Sustainable Construction Technologies

Sustainable Construction Technologies

Life-Cycle Assessment

Edited by

Vivian W.Y. Tam
School of Computing, Engineering and Mathematics,
Western Sydney University, Penrith, NSW, Australia
College of Civil Engineering, Shenzhen University,
Shenzhen, P.R. China

Khoa N. Le
School of Computing, Engineering and Mathematics,
Western Sydney University, Penrith, NSW, Australia

Butterworth-Heinemann is an imprint of Elsevier
The Boulevard, Langford Lane, Kidlington, Oxford OX5 1GB, United Kingdom
50 Hampshire Street, 5th Floor, Cambridge, MA 02139, United States

British Library Cataloguing-in-Publication Data
A catalogue record for this book is available from the British Library

Library of Congress Cataloging-in-Publication Data
A catalog record for this book is available from the Library of Congress

ISBN: 978-0-12-811749-1

Publisher: Matthew Deans
Acquisition Editor: Ken McCombs
Editorial Project Manager: Jennifer Pierce
Production Project Manager: Surya Narayanan Jayachandran
Cover Designer: Mark Rogers

Typeset by MPS Limited, Chennai, India

Contents

List of Contributors

Oladapo Adebayo Akanbi School of Engineering, University of Central Lancashire, Preston, United Kingdom

Ali Akbarnezhad School of Civil and Environmental Engineering, University of New South Wales, Sydney, NSW, Australia

Laura Almeida School of Computing, Engineering and Mathematics, Western Sydney University, Penrith, NSW, Australia

Rahman Azari College of Architecture, Illinois Institute of Technology, Chicago, IL, United States

Yapicioglu Balkiz Arkin University of Creative Arts and Design, Kyrenia, Cyprus

G.K.C. Ding School of Built Environment, University of Technology Sydney, NSW, Australia

Isidore C. Ezema Department of Architecture, Covenant University, Ota, Ogun State, Nigeria

Alireza Ahmadian Fard Fini School of the Built Environment, University of Technology Sydney, Ultimo, NSW, Australia

Cheng Siew Goh Department of Quantity Surveying and Construction Project Management, Heriot-Watt University Malaysia, Putrajaya, Malaysia

Md Mahmudul Haque EnviroWater, Sydney, NSW, Australia

I.M. Chethana S. Illankoon School of Computing, Engineering and Mathematics, Western Sydney University, Penrith, NSW, Australia

Ruoyu Jin Division of Built Environment, School of Environment and Technology, University of Brighton, Brighton, United Kingdom

Hoda Karimipour School of Computing, Engineering and Mathematics, Western Sydney University, Penrith, NSW, Australia

Ki Pyung Kim School of Natural and Built Environments, University of South Australia, Australia

Khoa N. Le School of Computing, Engineering and Mathematics, Western Sydney University, Penrith, NSW, Australia

Cazacova Liudmila Arkin University of Creative Arts and Design, Kyrenia, Cyprus

Loosemore Martin Faculty of the Built Environment, University of New South Wales, Sydney, Australia

Ogunbiyi Oyedolapo Department of Quantity Surveying, Yaba College of Technology, Lagos, Nigeria

Forsythe Perry University of Technology Sydney, Sydney, Australia

Ataur Rahman School of Computing, Engineering and Mathematics, Western Sydney University, Sydney, NSW, Australia

M Ashiqur Rahman School of Computing, Engineering and Mathematics, Western Sydney University, Sydney, NSW, Australia

Muhammad Muhitur Rahman Department of Civil and Environmental Engineering, King Faisal University, Al Hofuf, Saudi Arabia

Goulding Jack Steven Department of Architecture and Built Environment, Northumbria University, Newcastle, United Kingdom

Vivian W.Y. Tam School of Computing, Engineering and Mathematics, Western Sydney University, Penrith, NSW, Australia; College of Civil Engineering, Shenzhen University, Shenzhen, China

Oluyemi Toyinbo Department of Environmental and Biological Sciences, University of Eastern Finland, Kuopio, Finland

Cuong N.N. Tran School of Computing, Engineering and Mathematics, Western Sydney University, Penrith, NSW, Australia

Chapter 1

Introduction

I.M. Chethana S. Illankoon[1], Vivian W.Y. Tam[1,2], Hoda Karimipour[1] and Khoa N. Le[1]

[1]School of Computing, Engineering and Mathematics, Western Sydney University, Penrith, NSW, Australia, [2]College of Civil Engineering, Shenzhen University, Shenzhen, China

1.1 INTRODUCTION

This chapter introduces life cycle of a building. There are different stages in building life cycle and these can be classified into various aspects. There are managerial and technological decisions to be taken within the building life cycle. Therefore, this chapter provides a classification for building life cycle and the factors to be considered in each stage of building life cycle. There are many criteria considered for sustainable buildings or green buildings. With the development of green buildings, a yardstick was required to evaluate green buildings' performance, (Crawley and Aho, 1999) and therefore in 1990, Building Research Establishment Environment Assessment Method (BREEAM) was established as the first green building assessment tool (Building Research Establishment [BRE], 2018). Ever since there had been lot of tools developed. However, there are seven main criteria used in majority of the green assessments namely; Site, Energy, Water, Indoor Environment Quality (IEQ), Material, Waste and pollution, and Management (Illankoon et al., 2017). The remaining chapters of this book focus on these criteria and the sustainable technologies for respective criteria.

Chapter 2, Sustainability in Project management, of this book focuses on the management approaches for sustainability. This chapter discusses on the innovative approaches of sustainable management and life-cycle assessment. Although conventional buildings consume nearly 40% of primary energy production globally, green buildings substantially reduce energy consumption on a per square foot basis and they also focus on IEQ (MacNaughton et al., 2018). Therefore Chapter 3, Management, discusses the sustainable technologies for IEQ and Chapter 4, Indoor Environmental Quality, focuses on embodied and operational energy. Chapter 5, Life Cycle Energy Consumption of Buildings; Embodied 1 Operational, of this book looks into the novel approach of sustainable procurement and transport of construction

Sustainable Construction Technologies. DOI: https://doi.org/10.1016/B978-0-12-811749-1.00001-8

material. This chapter discusses on the current approaches, new combinatorial approach of procurement of construction material, and finally discusses on various case studies. Similarly, Chapter 7, Sustainable Procurement and Transport of Construction Materials, also focuses on material, specifically on cement and cement based material, steel, aluminum, insulation, bricks, and ceramic tiles. Chapter 6, Energy: Current Approach, and Chapter 8, Sustainable Water Use in Construction, focus on sustainable water use and emissions, respectively.

These novel and effective technologies need to be adopted into construction to derive results. Therefore, Chapter 9, Material, discusses on the sustainable construction technology adoption in detail. Lean principles in construction are discussed in Chapter 10, Emissions, of the book. Finally, Chapter 11, Sustainable Construction Technology Adoption, illustrates on bridging the sustainable construction technologies and heritage with a novel approach to conserve the built environment.

1.2 SIGNIFICANCE OF CONSTRUCTION AND BUILDING SECTOR

Construction industry plays a major role in a country's economy. The construction industry is one of the backbones of the economy of many countries (Cheung et al., 2001). It provides unique products with significant value. There are two main branches of construction industry, namely building sector and the infrastructure development. These two sectors consume many renewable and nonrenewable ecological resources were harvested, extracted, and productively used (Tatari and Kucukvar, 2012). Building sector satisfies certain social amenities such as houses, schools, hotels, religious and cultural centers, recreational facilities, etc. Infrastructure development provides base for economic activities and transactions by means of developing roads, airports, harbor bridges, etc.

Construction industry is different from other industries due to its unique characteristics such as different types of products, stakeholders, processes, and operating environments (Waidyasekara and Silva, 2014). The products are one-off in nature and usually it is developed based on the client's requirement. Even the procurement of the product is a lengthy process and not an off-the-shelf arrangement. The life span of buildings depend on series of factors such as quality of components, design level, work execution level, indoor environment, outdoor environment, usage conditions, and maintenance level (Langston, 2011). Considering all these facts when assessing the life cycle of a green building Green Building Council of Australia (2015), considers at least a 60-year period of time. Infrastructure development is also a section of construction industry. It has a "knock on" effect on the economic development of a country and it requires public financial resources as well (Kumaraswamy and Zhang, 2001). Further, government expenditure on infrastructure is enormous because it is capital-intensive (Babatunde, 2018).

The construction industry is influenced by the government taxes (Zainal et al., 2016) and the economic condition (Tumanyants, 2018).

The design of the building needs to be set up catering the client or the developer's requirements. Further, there are many decisions to be taken regarding the structure, material, and especially the cost of the building. As mentioned earlier, initial costs for buildings are higher. Therefore, except for iconic structures, all most all the building designs significantly focus on the cost impact. Developers always focus on the initial cost of the building.

Design stage is very crucial for the development of the building in an efficient and effective way. The design of the building has a severe impact toward the better functioning of the building. Apart from that, during this stage the materials are also selected. The selection of materials is very much important as it has a significant impact over the cost as well as the environment. Each year, as an example, building construction around the world alone consumes about 40% of the raw stone, gravel, and sand; about 25% of virgin wood and account for about 40% of the energy; and about 16% of water. (World Watch Institute, 2015). This signifies the impact on environment if the necessary material and better designs are not selected within the design stages.

As mentioned earlier, buildings have longer life spans. When a building functions, it consumes a lot of energy for lighting, air conditioning, water heating, and so on. Further, it also generates emissions in terms of foul air, waste water, carbon dioxide, etc. US Green Building Council Research Committee (2008) indicates that US buildings are responsible for about 38% of CO_2 emissions, about 71% of electricity consumption, about 39% of energy use, about 12% of water consumption, and about 40% of nonindustrial waste. This signifies the impact of the buildings toward the environment and the natural resources.

There are certain social impacts visible in ineffective building designs. Sick building syndrome is one of the mainly discussed topics. It includes series of symptoms such as fatigue, frequent headaches, and dryness in skin among workers in modern office buildings and the World Health It includes series of symptoms such as fatigue, frequent headaches, and dryness in skin among workers (Ghaffarianhoseini et al., 2018) and World Health Organization estimates that between 10% and 30% of the occupants of these buildings are affected by this (Lyles et al., 1991). Wong et al. (2009) carried out a survey to report the frequency of residents getting these symptoms for selected building. According to the results, more 30% of the respondents reported that the symptoms were related to their built environment (Wong et al., 2009). As the name itself suggests, this is medical condition affecting occupants of a building, attributed to unhealthy or stressful factors in the working environment such as fresh air ventilation rates, temperature, humidity, dust, and the microbial content of the air (Burge, 2004), which is basically due to a poor design. Therefore, by now there is a need to develop quality buildings, which are resource efficient, energy efficient, environmentally and socially sustainable.

1.3 GREEN BUILDINGS AND ITS LIFE CYCLE

There are a lot of definitions put forward to define green buildings. The most recent and widely used definitions are given in Table 1.1. According to Table 1.1, green buildings are structures that are environmentally friendly with efficient use of energy, water, and other resources and providing a better living and working environment for the occupants.

Green buildings focus on Energy (refer Chapter 4: Indoor Environmental Quality), Water (refer Chapter 6: Energy: Current Approach), Indoor Environment Quality (IEQ) (refer Chapter 3: Management), Material (refer Chapter 5: Life Cycle Energy Consumption of Buildings; Embodied 1 Operational and Chapter 7: Sustainable Procurement and Transport of Construction Materials), Emissions (refer Chapter 8: Sustainable Water Use in Construction), and Management (refer Chapter 2: Sustainability in Project management). As an example, sustainable material are selected within the initial decision-making process when finalizing the design specifications. Further, proper ventilation can be provided through a proper building design, which is possible in the design stage of the building life cycle. Emissions can be reduced throughout the operational life of the building by introducing various green building technologies, which are discussed in detail in the upcoming chapters. Better lighting controls and proper air conditioning are provided to the occupants to gain better working environment throughout the operational life. Environmentally and socially responsible deconstruction plan with minimum pollution is executed during the demolition or the disposal stage of the building. Each of these aspects required in a green building is achieved within the life cycle.

Usually, green building construction implies a process that starts in the planning stage and continues after the construction team has left the site (Hill and Bowen, 1997). According to Sterner (2002), there are many aspects to be considered throughout the green building life cycle such as responsibilities including managing the serviceability of a building during its lifetime, its possible deconstruction, and the recycling of resources to reduce the waste stream associated with demolition, choice of material, technical solutions, construction methods, types of services installation, responsibility for the development of environmentally aware processes by stipulating the requirements under which projects are designed by the clients and developers. These aspects occur at different stages in the green building life cycle.

Life cycle of a building commences with the initial decision-making stage. The life cycle of green buildings is also no exception; it also has a design stage with decision-making process. After the decision-making process, there is construction phase, whereas actual construction takes place according to the designs and specification. This is then followed by the operational stage of the building. Finally, at the end of the life cycle of the building there is a disposal stage. Green buildings also follow the same pattern.

TABLE 1.1 Recent Definitions on Green Buildings

References	Definitions
World Wildlife Fund (2015)	Physical structure that uses a design and planning process that is environmentally responsible and resource efficient
US Environment Protection Agency (2014a)	Practice of creating structures and using processes that are environmentally responsible and resource efficient throughout a building's life cycle from siting to design, construction, operation, maintenance, renovation and deconstruction. Further, this practice of green buildings and complement the classical building design concerns of economy, utility, durability, and comfort
Robichaud and Anantatmula (2011)	Green building is a philosophy and associated project and construction management practices that seek to: (1) minimize or eliminate impacts on the environment, natural resources, and nonrenewable energy sources to promote the sustainability of built environment; (2) enhance the health, well-being and productivity of occupants and whole communities; (3) cultivate economic development and financial returns for developers and whole communities; and (4) apply life cycle approaches to community planning and development
Hoffman and Henn (2008)	Green building is a term encompassing strategies, techniques and construction products that are less resource intensive or pollution producing than regular construction
US Green Building Council (2007)	Efficient building with savings in energy costs of 20%–50% are common through integrated planning, site orientation, energy-saving technologies, on-site renewable energy-producing technologies, light-reflective materials, natural daylight and ventilation, and downsized HVAC and other equipment
ASHRAE (2006, p. 4)	Green building is one that achieves high performance, over the full life cycle, in the areas such as minimizing natural resource consumption through more efficient utilization of nonrenewable natural resources, land, water, and construction materials, including utilization of renewable energy resources to achieve net zero energy consumption, minimizing emissions that negatively impact our indoor environment and the atmosphere of our planet, especially those related to IAQ, greenhouse gases, global warming, particulates, or acid rain, minimizing discharge of solid waste and liquid effluents, including demolition and occupant waste, sewer, and stormwater, and the associated infrastructure required to accommodate removal, minimal negative impacts on site ecosystem, maximum quality of indoor environment, including air quality, thermal regime, illumination, acoustics/noise, and visual aspects to provide comfortable human physiological and psychological perceptions
Cassidy et al. (2003)	Buildings which increase the efficiency of sites use energy, water, and materials, and reduce building impacts on human health and the environment, through better siting, design, construction, operation, maintenance, and removal throughout the complete building life cycle

Wyatt (1994) has deemed green building construction to include "cradle to grave" appraisal, which includes managing the serviceability of a building during its lifetime and eventual deconstruction and recycling of resources to reduce the waste stream usually associated with demolition. Further, Kibert (1994) identified seven principles of sustainable construction. These seven principles are: reduce resource consumption, reuse resources, use recyclable resources, protect nature, eliminate toxics, apply life costing, and focus on quality. These principles can be applied across the entire life cycle of construction, from planning to demolition (Kibert, 1994). These principles consider carrying out the construction by reducing resource consumption, reusing resources and using recycled resources, which tends to protect nature by eliminating toxics (Lavy and Fernández-Solis, 2009). However, in green buildings, there is a proper management put in place from the initial stages to guide through the process. Therefore, the essence of green concepts is put into practice throughout the green building life cycle.

According to Australian National Audit Office (2001), there are five main phases, which are design, purchase and construction, operations, maintenance, development and disposal. However, in this classification the first three phases occur in the earlier stages of the life cycle. The operation, maintenance, and development occur throughout the building life cycle, and finally, the disposal of the building occurs. Further, there are usually four stages in any product development; namely, design, production, usage, and disposal (Asiedu and Gu, 1998). Similarly, life cycle of green buildings can also be illustrated in these four stages. The design stage included the designing of the building and the production refers to the construction phase. Usage phase of the building refers to the operational phase which includes maintenance of the buildings and finally the disposal of the building. Cao and Folan (2012) identified three basic stages of product development; namely, the beginning of life, the middle of life, and the end of life. Further, the "beginning of life" stage includes the concept development and the manufacturing process and the "middle of life" process includes the actual usage of the product (Cao and Folan, 2012). In terms of green building life cycle perspective, these three stages can be identified as the design and construction phase, operational and maintenance phase, and finally, the disposal phase.

Considering these phases, Fig. 1.1 illustrates the green building life cycle and each of the significant aspects considered throughout the cycle.

According to Fig. 1.1, there are main four phases in green building life cycle. Design phase focuses on the decision-making regarding the green building. There are critical aspects discussed in this phase focusing on the design, procurement route, land use, accessibility, and so on. The second stage is construction phase. This is the phase where actual construction occurs. When compared to a conventional building, in a green building, there

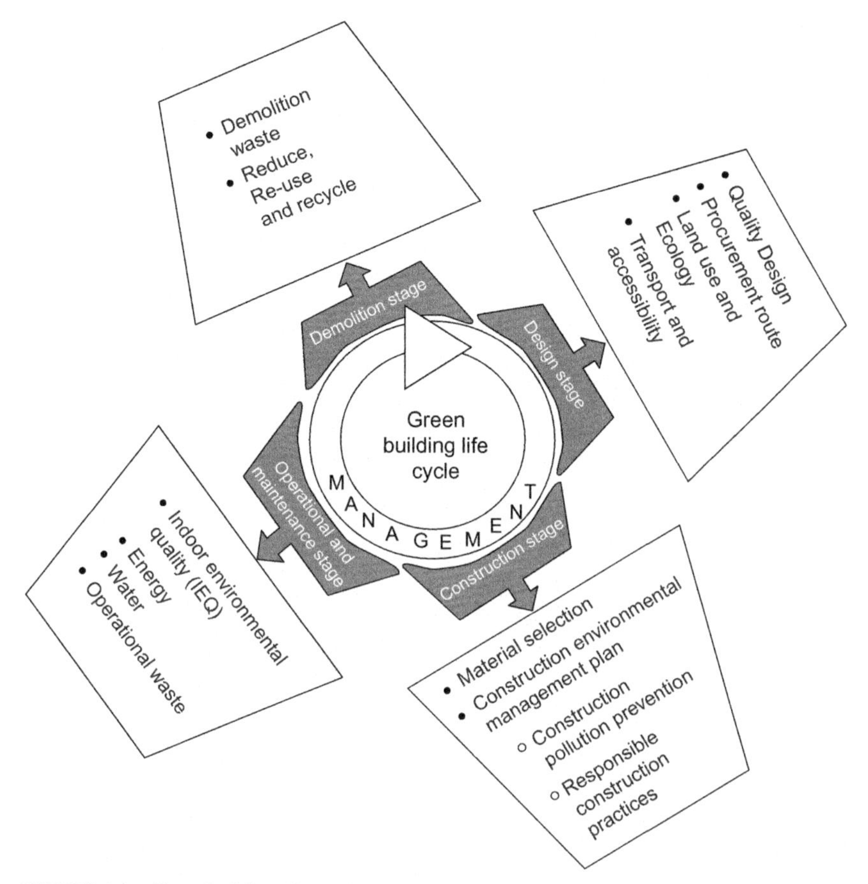

FIGURE 1.1 Green building life cycle stages.

is a more environmentally friendlier approach conducted during the construction. Next is the operational and maintenance stage. In other words, this is the period in which the building functions. Throughout this phase, it consumes energy, water, and many resources, and in turn emits many emissions to the environment. According to World Green Building Council (2015), in a green building, it is expected to incur an increase of 0.4%–12.5% of design and construction cost while experiencing an energy saving of up to 25%–30%. These energy savings are experienced during the operational stage of the building life cycle. Final stage is the disposal phase. This is the phase where the building is deconstructed. This marks the end of the life cycle of this green building. Further, for a green building, management function is applicable throughout the building life cycle.

1.3.1 Design Stage

As illustrated in Fig. 1.1, this stage includes certain aspects considered in the initial decision-making. Quality design, procurement route, accessibility and transport, land use and ecology, sustainable technology innovations are the main aspects identified in this phase.

1.3.1.1 Quality Design

The design itself contributes significantly toward the green features. According to Anastas and Zimmerman (2003), there are main 12 principles of green design. These principles provide a framework, which moves beyond baseline engineering quality and safety specifications to consider environmental, economic, and social factors. Fig. 1.2 illustrates the 12 principles in detail.

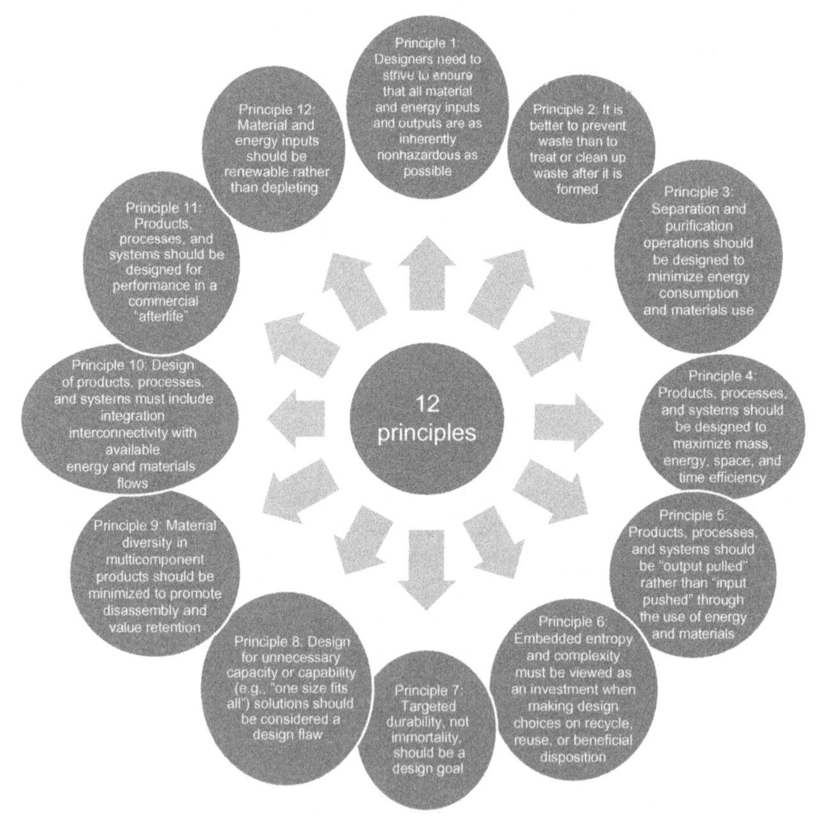

FIGURE 1.2 12 principles of green engineering design. *Adapted from Anastas, P.T., Zimmerman, J.B., 2003. Peer reviewed: design through the 12 principles of green engineering. Environ. Sci. Technol. 37 (5), 94A–101A.*

These principles given in Fig. 1.2 illustrate all the aspects of green building design. It discusses on the material selection, waste minimization, energy efficiency, recycling and reusing, and so on. Therefore, in green building design the initial design stage is critical. It lays the basis for the greener buildings considering the triple bottom line; that is, environmental, economical, and social sustainability.

1.3.1.2 Procurement Route

Project procurement has been described as an organized method or process and procedure for clients to obtain or acquire construction products (Abdul Rashid et al., 2006). Using the best-suited procurement route is very important for any construction. For green building development also it is not an exception. Procurement route is significant in managing the project objectives and to deal with the risks in construction in terms of cost fluctuations, time, quality concerns, and so on. Green buildings are resource and energy efficient structures compared to conventional buildings and therefore innovative technologies are put forth to achieve green building requirements. Therefore, the risk in green building construction is higher compared to the conventional buildings. Further, there is an initial cost premium for green buildings as well. As a result, selecting the suitable procurement route is very much important. Procurement is critical as it determines the overall framework embracing the structure of responsibilities and authorities for participants within the building process (Cheung et al., 2001).

There are main factors considered in procurement selection. The risk taking ability of the client, time for construction, cost, quality standards, complexity of the project, and the extent of client involvement are some of the factors affecting the selection of procurement route. In deciding the procurement route there are three aspects to be finalized. These are the working arrangement, tendering process, and the form of contract. Working arrangement defines the roles and duties of the client/consultant and the contractor and the payment methods. Green building projects demand open lines of communication between disciplines due to the more complex interdependencies between building systems and project organizations than traditional projects (Riley et al., 2003). Therefore, according to Riley et al. (2003), integrated project teams are more suitable for green buildings. As an example, there are several working arrangements such as traditional procurement, design and building, design and management path. There are variants to each of these working arrangements. Further depending on the requirements of the green building the working arrangement is established clearly defining the roles and responsibilities. The way in which environmental requirements are stipulated in procurement documents is significant for the development of a project's environmental features (Sterner, 2002).

The next is the contractor selection. Usually there is a main contractor for the building construction. It is a critical decision that involves several key project team members, including the owner and designer and requires to select the team that will deliver the project (Wardani et al., 2006). However, in green building construction there are a lot of specialized works involved, which required certain specialized contractors. There are methods for selecting these contractors in the initial decision-making stages. As an example, client/consultant can use variants of tendering to select the contractors or negotiate with contractors to select the best suited for the project. Finally, it is necessary to decide the form of contract for the project. Parties can either use standard international contracts such as FIDIC (The International Federation of Consulting Engineers), locally developed standard contracts or develop tailor made contracts depending on the requirements of the green building project.

1.3.1.3 Land Use/Ecology and Accessibility

Identification of possible impacts of building construction projects on the environment is a task that needs to be accomplished for the realization of an effective environment. The environment is fast deteriorating in its ability to support life forms, with every large stride taken toward developments, 10 strides are taken backward in our failure to protect the environment as well as save rare species from the danger of extinction (UN-Habitat, 2006).

According to Howard (2000), environmental impact can be classified into visual impact, material impact and resources use, energy impact, space condition and lighting impact and land use and ecological impact. Any building structure above the earth surface assumes visual impact on the environment. Large projects usually involve extensive land disturbance, involving removing vegetation and reshaping topography. Such activities make the soil vulnerable to erosion. Construction activities lead to serious erosion and bring harmful effects on the environment (Wu et al., 2012). Measures to address the impact of land disturbance on the environment should be included in the planning and design phase of the project, before any land is cleared (Environmental Protection Authority, 1996). Also, the presence of nature within the built environment can take many forms, depending upon the history and geography of the site and the density of buildings or infrastructure. The existence of natural features such as rivers, geological contours, and land unsuitable for building is another factor that allows nature to persist in the urban setting. In some cities, zoning regulations call for plenty of open space, which is often developed as green space. In other cities, there is an unbroken array of buildings, connecting roads, and parking areas, leaving little room for nature (Allacker et al., 2014).

Sustainable integration of buildings and infrastructure requires a major change in urban form and building design. In some new communities, elements of urban infrastructure systems are moving much closer to and even inside the buildings themselves. Increasingly, there is a blurring of the traditional boundaries that separate buildings from their civil infrastructure. In some communities, large distribution grids and remote treatment and generation facilities are giving way to a network of distributed or "on-site" infrastructure systems, with shared elements integrated into the fabric of the built environment (United Nations Environment Programme, 2001). More diverse land use and building types can complement these on-site infrastructure systems, creating self-reliant, mixed developments of housing, commercial space and industry. In these communities, each new housing development is seen simultaneously as a center of employment, communications, and food production, as well as a facility for power generation, water treatment, storm water management, and waste management. Apart from that, these site needs to be integrated to the public transport as well. According to Green Building Council of Australia (2015), provision of access to public transport leads to reduction in the carbon emissions arising from occupant travel to and from the project and also promotes the health and fitness of commuters, and the increased livability of the location.

In this regard, designers and engineers are faced with an increasing range of options. The life cycle impacts and energy and material flows need to be assessed for very diverse technologies, and for a greater variety of scales and locations. Decision makers need comprehensive models in order to combine the flows from different stocks (i.e., buildings, roads, pipes, wires, etc.) and allow meaningful comparisons between integrated and less integrated systems (Allacker et al., 2014). In summary, all those measures, which affect the environment during or after a construction, should be considered before any development. According to Intergovernmental panel on climate change (2014), due to all these various adverse environmental effects, it is necessary to focus on innovation and investments in environmentally sound technologies and infrastructure and sustainable livelihoods.

1.3.2 Construction Stage

This is the stage where the actual construction of the building takes place. Compared to conventional buildings the construction process focuses on the environmental impact in a wider perspective. Therefore, green building projects launch construction with kick-off meeting that includes a sustainable education component for on-site construction personnel; monthly on-site meetings are required by entire site workforce and include periodic education and training sessions on green building (Pulaski et al., 2004). Further, sustainability requirements are also reviewed with each subcontractor prior to commencing work (Pulaski et al., 2004).

As illustrated earlier, green building development is a process. It is not only the final product. Therefore, the construction process also should be environmentally and socially friendly. The selection of suppliers for materials and the type of material selected must be given a consideration. Further, the construction activities should be managed effectively to minimize any environmental impact. The aspects identified in Fig. 1.1 can be further illustrated as follows.

1.3.2.1 Material Selection and Sourcing

Usually, materials reach up to 20%–30% of the total building cost (Ross et al., 2007). Therefore, it has a considerable impact on the cost of the green building as well. Impacts considered from building materials could be addressed at different stages; from processing and manufacturing to logging and storage through installation and use (Ijigah et al., 2013). However, material selection is partially carried out in initial decision-making phase. Usually within the initial selection it is necessary to select materials with lower environmental impact. Further, the green building design is developed to consume less amount of material compared to a conventional building. The characteristics of the materials in green buildings focus on the contribution to the heat island effect, proportion of recycled content, distance from the supplier or producer to the project site, and emissions of indoor pollutants (Castro-Lacouture et al., 2009).

Material sourcing is carried out during the construction stage. In green buildings, sourcing of material is done in a responsible manner to have a minimal impact on the environment. According to Green Building Council of Australia (2015), for a green building, steel should be sourced from a responsible steel maker, timber used in the building and construction works is certified by a forest certification scheme and all permanent formwork, pipes, flooring, blinds, and cables in a project should not contain PVC and have an Environmental Product Declaration (EPD). Green building rating tools such as Leadership in Energy and Environmental Design (LEED) adapt to take into account a more-convenient distance from the place where materials are extracted, harvested, recovered, or manufactured to the project site (Castro-Lacouture et al., 2009). Further priority is given to recycled and reusable material. Raw materials are sourced mostly from rapidly renewable resources. Priority is given for regional raw material to reduce the transport distance leading to minimization of carbon emission from fossil fuel combustion. Supplementary cementitous materials (SCMs) are used for green buildings. Fly ash, ground granulated blast furnace slag and silica fume can be used as SCM (Meyer, 2009). Using earth bricks is also another option. These are the main aspects considered in green building construction considering material selection and sourcing. According to Spiegel (2012), the green building material selection process is slightly different from the conventional building material selection. The green building's material selection is illustrated in Fig. 1.3.

Step 1: Identify material categories
•General material categories are identified. As an example, if masonry is decided for the exterior walls, types of masonry—clay, concrete, or stone is explored
Step 2: Identify (Green) building material performance criteria
•Match the green performance criteria to the green goals of the projects. (e.g., energy efficiency, zero waste)
Step 3: Identify (Green) building material options
•In addition to the standard array of material subcategories, identify green subcategories. (e.g., Adobe masonry units, rammed earth, and straw bale might be viable options for exterior masonry walls)
Step 4: Gather technical information
•Technical information regarding the greenness of building products (Sources include: product representatives, governmental agencies, building codes, trade organizations, industry standards, material safety data sheets, green rating programs, and environmental nonprofit organizations.)
Step 5: Review submitted information for completeness
•The green building product information should provide all the necessary data for the professionals to take effective decisions.
Step 6: Evaluate (Green) materials
•Evaluation is the most important process (e.g., use of renewable resources, or nonrenewable resources, the quality of the resource (present and future), the cycle of the resource, consider local, regional, and global implication of choices.)
Step 7: Select and document choice
•Specify what is selected.

FIGURE 1.3 Green material selection process. *Developed from Spiegel, R., 2012. Green Building Materials: A Guide to Product Selection and Specification, third ed. Wiley, Hoboken, NJ.*

Fig. 1.3 provides a clear process of material selection. The stage 1 usually occurs in the design stage. However, giving details on required products can be experienced in other stages as well.

1.3.2.2 Construction Environmental Management Plan

On-site construction activities relate to the construction of a physical facility, resulting in air pollution, water pollution, traffic problems, and the generation of construction wastage (Shen and Tam, 2002). Therefore, these practices need to be managed throughout the construction process to minimize the adverse impact. There is an environmental management plan put in place to manage the construction practices in an environmental and socially friendly manner. Sources of pollution and hazards from construction sites include dust, harmful gases, noises, blazing lights, solid and liquid wastes, ground movements, messy sites, fallen items, etc. and these types of pollution and hazards can not only annoy residents nearby, but also affect the health and well-being of people in the surrounding area (Chen et al., 2000).

Usually a construction environmental management plan included two areas such as details of the project site and environment and the site management procedures. Details of site and environments provides information on environmental conditions such as topography, geology, sensitive ecological receptors, underground services, groundwater, and contaminant conditions (Auckland Council, 2015). Further, this provides details on site management goals, human health site management practices such as health and safety planning, site inductions, and personal protective equipment. Other than that, usually construction management plan illustrates on the environmental site management practices such as dust control and monitoring, odor control and monitoring, sediment control and monitoring, spoil stockpiling, soil disposal (Auckland Council, 2015). Apart from these, if the project is situated in a congested area, this included a traffic management plan and also if the site is of cultural heritage this includes a cultural management plan as well (Sanson and O'Mahony, 2010). There is no clear cut definition on the inclusions of the construction management plan, although it depends on the requirements of the site. These procedures and plans laid down by the construction management plan is followed throughout the construction phase and also during the demolition stage.

In certain instances construction management plan includes other subplans as well. These subplans include air quality management plan, soil and water management plan, erosion and sediment control management plan, flora and fauna management plan, noise management plan, traffic management plan, cultural heritage management plan, landscape and rehabilitation management plan, emergency response plan, and waste management plan (Sanson and O'Mahony, 2010).

Construction waste management is one of the main areas discussed in a construction environmental management plan. Building projects usually lead to a significant construction and demolition waste generation (Poon et al., 2004). In Australian context, construction and demolition waste has been estimated to contribute around 25.8% of overall landfill (Australian Bureau of Statistics, 2011). Therefore, waste management must be discussed in both construction and demolition stages of the project. In reducing waste management, two principles prevail: (1) reduction of the quantities of waste generated and (2) adopt an effective system for managing the unavoidable waste produced (Teo and Loosemore, 2001). In managing the unavoidable waste, there are three options in order of preference; namely, "reuse," "recycling," or "disposal" (Teo and Loosemore, 2001). These are considered in developing a waste management plan in construction process.

Implementing a construction management plan derives many benefits such as cost saving due to the reduction of fines associated with convictions, improving corporate image in environmental performance, contribution to the improvement of public environmental standards, contribution to environmental protection, increasing overall business competitiveness, reduction of environmental complaints, improving staff work environment, thus increasing their morale, reduction of environment-related sickness and injuries and reduction of environmental risks polluted air, land, and water (Shen and Tam, 2002).

1.3.3 Operational and Maintenance Stage

According to International Organization for Standardization [ISO] (2017), operational and maintenance phase of a building initiates from the end of construction phase and extended till the beginning of demolition phase. Within this period of time, the building needs to be maintained. Maintenance of a building can be identified as the processes and services undertaken to preserve, protect, enhance and care for the buildings' fabrics and services after completion, in accordance with the prevailing standards to enable the building and services to serve their intended functions throughout their entire life span without drastically upsetting their basic features and uses (Olanrewaju et al., 2010).

There are different types of maintenance required. Mainly there are three types of maintenance; namely, corrective maintenance, predictive maintenance, and condition-based maintenance (Horner et al., 1997). Other than that, there are replacements also. According to Stanford (2010), maximum performance life of a building component is the time over which the component serves its anticipated function over the range from 100% to 0% and once the performance level falls below some minimum, and the cost of continuing to maintain a failing component has reached the end of design service life. In such cases, the component is replaced.

There are a lot of factors which affect the building maintenance. According to Perera et al. (2016), there are main five factors that affect the building maintenance; namely, (1) building characteristics, such as building services, finishes, age of the building, area, material used, building design and construction, and location of the building elements; (2) maintenance factors, such as execution of maintenance, budgetary constraints, quality of parts and material, workmanship, and so on; (3) tenant factors, such as expectations of tenants, accessibility to property, delay, and failure in reporting problems; (4) regulatory and economic factors, such as price inflation, changes in standards and legislation; (5) other factors, such as building energy management systems, warranty, and after-sales services of material and changes in climate conditions.

In green buildings, there are certain aspects, which are discussed throughout the operational period. Most of the green buildings rating tools mainly focus on these criteria such as indoor environmental quality (IEQ), energy, water, and operational waste (Building Construction Authority, 2013; Building Research Establishment Environment Assessment Method, 2014; Green Building Council of Australia, 2015; US Green Building Council, 2014). Most of the benefits of green buildings such a better human conditions, efficiency in energy and water, life cycle cost savings and the lower negative impact to the environment are derived through the operational phase with effective operations of these aspects.

1.3.3.1 Indoor Environmental Quality

As the name suggests, IEQ refers to the quality environment within the building especially focusing on the occupants. According to the definitions of green buildings, it not only focuses on the environment but also considered the health and well-being of the occupants as well. IEQ, therefore, focuses on the aspects that affect the IEQ in buildings including temperature, humidity, ventilation systems characteristics, building envelope, and finish materials (Singh et al., 2011). There are main four categories for IEQ. These categories are thermal comfort, indoor air quality (IAQ), visual comfort, and acoustic comfort (Hui et al., 2010; Huizenga et al., 2006; Kim and de Dear, 2012). There are many strategies put into practice in order to achieve these comfort levels.

According to US Green Building Council (2011), IAQ is obtained by prohibiting smoking within the operational stage, locating air intakes away from likely exhaust sources, using low emitting green materials for construction, design for proper ventilation, conducting a flush out before occupancy, installing an entryway grate to remove pollutants carried by people, ensure adequate ventilation, monitoring CO_2 levels and identifying opportunities for improving building cleanliness. Most of these strategies are embedded to the design at the initial stages.

Visual comfort can be obtained by strategies such as providing a high level of lighting system control by individual occupants or by specific groups and providing individual lighting controls for 90% (minimum) of the building occupants, introduction of daylight and views (Dilrukshi et al., 2014). Basic strategies such as installing operable windows, giving occupants temperature and ventilation control (adjustable air diffusers and thermostats) and conducting occupant surveys to adjust the thermal comfort levels are adopted to enhance the thermal comfort in green buildings (US Green Building Council, 2011). By including appropriate acoustic design such as using soft surfaces and other strategies to ensure sound level remain comfortable (US Green Building Council, 2011) and using natural material such as bamboo or recycled rubber as an alternative to synthetic material for airborne sound insulation are basic strategies to embed acoustic comfort in green buildings (Asdrubali, 2011). There are certain other aspects, which need to be considered in IEQ as well. Frontczak et al. (2012) identified spatial comfort and building maintenance and cleanliness as IEQ aspects, which needs more attention. Further, according to Clements-Croome (2006), aspects such as color schemes, building materials, and radiation and electromagnetic fields also affect the IEQ of the building. However, in summary it suggests that improved IEQ has a positive correlation with reduced health issues such as asthma/respiratory allergies and improved productivity (Singh et al., 2010).

1.3.3.2 Energy

Energy use is a central issue as energy is generally one of the most important resources used in buildings over their lifetime (Thormark, 2002). Further, due to the rising energy costs and the awareness on the environmental impacts, energy consumption within buildings is largely considered from the design stages of the building life cycle. In green buildings, an energy criterion is widely discussed. Almost all the green building rating tools focus on energy efficiency significantly in evaluating green buildings. As an example, Green Star Australia (Green Building Council of Australia, 2015) allocates 22% of credits points for energy efficiency of the green building. Therefore, "energy" has become one of the significant aspects in green building life cycle.

Buildings demand energy in their life cycle, both directly and indirectly. Directly for their construction, operation (operating energy), rehabilitation and eventually demolition; indirectly through the production of the materials they are made of and the materials technical installations are made of (embodied energy) (Sartori and Hestnes, 2007). During the construction stage also energy is used for many construction activities. However, the use of energy within the operational life of the building is significant. Within the operational life, energy is used for numerous day-to-day activities. This operational energy comprises the energy used for space heating and cooling, hot water heating, lighting, refrigeration, cooking, and appliance and equipment operation (Fay et al., 2000). Therefore, efficiency in energy is essential during the operational life of the green building.

Green buildings usually focus on the target of reducing the operating energy. There are main two ways of achieving energy reduction, namely, using active design strategies or passive design strategies. Passive technologies include, for example, increased insulation, better performing windows, reduction of infiltration losses and heat recovery from ventilation air and/or waste water (Sartori and Hestnes, 2007). Active technologies include heat pumps coupled with air or ground/water heat sources, solar thermal collectors, solar photovoltaic panels and biomass burners (Sartori and Hestnes, 2007). Green Building Council of Australia (2015) encourages on-site energy generation, reduction in peak demand for energy, and reduction in the predicted energy consumption and GHG emissions. After thoroughly analyzing eight widely used green assessment tools, Illankoon et al. (2017) depict that all the assessments give priority to energy in green buildings and in some instance there is a possibility that a building might be given a green certification, although it does not focus on other green requirements such as water and IEQ. This signifies the importance of energy efficiency in green buildings.

Conventional energy efficiency technologies such as thermal insulation, low-emissivity windows, window overhangs, and daylighting controls can be used to decrease energy use in new commercial buildings by 20%–30% on average and up to over 40% for some building types and locations (Kneifel, 2010). Although increasing energy efficiency usually increases the first costs of a building, the energy savings over the service life of the building often offset these initial higher costs (Kneifel, 2010). Therefore, energy is one of the main aspects, which derives long-term cost savings compared to conventional buildings focusing on the building life cycle.

1.3.3.3 Water

Water is one of the main necessities of mankind. Therefore, throughout the operational phase of the building water resource is used. However, reducing water consumption and protecting water quality are key objectives of sustainable construction. Usually, water efficiency in green construction can be identified as minimizing and eliminating unnecessary water use and wastage with minimal damage to the environment, society, and economy (Waidyasekara and Silva, 2014). According to Crawford and Pullen (2011), buildings consume around 12% of global water consumption including the material production as well. During the operational stages, water is used for cleaning purposes, cooking, gardening etc. However, during the construction stage there is considerable water consumption as well. Therefore, Green Building Council of Australia (2015) award points for using reclaimed water for concrete mixing in green building construction.

There are major two types of water efficiency practices, namely, (1) practices, which are based on modifications in plumbing, fixtures, or water supply operating procedures, and (2) behavioral practices, which is based on changing water use habits (US Environment Protection Agency, 2014b). Further, water efficiency can be maintained through regular maintenance by identifying the leakages as well (US Environment Protection Agency, 2014b). Most of the green building rating tools always highlight on the water efficient fixates (Green Building Council of Australia, 2015; US Green Building Council, 2014) and the main idea is to save drinking water. Rainwater harvesting is one of the main methods used for water efficiency in buildings (Herrmann and Schmida, 2000).

In general, wastewater reuse or reclamation serves as an efficient and valuable way to cope with the scarcity of water resources and severity of water pollution (Chu et al., 2004). The substitution of drinking water with reusable gray water for toilet flushing and garden irrigation, helps to support the sustainability of valuable water resources (Nolde, 2000) in green buildings. Gray water is usually collected separately from clothes washers, bathtubs, showers and sinks, and laundry water, which constitutes to around 50%–80% of residential wastewater (Al-Jayyousi, 2003). By adopting these strategies, the water efficiency is achieved throughout the operational life of the green building.

1.3.3.4 Operational Waste

Handling operational waste within the buildings is one of the main issues during the operational stage if proper procedures are not put in place. Operational waste handling in and around buildings are treated as a service requiting proper planning, design, specification, and installation, with reference to the needs of the particular buildings, the environment of the surrounding area, and recycling (Wise and Swaffield, 2002). Provision of storage containers, provision of garbage chutes, and usage of compactors are the main established practices for waste management (Wise and Swaffield, 2002).

Green building rating tools also encourage the separation of waste material into general waste, paper and cardboard, glass and plastics (Green Building Council of Australia, 2015) into separated collection bins. Usually, the disposal falls within the domain of municipal councils.

However, this is a process clearly requires the contribution of the occupants as well.

1.3.4 Demolition Stage

Demolition stage represents the end of life of the green building. The structure of the building is demolished and the whole building structure, including the superstructure and the concrete foundations will end up as demolition waste. Demolition waste is 10–20 times by weight as much as waste generated from the construction of new buildings (Poon et al., 2001). Further, demolition waste usually consists of high percentage of inert materials like bricks, sand, and concrete, and lesser percentages of material like metals, timber, paper, glass, plastics, and other mixed materials (Poon et al., 2001). If these wastes are not properly managed, it directly goes to landfills causing significant adverse impacts to the environment. Further, there are certain materials such as bricks, which can be reused if properly demolished with extra care. Not only that, given appropriate sorting at the generation source, the inert material is the substitute for primary aggregate and can be used in lower grade applications such as base course in road beds and engineering fill or even in higher grade product recycling such as the manufacturing of reconstituted concrete (Chung and Lo, 2003). Therefore, all these must be taken into consideration in the process of demolition of the building. Similarly, building design and material selection at the early stages also presents a major scope for waste reduction (Chung and Lo, 2003).

There are many sustainable ways of dealing with construction and demolition waste. During the construction stage this issue is addressed through the construction environmental plan. However, from that phase onward there should be a plan put in place for demolition waste as well. According to Peng et al. (1997), there is a hierarchy of possibilities for disposing of demolition waste as given in Fig. 1.4.

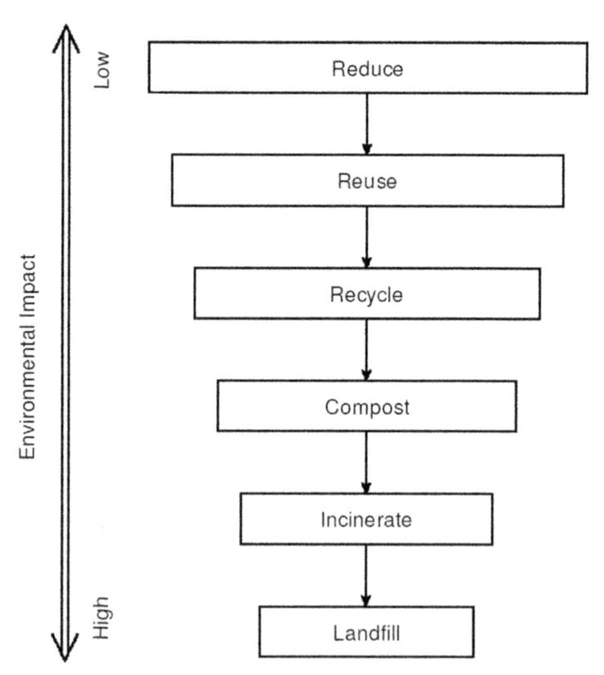

FIGURE 1.4 Hierarchy for construction demolition waste disposal. *Adapted from Peng, C.-L., Scorpio, D.E., Kibert, C.J., 1997. Strategies for successful construction and demolition waste recycling operations. Constr. Manage. Econom. 15 (1), 49–58.*

A given in Fig. 1.4, material usage reduction is the best or the most effective way. However, next is the reuse and recycle. According to Tam and Tam (2006), the most effective way to is to implement all the three reuse, recycling, and reduction ways of construction materials in construction activities. Reuse, refers to moving materials from one application to another (Peng et al., 1997). However in recycling, new materials are produced out of wastes which does not only fulfill the purpose of recycling but also generates economic benefits (Peng et al., 1997). The end of demolition stage marks the end of the entire life cycle of the green building.

This life cycle of building identified mainly four stages (Fig. 1.1). However, most of the decisions are interconnected and decisions taken in the initial stages affect the smooth process of the latter stages. Therefore, managing this process is essential throughout the building life cycle. According to Green Building Council of Australia (2015), management function improves sustainability performance of the green building by influencing areas where decision-making is critical and also promotes practices that ensure a project will be used to its optimum operational potential.

1.4 GEOGRAPHICAL INFORMATION SYSTEM (GIS) FOR GREEN BUILDING CONSTRUCTION

Over the past 30 years, key practices for shaping and managing the built environment within building and related industries have proved unsustainable. Available evidence demonstrates that these practices were not economically, environmentally, or socially viable in the long run. However, by now, there new dimensions come into existence to provide a better performance of green buildings. Geographical information system (GIS) is one of the examples. Using traditional approaches representing represents only three dimensional objects to create digital surface model is not efficient to get accurate geo-information covering urban areas and therefore, to avoid that issue GIS technology can be proposed (Dibs et al., 2018).

GIS is a way to achieve required stewardship, sustainability, and savings targets significantly enhances the ability for a triple bottom line approach in expanding, operating, and maintaining the built environment. At present, industry stakeholders most often use technologies such as building information models (BIM) and other more traditional computer-aided drafting (CAD) systems to design and store data about buildings. Usually, this information is queried and reported building by building. The challenge for facility managers is querying, analyzing, and reporting this information for all buildings across a site or an even broader geographic region. Facility age and the technologies used to design them contribute to this challenge. In many cases, building data are managed within spreadsheets and in hard and soft copy floor plans with no real organized system of data management. By importing and aggregating into a GIS the geometries and tabular data of the multiple BIM and/or CAD files required to accurately represent the built environment, the efficiencies and power of BIM can be leveraged, extended, and connected in geographic space to other relevant site, neighborhood, municipal, and regional data.

GIS technology can be exploited to provide key facility information for decision makers when they need it. In this context, it is used to answer questions regarding the best manner to develop and manage the built environment. This ability is largely a result of the relational database technology underlying it, as well as the capacity for GIS to identify spatially related objects. Spatial relationships allow GIS to merge different worlds of knowledge—it is significant and powerful because it unearths and exposes related patterns that would otherwise go undiscovered. It is a powerful system and enabling technology for shaping and managing the built environment—one that:

- Provides a common and coordinated view, thereby increasing collaboration and understanding while reducing risk and its associated costs.
- Enables visualization, analysis, and comparison of possible alternatives to optimize performance.
- Provides the analytic tools necessary for stakeholders to determine which strategy presents the best short- and long-term solutions to pursue.
- Can provide the support the building industry requires to realize more sustainable development practices and patterns (Wallis et al., 2012).

Recent innovation in GIS technology together with growing use of geomatics technologies for nontraditional surveying applications is leading toward new uses for these technologies. This is being prompted by the creation of new architectural designs, sustainable designs incorporating more complex geometries, and construction-operation requirements. Spatial technologies are capable of monitoring design workflows and can often be used to validate that projects are constructed within design specifications.

Many world-class design projects also involve the use and application of parametric design technology, 3D city models and advances in architectural design. These projects often include ecodesign elements that directly involve the use and application of spatial analysis and geoprocessing techniques.

The development of integrated building models that incorporate spatial data derived from geomatics technologies is part of the overlay project communication strategy. CAD users, GIS users, survey technologies, and engineers are all operating within one streamlined operational space, exchanging their data needs and analysis requirements through integrated business process chain. The GIS professionals and structural engineers might very well be exchanging data files and looking at the same BIM model on any given day. While one person pours the cement or prepares caldding and surface elements, their work is interchangeable and connected to that of others on the project (Sensors and Systems, 2011).

Enabling technology, such as GIS, provides industry managers and executives with the tools required to be better stewards of the built environment. The common and coordinated awareness that GIS delivers provides a better understanding of the present. Shared awareness enables stakeholders to visualize and analyze data regarding the built environment and its links to the world at large. This enables better collaboration among stakeholder disciplines, thereby reducing the unknowns and leading to lower project contingencies, risk, and cost (Wallis et al., 2012).

1.5 SUMMARY

The chapter provided a brief introduction to the construction industry on building's life cycle. Afterward, the green building concept is discussed and green buildings are identified as resource efficient structures, which are environmentally friendly providing better quality spaces to occupants. Then the green building's life cycle is illustrated in four stages: design stage, construction stage, operational and maintenance stages, and demolition stage. Each of these stages is discussed in terms of the main aspects to be considered in each of those stages. Finally, this chapter provided a brief introduction to the green building future development in terms of GIS applications.

REFERENCES

Abdul Rashid, R., Taib, I.M., Ahmad, W.B., Nasid, M.A., Ali, N., Zainordin, Z.M., 2006. Effect of procurement systems on the performance of constructn projects. International Conference on Construction Industry, Padang, Indonesia, 21-25th June.

Al-Jayyousi, O.R., 2003. Greywater reuse: towards sustainable water management. Desalination 156 (1), 181–192. Available from: https://doi.org/10.1016/S0011-9164(03)00340-0.

Allacker, K., Maia de Souza, D., Sala, S., 2014. Land use impact assessment in the construction sector: an analysis of LCIA models and case study application. Int. J. Life Cycle Assess. 19, 1799–1809. Available from: https://doi.org/10.1007/s11367-014-0781-7.

Anastas, P.T., Zimmerman, J.B., 2003. Peer reviewed: design through the 12 principles of green engineering. Environ. Sci. Technol. 37 (5), 94A–101A.

ASHRAE, 2006. Green/sustainable high-performance design. In ASHRAE (Ed.), The ASHRAE Greenguide: The Design, Construction, and Operation of Sustainable Buildings, second ed. Butterworth-Heinemann, Burlington, pp. 3–16.

Asdrubali, F., 2011. Green and sustainable porous materials for noise control in buildings: a state of the art. In Symposium on the Acoustics of Poro-Elastic Materials. 14–16 December, Ferrara.

Asiedu, Y., Gu, P., 1998. Product life cycle cost analysis: state of the art review. Int. J. Prod. Res. 36 (4), 883–908.

Auckland Council, 2015. Construction Environmental Management Plan.

Australian Bureau of Statistics, 2011. 8698.0 – Waste Management Services, Australia, 2009–10. Retrieved from: http://www.abs.gov.au/ausstats/abs@.nsf/Products/8698.0~2009-10~Main + Features~Waste + management + services#summary.

Australian National Audit Office, 2001. Life cycle costing – Better practice guide. Retrieved from: <http://www.anao.gov.au/uploads/documents/Life_Cycle_Costing.pdf>.

Babatunde, S.A., 2018. Government spending on infrastructure and economic growth in Nigeria. Econ. Res.-Ekon Istraž 31 (1), 997–1014. Available from: https://doi.org/10.1080/1331677X.2018.1436453.

Building Construction Authority, 2013. BCA Green Mark for new non-residential buildings version NRB/4.1. Singapore. Retrieved from: < https://www.bca.gov.sg/GreenMark/others/gm_nonresi_v4.1.pdf>.

Building Research Establishment Environment Assessment Method, 2014. BREEAM International New Construction Technical Manual (SD 5075 1.0:2013). Retrieved from: https://www.breeam.com/BREEAMInt2016SchemeDocument/#resources/output/10_pdf/a4_pdf/nc_pdf_printing/sd233_nc_int_2016_print.pdf

Building Research Establishment [BRE]. 2018. What is BREEAM? Retrieved from: https://www.breeam.com/

Burge, P.S., 2004. Sick building syndrome. Occupat. Environ. Med. 61 (2), 185–190. Available from: https://doi.org/10.1136/oem.2003.008813.

Cao, H., Folan, P., 2012. Product life cycle: the evolution of a paradigm and literature review from 1950–2009. Product. Plan. Contr. 23 (8), 641–662.

Cassidy, R., Wright, G., & Flynn, L., 2003. White paper on sustainability: a report of the green building movement. Building Design and Construction. Reed Business Information. Clearwater.

Castro-Lacouture, D., Sefair, J.A., Flórez, L., Medaglia, A.L., 2009. Optimization model for the selection of materials using a LEED-based green building rating system in Colombia. Build. Environ. 44 (6), 1162–1170. Available from: https://doi.org/10.1016/j.buildenv.2008.08.009.

Chen, Z., Li, H., Wong, C.T., 2000. Environmental management of urban construction projects in China. J. Constr. Eng. Manage. 126 (4), 320–324.

Cheung, S.-O., Lam, T.-I., Leung, M.-Y., Wan, Y.-W., 2001. An analytical hierarchy process based procurement selection method. Constr. Manage. Econom. 19 (4), 427–437. Available from: https://doi.org/10.1080/014461901300132401.

Chu, J., Chen, J., Wang, C., Fu, P., 2004. Wastewater reuse potential analysis: implications for China's water resources management. Water Res. 38 (11), 2746–2756. Available from: https://doi.org/10.1016/j.watres.2004.04.002.

Chung, S.-s, Lo, C.W.H., 2003. Evaluating sustainability in waste management: the case of construction and demolition, chemical and clinical wastes in Hong Kong. Resour. Conserv. Recycl. 37 (2), 119–145. Available from: https://doi.org/10.1016/S0921-3449(02)00075-7.

Clements-Croome, D., 2006. Creating the Productive Workplace. Taylor & Francis.

Crawford, R.H., Pullen, S., 2011. Life cycle water analysis of a residential building and its occupants. Build. Res. Inform. 39 (6), 589–602.

Crawley, D., Aho, I., 1999. Building environmental assessment methods: applications and development trends. Build. Res. Inf. 27 (4–5), 300–308. Available from: https://doi.org/10.1080/096132199369417.

Dibs, H., Al-Hedny, S., Abed Karkoosh, H.S., 2018. Extracting detailed buildings 3D model with using high resolution satellite imagery by remote sensing and GIS analysis; Al-Qasim Green University a case study. Int. J. Civ. Eng. Tech. 9 (7), 1097–1108.

Dilrukshi, H., Mallawarachchi, H., & Karunasena, G., 2014. Application of green building concept to enhance indoor environmental quality in hospital buildings in Sri Lanka. Paper Presented at the 3rd World Construction Symposium, Colombo, Sri Lanka.

Environmental Protection Authority, 1996. Environmental Guidelines for Major Construction Sites. Victoria, Australia. Retrieved from: <http://www.epa.vic.gov.au/~/media/Publications/480.pdf>.

Fay, R., Treloar, G., Iyer-Raniga, U., 2000. Life-cycle energy analysis of buildings: a case study. Build. Res. Inform. 28 (1), 31–41.

Frontczak, M., Schiavon, S., Goins, J., Arens, E., Zhang, H., Wargocki, P., 2012. Quantitative relationships between occupant satisfaction and satisfaction aspects of indoor environmental quality and building design. Indoor Air 22 (2), 119–131.

Ghaffarianhoseini, A., AlWaer, H., Omrany, H., Ghaffarianhoseini, A., Alalouch, C., Clements-Croome, D., et al., 2018. Sick building syndrome: are we doing enough? Archit. Sci. Rev. 61 (3), 99–121. Available from: https://doi.org/10.1080/00038628.2018.1461060.

Green Building Council of Australia, 2015. Green Star—Design & as built v1.1. Retrieved from: <https://www.gbca.org.au/green-star/green-star-design-as-built/the-rating-tool/>.

Herrmann, T., Schmida, U., 2000. Rainwater utilisation in Germany: efficiency, dimensioning, hydraulic and environmental aspects. Urban Water 1 (4), 307–316. Available from: https://doi.org/10.1016/S1462-0758(00)00024-8.

Hill, R.C., Bowen, P.A., 1997. Sustainable construction: principles and a framework for attainment. Construct. Manage. Econom. 15 (3), 223–239. Available from: https://doi.org/10.1080/014461997372971.

Hoffman, A.J., Henn, R., 2008. Overcoming the social and psychological barriers to green building. Organ. Environ. 21 (4), 390–419. Available from: https://doi.org/10.1177/1086026608326129.

Horner, R., El-Haram, M., Munns, A., 1997. Building maintenance strategy: a new management approach. J. Qual. Maint. Eng. 3 (4), 273–280.

Howard, N., 2000. Data for Sustainable Construction. UK. Retrieved from: <http://projects.bre.co.uk/sustainable/SusConstructionData.pdf>.

Hui, P., Wong, L., Mui, K., 2010. Occupant acceptance as a screening parameter for indoor environmental assessments. Facilities 28 (7/8), 338–347.

Huizenga, C., Abbaszadeh, S., Zagreus, L., & Arens, E.A., 2006. Air quality and thermal comfort in office buildings: results of a large indoor environmental quality survey.

Ijigah, E.A., Jimoh, R.A., Aruleba, B.O., Ade, A.B., 2013. An assessment of environmental impacts of building construction projects. Civil Environ. Res. 3 (1), 2222–2863. Retrieved from: <http://www.iiste.org/Journals/index.php/CER/article/viewFile/3871/3930>.

Illankoon, I.M.C.S., Tam, V.W.Y., Le, K.N., Shen, L., 2017. Key credit criteria among international green building rating tools. J. Clean. Prod. 164, 209–220. Available from: https://doi.org/10.1016/j.jclepro.2017.06.206.

Intergovernmental panel on climate change. 2014. Climate change 2014 synthesis report. Retrieved from https://www.ipcc.ch/pdf/assessment-report/ar5/syr/AR5_SYR_FINAL_SPM.pdf

International Organisation for Standardization [ISO], 2017. Buildings and constructed assests – Service life planning Part 5: Life cycle costing (Vol. ISO 15686-5). International Standard Organisation [ISO], Switzerland.

Kibert, C.J., 1994. Establishing principles and a model for sustainable construction. Paper presented at the Proceedings of the First International Conference on Sustainable Construction.

Kim, J., de Dear, R., 2012. Nonlinear relationships between individual IEQ factors and overall workspace satisfaction. Build. Environ. 49, 33–40. Available from: https://doi.org/10.1016/j.buildenv.2011.09.022.

Kneifel, J., 2010. Life-cycle carbon and cost analysis of energy efficiency measures in new commercial buildings. Energy Build. 42 (3), 333–340.

Kumaraswamy, M.M., Zhang, X.Q., 2001. Governmental role in BOT-led infrastructure development. Int. J. Proj. Manage. 19 (4), 195–205. Available from: https://doi.org/10.1016/S0263-7863(99)00069-1.

Langston, C., 2011. Estimating the useful life of buildings. Paper presented at the 36th Australasian University Building Educators Association (AUBEA) Conference. https://epublications.bond.edu.au/cgi/viewcontent.cgi?article = 1029&context = aubea_2011

Lavy, S., Fernández-Solis, J.L., 2009. LEED accredited professionals' perceptions affecting credit point adoption. Facilities 27 (13/14), 531–548. Available from: https://doi.org/10.1108/02632770910996360

Lyles, W.B., Greve, K.W., Bauer, R.M., Ware, M.R., Schramke, C.J., Crouch, J., et al., 1991. Sick building syndrome. South. Med. J. 84 (1), 65–71, 76. Retrieved from: <http://europepmc.org/abstract/MED/1986430>.

MacNaughton, P., Cao, X., Buonocore, J., Cedeno-Laurent, J., Spengler, J., Bernstein, A., et al., 2018. Energy savings, emission reductions, and health co-benefits of the green building movement review-article. J. Expo. Sci. Environ. Epidemiol. 28 (4), 307–318. Available from: https://doi.org/10.1038/s41370-017-0014-9.

Meyer, C., 2009. The greening of the concrete industry. Cem. Concr. Compos. 31 (8), 601–605. Available from: https://doi.org/10.1016/j.cemconcomp.2008.12.010.

Nolde, E., 2000. Greywater reuse systems for toilet flushing in multi-storey buildings – over ten years experience in Berlin. Urban Water 1 (4), 275–284. Available from: https://doi.org/10.1016/S1462-0758(00)00023-6.

Olanrewaju, A.L., Khamidi, M.F., Idrus, A., 2010. Building maintenance management in a Malaysian university campuses: a case study. Austr. J. Constr. Econom. Build. 10 (1/2), 101.

Peng, C.-L., Scorpio, D.E., Kibert, C.J., 1997. Strategies for successful construction and demolition waste recycling operations. Constr. Manage. Econom. 15 (1), 49–58. Available from: https://doi.org/10.1080/014461997373105.

Perera, B., Illankoon, I.M.C.S., Perera, W., 2016. Determinants of operational and maintenance costs of condominiums. Built-Environ. Sri Lanka 12, 1.

Poon, C.S., Yu, A.T.W., Ng, L.H., 2001. On-site sorting of construction and demolition waste in Hong Kong. Resour. Conserv. Recycl. 32 (2), 157–172. Available from: https://doi.org/10.1016/S0921-3449(01)00052-0.

Poon, C.S., Yu, A.T.W., Wong, S.W., Cheung, E., 2004. Management of construction waste in public housing projects in Hong Kong. Constr. Manage. Econom. 22 (7), 675–689. Available from: https://doi.org/10.1080/0144619042000213292.

Pulaski, M.H., Horman, M., Riley, D., Dahl, P., Hickey, A., Lapinski, A., et al., 2004. Field Guide for Sustainable Construction. Pentagon Renovation and Construction Program Office, Washington.

Riley, D., Pexton, K., Drilling, J., 2003. Procurement of sustainable construction services in the United States: the contractor's role in green buildings. Industry Environ. 26 (2), 66–69.

Robichaud, L.B., Anantatmula, V.S., 2011. Greening project management practices for sustainable construction. J. Manage. Eng. 27 (1), 48–57. Available from: https://doi.org/10.1061/(ASCE)ME.1943-5479.0000030.

Ross, Mario L.ópez-Alcalá, Arthur, A., Small, I., 2007. Modeling the private financial returns from green building investments. J. Green Build. 2 (1), 97–105. Available from: https://doi.org/10.3992/jgb.2.1.97.

Sanson, K., & O'Mahony, G.M.G, 2010. Construction Environmental Management Plan. APA group, Australia.

Sartori, I., Hestnes, A.G., 2007. Energy use in the life cycle of conventional and low-energy buildings: a review article. Energy Build. 39 (3), 249–257.

Sensors and Systems, 2011. Is the role of GIS and geomatics technology changing for architecture and infrastructure construction? Retrieved from: <http://sensorsandsystems.com/his-the-role-of-gis-and-geomatics-technology-changing-for-architecture-and-infrastructure-construction/>.

Shen, L.Y., Tam, V.W.Y., 2002. Implementation of environmental management in the Hong Kong construction industry. Int. J. Project Manage. 20 (7), 535–543. Available from: https://doi.org/10.1016/S0263-7863(01)00054-0.

Singh, A., Syal, M., Korkmaz, S., Grady, S., 2010. Costs and benefits of IEQ improvements in LEED office buildings. J. Infrastr. Syst. 17 (2), 86–94.

Singh, A., Syal, M., Korkmaz, S., Grady, S., 2011. Costs and benefits of IEQ improvements in LEED Office Buildings. J. Infrastr. Syst. 17 (2), 86–94. Available from: https://doi.org/10.1061/(ASCE)IS.1943-555X.0000046.

Spiegel, R., 2012. Green Building Materials: A Guide to Product Selection and Specification, third ed. Wiley, Hoboken, NJ.

Stanford, H.W., 2010. Effective Building Maintenance: Protection of Capital Assets. Fairmont Press, Boca Raton, FL/Lilburn, GA.

Sterner, E., 2002. Green Procurement' of Buildings: A Study of Swedish Clients' Considerations. Constr. Manage. Econom. 20 (1), 21–30. Available from: https://doi.org/10.1080/01446190110093560.

Tam, V.W.Y., Tam, C.M., 2006. A review on the viable technology for construction waste recycling. Resour. Conserv. Recycl. 47 (3), 209–221. Available from: https://doi.org/10.1016/j.resconrec.2005.12.002.

Tatari, O., Kucukvar, M., 2012. Sustainability assessment of US construction sectors: ecosystems perspective. J. Constr. Eng. Manage. 138 (8), 918–922.

Teo, M.M.M., Loosemore, M., 2001. A theory of waste behaviour in the construction industry. Constr. Manage. Econom. 19 (7), 741–751. Available from: https://doi.org/10.1080/01446190110067037.

Thormark, C., 2002. A low energy building in a life cycle—its embodied energy, energy need for operation and recycling potential. Build. Environ. 37 (4), 429–435.

Tumanyants, K., 2018. Economic impact of the change in tax rate on small enterprises of manufacturing and construction sectors: evidence from Russia 2006-2014. Bus. Econ. Horiz. 14 (3), 642–658. Available from: https://doi.org/10.15208/beh.2018.45.

United Nations Environment Programme. 2001. Energy and Cities: Sustainable Building and Construction: Summary of Main Issues. Retrieved from http://www.unep.or.jp/ietc/focus/EnergyCities1.asp

UN-Habitat, 2006. The Global Campaign on Urban Governance. Retrieved from: <www.unhabitat.org>.

US Environment Protection Agency, 2014a. Green Building. Retrieved from: <http://archive.epa.gov/greenbuilding/web/html/about.html>.

US Environment Protection Agency, 2014b. Sustainable Water Infrastructure. Retrieved from: <https://www.epa.gov/sustainable-water-infrastructure>.

US Green Building Council, 2007. Making the Business Case for High Performance Green Buildings. Washington, DC. Retrieved from: <http://www.usgbc.org/Docs/Member_Resource_Docs/makingthebusinesscase.pdf>.

US Green Building Council, 2011. Green Building and LEED Core Concepts Guide. US Green Building Council.

US Green Building Council, 2014. LEED V4 for Building Design and Construction. US Green Building Council, Washington, DC.

US Green Building Council Research Committee, 2008. A National Green Building Research Agenda. Retrieved from: <http://www.usgbc.org/resources/national-green-building-research-agenda>.

Waidyasekara, K.G.A.S., & Silva, M.L.D, 2014. A critical review of water studies in construction industry. Paper presented at the 3rd World construction symposium, Colombo, Sri Lanka.

Wallis, P., AICP, A.P., GISP, L., 2012. The Role of GIS Technology in Sustaining the Built Environment. Esri, New York.

Wardani, M.A.E., Messner, J.I., Horman, M.J., 2006. Comparing Procurement Methods for Design-Build Projects. J. Constr. Engin. Manage. 132 (3), 230–238. Available from: https://doi.org/10.1061/(ASCE)0733-9364(2006)132:3(230).

Wise, A.F.E., & Swaffield, J.A., 2002. 6—Solid Waste Storage, Handling and Recovery Water, Sanitary and Waste Services for Buildings, fifth ed. Butterworth-Heinemann, Oxford, pp. 81–93.

Wong, S.-K., Wai-Chung Lai, L., Ho, D.C.-W., Chau, K.-W., Lo-Kuen Lam, C., Hung-Fai Ng, C., 2009. Sick building syndrome and perceived indoor environmental quality: A survey of apartment buildings in Hong Kong. Habitat Int. 33 (4), 463–471. Available from: https://doi.org/10.1016/j.habitatint.2009.03.001.

World Green Building Council, 2015. Retrieved from: <http://www.worldgbc.org/index.php?cID = 1>.

World Watch Institute, 2015. Vision for a sustainable world. Retrieved from: <http://www.worldwatch.org/>.

World Wildlife Fund, 2015. Green building design. Retrieved from: <http://www.wwf.org.au/get_involved/change_the_way_you_live/green_building_design/>.

Wu, Q., Wang, L., Gao, H., Chen, Y., 2012. Current situation analysis on soil erosion risk management in development and construction project. Adv. Mater. Res. vol. 430–432, 1967–1969.

Wyatt, D.P., 1994. Recycling and serviceability: the twin approach to securing sustainable construction. Paper presented at the Proceedings of First International Conference of CIB TG.

Zainal, R., Teng, T.C., Mohamed, S., 2016. Construction costs and housing prices: impact of goods and services tax. Int. J. Econ. Finan. Issues 6 (7Special Issue), 16–20.

Chapter 2

Current Management Approach

Ruoyu Jin
Division of Built Environment, School of Environment and Technology, University of Brighton, Brighton, United Kingdom

2.1 DEFINITION OF SUSTAINABILITY AND SUSTAINABLE DEVELOPMENT

Sustainability was defined in the World Commission Report (also known as Brundtland Commission, 1987):

> *Development that meets the needs of the present without compromising the ability of future generations to meet their own needs.*

The triple bottom line of sustainability originally proposed by Elkington (1994) to cover the three major categories in terms of social, environmental, and economical aspects has been widely applied crossing fields, including business (Slaper and Hall, 2011), manufacturing (Gimenez et al., 2012), and ecology (Milne and Gray, 2013). Compared to sustainability, sustainable development (SD) recognizes the involvement of people, the planet, and wealth (Institute for International Urban Development, 2017). SD may cover more categories than sustainability beyond social, environmental, and economical aspects. SD can also be understood as the process to reach the goal of sustainability as indicated by Robèrt et al. (2002). Fig. 2.1 illustrates the key breakdowns that have been widely recognized within SD.

Besides the triple dimensions of SD described in Fig. 2.1, extra dimensions have also been proposed in SD studies. These extra dimensions include institutional and cultural components from various studies such as Valentin and Spangenberg (2000), Nurse (2006), and Herva et al. (2011). Five-dimensional SD was proposed by Ilskog (2008) as technical, economical, social/ethical, environmental, and institutional sustainability. Seven-dimensional SD was provided by Perlas (1994) covering aspects of social, cultural, ecological, economical, scientific, technological, and human oriented.

Sustainable Construction Technologies. DOI: https://doi.org/10.1016/B978-0-12-811749-1.00002-X

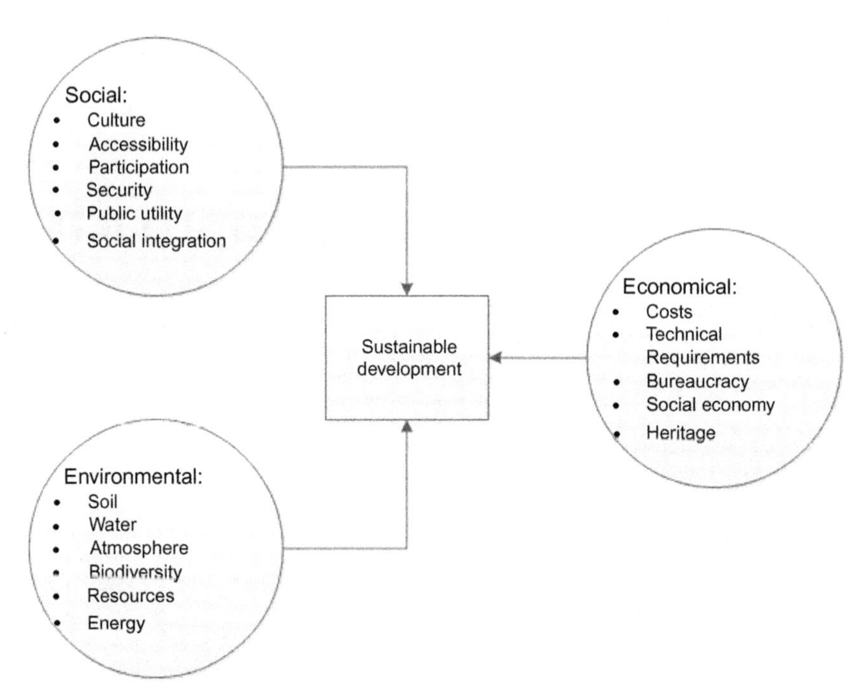

FIGURE 2.1 Breakdown structure of sustainable development. Source*: Adapted from Fernández-Sánchez, G., Rodríguez-López, F., 2010. A methodology to identify sustainability indicators in construction project management—Application to infrastructure projects in Spain. Ecol. Indicat. 10, 1193–1201.*

Human beings need buildings for sustaining their lives during civilization (Yılmaz and Bakış, 2015). Buildings consume a huge amount of resources and energy. They impact the climate change by affecting quality of urban air and water (Vyas et al., 2014). According to Dixon (2010), about 23% of air pollution, 50% of greenhouse gas production, 40% of water pollution, and 40% of solid waste in cities are caused by buildings. The high environmental impacts of buildings can be caused by the consumption of nonrenewable resources, decrease in biological diversity, destruction of forest areas, loss of agricultural areas, air, water and soil pollution, destruction of natural green areas, and global warming (Yılmaz and Bakış, 2015). On the other hand, the building construction industry could increase economical sustainability (CIB and UNEP-IETC, 2002) and provide job opportunities to low-income people due to its labor-intensive nature (Yılmaz and Bakış, 2015). It could contribute to the social sustainability for preventing poverty in the society (CIB and UNEP-IETC, 2002). The comprehensive measurement and assessment of sustainability in the building industry is being developed, tested, and updated.

2.1.1 Building Sustainability Rating Systems

Multiple tools providing indicators to measure and evaluate the SD within the building sector are available (Fernández Sánchez, 2008). Examples of sustainability indicator within the building sector include LEED (Leadership in Energy and Environmental Design, United States), BREEAM (The Building Research Establishment's Environmental Assessment Method, United Kingdom), and Green Star (Green Building Council, Australia and New Zealand), etc. The comparisons of these three sustainability evaluation systems in terms of history, certification level, number of certified projects, types of projects applied, evaluation criteria, and available versions are listed in Table 2.1.

TABLE 2.1 Comparisons Among LEED, BREEAM, and Green Star in the Building Sector

	LEED (U.S. Green Building Council, 2016a)	BREEAM	Green Star (Green Building Council of Australia, 2016)
Background	LEED was developed by USGBC and the first version was released in 2000	First launched by Building Research Establishment (BRE) in 1990, the world's first sustainability assessment method for buildings	Green Star was developed by GBCA in 2002 and its environmental rating system for buildings was launched in 2003
Historical versions	LEED v1.0 in 1998, LEED Green Building Rating System v2.0 in 2000, LEED New Construction (NC) v2.0 in 2005, LEED v3 in 2009, and LEED v4 in 2013	Since the launch of the first version in 1990, BREEAM Offices standard was released in 1998; the version of BREEAM for new homes was launched in 2000. In 2008, both the BREEAM scheme and its international version were launched. BREEAM New Construction was launched in 2011 and updated in 2014	Design & As Built v 1.0 was released in October 2014. The v1.1 was released in July 2015, and has been kept updated since then

(Continued)

TABLE 2.1 (Continued)

	LEED (U.S. Green Building Council, 2016a)	BREEAM	Green Star (Green Building Council of Australia, 2016)
Types of projects applied	Building design and construction, interior design and construction, building operations and maintenance, neighborhood development, homes	Communities, infrastructure, new construction, in-use buildings, refurbishment and fit-out buildings	Communities, design and as built, interiors, performance
Building categories[a]	New construction, core and shell, schools, retail, data centers, warehouses and distribution centers, hospitality, healthcare	Data centers, education, healthcare, Industrial, mixed use, office, residential, retails, other buildings	Design and construction of new buildings or major refurbishments
Evaluation criteria[a]	Location and transportation (LT), sustainable sites (SS), water efficiency (WE), energy and atmosphere, materials and resources (MR), indoor environmental quality (IEQ), Innovation (IN), regional priority (RP)	Energy, health and well-being, Innovation, land use, materials, management, pollution, transport, waste, and water	Management, indoor environment quality, energy, transport, water, materials, land use and ecology, emissions, innovation
Certification level	Certified, silver, gold, and platinum according to credit points earned from all criteria	Pass, good, very good, excellent, and outstanding according to the weighting system within the evaluation criteria	Communities, design & as built, and interiors projects could achieve from 4 to 6 star in the Green Star system, and performance projects rates from 1 to 6 star

(*Continued*)

TABLE 2.1 (Continued)

	LEED (U.S. Green Building Council, 2016a)	**BREEAM**	**Green Star (Green Building Council of Australia, 2016)**
Certified buildings worldwide	Up to April 2016, more than 79,100 LEED projects in over 160 countries and territories have gained LEED certification (USGBC, 2016c)	Up to August 2016, there have been more than 548,400 BREEAM certified developments and around 2,248,000 BREEAM registered buildings for assessments in 77 countries worldwide. Over 260,000 buildings had been certified in over 50 countries up to February 2014 (BREEAM, 2014)	More than 600 projects in Australia have achieved Green Star ratings. Green Star has also been adopted in New Zealand and South Africa
Achievements	LEED buildings achieve 20% lower maintenance costs (USGBC, 2016c), have 34% lower CO_2 emissions, consume 25% less energy and 11% less water, and have diverted more than 80 millions tons of waste from landfills (U.S. Department of Energy, 2011)	Limited information has been found on the quantified measurements of BREEAM-certified buildings. Instead, questionnaire-based survey results such as BSRIA report (2012) indicated positive feedback from BREEAM users. For example, nearly 60% of respondents believed that BREEAM could reduce construction waste and materials use	Compared to average Australian buildings, Green Star-certified buildings could produce 62% fewer greenhouse gas emissions, 66% less electricity, use 51% less portable water, and higher recycling rate of construction and demolition wastes (Green Building Council of Australia, 2013)

[a] *The building categories and evaluation criteria for LEED is based on the project type of Building Design and Construction from USGBC (2016b), equivalent to Design & As Built certification in the Green Star system (Green Building Council of Australia, 2016).*

Among the three rating systems within the building sustainability, LEED and BREEAM might be the two more widely adopted systems worldwide according to the number of certified buildings. However, compared to LEED and Green Star systems, which have quantitative data measuring benefits achieved from sustainability certification, limited data have been found in light of the BREEAM-certified projects, such as the reduction of greenhouse gas emission. All the three rating systems cover a comprehensive list of building categories, both new construction and refurbishment, crossing different stages of project life from design, construction, to operation and maintenance. To determine the certification level for a given project, LEED and Green Star adopt the scoring approach with certain credit points assigned to each individual scoring item under different evaluation criteria. The BREEAM uses the weighting method by assigning predetermined percentages distributed among the evaluation criteria. Similar major evaluation criteria can be found in these three rating systems, for instance, energy, materials, and waste management, etc. The detailed evaluation criteria and scoring/weighting methods are listed in Tables 2.2–2.4 using the new building construction as the example.

TABLE 2.2 Evaluation Criteria of LEED v4 New Construction

	Evaluation category	Evaluation item
LEED v4 New Construction and Major Renovation (U.S. Green Building Council, 2016b)	Integrative process (one credit)	
	Location and transportation (16 points)	Sensitive land protection (one point)
		High priority site (two points)
		Surrounding density and diverse uses (five points)
		Access to quality transit (five points)
		Bicycle facilities (one point)
		Reduced parking footprint (one point)
		Green vehicles (one point)
	Sustainable sites (10 points)	Construction activity pollution prevention (prerequisite)
		Site assessment (one point)
		Site development—protect or restore habitat (two points)
		Open space (one point)
		Rainwater management (three points)
		Heat island reduction (two points)
		Light pollution reduction (one point)

(*Continued*)

TABLE 2.2 (Continued)

	Evaluation category	Evaluation item
	Water efficiency (11 points)	Outdoor water use reduction (prerequisite)
		Indoor water use reduction (prerequisite)
		Building-level water metering (prerequisite)
		Outdoor water use reduction (two points)
		Indoor water use reduction (six points)
		Cooling tower water use (two points)
		Water metering (one point)
	Energy and atmosphere (33 points)	Fundamental commissioning and verification (prerequisite)
		Minimum energy performance (prerequisite)
		Building-level energy metering (prerequisite)
		Fundamental refrigerant management (prerequisite)
		Enhanced commissioning (six points)
		Optimize energy performance (18 points)
		Advanced energy metering (one point)
		Demand response (two points)
		Renewable energy production (three points)
		Enhanced refrigerant management (one point)
		Green power and carbon offsets (two points)

(Continued)

TABLE 2.2 (Continued)

	Evaluation category	Evaluation item
	Materials and resources (14 points)	Storage and collection of recyclables (prerequisite)
		Construction and demolition waste management planning (prerequisite)
		Building life-cycle impact reduction (six points)
		Building product disclosure and optimization—environmental product declarations (two points)
		Building product disclosure and optimization—sourcing of raw materials (two points)
		Building product disclosure and optimization—material ingredients (two points)
		Construction and demolition waste management (two points)
	Indoor environmental quality (10 points)	Minimum indoor air quality performance (prerequisite)
		Environmental tobacco smoke control (prerequisite)
		Enhanced indoor air quality strategies (two points)
		Low-emitting materials (three points)
		Construction indoor air quality management plan (one point)
		Daylight (three points)
		Quality views (one point)
	Innovation (six points)	Innovation (five points)
		LEED accredited professional (one point)
	Regional priority (four points)	

TABLE 2.3 Evaluation Criteria of BREEAM New Construction 2014 Version

	Evaluation category	Evaluation item
BREEAM New Construction (BREEAM, 2014)	Management	Project brief and design (four credits)
		Life cycle cost and service life planning (four credits)
		Responsible construction practices (six credits)
		Commissioning and handover (four credits)
		Aftercare (three credits)
	Health and well-being	Visual comfort (credits varying from three to six depending on building type)
		Indoor air quality (four to five credits depending on building type)
		Safe containment in laboratories (up to two credits depending on building type
		Thermal comfort (three credits available)
		Acoustic performance (up to three or four credits depending on building type)
		Safety and security (two credits available)
	Energy	Reduction of energy use and carbon (12 credits)
		Energy monitoring (one to two credits depending on building type)
		External lighting (one credit)
		Low carbon design (three credits)
		Energy efficient cold storage (two credits)
		Energy efficient transportation systems (three credits)
		Energy efficient laboratory systems (one to five credits depending on building type)
		Energy efficient equipment (two credits)
		Drying space (one credit)

(Continued)

TABLE 2.3 (Continued)

	Evaluation category	Evaluation item
	Transport	Public transport accessibility (one to five credits depending on building type)
		Proximity to amenities (up to two credits depending on building type)
		Cyclist facilities (up to three credits depending on building type)
		Maximum car parking capacity (up to two credits depending on building type)
		Travel plan (one credit)
	Water	Water consumption (five credits)
		Water monitoring (one credit)
		Water leak detection (two credits)
		Water efficient equipment (one credit)
	Materials	Life cycle impacts (up to six credits depending on building type)
		Hard landscaping and boundary protection (one credit)
		Responsible sourcing of materials (four credits)
		Insulation (one credit)
		Designing for durability and resilience (one credit)
		Material efficiency (one credit)
	Waste	Construction waste management (four credits)
		Recycled aggregates (one credit)
		Operational waste (one credit)
		Speculative floor and ceiling finishes (one credit)
		Adaptation to climate change(one credit)
		Functional adaptability (one credit)

(*Continued*)

TABLE 2.3 (Continued)

	Evaluation category	Evaluation item
	Land use and ecology	Site selection(two credits)
		Ecological value of site and protection of ecological features(two credits)
		Minimizing impact on existing site ecology(two credits)
		Enhancing site ecology (one to four credits depending on building type)
		Long-term impact on biodiversity (two credits)
	Pollution	Impact of refrigerants (three credits)
		NO_x emissions (two or three credits depending on building type)
		Surface water run-off (five credits)
		Reduction of night time light pollution (one credit)
		Reduction of noise pollution (one credit)

TABLE 2.4 Evaluation Criteria of Green Star Design & As Built v1.1 Updated in February 2016

Green Star Design & As Built v1.1	Evaluation category	Evaluation item
	Management (14 credit points)	Green Star accredited professional (one point)
		Commissioning and tuning (four points)
		Adaptation and resilience (two points)
		Building information (two points)
		Commitment to performance (two points)
		Metering and monitoring (one point)
		Construction environmental management
		Operational waste (one point)

(Continued)

TABLE 2.4 (Continued)

Green Star Design & As Built v1.1	Evaluation category	Evaluation item
	Indoor environment quality (17 credit points)	Indoor air quality (four points)
		Acoustic comfort (three points)
		Lighting comfort (three points)
		Visual comfort (three points)
		Indoor pollutants (two points)
		Thermal comforts (two points)
	Energy (22 credit points)	Greenhouse gas emissions (20 points)
		Peak electricity demand reduction (two points)
	Transport (10 points)	Sustainable transport (10 points)
	Water (12 points)	Potable water (12 points)
	Materials (14 points)	Life cycle impacts (seven points)
		Responsible building materials (three points)
		Sustainable products (three points)
		Construction and demolition waste (one point)
	Land use and ecology (six points)	Ecological value (three points)
		Sustainable sites (two points)
		Heat island effect (one point)
	Emissions (five points)	Stormwater (two points)
		Light pollution (one point)
		Microbial control (one point)
		Refrigerant impacts (one point)

The evaluation items within Table 2.2 for LEED v4 only apply to new construction. Depending on building type (e.g., Core and Shell, Schools, Retail, Data Centers, Warehouses and Distribution Centers, Hospitality, and Healthcare), the detailed evaluation items or the assigned credit points may slightly vary. LEED v4 sets prerequisites or required evaluation items to be met in its sustainability rating system. For example, outdoor and indoor water use reduction as listed in Table 2.2. A project will be evaluated according to the total credit points earned out of totally 110 points. The Platinum certification will be awarded to projects earned between 80 and 110 points. Following Platinum certification are Gold (60–79 points), Silver (50–59 points), and Certified (40–49 points).

Similar to LEED v4, the BREEAM (2014) also defines minimum standards for projects to achieve certain rating level. For example, a minimum of five credit points within the item of "Reduction of energy use and carbon emissions" are demanded in order to obtain the certification level of Excellent within BREEAM New Construction 2014 version. The detailed breakdown of evaluation items within BREEAM (2014) is summarized in Table 2.3.

Besides the nine defined categories which constitute the total weight of 100% within the sustainability criteria for new construction projects, up to an extra 10% innovation credits could be awarded "to support innovation within the construction industry through the recognition of sustainability related benefits which are not rewarded by standard BREEAM issues" as defined in (BREEAM, 2014). The sustainability credit points listed in Table 2.3 are not simply summed up to determine the rating level for projects under evaluation. Instead, each major evaluation category is weighted in the predefined BREEAM rating system. For example, the Management category is weighted at 12% for fully fitted out type of buildings. Based on the 100% weighting plus an extra 10% assigned in the innovation part, buildings that earn over 85% out of totally 110% would be certified as Outstanding, followed by Excellent (over 70%), Very Good (over 55%), Good (over 45%), and Pass (over 30%).

Compared to BREEAM New Construction 2014 version in which the credits assigned to some subcategories within the evaluation criteria are building type dependent, the Green Star Design & As Built v1.1 does not emphasize the linkage of building types to credit points assigned to each evaluation item. Table 2.4 summarizes the rating system of Green Star Design & As Built v1.1 updated by GBCA in February 2016.

Similar to BREEAM (2014), Green Star (Green Building Council of Australia, 2016) also specifies an extra 10% of credits in weighting the innovation criteria in terms of technology, market transformation, benchmarks, and addressing sustainability issues from other global rating tools outside the scope of Green Star. It is indicated that Green Star system encourages the update of sustainability credits by recruiting more sustainable evaluation items from other rating systems, since it is a relatively new system. Compared to LEED and BREEAM ratings, limited information can be found in light of the minimum requirements in the Green Star scoreboard list. Based on the overall score earned from the rating scale, a project gained over 75 points out of 100 is certified as Six Star within the category of World Leadership, followed by Five Star corresponding to the Australian

Excellence category from the range of 60–74 points, and Four Star representing the Best Practice category with overall credit points between 45 and 59.

It is worth noticing that the LEED (USGBC, 2016a) and Green Star (Green Building Council of Australia, 2016) rating systems in Tables 2.2–2.4 for building design and development cover buildings under new construction and renovation, while the BREEAM (2014) excludes the existing buildings' refurbishment and operation. In another word, BREEAM has a separate scheme (i.e., In-Use scheme) for the evaluation of existing building renovation. Besides the three introduced sustainability rating systems (i.e., LEED, BREEAM, and Green Star), there are other widely adopted multicriterion systems worldwide, such as the Japanese rating system CASBEE (Comprehensive Assessment System for Building Environmental Efficiency), the Chinese Three Star, and the U.S. Assessment and Rating System (STARS).

Nine major sustainability rating systems, including land use and ecology, energy, water, materials, indoor environmental quality (IEQ) and human comfort, management, pollution, transport, and innovation, are compared of their weightings among the three rating systems for new construction projects. The percentages of these major assessment categories within the three rating systems are displayed in Fig. 2.2.

Among the nine evaluation categories compared in Fig. 2.2, energy is highest weighted sustainability assessment criteria, with an average weight percentage at 22.3%. The rankings of average weightings for the remaining eight categories from highest to lowest are IEQ and human comfort (13.7%), materials (13.4%), transport (13.0%), water (9.7%), land use and ecology

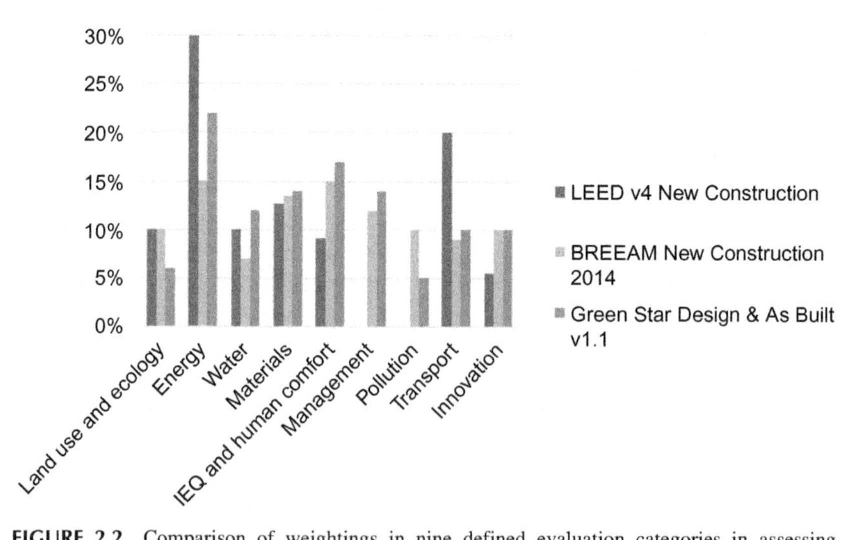

FIGURE 2.2 Comparison of weightings in nine defined evaluation categories in assessing building sustainability.

(8.7%), management (8.7%), innovation (8.5%), and pollution (5.0%). It is noticed that LEED v4 does not include separate categories in management and pollution. However, pollution and management related items fall into other categories in LEED v4. For example, construction activity pollution prevention is set as a prerequisite in the category of sustainable site, and the fundamental refrigerant management is a prerequisite in the category of energy. A similar comparison of major sustainability assessment categories can be found in Berardi's (2012) study in which more rating systems such as CASBEE were included.

2.1.2 Sustainability in Infrastructure

Although tending to be centered on the building sector, sustainability in construction has been introduced into infrastructure projects (Fernández-Sánchez and Rodríguez-López, 2010). There are existing rating systems for infrastructure as well. For example, the BREEAM's (2016) New Construction Infrastructure Technical Manual covers major categories for civil engineering project sustainability in terms of integrated design, resilience, stakeholders, local well-being, transport, land use and ecology, landscape and heritage, pollution, materials, carbon and energy, waste, and water. Other existing rating systems for infrastructure projects include BEST-in-Highways, Envision, GreenLITES, Greenroads, I-LAST and Invest, etc. Compared to the building sector, the sustainability in the infrastructure factor may cover a more comprehensive list of factors within the triple bottom line of sustainability. These factors include cost, energy consumption, resource requirements, capacity, service quality, safety, impacts on society, and impacts on the environment (Lee et al., 2011; Martland, 2012).

Summarized from these existing rating systems of infrastructure, the sustainability indicators can be further identified from the breakdown structure shown in Fig. 2.1. Table 2.5 lists these indicators identified by Fernández-Sánchez and Rodríguez-López (2010). In recent years, multiple stakeholders (e.g., governmental authority and developers) have raised more concerns of infrastructure resilience towards natural disasters or extreme weather conditions. Many developed high density cities worldwide are facing the unprecedented challenges including that their infrastructure systems need to make the cities sustainable and resilient against climate change and manmade threats (Yang et al., 2018).

2.1.3 Corporate Sustainability

Sustainability could be assessed at different levels, including the project level introduced in previous sections within building and infrastructure sectors, and the organization level. The adoption of SD at the firm level has been a success (Dyllick and Hockerts, 2002). Since mid-1990s, SD has been shifted

TABLE 2.5 Identified Sustainability Indicators for Infrastructure Projects (summarized from Fernández-Sánchez and Rodríguez-López, 2010)

Sustainability category	Breakdown	Indicators
Environment	Soil	Ecological value; erosion and sedimentation control plan; soil consumption
	Water	Water saving; water consumption; protection of water resources; control and monitor consumption
	Atmosphere	Ventilation; noise; GHG emissions; particulate emission and dust; odors; air quality; NO*x* and SO_2 emission
	Biodiversity	Impacts on the environment; protection of flora and fauna; Barrier effect of the project; natural heritage; ecological footprint
	Landscape	Visual impact
	Resources	Optimization of resources; civil engineering materials; equipment and materials with ecologic label; use of regional materials; materials with low health risks; use of durable materials
	Waste	Waste management
	Energy	Energy consumption; renewable energy; savings and energy efficiency; light pollution
	Risks	Mitigating the effects of floods and droughts; adaptation and vulnerability to climate change; infrastructure control-risk management
Social	Culture	Cultural heritage; built heritage; respect customs and beauty of the place
	Accessibility	Public access; human biodiversity access
	Participation of all actors	Public participation and control over the projects; public information; participation of associations and organizations; multidisciplinarity
	Security	Safety and health of workers; user security; impact on the global community; technical and environmental training; security of the infrastructures
	Public utility	Project declared of general interest; satisfaction of society; happiness
	Social integration	Local workers during construction, operation, and maintenance; raising levels of training and information; environmental campaigns; integration into the society
	Responsibility	Corporate social responsibility to the sponsor; environmental and sustainable awareness; necessity/urgency of the work

(*Continued*)

TABLE 2.5 (Continued)

Sustainability category	Breakdown	Indicators
Economy	Costs	Direct costs; indirect costs; cost/benefit of society; life cycle cost; cost incurred to users; local economy
	Technical requirements	Constructability; quality control; durability; functionality; research, development, and innovation in design; plain maintenance; operating manual; design for disassembly or change of use; environmental management accreditation; quality management accreditation; synergies with other projects
	Bureaucracy	Types of contracts; synergies with actors; product warranties, installations and set; project management; governance and strategic management of projects

Source: From Fernández-Sánchez, G., Rodríguez-López, F., 2010. A methodology to identify sustainability indicators in construction project management—Application to infrastructure projects in Spain. Ecol. Indicat. 10, 1193–1201.

toward business, and eco-efficiency became a guiding principle in corporate strategies (Dyllick and Hockerts, 2002). Transposing the concept of sustainability to the business context, corporate sustainability was defined by Dyllick and Hockerts (2002) as "meeting the needs of a firm's direct and indirect stakeholders (such as shareholders, employees, clients, pressure groups, communities, etc.), without compromising its ability to meet the needs of future stakeholders as well." Other definitions of corporate sustainability (e.g., Shrivastava, 1995; Starik and Rands, 1995) in the mid-1990s were highly weighted on ecological sustainability, while some other definitions of corporate sustainability in 2000s (e.g., Banerjee, 2003; Bansal, 2005) tended to be more comprehensive covering other criteria such as economical, social, and environmental aspects. The relationship between business and social sustainability has been receiving more attention (Dyllick and Hockerts, 2002).

Baumgartner and Ebner (2010) perceived corporate sustainability as SD incorporated by the organization. Similar to Dyllick and Hockerts (2002), Baumgartner and Ebner (2010) also considered corporate sustainability containing the three pillars of SD (i.e., economical, ecological, and social aspects). The triple dimensions within corporate sustainability were described by Baumgartner and Ebner (2010) with details as summarized in Table 2.6.

TABLE 2.6 Strategies of Corporate Sustainability

Dimensions within corporate sustainability	Strategies	Description
Economical aspect	Innovation and technology	Adoption of sustainable research and development to reduce environmental impacts in new products and business activities; integration of environmental technologies and concentration on cleaner production
	Collaboration	Cooperation and collaboration with various business partners; exchange of information and knowledge; working in common programs and networks on innovative products and technologies
	Knowledge management	Activities and approaches to keep sustainability related knowledge in the organization; methods to plan, develop, organize, maintain, transfer, apply and measure specific knowledge and to improve the organizational knowledge base
	Processes	Clear processes and roles defined for business activities to be efficiently conducted; adaptation of process management on sustainability necessities to implement corporate sustainability systematically; integration of sustainability into daily business life
	Purchase	Consideration of sustainability issues in purchase; awareness and consideration of sustainability related issues in the organization as well as alongside the supply chain; relationship with suppliers focusing also on sustainability
	Sustainability reporting	Consideration and reporting of sustainability issues within company reports
Ecological dimension	Evaluation of environmental impacts by corporate activities	Environmental impacts including: use of renewable and nonrenewable resources and energy

(*Continued*)

TABLE 2.6 (Continued)

Dimensions within corporate sustainability	Strategies	Description
		through the company including recycled resources; emissions into the air, the water, the ground, and wastes including hazardous wastes due to corporate activities; impact on biodiversity; environmental aspects of the product over the whole life cycle
Social aspects	Corporate governance	Transparency in all its activities in order to ameliorate relationship toward its stakeholders; giving insight into all relevant data; following rules of markets on corporate governance and defining responsibilities and behavior of the board
	Motivation and incentives	Active involvement and exemplary function of management on sustainability topics for employees; awareness of needs, claims, and motivation factors of employees in order to implement sustainability sufficiently into the organization; development of incentives and reward systems
	Health and safety	Ensuring no health and safety risks occur when working within the organization; no negative impact of employees' physical health at any time; operation of programs for employees to prevent dangers and to stay generally fit and healthy
	Human capital development	Development of human capital for sustainability related issues through specific programs; broad cross-working education in order to become aware of the different challenges and issues of corporate sustainability

Source: Adapted from Baumgartner, R.J., Ebner, D., 2010. Corporate sustainability strategies: sustainability profiles and maturity levels. Sustain. Develop. 18(2), 76–89.

2.1.4 Implementation of Sustainable Construction Management

The sustainability indicators in Tables 2.2–2.5 need to be addressed to implement sustainability in the construction industry. Tables 2.2 to 2.4 showcase three international rating systems for sustainability in the building sector. Although detailed breakdown of sustainability measurements may vary among these rating systems, a general list of main indicators can be found with focuses on energy consumption, indoor comfort, materials, and other key indicators. In comparison, there have not been sufficiently developed sustainability rating systems for infrastructure facilities. There is a need to establish an internationally recognized sustainability rating system for infrastructure projects. Among these indicators, energy consumption, one of the highest weighted evaluation categories in existing sustainability rating systems, is a major sustainability criteria in both building and infrastructure sectors. In the United States, buildings account for a large proportion of all energy resources including electricity and natural gas (Schlueter and Thesseling, 2008). Residential and commercial buildings in the United States currently consume about 40% of all primary energy (Costa et al., 2013). Potential energy savings could be achieved between 20% and 40% in the building sector (Department of Energy & Climate Change (DECC), 2012; International Energy Agency (IEA), 2012; US Energy Information Administration (EIA), 2012). Public institutions worldwide have been founded to achieve energy-efficient and low-emission buildings (Hong et al., 2015). For example, the Energy Performance of Building Directive (EPBD), established in 2012 by the European Union to require the energy performance certificate in the building sale or rental process as the way to achieve energy saving. Due to the rise of global sustainability, managing minimizing the energy consumption and carbon emissions over the full life cycles of buildings is undergoing a fast growth in the fields of construction and engineering (Wong and Zhou, 2015).

Besides energy performance, environmental impact assessment (EIA) should be embraced in sustainable construction, as construction is not an environmentally friendly process (Li et al., 2010). EIA was defined by Glasson et al. (2013) as a systematic process to examine the environmental consequences of development actions with emphases on prevention. Impact assessment practice worldwide has been widely used at the project level (Wood, 2003). Li et al. (2010) proposed an EIA model for construction processes intending to serve as an assessment tool to support contractors and other decision-makers to identity major environmental impact factors, and to optimize construction plans in early construction stages. In the typical construction management, both the designer and contractors have major influences on construction's environment impacts as indicated by Li et al. (2010), specifically, designers are responsible for choosing the building materials, and contractors would be involved in construction equipment and ancillary materials. Three categories of environmental impacts were defined by Li et al. (2010), namely ecosystems, natural resources, and human health.

2.2 INNOVATIVE APPROACH IN SUSTAINABLE MANAGEMENT

The innovative approach in sustainable management would be divided into construction method, digital technology, and project delivery process.

2.2.1 Construction Method

Traditional on-site construction is blamed on its generation of wastes in steel and cement, high water consumption, environmental pollution, and labor intensiveness. Modern methods of construction, or MMC, as defined by Alwan et al. (2016), represents new construction technologies where building components are manufactured and assembled in a factory in a controlled environment leading to cleaner production within the construction industry. The MMC is also defined as prefabrication construction, kit-house, or off-site construction in literature. MMC, for example, the prefabricated steel construction, leads to more sustainable construction due to the shortened construction period, reduced on-site labor, improved productivity, enhanced product quality, less land resource occupied in construction, water saving, reduced site noise, and decreased dust or site pollution. Prefabrication construction provides more control over the quality of components and construction safety (Molavi and Barral, 2016). The recyclable nature of materials could also decrease construction waste and reduce the environmental impact. Similar impacts of prefabrication on infrastructure projects (i.e., bridge) on sustainability can be found in the study of Khan (2015) in preventing project schedule delays and reducing environmental impact through off-site production and on-site installation. Prefabrication products, such as precast concrete structural members, offer designers and contractors significant advantages in terms of construction time, safety, environmental impact, constructability, and cost (Khan, 2015).

Prefabrication has been developed since 1970s and have been further improved in the past 30 years (Tam et al., 2015). The governmental policy could drive the prefabrication construction to move forward. Since 2001, Hong Kong Buildings Department joint with Planning and Lands Department have provided incentives to encourage the green-featured building components, including balconies, wider common corridors and lift lobbies, communal sky gardens, podium gardens, acoustic fins, sunshades and reflectors, wing walls, wind catchers and wind funnels (Tam et al., 2015). Further promoted prefabrication members in Hong Kong include precast wall panels, utility platforms, and noise barriers, etc. Prefabrication is viewed as a construction method to enhance sustainability in terms of reducing environmental impact (e.g., site waste and noise). Since recent years, giant construction markets, such as mainland China, is undergoing a rapid policy development of encouraging and demanding the implementation of prefabrication construction. The State Council of China announced the goal that the prefabrication rate in new building construction should reach 30% within 10 years starting in 2016. Chinese Government's *Development Guidelines of Construction Industry Modernization* released in 2016 requested that prefabrication buildings should account for higher than 20% of all new construction projects by 2020. Following the recently announced central government's strategic goals, more than 30 provinces or municipalities in mainland China has published the guidelines on prefabrication construction, including Shanghai and Zhejiang representing economically active and developed regions. Shanghai municipal government provides financial incentives for building projects with higher prefabrication percentage (i.e., over 45%). Zhejiang provincial government has incorporated prefabrication as part of the green building development plan. Legislation is under development to drive the implementation of prefabrication construction. It can be expected that more detailed standards and industry codes would be further developed in the near future to widen the applications of precast concrete and steel construction in mainland China. Jin et al. (2018) provided the big picture of countries that had been active in prefabricated construction (i.e., off-site construction) research in the recent decade. Besides China, the movement of prefabricated construction approach can be seen in multiple other regions including both developing and developed countries, such as Brazil (Matoski and Ribeiro, 2016), Malaysia (Akmam Syed Zakaria et al., 2018), and United Kingdom (Charted Institute of Buildings, 2018). For example, UK's industry report titled "Modernise or Die" strongly recommended the update of off-site construction as one of the two keys (the other is BIM or Building Information Modeling) to improve the performance of the construction industry (Farmer, 2016).

Currently, one of the major challenges in prefabrication construction is the lack of comprehensive cognition of the cost as identified by Mao et al. (2016). The total cost of prefabrication construction could be much higher than that of the conventional construction in China due to the production of prefabricated component, transportation, and design consultancy (Mao et al., 2016). However, these higher cost items are largely due to the lack of experience, skills, and market demand as China is still in the early stage of developing the prefabrication construction market, and the overall cost of prefabrication projects could be lower from the long-term perspective (Mao et al., 2016). Other case studies performed by Tam et al. (2015) showed that prefabrication could save up to 100% in external scaffolding, 50% of material usage, 30% reduction for site manpower, and extra 4%–6% saving from gross floor area exemption. With the cost being one of the major concerns in adopting prefabrication, it is fair to summarize that the long-term benefits of prefabrication construction could outweigh its drawbacks compared to in-situ construction. However, the requirements in management experience, multi-party collaboration, sophisticated industry standards in prefabrication members (e.g., precast concrete panels), governmental support, the market condition, and other external factors would influence the implementation of prefabrication construction.

Other challenges within the prefabrication construction are the procurement management as proposed by Molavi and Barral (2016). According to Molavi and Barral (2016), contractors and suppliers should be involved in the project design stage to ensure that fabrication, transportation, storage, and installation occur in a cohesive manner. It is indicative of the importance of collaboration among different trades such as modular site assembly integrated with electrical, mechanical, plumbing, and other building service facilities. An early communication among contractors, suppliers, and other project parties, together with qualified modular design and intensive administrative work, would be imperative in the modular construction as indicated by Molavi and Barral (2016). The experience of stakeholders, project designers, engineers, and contractors in prefabrication construction would be important in the successful delivery of prefabrication construction projects.

2.2.2 Digital Technology: Building Information Modeling

Building Information Modeling, or BIM, the newly emerged digital technology worldwide, is undergoing its fast growth in the architecture, engineering, and construction (AEC) industries. BIM is defined by Eastman et al. (2011) as one of the most promising developments in the AEC industries by providing the digital construction of accurate virtual models.

The demand for sustainable building facilities with minimal environmental impact is increasing (Biswas et al., 2008; Autodesk, Inc, 2009; Azhar, 2010; Azhar et al., 2010a). The strategic plan released by China State Ministry of Housing and Urban-Rural Construction (State Ministry of Housing and Urban-Rural Construction (SMHURC), 2013) required that green building projects in the provincial level should adopt BIM in both design and construction by 2016. A recent questionnaire survey by Jin et al. (2017) to AEC professional in mainland China released the information that BIM would have an increased application in China's green building projects in the follow-up years (Poirier et al., 2017). The early design and preconstruction of a building are the most critical phases for decision-making in sustainability (Azhar, 2010b). Traditional computer-aided design (CAD) lacks the capacity to perform sustainability analyses in these critical phases (Azhar et al., 2010a). In comparison, BIM, the tool that enables the process of collaborative design by using one universal computer model rather than separate sets of drawings (Alwan et al., 2016), drives the multidisciplinary information to be integrated within one model and motivates environmental performance analysis and sustainability measurements (Schlueter and Thesseling, 2008; Azhar et al., 2010b). The technical core of BIM is software tools that enable 3D modeling and information management (WSP, 2013). Additional information, such as sustainability and maintenance information, can be inserted into the BIM framework (Wong and Zhou, 2015). Kriegel and Nies (2008) indicated that BIM could aid in sustainable design in these areas:

- building orientation;
- building form and envelope;
- daylighting analysis;
- water harvesting;
- energy modeling;
- sustainable materials;
- site and logistics management.

From the technical and scientific aspects, BIM, as the trend of the AEC industries in the near future, has its potential of being applied in the "green" building project life cycle. Green BIM is defined by McGraw-Hill Construction (2010) as the application of BIM tools to achieve sustainability and/or improved building performance. Alawini et al. (2013) stated that Green BIM is a tool to help the integration of sustainable components, such as energy efficiency application, into the building project life cycle. While the definition of Green BIM is being kept updated in both academia and industry, Wong and Zhou (2015) summarized Green BIM as "a model-based process of generating and managing coordinated and consistent building data during its project life cycle that enhance building energy-efficiency performance, and facilitate the accomplishment of established sustainability goals."

A key technological issue when linking BIM into building sustainability design is the digital data sharing and exchange in certain format such as Industry Foundation Class (IFC). The capacity of data exchange within the BIM-based platform would be related to the BIM maturity levels defined by BIM Industry Working Group (2011)'s Strategy Paper. Table 2.7 summarized the BIM maturity levels.

BIM Level 2, as described in Table 2.7, is required to be implemented in the UK industry by 2016 according to BIM Industry Working Group (2011). It is expected that the BIM maturity level will move toward web-based integrated data that enable building information exchange in the near future. BIM would play its higher potential in green building design as well as the building life cycle assessment (LCA).

COBie, representing Construction Operations Building Information Exchange, is a data format (i.e., IFC) for the publication of a subset of building information focusing on delivering building data (coBuilder, 2015). As a

TABLE 2.7 BIM Maturity Levels

BIM maturity level	Key feature	Standard or digital platform	Data exchange format
Level 0	2D-based computer-aided design without data sharing or integration among different disciplines	CAD	Paper drawings or electronic paper
Level 1	2D or 3D CAD with a collaboration tool	BS 1192:2007	A common data environment possibly with some standard data structures and formats but without integration
Level 2	3D environment held in separate discipline BIM tools with certain integration based on proprietary interfaces, possibly with 4D program data and 5D cost elements	pBIM (proprietary)	Files-based collaboration and BIM library management
Level 3	Fully open process and data integration enabled by "web services"	iBIM compliant with IFC standards	Integrated and interoperable data in web services

Source: Summarized from BIM Industry Working Group, 2011. Strategy Paper for the Government Construction Client Group.

means of information exchange among parties, COBie is a vehicle for sharing nongraphic data about a facility to ensure that the owner, operator, and occupants receive the information about the facility (BIM Industry Working Group, 2011). Linked into the green building design and follow-up life cycle stages in the operation, COBie can build, store, and provide the building information of green building products (e.g., double-glazed low-e window). Fig. 2.3 described the product information exchange process in COBie.

BIM Level 3 incorporated with COBie would enhance the sustainable design, construction, and operation of buildings through integrated building information crossing the major building life cycle stages. It is since the late 1990s that the potential of computational assessment methods and tools allowing actual environmental performance assessment of buildings have

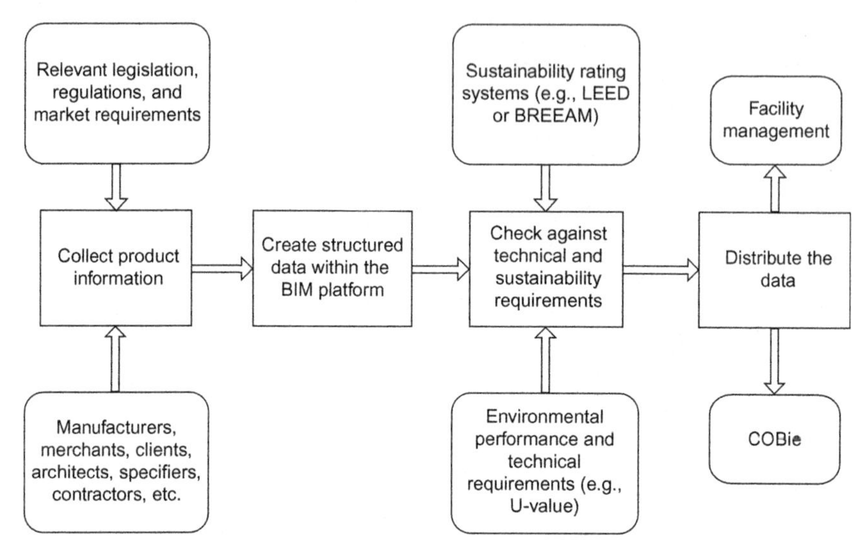

FIGURE 2.3 The building product exchange process. Source: *Adapted from coBuilder, 2015. BIM data (COBie/IFC). NEC Birmingham.*

been emphasized (Curwell et al., 1999). For example, Alwan et al. (2015) examined the feasibility of applying BIM in the building design against criteria in LEED. The technological approach through streamlined, architectural and data exchange in the BIM platform was found effective in assisting achieving LEED certification and general low carbon design (Alwan et al., 2015). Other recently performed BIM-driven building sustainability studies include the BIM used in LEED energy simulation by Ryu and Park (2016), whose research involved the data format verification when transporting the architectural geometry model into the follow-up energy simulation. More existing studies of BIM applications in building sustainability conducted between 2004 and 2014 are summarized in Table 2.8.

These previous studies listed in Table 2.8 covered BIM applications in different building project stages, from design, construction, operation, retrofitting, to demolition. Despite the potential of BIM capacities in enhancing building sustainability, the Green BIM's application in sustainable design and construction is still undergoing the development. Wu and Issa (2013) concluded that the adoption rate of BIM in green building projects is still low and its potential needs further exploration because of the limited knowledge of BIM from practitioners. More practices of digital technologies' application in green building management are expected in the future construction management, especially over the building life cycle as recommended by Wong and Zhou (2015), although conventional BIM is more commonly applied in the early stages of building life cycles with fewer applications in construction, maintenance, and demolition stages

TABLE 2.8 BIM Applications in Multiple Project Delivery Stages

Project stage	Implemented or suggested BIM application(s)	References
Building design	BIM-extended tool enabling sustainable design decision making	Inyim et al. (2015)
	BIM-extended "Evolutionary Energy Performance Feedback for Design" tool that guides energy performance at the early design stage by offering rapid iteration with performance feedback	Lin and Gerber (2014)
	BIM suggested to improve the building delivery performance and minimize unnecessary environmental impact through enhanced multiparty communication and reduced design errors	Clevenger and Khan (2014)
	BIM applied in the sustainability rating and certification of green buildings with BIM incorporated to assist environmental performance evaluation	Biswas et al. (2008)
	Autodesk Revit BIM tool suggested to assess and document the LEED rating system	Barnes and Castro-Lacouture (2009); Azhar et al. (2010a; 2010b)
	BIM's potential application for the Australian Green Star Building certification	Gandhi and Jupp (2014)
	BIM explored in Hong Kong's sustainable building certification process	Wong and Kuan (2014)
	Integrated system incorporating BIM, energy performance analysis, and cost estimate to assess a variety of sustainable building design options	Jalaei and Jrade (2014)
	Existing commercially available BIM tool (e.g., Autodesk Revit) developed to help designers in converting conceptual designs into energy analytical models; other software tools developed to support the sustainability design decisions such as daylighting and solar access; tools suggested to automate the material quantity takeoffs	Wong and Zhou (2015)

(Continued)

TABLE 2.8 (Continued)

Project stage	Implemented or suggested BIM application(s)	References
Building construction	BIM systems adopted to monitor site carbon emissions	Hajibabai et al. (2011); Stadel et al. (2011)
	BIM tool developed to assist project team members in estimating and visualizing carbon emission in construction	Wong et al. (2013)
	BIM integrated with geographic information system (GIS) to optimize concrete truck mixer routes and minimize emissions	Artenian et al. (2010)
Building operation	Analysis of heating and cooling requirements	Tzivanidis et al. (2011)
	Finding daylighting opportunities and reducing artificial lighting load	Novitski (2009)
	Selection of proper building equipment to reduce energy use	Novitski (2009)
	Existing BIM tools (e.g., Autodesk Green Building Studio) expected to compute energy consumption, to promote energy efficiency, and to reduce resource waste	Wong and Zhou (2015)
Building retrofitting	BIM-based knowledge-sharing system as a platform for facility managers and maintenance teams to learn from preceding experience and record	Motawa and Almarshad (2013)
	BIM integration in implementing sustainable design principles into the retrofitting of existing buildings	Hammond et al. (2014)
	BIM platform containing data access and exchange (e.g., security and data privacy issues) for energy efficient retrofitting work	Jiang et al. (2012)
Building demolition	BIM tool developed to identify and estimate construction and demolition (C&D) wastes by allowing practitioners to develop a more efficient material recycling plan before the actual demolition or renovation	Associated General Contractors of America (2005)
	BIM system established to extract the information on volumes and materials	Cheng and Ma (2012; 2013)

(*Continued*)

TABLE 2.8 (Continued)

Project stage	Implemented or suggested BIM application(s)	References
	for building elements by incorporating the information for detailed waste estimate and planning	
	BIM-based model for assessing the impacts of various building demolition options in terms of their economical costs and environmental benefits	Akbarnezhad et al. (2014)

(Eadie et al., 2013). Future green BIM should incorporate three R's concept (i.e., reduce, reuse, and recycle) in the sustainability studies for both new development and retrofitting projects, and it is necessary to use cloud-based BIM technology to enable the management of building sustainability through "big data" (Wong and Zhou, 2015).

The integration of cloud-based technology and BIM provides an innovative approach through information exchange during the construction, and also a potential for better sustainability management over the building life cycle (Wong and Zhou, 2015). Currently there are a few challenges of implementing Green BIM, including but not limited to

- lack of suitable green BIM software tools and complicated models (McGraw-Hill Construction, 2010);
- technical compatibility issues such as interoperability issue when transporting data stored in BIM into energy simulation (Moon et al., 2011);
- cost of software licenses (Pelsmakers, 2013);
- cultural resistance to adopt BIM in the industry (Dawood and Iqbal, 2010);
- insufficient guidelines and standards from authorities (Smith and Tardif, 2009);
- insufficient BIM education and training (Tang et al., 2015).

2.2.3 Project Delivery Methods

BIM represents a new paradigm that motivates the integration of the roles of all stakeholders on a project by assisting architects, engineers, and constructors to visualize the simulated built environment and to identify any potential design, construction, and operational issues (Azhar, 2011). From the perspective of project delivery method, a collaborative process would be considered important in enabling and motivating the adoption of BIM in construction projects. A more recently emerged project delivery method, commonly referred as integrated project delivery (IPD), could more effectively facilitate the use of BIM in project management (Kent and Becerik-Gerber, 2010). IPD was defined by American Institutes of Architects (2007) as

> *A project delivery approach that integrates people, systems, business structures and practices into a process that collaboratively harness the talents and insights of all participants to optimize project results, increase value to the owner, reduce waste, and maximize efficiency through all phases of design, fabrication, and construction.*

According to American Institutes of Architects (2007), integrated processes are being acknowledged and encouraged in rating systems such as LEED, and a complex interactions of systems must be taken into account in order to achieve higher reduction in energy and carbon. Jones (2014) claimed that IPD and BIM formed essential tools and strategies in decision-making, and in particular, the collaborative system would assist the design process to deliver sustainable buildings. In the infrastructure sector, integrated project contracts, defined as combining the stages of design, construction, and maintenance into one arrangement, could lead to more SD in the project life cycle (Lenferink et al., 2013). The effects of IPD and BIM to building sustainability could be illustrated in Fig. 2.4.

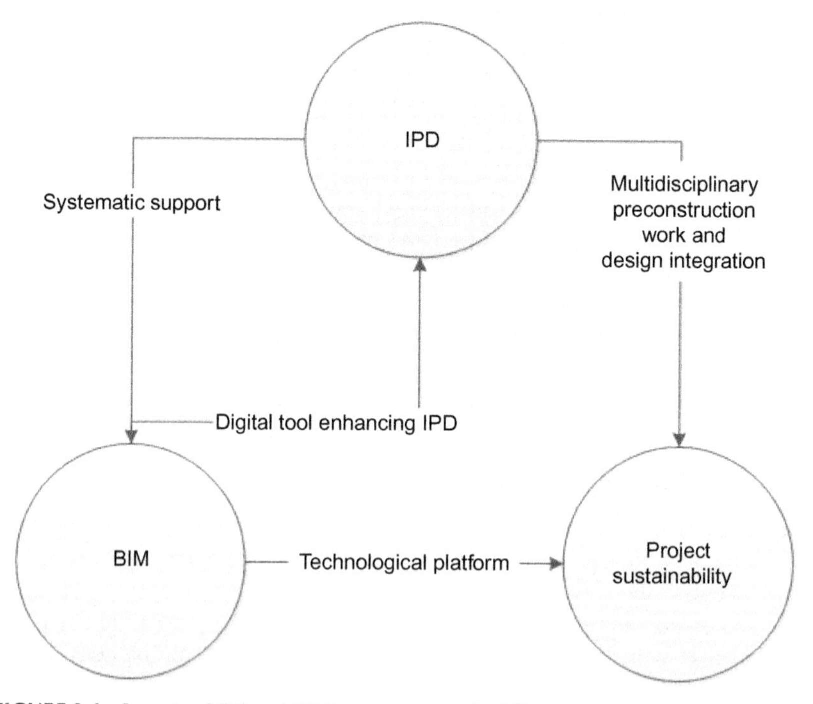

FIGURE 2.4 Impacts of IPD and BIM on project sustainability.

Fig. 2.4 can be understood as the relationships among the innovative project delivery method (i.e., IPD), digital technology (i.e., BIM), and project goals (i.e., sustainable design and construction). BIM and IPD could provide mutual support to each other in that BIM enhances the implementation of IPD as a digital tool enabling multidisciplinary collaboration in project design and construction. Specifically, BIM combines design, fabrication information, erections, and project management logistics in one database and provides the collaboration platform (American Institutes of Architects, 2007). BIM can also be used as the facility management tool by the owner in monitoring long-term energy performance, maintenance, and remodeling following construction completion (American Institutes of Architects, 2007). IPD, on other hand, could provide the collaborative environment and systematic support to BIM adoption in the project. It was found by Kent and Becerik-Gerber (2010) that although BIM was not a prerequisite in IPD process, those industry practitioners with more IPD experience were more liked to adopt BIM and use more capabilities of BIM. Sustainability-based decisions can be achieved and assessed in early project stages (i.e., preconstruction and design) with multiparty involvement. The success of IPD is measured by expressly stated shared goals, one of them being the sustainability (American Institutes of Architects, 2007). Fig. 2.5 illustrates the

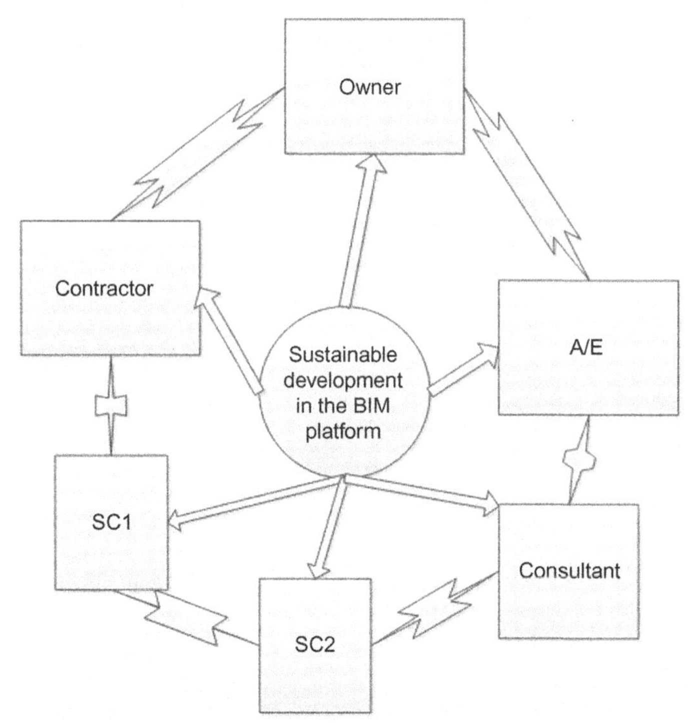

FIGURE 2.5 Working mechanism of BIM-enabled sustainability within the IPD-based multiparty collaboration.

working mechanism of BIM-driven sustainability in the multiparty-involved IPD process. These multiple project parties shown in Fig. 2.5 include A/E (i.e., architects and engineers) as well as various SCs (i.e., subcontractors).

The IPD process would enable all key project participants, including the owner, to be involved in the collaborative work in the same "Big Room" to define the sustainability and cost goals of the project (Jones, 2014). The key features of BIM and IPD integrated sustainability process can be summarized in Table 2.9 according to Jones (2014).

It is worth noticing that despite the promising development of BIM-incorporated IPD in assisting project sustainability, the applications of IPD projects in the real world may still be limited. The challenges encountered when implementing IPD equipped with BIM platform in the green building design and construction could be further explored through more case studies. As BIM is adopting more dimensions, such as scheduling-based 4D and cost-based 5D, BIM's potential in enhancing the collaborative IPD workflow could be expected. However, implementation of IPD is still relatively uncommon in the construction industry worldwide compared to more traditional approaches, for example, design–bid–build (i.e., hard-bid). Similar to IPD, other fast-track-featured project delivery methods, such as design-build and construction management at-risk, could allow more opportunities in incorporating BIM as the digital platform, and also the

TABLE 2.9 Sustainability Driven by BIM and IPD

Features	Description
"Big Room" concept	Entire planning and design carried out in the early project stage in light of construction management, MEP engineering, energy technology, building physics, acoustics, façade construction, and other specialists
Evaluation and procurement	Evaluation of sustainability goals through using local resources, and procurement of materials with minimal pollutions and recyclable
Review of functional units	Sustainability analysis in the areas of efficient envelope design, solar and efficient lighting, construction systems, energy requirements, life-cycle maintenance costs, air quality health impact, and design for safety
Collaborative data model	The integrated and transparent construction process enabling the shared building model and data among all project team members

Source: From Jones, B., 2014. Integrated project delivery (IPD) for maximizing design and construction considerations regarding sustainability. Integrated project delivery (IPD) for maximizing design and construction considerations regarding sustainability. Proced. Eng. 95, 528–538.

sustainability-oriented collaboration. These collaborative design and construction environment in IPD or other fast-track project delivery processes are largely based on the trust working relationship among project parties. Long-term business relationship and previous collaboration experience could make it easier to implement IPD or other fast-track processes. Similar statements could be found from other studies such as Kent and Becerik-Gerber (2010) that "trust, respect, and good working relationships are the key to successful IPD projects."

2.3 LIFE CYCLE ASSESSMENT

2.3.1 Definition of Life Cycle Assessment

LCA is defined by ISO 14040 (2016) as a tool for environmental management of product and services systems, and it encompasses the assessment of impacts on the environment from the extraction of raw materials to the final disposal of waste. European Commission (2016) defines LCA as "social and environmental assessments of supply chains and end-of-life waste management." U.S. EPA, SAIC (2006) defined LCA as "a cradle-to-grave approach for assessing industrial systems that evaluates all stages of a product's life." LCA consists of four major stages: (1) definition of goal and scope, (2) Life cycle inventory (LCI) analysis, (3) life cycle impact assessment (LCIA), and (4) Interpretation (U.S. Energy Information Administration (EIA), 2012). The detailed descriptions of these steps are provided below according to American Institute of Architects (2010) and U.S. EPA (2012).

1. *Definition of goal and scope:* intended application, the functional unit, system boundaries and life cycle phases for analysis are established at this stage. The type of analysis and impact categories are to be evaluated. The functional unit is the important basis to compare and analyze inputs and outputs.
2. *Life cycle inventory (LCI) analysis:* the energy and raw materials consumed and the emissions to atmosphere, energy consumption, liquid discharges, and solid wastes are quantified in these processes including raw material extraction and processing, materials production, manufacturing of finished products, transportation, lifetime operation/use, and disposal/recycling. An inventory of all the inputs and outputs of the system are identified and quantified throughout the life cycle. Software tools and databases are critical in this inventory analysis stage. Software tools can be as simple as spreadsheet, or be more complexed such as cost-estimate tool to automate tabulation of material quantities from assemblies.
3. *Life cycle impact assessment (LCIA):* this stage provides indicators to analyze the potential environmental impacts of resource extractions and emissions in an inventory (Pennington et al., 2004). It provides an evaluation of the magnitude and significance of a product life cycle quantified

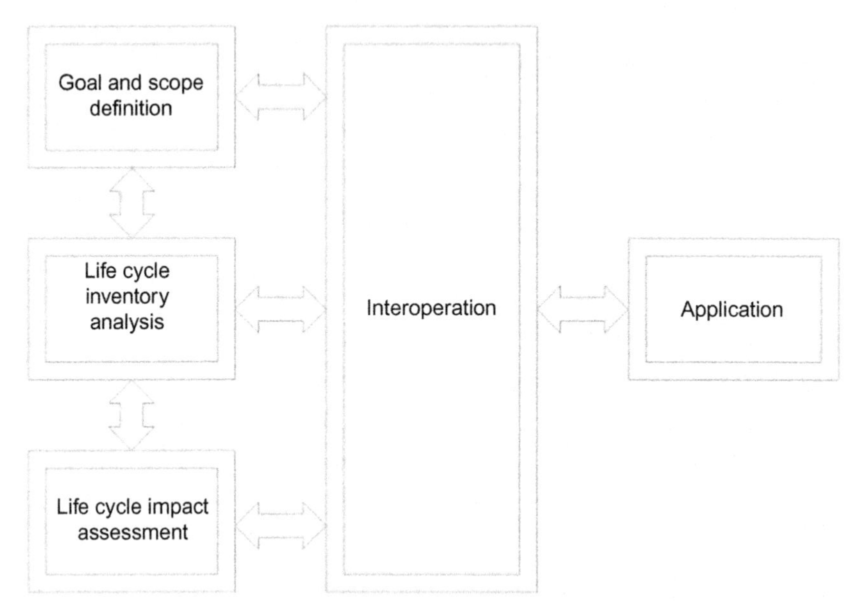

FIGURE 2.6 Four major stages of LCA according to ISO 14040, 2006.

in functional units, in terms of various influence factors such as land use, noise, climate change, etc. These categories may be weighted for importance. In other words, data from the inventory analysis in Step (2) is attributed to impact categories defined in Step (1). Impact assessment differ among LCA tools, one of them could be BEES LCA tool listed as an example by American Institute of Architects (2010).

4. *Interoperation:* interoperation occurs at all stages of the LCA with the purpose of identifying the potential areas, influences and major burdens for reducing the environmental impacts.

The interconnections among these four major stages of LCA can be demonstrated in Fig. 2.6.

Relevant terminologies within LCA are defined by American Institute of Architects (2010) as summarized in Table 2.10.

LCA can be applied in the sustainable design and construction of building projects. In order to deliver environmentally sustainable buildings, the environmental impacts at each stage of their life cycle must be assessed, and among these stages, raw material extraction, manufacturing, transportation, deconstruction and recycling are of great importance (Australian Government & Forest and Wood Products Research and Development Corporation, 2006). The typical life cycle of a building from raw material supply to demolition is illustrated in Fig. 2.7.

According to American Institute of Architects (2010), compared to the baseline building, a green building containing high-performance insulation

TABLE 2.10 Definitions of LCA-Related Terminologies

Terminology	Definition according to American Institute of Architects (2010)
Impact category	Each category is an indicator of the contribution of a product to a specific environmental problem. These categories are defined by LCIA methods. The impact categories of LCA methodologies vary from system to system
Functional unit	It can be defined as the unit of comparison that assures that products are being compared based on an equivalent level of function or service
System boundary	It is defined as interface between a product system and the environment or other product systems. It defines the activities and processes that will be included or excluded in each life-cycle stage for the LCA analysis
LCI database	LCI data is the key in any LCA analysis. A database may contain inputs and outputs of for each unit process within a product system. The data in the LCI databases generally account for raw material extraction, transportation to manufacturing unit, manufacturing process, and packaging and distribution
Life cycle management	It is a framework that utilizes methods like LCA and life cycle costing to support decisions leading to sustainable development
Life cycle costing (LCC)	Compared to LCA, which is based on environmental benefits of a system or product, LCC provides decision support on financial benefits, with the basis for contrasting initial investments with future costs over a specified period of time. It involves the systematic consideration of all relevant costs and revenues associated with the acquisition and ownership of an asset

and glazing or photovoltaics, although with higher initial embodied energy, could turn out with less overall impact of the embodied energy from the long-term building operation.

2.3.2 Life Cycle Assessment in Construction Management

Generally, a project delivery process would cover these major stages: predesign, design, procurement, construction, closeout, maintenance and operation, and the final demolition. Literature has more focused on reducing operational impacts of buildings and their emission to the environment as the cradle-to-gate impacts related to buildings including material extraction, manufacturing and transportation are often ignored and outweighted by building operation (Russell-Smith and Lepech, 2015). On the other hand, the need to understand embodied energy becomes more important and an LCA including the materials

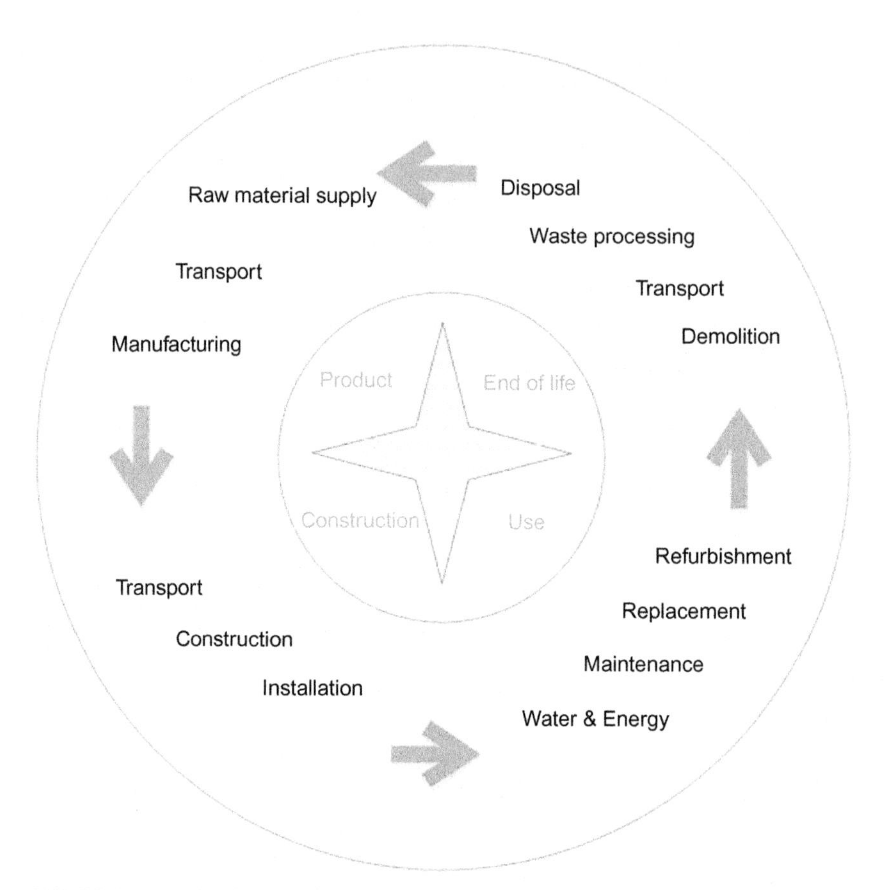

FIGURE 2.7 A description of building life cycle. Source*: From Bribián I.Z., Usón, A.A., Scarpellini, S., 2009. Life cycle assessment in buildings: state-of-the-art and simplified LCA methodology as a complement for building certification. Build. Environ. 44, 2510–2520.*

manufacturing and construction phase of a project is the primary means of computing the embodied energy (American Institute of Architects, 2010). Decisions made on early stages of project planning and design, for example, a site waste management plan as recommended by Mcdonald and Smithers (1998), could have a higher influence over the later project stage. The project management research, back to 1970s, has shown that management activities in the earliest stage of the project exert a significant influence over a project's outcome (Paulson, 1976; Baker et al., 1988). The well-known cost influence curve, widely used in the construction management field, such as Rocque (2003), Lu et al. (2014), and Osborn Engineering (2015), demonstrating that the early stage decision tends to have a higher degree of impact over the project cost performance. The planning and design stages are the point when most basic decisions on sustainability, energy use, and environmental design are made

(Liu et al., 2011). It can also be inferred that the earlier planning and design in the project life cycle stage could influence more on the project sustainability outcomes. As architects are generally involved in the project life cycle at an earlier stage, American Institute of Architects (2010) claimed that "the greatest incentive for the use of LCA in the design process is the ability of an architect to show to the client that the use of LCA will improve and demonstrate the "greenness" of the project and help significantly increasing long-term paybacks by better decision-making."

Compared to the traditional construction management methodologies, it is suggested that adjustments in motivating cross-disciplinary coordination, construction techniques and building systems should be made early in the project life cycle within the sustainable project management (Robichaud and Anantatmula, 2011). The incorporation of a charrette in the early stage of a construction project is recommended by Robichaud and Anantatmula (2011) as a solution to improving communication within a project stakeholder group. Charrette is defined by The National Charrette Institute (2016) as "Engage the creative moment with all key participants collaborating on a feasible action plan."

Several key strategies defining the process of charrette was summarized by Lennertz (2003) as listed below.

- *Work collaboratively:* all interested parties should be involved from the beginning.
- *Design Cross-functionally:* all project parties meet for a public briefing, discussion or presentation periodically enhancing the collaboration among representatives of different disciplines such as architects, planner, engineers, market experts, and public agency staff.
- *Compress work sessions:* a typical 4-to-7-day charrette process enables time compressive by accelerating decision-making and reducing unconstructive negotiation tactics.
- *Communication in short feedback loops:* simultaneous brainstorming and negotiation during a charrette enables changing minds and encouraging unique solutions to problems. The process of design and feedback is shortened.
- *Study the details and the whole:* designed details can be accomplished by looking at both details (e.g., building types, block sizes, and public space) and the big picture (site circulation, land use, and public amenities) in a concurrent approach.
- *Confirm progress by measuring outcomes:* measuring progress through agreed-upon outcomes assures the transparency of the decision-making process.
- *Produce feasible plans:* every decision point must be fully informed by the legal, financial and engineering disciplines. The implementation tools such as codes and regulations form the basis of the success of the plan.

- *Use design to achieve a shared vision and create holistic solutions:* the charrette design team specializes in capturing ideas quickly in drawings to enhance the discussion.
- *Hold the charrette on-site or near the site:* a charrette studio either in the neighborhood or on the project site can be set up to allow project participants to have quick access to the site.

Table 2.11 compares the differences between sustainable project management approach and conventional construction in the project life cycle adjusted from Robichaud and Anantatmula (2011).

It is indicated from Table 2.11 that the integrated approach of involving multiple parties in the early project life cycle stage, including stakeholder, regulation authorities, the owner, the design firm, contractors, and extra consultants, is a key feature for sustainable project management. Compared to the traditional project, sustainable projects could provide more supports for designers or project team members to visualize the feasibility of early design decisions according to Wong and Zhou (2015). LCA can assist in the project design stages, namely predesign, schematic design, and design development (American Institute of Architects, 2010). American Institute of Architects (2010) listed the LCA approach in enhancing the design process:

- Making choices among different design options;
- Making choices among different structural systems, assemblies and products;
- Identifying products or assemblies causing the maximum and minimum contribution to the overall environmental impact throughout the building's life cycle;
- Identifying stages of the building life cycle causing the maximum and minimum impact;
- Mitigating impacts targeted at a specific environmental issue.

The newly emerged digital technology, such as BIM, could enhance the application of LCA. For example, LCADesign™ tool, developed by Cooperative Research Center for Construction Innovation in Australia, is fully automated from 3D Auto CAD drawings to enable the computation of environmental impacts resulting from the selection of materials to be reflected in the design assessment (Seo et al., 2007). The LCADesign™ tool can read the information in the IFC format of the 3D drawing file (American Institute of Architects, 2010).

It is expected by American Institute of Architects (2010) that the availability of BIM integrated tools with LCA or other similar tools could change the way LCA is perceived by architects. Since then, more recent studies on applying BIM for the early design stages of building projects for BIM considering LCA. For example, Nizam et al. (2018) developed a framework and prototype to estimate building embodied energy within the native BIM

TABLE 2.11 The Comparison Between Sustainable Project Management and Conventional Construction

Project stage		Conventional project management	Sustainable project management
Project planning	Feasibility studies and project need assessment	Needs are defined based on market conditions and physical needs	Besides market conditions and physical needs, early project feasibility study should involve environmental goal, sustainability certification (e.g., LEED in the United States), and the budget for sustainability
	Project manager selection	May not happen at this early stage, or an in-house manager may be hired	A manager or green building consultant experienced in sustainable construction and familiar with product and market would be hired. A sustainability accredited (e.g., LEED) professional is preferred
	Preliminary site analysis and plan	A preliminary budget would be developed based on benchmarked traditional projects; unit costs are applied to a preliminary scope of work	Economical and ecological goals would be finalized based on cost/benefit analysis. Site characteristics should be considered. Building needs and ecological issues should be weighed. A LEED checklist and documenting system are involved in the feasibility stage
	Design charrette	Charrette, often perceived as economical waste or schedule inhibitors, may not be implemented in a conventional project	Must include all key external stakeholders. Other project participants including designer, architects, contractors, environmental engineers, and real estate consultants may also be involved
	Final site selection	Conventional projects usually do not have stakeholder involvement in site selection	Stakeholder would be involved in site selection. The owner, project manager, architect, and the contractor, and all other parties have a stake in site selection

(*Continued*)

TABLE 2.11 (Continued)

Project stage		Conventional project management	Sustainable project management
Project procurement	Initial budget and schedule	Budgets are typically developed by an architect based on a formula or unit cost estimate varying as much as 15% from actual costs	Builder, project manager, architect, and real estate consultant are all involved in preconstruction estimates. Experience is required in cost estimate related to green building products. Life cycle cost may also be included in the budget. Long-term gains from operational savings would be focused on instead of short-term return on investment
	Zoning approval	Rework in the planning and feasibility studies may be required if the earlier concepts do not fit zoning ordinances or local land use goals	The zoning approval process can often go more smoothly with early involvement of local government planners and other regulatory agencies
	Design team procurement	The architect or general contractor is selected depending on the contract type	Usually the core design team has already been procured. Extra experts for technical systems may also be procured
	Construction document development	Sustainability initiatives may cause rework in the existing design	Construction documents can be developed more efficiently with less design modifications due to the early participation of the integrated team
	Government permitting review	Plans are often reviewed for the first time for engineering compliance (e.g., grading, environmental control), building codes, etc.	Government stakeholders are involved at earlier stages to ensure compliance with regulations. The relevant environmental systems are significantly connected to LEED requirements
	Project bidding	Design–bid–build is the most common bidding method. Subcontractors are usually selected on a closed-book basis	Open-book subtracting may suit more on green projects and allows the owner to have access to the estimators and pricing from subcontractors

Construction	Implementation contracting	Contracting methods such as cost-plus-fee or cost-plus-percentage are applied according to associated project risks	A different type of contract may be applied in green projects incorporating terms for sustainability practices. Provisions should also be provided in the contract, such as LEED criteria, the use of recyclable materials, and agreement to return unused materials
	On-site construction	There is limited cross-communication among the site workforce, including subcontractors	Kickoff meeting that includes a sustainable education component for on-site construction is launched. Periodical on-site meetings and training sessions on green building are required by the entire workforce
	Inspections	Fields caused by fragmentation in the owner–architect–contractor relationship may require additional government inspections, causing cost and schedule inefficiencies	Government regulators are working as a partner in the project. Less rework and field adjustments decrease the chances of having reinspections
	Sustainability certification	Green certification is typically nonapplicable for conventional projects	The ongoing work of the project manager with the benefits of an integrated team and specialized technology improves the efficiency of documentation submission
Operation	Close out; Occupancy and operations	Limited testing is performed before the building is turned over for operations	Qualified commissioning authority is hired to ensure that building systems function as intended and set forth in the project criteria

Source: Adapted from Robichaud, L.B., Anantatmula, V.S., 2011. Greening project management practices for sustainable construction. J. Manage. Eng. 27(1), 48–57.

environment. The framework addressed the existing gap of ignoring the importance of retaining the LCA results in BIM. Najjar et al. (2017) integrated BIM and LCA by evaluating the environmental impacts of building materials in the early stage of building design. It was further recommended by Najjar et al. (2017) that future work could be performed to investigate elementary flow integration between BIM tools and LCA methodology to address the challenges of comparing different scenarios. More BIM applications in assisting LCA can be found in other studies, such as BIM enabled LCA for carbon footprint analysis (Yang et al., 2018). However, incorporating SD principles in conducting the project feasibility stage may have not been effectively understood by project stakeholders (Shen et al., 2010). Generally, larger stakeholders are more likely to adopt LCA approach due to their financial affordability compared to smaller companies (American Institute of Architects, 2010). It was indicated by Shen et al. (2010) of the need to shift the traditional management approach in the project feasibility study to a new approach embracing the principles of SD, specifically the need to have more weightings on social and environmental factors besides the economical attributes. The totally 18 economical, 9 social, and 8 environmental performance attributes provided by Shen et al. (2010) are listed in Table 2.12.

Generally speaking, economical performance is weighed more than social and environmental attributes in most project practice, although all the three aspects should be fully considered, in particular, project quality, safety performance and environmentally friendly practice (Shen et al., 2010). It is not uncommon that stakeholders are more concerned about the budget than social and environmental impacts of construction projects, especially for private stakeholders, whose major motivation in project investment may be to gain the capital benefits. The higher cost associated with sustainability was identified as one major challenge in green projects by multiple researchers including Tagaza and Wilson (2004), Zhang et al. (2011), and Hwang and Tan (2012). Other challenges in implementing green projects include:

- Technical difficulties during the construction process (Ling, 2003; Tagaza and Wilson, 2004; Zhang et al., 2011; Hwang and Tan, 2012).
- Risks due to different contract forms of project delivery (Tagaza and Wilson, 2004; Hwang and Tan, 2012).
- Longer approval process for new green technologies and recycled materials (Eisenberg et al., 2002; Ling, 2003; Tagaza and Wilson, 2004; Zhang et al., 2011; Hwang and Tan, 2012).
- Unfamiliarity with green technologies (Eisenberg et al., 2002; Ling, 2003; Tagaza and Wilson, 2004; Zhang et al., 2011).
- Greater communication and interest required among project team members (Tagaza and Wilson, 2004).
- More time required to implement green construction practices (Tagaza and Wilson, 2004).

TABLE 2.12 Attributes Within Project Feasibility Study

Attributes		
Economical	**Social**	**Environmental**
• Governmental strategic development policy • Tax policy • Demand and supply analysis • Market forecast • Project function and size • Market competition • Location advantage • Technology advantage • Budget estimate • Financing channels • Investment plan • Life cycle cost • Life cycle profit • Finance risk assessment • Return of investment (ROI) • Net present value (NPV) • Pay-back period • Internal rate of return (IRR)	• Influence to the local social development • Provision capacity of employment • Provision capacity of public services • Provision capacity of public infrastructure facilities • Provision of the infrastructures for other economical activities • Safety standards • Improvement to the public health • Cultural and heritage conservation • Development of new settlement and local communities	• Eco-environmental sensitivity of the project location • Air impacts • Water impacts • Noise impacts • Waste impact • Environmentally friendly design • Energy consumption performance • Land consumption

Source: From Shen, L., Tam, V.W.Y., Tam, L., Ji, Y., 2010. Project feasibility study: the key to successful implementation of sustainable and socially responsible construction management practice. J. Clean. Product. 18, 254–259.

A competent project manager is vital to project success (Frank, 2002; Ahadzie, 2007). In green-featured construction, the capacity of the project manager to address these aforementioned challenges matters to successful project outcomes. The procurement of project manager is listed by Robichaud and Anantatmula (2011) as one key work that differentiates the conventional construction project and the green construction in the project planning stage. As studied by Hwang and Ng (2013), green project requires different proficiencies from the project manager.

To ensure the implementation of sustainable construction practice, Shen et al. (2010) recommended the roles of multiple project participants including government, clients, architects and engineering consultants, as well as contractors and suppliers.

- *Government:* government has an important role in promoting the sustainable construction management in the project feasibility study stage through multiple ways including but not limited to laws, regulations, industry specifications, administrative examination and approval, tax fine and incentives, etc.
- *Clients:* the owner of a project has the key influence in the sustainability performance. The owner could work closely with other project parties to incorporate advices from authorities, planning professionals, and architects/engineers into the project feasibility study.
- *Architects and engineering consultants:* they should be equipped with knowledge of sustainable construction principles and be able to apply sustainability into project practice. It is recommended that architects and engineering consultants be involved in the project feasibility study.
- *Contractors and suppliers:* they can also be involved in the project feasibility study stage by providing information and suggestions on the environmental effects of construction activities and various materials. Advices on alternative construction methods and plants on the project sustainability may be provided from contractors and suppliers in the feasibility study.

2.4 INTERNATIONAL CASE STUDIES

Literatures in the past may have more focused on sustainable innovation in the company or national level with limited studies in the project level (Li and Wang, 2016). In this section, case studies of sustainable project management from international experience are presented.

2.4.1 Sustainable Earthwork Construction in Australia (Li and Wang, 2016)

Li and Wang (2016) reported a sustainable earthwork construction located in Melbourne, Australia, with approximately 200,000 m^2 land, involved in processing the existing contaminated site materials, cut and fill of over 600,000 m^3 site materials, and transporting over 200,000 m^3 clean fill from external sources to construction surface. In the busiest period of construction, there could be more than 200 regular trucks working on-site. The site was adjacent to a major road network on the western boundary, close to residents on the eastern boundary, and secured by permanent with mesh fence along all border. The clients had high expectations on environmental standard and sustainable construction in the preconstruction stage. Modifications of the traditional project management processes and practices were made to deliver sustainable construction based on clients' expectations and cost restriction.

According to Li and Wang (2016), the most efficient way to achieve sustainable construction project was starting from elimination of environmental impacts. Specifically, these environmental impacts in this project included transport traffic causing road pollution, hazardous materials (e.g., asbestos), and construction vibration. The triple-bottom line of sustainability as described in Fig. 2.1, when applied to this earthwork project, can be illustrated in Fig. 2.8.

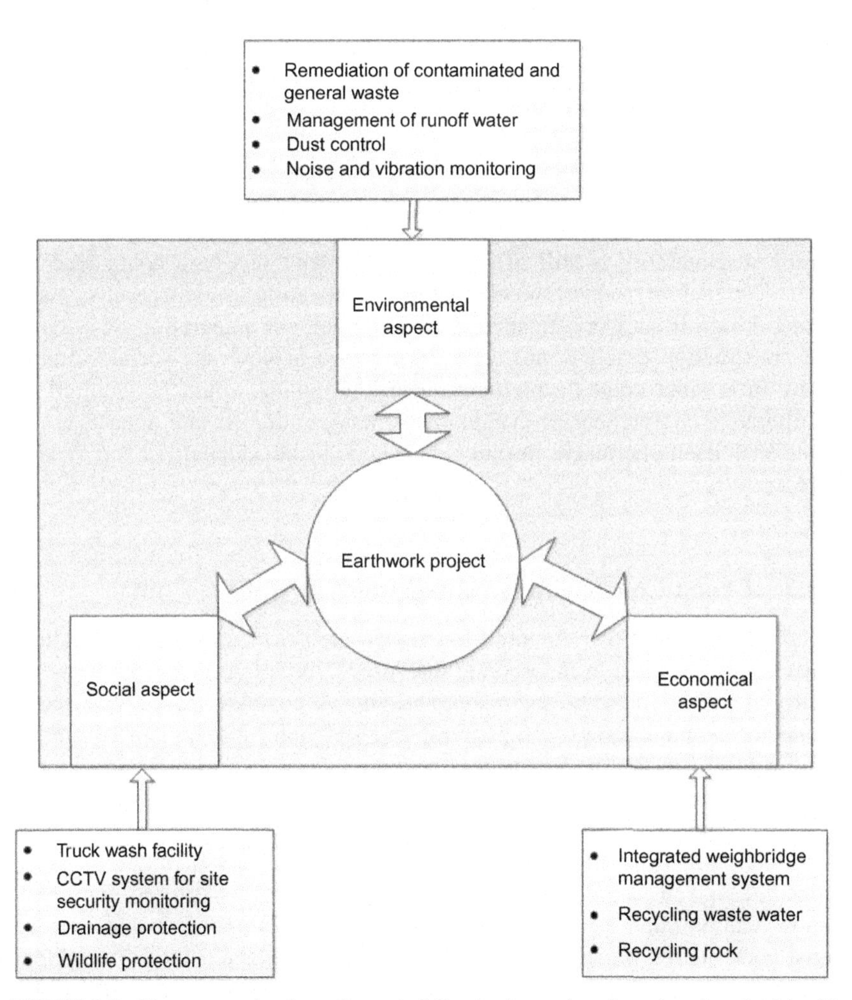

FIGURE 2.8 The systematic view of sustainability in the earthwork project described by Li and Wang (2016).

Fig. 2.8 lists the major sustainable construction management actions for the earthwork project. The sustainable construction framework covering these actions in Fig. 2.8 is listed in Table 2.12, which provides more detailed descriptions of sustainable construction actions. Corresponding sustainability categories of infrastructure from Table 2.5 are also listed in Table 2.13.

Some actions listed in Table 2.13 may meet multiple infrastructure sustainability indicators from Fernández-Sánchez and Rodríguez-López (2010). For example, wheel wash facility and static washing bath meets the cost sustainability by recycling and reusing dirty water when also addressing the social responsibility of keeping public vehicles clean. Li and Wang (2016) summarized this project that minimizing negative environments and maximizing positive impacts are a major objective of sustainability because the environment consists of ecosystems highly related to human health. The environment impacts addressed in this earthwork project, according to Table 2.13, included water usage, hazardous materials (e.g., asbestos fiber), wildlife preservation, material recycling, existing stormwater, construction dust, and underground water, etc.

While this earthwork project focused on sustainable practices to improve project environmental performance, Li and Wang (2016) stated that journey toward sustainability is still in an early stage, and this case could lead to future sustainability opportunities including innovation of evaluation system, project design framework, finance analysis model, and integration of sustainable construction practice into code. In order to achieve the social sustainability in a wider context, intelligent decision-making would be required to incorporate a comprehensive evaluation of many trade-offs and impacts associated with each alternative resource or technic to be adopted (Li and Wang, 2016).

2.4.2 Green Construction Management in Dalian, China

Dalian Donggang C05, the multifunctional high-rise commercial building project, was designed to incorporate the "green" building feature in terms of environmental friendliness and resource saving. It contains spaces for electrical device manufacturing, office areas, five-star hotel rooms, and basement parking lots. The project footprint covers an area of 11,295 m^2, with a total gross floor area around 124,000 m^2. Dalian Donggang C05 project consists of four basement floors, and 42 superstructure floors. All these basement floors are built for parking lots. The first 20 superstructure floors are used as office areas, and the floors from 21 to the top are constructed for hotel rooms. The building has a total height of 220.8 m. Compared to other traditional construction projects in mainland China, more sustainable construction methods or management can be summarized focusing on five areas including environmental protection, material saving, energy saving, water saving, and

TABLE 2.13 Sustainability Actions in the Earthwork Construction Project

Section	Sustainability actions	Descriptions	Corresponding sustainability indicators from Fernández-Sánchez and Rodríguez-López (2010)
Elimination of environmental impacts	System and audit	Adoptions of company policies and public sustainability guidelines, and auditing of environmental management plan	Bureaucracy; responsibility
	Wheel wash facility and static washing bath	Mobile truck wheel wash system to reduce the transport traffic impact by washing vehicles and a static washing bath to assist the wheel wash system; reuse of waste water after wash for site dust suspection	Costs; water; responsibility
	Pump and sprinklers	Reduction of asbestos and decreasing workers' exposure to asbestos	Atmosphere; social integration
	Static compaction technic	Static compaction mode applied in most of the major earthwork, and no damages to surrounding properties	Atmosphere
Substitution of the traditional process with alternative environmental friendly method	Trees removal with specialized excavator and wildlife management	Protection to wildlife allowing them enough time to be relocated; specialized excavator used to prevent dust and erosion	Atmosphere; biodiversity
	Mobile jaw crusher	Jaw crusher used to sort out all oversize rock, crush them for reuse	Resources; costs
	GPS controlled plant operation	GPS controlled technic to track heavy equipment operation and reduce fuel waste and rework, when maintaining work efficiency	Energy; atmosphere; technical requirements

(*Continued*)

TABLE 2.13 (Continued)

Section	Sustainability actions	Descriptions	Corresponding sustainability indicators from Fernández-Sánchez and Rodríguez-López (2010)
Engineering control via newly developed technic and implementation	Drain warden filter	A drain warden filter adopted to protect the existing drainage system on site	Culture; responsibility
	CCTV	The advanced intruder detection CCTV system to monitor the site and support live view from mobile facilities	Security; technical requirements
	Dust monitor	Monthly deposition dust monitoring work was performed to measure the dust emission level	Atmosphere
	Underground water monitor	A number of underground water wells installed to ensure that underground water stay consistent with their original condition	Culture; responsibility
Administration control by better environmental focus system	Materials tracking system	Various types of wastes and unsuitable materials removed offsite with destination recorded forming the project statistics	Technical requirements
	Contaminated materials identify and stockpile management	All earthwork cut and fill cells to be pretested; unknown materials to be separated for further decision; prescribed industrial wastes treated and disposed following procedures	Resources

Source: Developed from Li, W., Wang, X., 2016. Innovations on management of sustainable construction in a large earthwork project: an Australian case research. International Conference on Sustainable Design, Engineering and Construction. Proced. Eng. 145, 677–684.

land resource saving. Table 2.14 summarizes the key features of sustainability in the Donggang C05 project.

These five focus areas listed in Table 2.14 represent the key sustainability indicators is China's construction industry. The development of green building in terms of "Four Savings and One Protection" (i.e., savings in land, energy, water, and materials, as well as environmental protection) has been listed as one major strategy in the published government document in China since 2012 according to Zhejiang Provincial Government Office (2012). These detailed actions described in Table 2.14 fall exactly into five categories related to land resource, energy, water, materials, and environmental protection. They could serve as jobsite sustainability examples for other construction projects in China.

Safety and security management is integrated with sustainability on the construction jobsite. The site computer devices are set with the face recognition function to permit employees' entrance to the jobsite. The mock safety experience zone is set on site for the experiential safety education to employees on the awareness of fall hazard. The jobsite monitoring system is established on site and TV monitoring screen is installed in site management offices.

BIM was adopted in the Donggang C05 project, for example, BIM-based visualization and guide on the site assembly of security check facilities, and the BIM-based design for the site project management facilities. BIM was also applied in the design and fabrication of steel and reinforced concrete members, as well as the detailed design of floor pipelines and utility rooms. BIM provided the digital data in the design for further construction stage use, and reduced the materials waste from rework. BIM was also used as the communication tool to show workers on construction details, such as foundation work, stair connection with the floor, and the waterproof construction.

It could be found from the Donggang C05 project that sustainability was not limited to resources, energy, and environmental protection, but may also cover site safety security, as well as adoption of the newly emerging digital technologies (e.g., BIM) to improve project deliver efficiency. It could be expected that BIM and other newly developed digital technologies (e.g., augmented reality), along with the movement of relevant industry standards to embrace IT technology, could be further adopted in construction projects to improve the site multitrade communication, to reduce rework, and to enhance the information exchange among multiple disciplines.

2.4.3 Integrated Project Delivery Approach in the Solar Decathlon Competition

Initiated by the U.S. Department of Energy, the Solar Decathlon competition (SDC) challenges college academic teams to design and build full-size solar-powered homes to optimize the 10 defined evaluation criteria including

TABLE 2.14 Sustainability Actions in Donggang C05 Project

Sustainable criteria	Detailed actions
Environmental protection	• Site temporary commercial flier board that could absorb dust and that is attached to the site railing protection • Flier board placed on site with the content of environmental protection education • Mobile and continuous enclosure • Reduced noise level and daily monitor of site noise • Storage areas for different types of site-generated wastes • Illuminating lamp shade to prevent light pollution • Site-installed monitoring system to display on-time noise, dust, temperature, and humidity
Material saving and efficient use	• Housekeeping for waste material storage • Wooden plank road on site made from recycled and reused wood form wastes • BIM adopted to reduce rework-caused material waste
Energy saving	• LED lighting facilities used to reduce electricity consumption • Lowered electrical voltage in site housekeeping and artificial lighting • Solar hot water facilities installed on site to save electricity consumption from the site showering activities • The fire protection system with adjustable frequency to reduce the pump energy consumption with an automatic start at insufficient pressure and stop upon achieving the set pressure
Water saving	• Foundation pit dewatered, purified, pumped, and stored for reuse in fire control, sprinkler, and flushing systems • Both automatic and manual setups are installed in the sprinkler system using water from fire protection water tank, and connected to the environmental monitoring data • Mobile dust control facilities operated in a flexible way to save water • Sensor-equipped inductive water faucet to reduce water waste • Infrared induction applied in the flushing system of site toilets to reduce water usage in flushing • Prefabricated concrete septic tank assembled and cleaned periodically
Land resource saving	• Proper planning and layout of site storage areas for different construction materials in order to optimize space usage

Note: Generally workers are offered their temporary accommodation on the jobsite during the construction in China. Therefore, the electrical voltage and solar hot water devices are adopted on site to reduce workers' energy consumption.

innovation, water use and reuse strategies, smart energy use, market potential, cost-effective architectural and engineering design, energy-efficient heating and cooling systems, appliances, and electronics, occupant health and comfort, and communications. Started in 2002 and promoted as an international architecture, engineering, and construction (AEC) competition held in multiple continents, including Africa, Asia, Europe, Latin America, and North America, SDC attracts international institutions including joint teams and sponsors within the AEC industry to adopt the integrated project delivery (IPD) approach to design and build sustainable houses with most building elements manufactured off-site and shipped to the competition site for an intense on-site assembly and follow-up energy efficiency tests. This section is focusing on only the case of National University of Singapore's SDC project. The main challenge, as described in this section, was adopting the IPD approach with BIM as the digital technology for the prefabrication construction project.

IPD method is typically incorporated in institutional teams participating in SDC, with academic disciplines from multiple AEC fields and industry partners working closely to deliver the design documents crossing different stages and construction related activities (e.g., scheduling, estimate, and site logistics). The team of National University of Singapore (TeamNUS, 2013) worked in the 2013 SDC held in Datong China is used as a case study. Similar to other institutional teams participating in SDC, TeamNUS was also an interdisciplinary team involving in total 59 students from 10 departments across the School of Design & Environment, Faculty of Engineering, Faculty of Arts and Social Sciences, and NUS Business School. This cross-disciplinary team was supported by other NUS public resources such as NUS Energy Office and Integrated Sustainability Solution Cluster. According to Huang (2013), the TeamNUS designed and prepared house prefabrication elements (e.g., structural steel) in Singapore, packed in container, and shipped to China. All the design work was performed in Singapore the year prior to the test. Industry consultancies, such as structural design, HVAC, and solar thermal consultancies involving AEC professions were organized in Singapore. Off-site prefabrication was performed by NUS students in Singapore, for example, composite floor off-site construction, masterclad wall, and secondary steel frames. Other activities conducted in Singapore prior to SDC included building system tests (e.g., PV panels). The project delivery timeline is presented in Table 2.15.

The solar-powered house assembled on the competition site in Datong China by TeamNUS was relocated back in Singapore after the SDC ended in August 2013. SDC serves as an academic-industry joint platform to involve academics, AEC students, industry consultants and sponsors to work in a collaborative environment to deliver a sustainable building, usually ranging from 100 to 200 m^2 of the gross floor area. Innovation in sustainable design is a key in SDC, and this would usually involve architects and engineers

TABLE 2.15 Workflow of TeamNUS's SDC Project From 2012 to 2013

Timeline	Project activities
March 2012	Student team building including communication team, site management team, HVAC team, solar PV team, architecture team, and admin team
March–April 2012	Quantity survey and health and safety studies
May 2012	Site visit in Datong China and workshop training
June 2012	Start of seeking industrial sponsorship (e.g., lighting and home appliances)
July 2012	Student teams receiving professional advices in architecture, structure, and other building design disciplines
September 2012	Design Development submission to SDC committee
January–March 2013	Mock site test from construction to commissioning in Singapore
March 2013	Construction document submission
May 2013	Dissembling of site components, packing, loading, and shipping to China for SDC
June 2013	Laying shipped building components on SDC site in Datong China
July 2013	Eight days spent on site assembling from foundation blocks, hoisting beams and columns, hooking up building service systems, interior work, furniture moving-in, flooring setup, to the final finish
August 2013	Public visit of the completed solar-powered house

Sources: TeamNUS, 2013. Retrieved from: <http://www.nus.edu.sg/SD2013/?page_id = 579> (accessed 04.10.16); Huang, Y.C., 2013. TeamNUS Solar Decathlon. National University of Singapore.

from multiple disciplines (all these roles played by AEC students guided by academic or industry advisors) to have early communication in the project feasibility study and design stages. For example, the TeamNUS provided the design concept of secondary house skin or protection skin as a passive strategy to relieve solar glare on the built house. The TeamNUS also designed and practiced other sustainable engineering methods, including low voltage direct current microgrid to achieve higher energy efficiency, graywater filtration adopting 3 R strategies (i.e., reduce, reuse, and recycle), and solar thermal collectors to heat cold water.

Other state-of-the-art technologies or sustainability devices are being applied in SDC, for example, using the energy generated from solar PV panels to charge the electrical car, and adoption of BIM in the life cycle analysis of the built solar house, etc. All these innovative strategies introduced in Section 2.2 are being applied in SDC, including prefabrication construction, digital technology (i.e., BIM), and IPD method. LCA can also be applied to the solar-powered houses built in SDC. Teams participating in SDC could market the project to their local communities and use the project as a pedagogical example to enhance the experiential learning for students. AEC students gain the field experience and project management practice through the project conceptual design to construction completion. SDC activities are incorporating the sustainability in multiple aspects, including social, economical, and technical areas. For example, Solar Decathlon Europe (2017) addresses the increased urban density issue by proposing the theme of prefabricated roof-top apartment. Sustainability has been extended to a more holistic approach, including financial affordability, usage of recycled materials, indoor human comfort, and implementation of emerging technologies (e.g., virtual reality, augmented reality, smart phone application in monitoring building performance) according to Solar Decathlon Europe (2017). It is believed that the IPD, as the innovative project delivery method, could enable the early stage communication among project teams from different disciplines and to further improve the design efficiency.

2.4.4 Life-Cycle Assessment of the New Jersey Meadowlands Commission Center (Source: From Krogmann et al., 2008)

The New Jersey Meadowlands Commission (NJMC) Center is an educational facility consisting of classrooms, laboratory, administrative offices along with an observatory. It has a total area of 891 m^2, with chromate copper arsenate treated wood piles as the foundation. The building structure is wood columns, concrete masonry units and glued-laminated wood beams. Floors are made of cast-in-place reinforced concrete slab. The building was being designed to meet the requirements of LEED Gold certification (Krogmann et al., 2008). The owner of the project, NIMC, contracted the Rutgers Center for Green Building to conduct the LCA using the building plans and specifications as inputs. LCA experts, general contractor, product vendor/manufacturer, the architect, owner, and energy modeler were involved in the LCA process. The project embraced these SD features listed by Dyszkiewicz (2008):

- 165-unit rooftop solar panel array;
- ceiling solar tubes;
- recycled building materials;
- recyclable and locally manufactured standing-seam metal roof;
- energy-efficient heating, lighting, and water system.

The major stages of LCA applied in the NJMC Center are summarized.

1. Step 1: Goal and scope definition

   The goal of the LCA in NJMC Center project was to evaluate and compare its life-cycle environmental impacts to that of conventional buildings, specifically, to focus on the primary energy consumption and global warming potential, and to compare the results with data from the literature. Table 2.16 defines these LCA related terminologies within the context of the NJMC Center.

   More detailed description of system boundary can be found in Table 2.17.

2. Step 2: Inventory analysis

   Inputs and outputs within the inventory were prepared using existing database including EcoInvent 2.0 database, Franklin US LCI, and Industry 2.0, etc. Generally the data collected for the material placement phase were based on contract documents and communication with the architect, the owner, manufacturers, and trade organizations. The data collection and assumption process in the NJMC Center project can be summarized in Table 2.18.

   Based on the data collection and assumption described in Table 2.18, material compositions in the building life cycle were computed using SimaPro 7.1 software. Inventory results showed the percentages of materials contributing to the NJMC building. For example, crushed concrete was accounted for 37.1% of the building mass. The material consumption for renovations and replacements was added to the life cycle building mass.

TABLE 2.16 Definitions of LCA Terminologies in the NJMC Center Project

LCA terminologies	Definition in the NJMC Center project
Scope	The scope of NJMC Center project focused on energy consumption and global warming. It excluded those building elements which were either with little energy use or out of the scope of building footprint, such as laboratory facilities, site work, and bathroom supplies
Impact category	Primary energy consumption, global warming potential, ozone depletion, acidification, and eutrophication potential
Functional unit	The functional unit in this project was calculated on the basis of per square foot
Building lifespan	The architect considered the building lifespan as 50 years
System boundary	The building life cycle consisted of three stages: material placement, operation, and decommissioning

TABLE 2.17 Description of System Boundary for LCA Study of NJMC Center Project

LCA stage	Activities involved in each stage
Material placement	Raw material extraction; avoided burden due to use of recycled and reused materials
	Refinement of raw materials; manufacturing of construction materials and components
Construction	Building construction; replacement of building materials and components
Operation	Building operation (electricity, heating, avoided burden due to the use of solar power)
Decommissioning	Material decommissioning and demolition
	Recycling of building materials; reuse of building materials; landfilling of building materials

Source: From Krogmann, U., Minderman, N., Senick, J., Andrews, C., 2008. Life Cycle Assessment of New Jersey Meadowlands Commission's Center for Environmental & Scientific Education Building. Rutgers Edward J. Bloustein School of Planning & Public Policy, New Jersey.

3. Step 3: Impact assessment

   Using the impact categories defined in Table 2.16, two different environmental impact methods (i.e., Building for Environmental and Economic Sustainability (BEES) and IMPACT 2002 +) were adopted in the study of Krogmann et al. (2008) to covert the inventory data to environmental impacts. Results from these two methods were compared. These two methods yielded similar results in energy use, global warming, and acidification potentials, but diverged in estimates of eutrophication and ozone depletion potential. Krogmann et al. (2008) stated that it was necessary to scrutinize these inconsistent results from these two methods by providing possible explanations.
4. Step 4: Interpretations

   The impact assessment results from Step 3 indicated that the primary energy consumption, global warming potential, and acidification potential for NJMC Center was lower than conventional education buildings benchmarked. Although NJMC Center had a higher impact in the material placement stage due to materials consumed in the construction of foundation work, roof elements, and photovoltaic cells, the higher initial impact could be offset in the later operation stage. The decommissioning stage had a relatively lower impact on the building life cycle compared to material placement and operation.

   It was inferred that the early sustainability design decisions, such as energy efficiency design and usage of renewable energy facilities in the

TABLE 2.18 Descriptions of Data Collection and Assumption in the LCA Study for the NJMC Center Project

Data or information	Detailed actions
Materials	Inputs were based on US-average electricity for materials produced in other states and composition of the New Jersey electricity grid for materials produced in-state
	Material losses during manufacturing and construction were assumed
	Information on replacement frequencies was obtained either from the architect or from published sources
	Transportation of raw materials to refinement and manufacturing was included in the inventory. Also included was the transportation from the manufacturing facility to the construction site
	Credits for recycling or reuse were not accounted in the inventory calculation
	Transportation to local recycling facilities was accounted
Energy	Energy consumption included the construction stage from the use of power tools, lighting, and heavy equipment
	Only heating, cooling, ventilation, lighting, and water heating were included in the operation stage
	Adjustments for solar photovoltaic panels were made in the calculation

Source: From Krogmann, U., Minderman, N., Senick, J., Andrews, C., 2008. Life Cycle Assessment of New Jersey Meadowlands Commission's Center for Environmental & Scientific Education Building. Rutgers Edward J. Bloustein School of Planning & Public Policy, New Jersey.

NJMC Center project, could lead to a more SD in the building life cycle (American Institute of Architects, 2010). Through the comparison of design choices using LCA approach, the environmental impacts during material placement, operation, and decommissioning of buildings can be evaluated (Krogmann et al., 2008). Choices imposing higher impacts during the material placement stage could yield significantly lower impacts during operation and these findings indicated the benefits that could be expected from green building practices (Krogmann et al., 2008). It could also be inferred from this case study that different impact assessment methods may lead to inconsistent results, possibly due to the different preassumptions contained within these methods. The researcher and practitioners should be able to identify the gaps among different methods and to explore causes of these divergences.

2.5 SUMMARY

This chapter studied on the sustainability and sustainability performance focusing on the management aspect. Sustainability rating systems worldwide (e.g., BREEAM) were reviewed and compared of their evaluation criteria. Energy efficiency was found as the key indicator crossing these rating systems. EIA in construction project management was also emphasized. SD in both project and organizational levels were introduced. More state-of-the-art movements in sustainable construction project management were described and discussed, including prefabrication construction, BIM, and IPD method. LCA formed part of project sustainability. Its major features and challenges were discussed. The chapter also provided case studies relevant to prefabrication construction, IPD method, and LCA. Sustainability can be incorporated into project management in multiple aspects, using rating systems as the technical guideline, BIM as the digital platform, prefabrication as the construction technique, and IPD as the system. It is expected that more international case studies would provide the practice and knowledge on how these aspects are interlinked to each other; for example, how BIM could provide the digital information platform to enable more efficient decisions in early project delivery stages which have a higher impact on the project sustainability performance. Future studies could be expanded to case studies of project sustainability in the organizational level and knowledge management of sustainability within the organization.

REFERENCES

Ahadzie, D.K. A model for predicting the performance of project managers in mass house building projects in Ghana. Ph.D. thesis, University of Wolverhampton, 2007.

Akbarnezhad, A., Ong, K., Chandra., L., 2014. Economic and environmental assessment of deconstruction strategies using building information modelling. Automat. Constr. 37, 131–144.

Akmam Syed Zakaria, S., Gajendran, T., Skitmore, M., Brewer, G., 2018. Key factors influencing the decision to adopt industrialised building systems technology in the Malaysian construction industry: an inter-project perspective. Architect. Eng. Des. Manag 14, 27–45.

Alawini, A., Tanatanmatorn, N., Tucker, D., 2013. Technology adoption: building IT. In: Daim, Tugrul Unsal, Oliver, Terry, Kim, Jisun (Eds.), Research and Technology Management in the Electricity Industry: Methods, Tools and Case Studies.

Alwan, Z., Greenwood, D., Gledson, B., 2015. Rapid LEED evaluation performed with BIM based sustainability analysis on a virtual construction. Constr. Innov. 15 (2), 134–150.

Alwan, Z., Jones, P., Holgate, P., 2016. Strategic sustainable development in the UK construction industry, through the framework for strategic sustainable development, using Building Information Modelling. J. Clean. Product. Available from: https://doi.org/10.1016/j.jclepro.2015.12.085.

American Institute of Architects. AIA guide to building life cycle assessment in practice. Washington, DC, 2010.

American Institutes of Architects. 2007, Integrated Project Delivery: A Guide. Version 1.

Artenian, A., Sadeghpour, F., and Teizer, J., 2010. A GIS framework for reducing GHG emissions in concrete transportation. In Proceedings of Construction Research Congress, Canada, May 2010, pp. 1557–1566.

Associated General Contractors of America. The Contractor's Guide to BIM, first ed. Associated General Contractors of America Research Foundation, Las Vegas, NV, 2005.

Australian Government & Forest and Wood Products Research and Development Corporation. Technical Evaluation of Environmental Assessment Rating Tools. Project PN05.IOI9. 2006.

Autodesk, Inc. Autodesk® Ecotect™ Analysis. Available online at: <http://usa.autodesk.com/adsk/servlet/pc/index?siteID = 123112&id = 12607162>. 2009.

Azhar, S., 2010. BIM for sustainable design: results of an industry survey. J. Build. Inform. Model. 4 (1), 27–28.

Azhar, S., 2011. Building Information Modeling (BIM): trends, benefits, risks, and challenges for the AEC industry. Leader. Manage. Engin. 11 (3), 241–252.

Azhar, S., Brown, J., and Sattineni, A. A case study of building performance analyses using building information modeling. Proceedings of the 27th International Symposium on Automation and Robotics in Construction (ISARC-27), Bratislava, Slovakia, June 25–27, 2010b.

Azhar, S., Carlton, W.A., Olsen, D., Ahmad, I., 2010a. Building Information Modeling for Sustainable Design and LEED Rating Analysis. Autom. Constr. 20 (2011), 217–224.

BIM Industry Working Group. Strategy paper for the government construction client group. 2011.

BREEAM, BREEAM UK New Construction, Technical Manual SD5076:0.1 (Draft)-2014.

BREEAM New Construction Infrastructure Technical Manual: Version: SD219—Issue: 2.0—Issue Date: 12/11/2015, 2016.

BSRIA Report in Association with Schneider Electric, The Value of BREEAM, 2012.

Baker, B.N., Murphy, D.C., Fisher, D., 1988. Factors affecting project success. In: Cleland, David, King, William (Eds.), Project Management Handbook, edited by Van Nostrand Reinhold.

Banerjee, S.B., 2003. Who sustains whose development? Sustainable development and the reinvention of nature. Organ. Stud. 24 (1), 143–180.

Bansal, P., 2005. Evolving sustainably: a longitudinal study of corporate sustainable development. Strat. Manage. J. 26 (3), 197–218.

Barnes, S., Castro-Lacouture, D., 2009. BIM-enabled integrated optimization tool for LEED decisions. Proceedings of the 2009 ASCE International Workshop on Computing in Civil Engineering 258–268.

Baumgartner, R.J., Ebner, D., 2010. Corporate sustainability strategies: sustainability profiles and maturity levels. Sustain. Develop. 18 (2), 76–89.

Berardi, U., 2012. Sustainability assessment in the construction sector: rating systems and rated buildings. Sustain. Develop. 20.6 411–424.

Biswas, T., Wang, T., and Krishnamurti, R. Integrated sustainable building rating systems with building information modeling. Master's thesis, Carnegie Mellon University, Pittsburgh, PA. 2008.

Bribián, I.Z., Usón, A.A., Scarpellini, S., 2009. Life cycle assessment in buildings: state-of-the-art and simplified LCA methodology as a complement for building certification. Build. Environ. 44, 2510–2520.

CIB & UNEP-IETC, 2002. Agenda 21 for Sustainable Construction in Developing Countries: a discussion document. Published by the CSIR Building and Construction Technology, Pretoria.

Charted Institute of Buildings (CIOB), 2018. Construction Manager. Available via. http://www.constructionmanagermagazine.com/. (accessed 05.02.18).

Cheng, J., and Ma, L. A BIM-based system for demolition and renovation waste quantification and planning. Proceedings of the 14th International Conference on Computing in Civil and Building Engineering, Moskow, 2012, 2012.

Cheng, J.C., Ma, L.Y., 2013. A BIM-based system for demolition and renovation waste estimation and planning. Waste Manage. 33, 1539–1551.

Clevenger, C.M., Khan, R., 2014. Impact of BIM-enabled design-to-fabrication on building delivery. Pract. Period. Struct. Design Constr. 19, 122–128.

coBuilder BIM data (COBie/IFC). NEC Birmingham, 2015.

Costa, A., Keane, M.M., Torrens, J.I., Corry, E., 2013. Building operation and energy performance: monitoring, analysis and optimisation toolkit. Appl. Ener. 101, 310–316.

Curwell, S., Yates, A., Howard, N., Bordass, B., Doggart, J., 1999. The green building challenge in the UK. Build. Res. Inform. 27, 286–293.

Dawood, N., and Iqbal, N. Building information modelling (BIM): a visual and whole life cycle approach, CONVR2010, Sendai, Japan, November 4–5, 2010. In K. Makanae, N. Yabuki, K. Kashiyama (Eds.), Proceedings of the 10th International Conference on Construction Applications of Virtual Reality, CONVR2010 Organizing Committee, 2010, pp. 7–14.

Department of Energy & Climate Change (DECC). UK Emissions Statistics: Frequently Asked Questions. DECC, London. 2012.

Dixon, W., 2010. The Impacts of Construction and the Built Environment. Briefing Notes. Willmott-Dixon Group. Letchworth Garden City, Hertfordshire, UK. Retrieved from: <https://www.willmottdixon.co.uk/asset/9462/download>.

Dyllick, T., Hockerts, K., 2002. Beyond the business case for corporate sustainability. Busin. Strat. Environ. 11 (2), 130–141.

Dyszkiewicz, T., 2008. Sustainability in Action: A New Jersey-Based Environmental Center Practices What It Preaches.

Eadie, R., Browne, M., Odeyinka, H., McKeown, C., McNiff, S., 2013. BIM implementation throughout the UK construction project lifecycle: an analysis. Automat. Constr. 36, 145–151.

Eastman, C., Teicholz, P., Sacks, R., Liston, K., 2011. BIM Handbook, A Guide to Building Information Modeling for Owners, Managers, Designers, Engineers, and Contractors. John Wiley & Sons, Inc, Hoboken, New Jersey.

Eisenberg, D., Done, R., Ishida, L. Breaking down the barriers: Challenges and solutions to code approval of green building. Research report by the Development Center for Appropriate Technology. 2002.

Elkington, John, 1994. Towards the sustainable corporation: win-win-win business strategies for sustainable development. Calif. Manage. Rev. 36.2 90–100.

European Commission. European platform on life cycle assessment. Retrieved from: <http://eplca.jrc.ec.europa.eu/> (accessed 29.09.16).

Farmer, M. (2016). Farmer Review 2016: Modernise or Die, Commissioned and published by the Construction leadership Council (CLC), UK.

Fernández Sánchez, G. Análisis de los Sistemas de Indicadores de Sostenibilidad. Planificación urbana y proyectos de construcción. Escuela Técnica Superior de Ingenieros de Caminos, Canales y Puertos, Universidad Politécnica de Madrid, 2008, Spain.

Fernández-Sánchez, G., Rodríguez-López, F., 2010. A methodology to identify sustainability indicators in construction project management—application to infrastructure projects in Spain. Ecol. Indicat. 10, 1193–1201.

Frank, T., 2002. The Superior Project Manager. Marcel Dekker, New York.
Gandhi, S., Jupp, J., 2014. BIM and Australian green star building certification. Comput. Civil Build. Eng. ASCE 275–282.
Gimenez, Cristina, Sierra, Vicenta, Rodon, Juan, 2012. Sustainable operations: their impact on the triple bottom line. Int. J. Product. Econom. 140.1 149–159.
Glasson, J., Therivel, R., Chadwick, A., 2013. Introduction to Environmental Impact Assessment. Routledge.
Green Building Council of Australia, The Value of Green Star—A Decade of Environmental Benefits. 2013.
Green Building Council of Australia, 2016. Retrieved from: <http://new.gbca.org.au> (accessed 02.08.16).
Hajibabai, L., Aziz, Z., Peña-Mora, F., 2011. Visualizing greenhouse gas emissions from construction activities. Constr. Innov. Inform. Process Manage. 11, 356–370.
Hall, T., J., 2011. The triple bottom line: what is it and how does it work? Indiana Busin. Rev. 86.1 4.
Hammond, R., Nawari, N., Walters, B., 2014. BIM in sustainable design: strategies for retrofitting/renovation. Comput. Civil Build. Eng. ASCE. pp. 1969–1977.
Herva, M., Franco, A., Carraso, E.F., Roca, E., 2011. Review of corporate environmental indicators. J. Clean. Product. 19, 1687–1699.
Hong, T., Koo, C., Kim, J., Lee, M., Jeong, K., 2015. A review on sustainable construction management strategies for monitoring, diagnosing, and retrofitting the building's dynamic energy performance: Focused on the operation and maintenance phase. Appl. Energ. 155, 671–707.
Huang, Y.C., 2013. TeamNUS Solar Decathlon. National University of Singapore.
Hwang, B.G., Ng, W.J., 2013. Project management knowledge and skills for green construction: overcoming challenges. Int. J. Project Manage. 31.2 272–284.
Hwang, B.G., Tan, J.S., 2012. Green building project management: obstacles and solutions for sustainable development. Sustain. Develop. 20.5 335–349.
ISO 14040, 2006. Environmental management—life cycle assessment—principles and framework. Retrieved from: <http://www.iso.org/iso/catalogue_detail?csnumber = 37456> (accessed 06.10.16).
Ilskog, E., 2008. Indicators for assessment of rural electrification e an approach for the comparison of apples and pears. Energ. Policy 36 (7), 2665–2673.
Institute for International Urban Development (2017). Sustainable Development vs. Sustainability. Available from: <http://i2ud.org/2012/06/sustainable-development-vs-sustainability/> (accessed 17.09.17).
International Energy Agency (IEA), 2012. Energy Technology Perspectives 2012: Scenarios and Strategies to 2050. IEA, Paris.
Inyim, P., Rivera, J., Zhu, Y., 2015. Integration of building information modeling and economic and environmental impact analysis to support sustainable building design. J. Manage. Eng. 31, (Special Issue: Information and Communication Technology (ICT) in AEC Organizations: Assessment of Impact on Work Practices, Project Delivery, and Organizational Behavior, A4014002).
Jalaei, F., Jrade, A., 2014. Integrating BIM with green building certification system, energy analysis, and cost estimating tools to conceptually design sustainable buildings. Constr. Res. Congr. 140–149.
Jiang, Y., Ming, J., Wu, D., Yen, J., Mitra, P., Messner, J., et al., 2012. BIM server requirements to support the energy efficient building lifecycle. Published in ASCE International

Conference on Computing in Civil Engineering - Clearwater Beach, FL, United States, 365–372.

Jin, R., Gao, S., Cheshmehzangi, A., Aboagye-Nimo, E., 2018. A holistic review of off-site construction literature published between 2008 and 2018. J. Clean. Prod. 202, 1202–1219.

Jin, R., Hancock, C.M., Tang, L., Wanatowski, D., 2017. Investigation of BIM investment, returns, and risks in China's AEC industries. J. Constr. Eng. Manage 143 (12), 04017089, http://dx.doi.org/10.1061/(ASCE)CO.1943-7862.0001408.

Jones, B., 2014. Integrated project delivery (IPD) for maximizing design and construction considerations regarding sustainability. Integrated Project Delivery (IPD) for Maximizing Design and Construction Considerations Regarding Sustainability. Proced. Eng. 95, 528–538.

Kent, D.C., Becerik-Gerber, B., 2010. Understanding construction industry experience and attitudes toward integrated project delivery. J. Constr. Eng. Manage. 136 (8), 815–825.

Khan, M.A., 2015. Chapter 8-Prefabrication of the superstructure. Accel. Bridge Constr. 353–398.

Kriegel, E., Nies, B., 2008. Green BIM. Wiley Publishing, Indianapolis, IN.

Krogmann, U., Minderman, N., Senick, J., Andrews, C., 2008. Life Cycle Assessment of New Jersey Meadowlands Commission's Center for Environmental & Scientific Education Building. Rutgers Edward J. Bloustein School of Planning & Public Policy, New Jersey.

Lee, J.C., Edil, T.B., Benson, C.H., Tinjum, J.M., 2011. Evaluation of variables affecting sustainable highway design with BE2ST-in-Highways system. J. Transport. Res. Board 2233.

Lenferink, S., Tillema, T., Arts, J., 2013. Towards sustainable infrastructure development through integrated contracts: experiences with inclusiveness in Dutch infrastructure projects. Int. J. Project Manage. 31, 615–627.

Lennertz, B., 2003. The charrette as an agent for change. In New Urbanism: Comprehensive Report J.I. Jackson & Best Practice Guide, third ed New Urban Publications, Ithaca, NY.

Li, W., Wang, X., 2016. Innovations on management of sustainable construction in a large earthwork project: an Australian case research. International Conference on Sustainable Design, Engineering and Construction. Proced. Eng. 145, 677–684.

Li, X., Zhu, Y., Zhang, Z., 2010. An LCA-based environmental impact assessment model for construction processes. Build. Environ. 45 (3), 766–775.

Lin, S.H.E., Gerber, D.J., 2014. Designing-in performance: a framework for evolutionary energy performance feedback in early stage design. Auto. Constr. 38, 59–73.

Ling, J.U. The project manager's personal characteristic, skills and roles in local construction industry. Published Master's dissertation, Faculty of Civil Engineering, University Technology Malaysia. 2003.

Liu, Z., Osmani, M., Demian, P., Baldwin, A.N., 2011. The potential use of BIM to aid construction waste minimalisation. Proceedings of the CIB W78-W102. International Conference-Sophia Antipolis, France.

Lu, W., Fung, A., Peng, Y., Liang, C., Rowlinson, S., 2014. Cost-benefit analysis of building information modeling implementation in building projects through demystification of time-effort distribution curves. Build. Environ. 82, 317–327.

Mao, C., Xie, F., Hou, L., Wu, P., Wang, J., Wang, X., 2016. Cost analysis for sustainable off-site construction based on a multiple case study in China. Habitat Int. 57, 215–222.

Martland, C.D., 2012. Toward More Sustainable Infrastructure: Project Evaluation for Planners and Engineers. Wiley, Hoboken, NJ.

Matoski, A., Ribeiro, R.S., 2016. Evaluation of the acoustic performance of a modular construction system: case study. Appl. Acoust. 106, 105–112.

McGraw-Hill Construction. Green BIM: How Building Information Modelling is Contributing to Green Design and Construction. SmartMarket Report, McGraw-Hill Construction, 2010.

Mcdonald, B., Smithers, M., 1998. Implementing a waste management plan during the construction phase of a project: a case study. Constr. Manage. Econom. 16, 71–78.

Milne, M.J., Gray, R., 2013. W (h)ither ecology? The triple bottom line, the global reporting initiative, and corporate sustainability reporting. J. Busin. Ethics 118.1 13–29.

Molavi, J., Barral, D.L., 2016. A construction procurement method to achieve sustainability in modular construction. Proced. Eng. 145, 1362–1369.

Moon, H.J., Choi, M.S., Kim, S.K., and Ryu, S.H. Case studies for the evaluation of interoperability between a BIM-based architectural model and building performance analysis programs. Proceedings of 12th Conference of International Building Performance Simulation Association, 2011.

Motawa, I., Almarshad, A., 2013. A knowledge-based BIM system for building maintenance. Automat. Constr. 29, 173–182.

Najjar, M., Figueiredo, K., Palumbo, M., Haddad, A., 2017. Integration of BIM and LCA: evaluating the environmental impacts of building materials at an early stage of designing a typical office building. J. Build. Eng. 14, 115–126.

National Charrette Institute. Collaboration by Design. Retrieved from: <http://charretteinstitute.org/> (accessed 14.09.16).

Nizam, R.S., Zhang, C., Tian, L., 2018. A BIM based tool for assessing embodied energy for buildings. Energ. Buildings 170, 1–14.

Novitski, B.J. BIM promotes sustainability, practitioners are finding paths to green through interoperable software. 2009. Available at: <http://continuingeducation. construction.com/article.php?L = 5&C = 516> (accessed 04.08.16).

Nurse, K., 2006. Culture as the fourth pillar of sustainable development. ISmall States Econom. Rev. Basic Stat. 11, 28–40.

Osborn Engineering, 2015 Progressive construction: no quick fixes in project delivery. Retrieved from: <http://www.osborn-eng.com/Blog/Post?ID = 20> (accessed 14.09.16).

Paulson Jr, B.C., 1976. Designing to reduce construction costs. J. Constr. Div. 102, C04.

Pelsmakers, S., 2013 BIM and its potential to support sustainable building. In Presented in the NLA/NBS "Green BIM" Conference, London, UK, 31 January 2013.

Pennington, D.W., Potting, J., Finnvenden, G., Lindeijer, E., Jolliet, O., Rydberg, T., 2004. Life cycle assessment—Part 2: Current impact assessment practice. Environ. Int. 30 (5), 721–739.

Perlas, N. Overcoming illusions about biotechnology. Third World Network, Penang, Malaysia, 1994.

Poirier, E.A., Forgues, D., French, S.S., 2017. Understanding the impact of BIM on collaboration: a Canadian case study. Build. Res. Inf 45 (6), 681–695. Available from: https://doi.org/10.1080/09613218.2017.1324724.

Robèrt, K.-H., Schmidt-Bleck, B., Aloisi de Laderel, J., Basile, G., Jansen, J.L., Kuehr, R., et al., 2002. Strategic sustainable development – selection, design and synergies of applied tools. J. Clean. Product. 10 (3), 197–214.

Robichaud, L.B. and Anantatmula, V.S., 2011. Greening project management practices for sustainable construction. J. Manage. Eng. 27 (1), 48–57.

Rocque, B.L., 2003. Enabling effective project sponsorship: a coaching framework for starting projects well. Three Houses Consulting LLC. Trumbull, CT.

Russell-Smith, S.V., Lepech, M.D., 2015. Cradle-to-gate sustainable target value design: integrating life cycle assessment and construction management for buildings. J. Clean. Product. 100, 107–115.

Ryu, H.S., Park, K.S., 2016. A study on the LEED energy simulation process using BIM. Sustainability 8, 138.

Schlueter, A., Thesseling, F., 2008. Building information model based energy/exergy performance assessment in early design stages. Automat. Constr. 18, 153–163.

Seo, S., Tucker, S., Newton, P., 2007. Automated material selection and environmental assessment in the context of 3D building modelling. J. Green Building 2 (2).

Shen, L., Tam, V.W.Y., Tam, L., Ji, Y., 2010. Project feasibility study: the key to successful implementation of sustainable and socially responsible construction management practice. J. Clean. Product. 18, 254–259.

Shrivastava, P., 1995. Environmental technologies and competitive advantage. Strategic Manage. J. 16, 183–200.

Smith, D.K., Tardif, M., 2009. Building Information Modeling: A Strategic Implementation Guide for Architects, Engineers, Constructors, and Real Estate Asset Managers. John Wiley & Sons.

Solar Decathlon Europe (2017). sde19 call for teams & deadline. Available at: <http://solardecathlon.eu/deadline-for-submissions-_-2018-sde-call-for-teams-_15_10_16/> (accessed 17.09.17).

Stadel, A., Eboli, J., Ryberg, A., Mitchell, J., Spatari, S., 2011. Intelligent sustainable design: integration of carbon accounting and building information modelling. J. Prof. Issues Eng. Edu. Pract. 137, 51–54.

Starik, M., Rands, G.P., 1995. Weaving an integrated web: multilevel and multisystem perspectives of ecologically sustainable organizations. Acad. Manage. Rev. 20 (4), 908–935.

State Ministry of Housing and Urban-Rural Construction (SMHURC). Request for Proposal on BIM Application in the Construction Industry. Beijing, China (in Chinese), 2013.

Tagaza, E., Wilson, J.L. Green buildings: drivers and barriers e lessons learned from five Melbourne developments. Report Prepared for Building Commission by University of Melbourne and Business Outlook and Evaluation. 2004.

Tam, V.W.Y., Fung, I.W.H., Sing, M.C.P., Ogunlana, S.O., 2015. Best practice of prefabrication implementation in the Hong Kong public and private sectors. J. Clean. Product. 109, 216–231.

Tang, L., Jin, R., and Fang, K. Launching the innovative BIM module for the architecture and built environment programme in China. In BIM 2015 First International Conference on Building Information Modelling (BIM) in Design, Construction and Operations, WIT Transactions on the Built Environment, vol. 149 (2015), pp. 145–156.

TeamNUS, 2013. Available at: <http://www.nus.edu.sg/SD2013/?page_id = 579> (accessed 04.10.16).

Tzivanidis, C., Antonopoulos, K.A., Gioti, F., 2011. Numerical simulation of cooling energy consumption in connection with thermostat operation mode and comfort requirements for the Athens buildings. Appl. Ener. 88 (8), 2871–2884.

U.S. Department of Energy, 2011. Re-Assessing Green Building Performance: A Post Occupancy Evaluation of 22 GSA Buildings. Pacific Northwest National Laboratory Richland, Washington.

U.S. EPA, SAIC. Life Cycle Assessment: Principles & Practice. 2006.

U.S. Energy Information Administration (EIA), 2012. Annual Energy Review 2011. EIA, Washington, DC.

U.S. Environmental Protection Agency (U.S. EPA). Life Cycle Assessment (LCA). 2012.

U.S. Green Building Council, 2016a. LEED. Available from: <http://www.usgbc.org/leed> (accessed 02.08.16).

U.S. Green Building Council, 2016b. LEED v4 for Building Design and Construction. Washington, DC, April

U.S. Green Building Council, 2016c. Benefits of Green Building. Available from: <http://www.usgbc.org/articles/green-building-facts> (accessed 02.08.16)

Valentin, A., Spangenberg, J.H., 2000. A guide to community sustainability indicators. Environ. Impact Assess. Rev. 20, 381–392.

Vyas, S., Ahmed, S., Parashar, A., 2014. BEE (Bureau of energyefficiency) and Green Buildings. Int. J. Res. 1, 23–32.

WSP, 2013. What is BIM? Available from: <http://www.wspgroup.com/en> (accessed 21.09.16).

Wong, J.K.W., Kuan, K.L., 2014. Implementing 'BEAM Plus' for BIM-based sustainability Analysis. Automat. Constr. 44, 163–175.

Wong, J.K.W., Zhou, J., 2015. Enhancing environmental sustainability over building life cycles through green BIM: a review. Automat. Constr. 57, 156–165.

Wong, J.K.W., Li, H., Wang, H., Huang, T., Luo, E., Li, V., 2013. Toward low-carbon construction processes: the visualization of predicted emission via virtual prototyping technology. Automat. Constr. 33, 72–78.

Wood, C., 2003. Environmental Impact Assessment: A Comparative Review, second ed Prentice Hall, Harlow.

World Commission on Environment and Development's (the Brundtland Commission) Report Our Common Future. Oxford University Press, Oxford, 1987.

Wu, W. and Issa, R. Integrated processmapping for BIM implemenation in green building project delivery. In N. Dawood, M. Kassem (Eds.), Proceedings of the 13th International Conference on Construction Applications of Virtual Reality, October 30–31, 2013, London, UK, 2013.

Yang, X., Hu, M., Wu, J., Zhao, B., 2018. Building-information-modeling enabled life cycle assessment, a case study on carbon footprint accounting for a residential building in China. J. Clean. Prod. 183, 729–743.

Yang, Y., Ng, S.T., Xu, F.J., Skitmore, M., 2018. Towards sustainable and resilient high density cities through better integration of infrastructure networks. Sustainable Cities and Society 42, 407–422.

Yılmaz, M., Bakış, A., 2015. Sustainability in construction sector. Proced. SocialBehav. Sci. 95, 2253–2262.

Zhang, X.L., Shen, L.Y., Wu, Y.Z., 2011. strategy for gaining competitive advantage in housing development: a China study. J. Clean. Product. 19 (1), 157–167.

Zhejiang Provincial Government Office, 2012. Suggestions to Enhance the Industrialization of the Building Fields, No. 152 (in Chinese).

FURTHER READING

ISO/TC 207/SC 5. Life cycle assessment. Available from: <http://www.iso.org/iso/home/standards_development/list_of_iso_technical_committees/iso_technical_committee.htm?commid=54854> (accessed 29.09.16).

Wu, W., Issa, R.R., 2012. Leveraging cloud-BIM for LEED automation. J. Inform. Technol. Constr. 17, 367–384.

Chapter 3

Management

Cheng Siew Goh
Department of Quantity Surveying and Construction Project Management, Heriot-Watt University Malaysia, Putrajaya, Malaysia

3.1 MANAGEMENT APPROACH

Sustainability in the built environment needs to incorporate numerous elements such as material technical performance, people, environment, urban planning, financial sense. The application requires multi- and interdisciplinary efforts to address sustainable development issues at various project stages including inception, design, construction, operation, and maintenance. In view of the breadth, sustainable projects should be regarded as a process instead of a final product to ensure the overarching goals of sustainability are always attained.

Several challenges in the transition of sustainable practice are found to be related to management issues such as stakeholders' attitudes, supply chain deficiencies, poor coordination and collaboration among stakeholders, and discouragement of existing systems (Goh, 2014). Ochieng et al. (2014) also pointed out that lack of consideration of clients and stakeholders, resistance to change, and fragmentation are barriers of sustainable construction. This shows that implementing effective and efficient management strategies has great potentials to improve sustainability in the built environment. The important role of project management in sustainable practice is also recognized by Wang et al. (2014). They asserted that sustainable development is more easily implemented at the strategic level than the operation level of a business. As compared to advanced energy efficiency measures and technological innovation, project management can offer a low-cost low-barrier avenue to achieve sustainable construction (Ball, 2002; Wu and Low, 2010).

Professionals in the built environments including clients, investors, architects, surveyors, engineers, sustainability consultant, contractors, and facilities managers have to share their respective knowledge and skills in working toward sustainable solutions. Cooperation and communication between

Sustainable Construction Technologies. DOI: https://doi.org/10.1016/B978-0-12-811749-1.00017-1

various project parties must be considered in project management for sustainable building projects (Wang et al., 2014). However, the traditional management approach is rather fragmented and contains extensive layers in the project organizational structure. This fragmentation prevents effective and efficient communication between various project parties in different stages of projects. A more integrated approach should therefore be employed for close interaction of suppliers, professionals, and users to deliver sustainable construction practice (Häkkinen and Belloni, 2011; Yao, 2013).

Literature shows that it is essential to make a change to conventional construction practice to deliver sustainable projects successfully. On one hand, Wang et al. (2014) considered that conventional project management approach in the construction industry lacks a whole life cycle perspective. On the other hand, Shen et al. (2010) also viewed that the traditional practice allows project clients to focus on the analysis on project economical performance in project inception and design stages by giving limited or no consideration to social and environmental issues. Contractors and suppliers have no or very little involvement in project feasibility stage even though their inputs are valuable on improving project buildability and performance to contribute to sustainability (Shen et al., 2010). These have demonstrated a need to shift the traditional method of project management to a new approach that embraces the principles of sustainable development.

The following section introduces five management approaches that are commonly employed by construction organizations. The approaches are sustainable procurement, integrated project delivery, environmental management system, building sustainability assessment and post-occupancy management.

3.2 SUSTAINABLE PROCUREMENT

Procurement is one of the key management processes in construction project management, which involves contractual agreement, technical performance system, culture, procedures, environmental sustainability, organization, conflicts, and building economies (Rowlinson, and McDermott, 2005). It offers great potentials to improve sustainable performance in construction practice.

Selecting an appropriate procurement approach is significant to integrate sustainability into construction and its operation. As described by Häkkinen and Belloni (2011), the difficulty to define measurable sustainability requirements and targets is one of the most important obstacles for sustainable buildings. It is therefore important to integrate sustainability goals and requirements into procurement and contract documents since the beginning of a development to avoid too much attention paid on merely cost factors.

Design–bid–build is the traditional procurement strategy that dominates the current construction industry. The conventional procurement system segregates the design, construction, and operation of engineering project, lacking a continuous management of the project (Wang et al., 2014). Little formally structured information is available to include environmental issues in the construction procurement process (Cole, 2000). However, sustainable construction practices and measures require much attention to be placed on green specifications such as the type of materials and equipment (Hwang and Tan, 2012). To work toward sustainability goals, sustainability requirements should be embedded in the supply chain and contract documents of construction practice. To align with it, the UK government published "Achieving Excellence in Construction Procurement Guide" and a set of guidance notes to inform sustainable construction (Department of Finance, n.d.).

A concept of "Sustainable Procurement" has recently emerged to bring promising implementation of sustainability in construction projects. It was first introduced following the Rio Earth Summit in 1992 and started growing into the government policies (ECO-Buy, 2013). The United Kingdom, United States, and Australian governments have now incorporated sustainability into their procurement decisions and operations. Sustainable procurement is defined as "a process whereby public organisations meet their needs for goods, services, works and utilities in a way that achieves value for money on a whole life cycle basis in terms of generating benefits not only to the organisation but also to society and economy, whilst significantly reducing negative impacts on the environment" (UNEP, 2012). This definition is widely adopted by United Nations Environmental Programme, Australian Government, UK Government, International Council for Local Environmental Initiatives, and European Commission (DG-Environment).

The noteworthy aspect of sustainable procurement is its focus on the entire life cycle of the goods and services rather than the up-front cost, underlying with the principle of value for money (ECO-Buy, 2013). It seeks to achieve the appropriate balance between the three pillars of sustainable development. Table 3.1 shows the economical, social, and environmental factors to be considered in sustainable procurement.

As outlined by BS 8903:2010 Principles and Framework for Procuring Sustainably—Guide, there are seven key stages of sustainable procurement processes: (1) identify the business need, (2) defining sourcing strategy, (3) identify suppliers and tender, (4) evaluation and award, (5) implement, (6) manage performance and relationship, and (7) review and learn. The use of sustainable procurement streamlines the delivery of sustainability in construction projects by offering consistent communication to key internal and external stakeholders on the development of organizational policy, strategy, action plans, and initiatives.

In United Kingdom, several policy initiatives related to sustainable procurement have been developed and these include Equality of Opportunity and Sustainable Development in Public Sector Procurement (2008), Sustainable Procurement Action Plan (2008), Government Construction Clients Sustainability, Sustainable Construction Group Guidance Notes, and CIFNI Sustainability Requirements for Construction Work Contracts.

TABLE 3.1 Economical, Social and Environmental Factors Considered in Sustainable Procurement

Economical factors	Social factors	Environmental factors
• Acquisition cost • Maintenance cost • Operation cost • End-of-life management cost • Whole life costs and value for money • Business continuity • Fair and viable margin in suppliers' agreement	• Social justice and equality • Safety and security • Human rights • Employment conditions and workforce welfare • Diverse base of suppliers • Fair trade and ethical sourcing • Community benefits	• Emission to air, land and water • Climate change • Biodiversity • Natural resource use • Water scarcity • Energy use and emission • Waste and by-products

Sources: UNEP, 2012. Sustainable Public Procurement Implementation Guidelines: Introducing UNEP's Approach. Paris: UNEP; BSI, 2010. BS 8093: 2010 Principles and Framework for Procuring Sustainably—Guide.

The UK Olympic Delivery Authority also integrated sustainability into its Procurement Policy by embedding eight sustainability related principles in informing its procurement approach (UK DEFRA, 2013). Similarly, Australia also commits to sustainable procurement by embodying it in National Waste Policy, the Australian Government ICT Sustainability Plan 2010–2015 and the Australian Government Financial Management Framework (ECO-Buy, 2013).

Nevertheless, Ruparathna and Hewage (2015) found limited number of sustainable procurement initiatives used in the Canadian construction industry. A slow progress with regard to sustainable procurement is noticed in Canada due to a lack of policies, tools, or databases for incorporating sustainability in procurement (Ruparathna and Hewage, 2015). Similarly, McMurray et al. (2014) also spotted very few aspects of sustainable procurement embedded in the Malaysian public and private organizational practices, due to a lack of guidance and awareness for sustainable procurement and "money politics." As addressed by McMurray et al. (2014), sustainable guidance criteria do not always reach the people responsible for procurement on a day-to-day basis.

Sustainable procurement is an enabler to embed sustainability in governance processes across an organization as well as throughout the whole supply chains. It gives active and visible leadership to organizations in managing internal operation and external practice with influences on her associated suppliers and external stakeholders. The stance and expectation of an organization on sustainability are communicated clearly to stakeholders which in turn lead to more effective engagement (Goodhew, 2016). Sustainable procurement could offer a well-suited mechanism to sustainable construction initiatives. However, a robust sustainable solution can only be attained with full accountability and involvement of all key stakeholders. More efforts should therefore be made to increase the industry acceptance toward sustainable procurement.

3.3 INTEGRATED PROJECT DELIVERY/INTEGRATED DESIGN APPROACH/INTEGRATIVE PROCESS

As discussed in the previous section, sustainable construction projects require extensive interdisciplinary collaboration to ensure sustainability principles in the building systems are synergistic and are in appropriate size. Several considerations are involved and they include resource efficiency, energy efficiency, environmental performance, social responsibility, design and management of buildings, innovation, building use flexibility, operation and maintenance, economical issues, site connectivity, and microclimate. Due to disciplinary boundaries and linear planning and design sequences, the conventional planning, design, construction and operation processes fail to view the entire building as a whole (USGBC, 2014). An integrated design approach (or integrate design approach or integrative process) could change the domain and make a collaborative environment.

An integrated design approach offers a more direct and active manner to work collaboratively with team players (Kubba, 2010). It involves active and continuing participation of project clients, project manager, building technologists, contractors, architects, civil and structural engineers, mechanical and electrical engineers, quantity surveyors, sustainability consultant and users. In an integrated design approach, all key team players are fully involved from project brief, predesign, design to construction, and commissioning stage (Kubba, 2010). They work collaboratively in defining the sustainability goals for the project. The key project participants would not work in isolation on the project parts they have control but think and act holistically about the whole project. It emphasizes connections and improves communications among stakeholders throughout the project phases. A "new team" is formed from stakeholders who traditionally work as separate entities. More time and energy are invested upfront early in the design phase to allow maximum flexibility in the delivery of sustainable projects (USGBC, 2014).

As described by Lewis (2004), the most critical element of integrated design approach is excellent communications and the following activities are generally included in an integrated design approach:

- Setting sustainability performance goals;
- Getting all team members involved early;
- Conducting design charrettes;
- Performing project reviews/peer reviews;
- Establishing working groups on sustainability issues; and
- Using building systems commissioning as a project process.

In the integrated design approach, there are a range of follow-through actions in place to implement sustainability strategies effectively and completely and these actions embrace peer reviews, documentation of sustainability measures, commissioning of building systems, post-occupancy evaluation, and tuning system performance (Lewis, 2004). In light of great contribution of the integrated approach, USGBC makes "integrative process/thinking" as part of credits to be scored in its latest LEED version—LEED v4.

3.4 ENVIRONMENTAL MANAGEMENT SYSTEM

The construction industry has been accounted for massive detrimental impacts on the environment. Environmental sustainability is therefore often given greater emphasis than the other two sustainability dimensions. The use of environmental management systems such as ISO 14000 family is one of the popular approaches used by construction organizations in their sustainability efforts.

The ISO 14000 family of standards gives tools for organizations across different background and sectors to manage their environmental responsibilities. The standards map out a series of framework for environmental management system by considering audits, communications, labeling, life cycle analysis, and environmental challenges. They offer guidance without giving fixed prescriptions for environmental performance criteria, and thus are applicable to all kinds of organizations regardless of size, type, nature, culture, social factors, and geographical characteristics.

The ISO 14000 standards offer a framework for construction enterprises to attain sustainable construction in a structured manner (Ofori et al., 2000). ISO 14001 is the standard widely implemented in the industry among the ISO 14000 family of standards. Organizations can demonstrate their conformity with the standard by making self-declaration, or seeking for third party confirmation of its conformance, or seeking certification by an external organization. ISO 14001 employs a Plan-Do-Check-Act approach to continually improve and document environmental management practice. However, adoption of ISO 14001 does not guarantee optimal environmental outcomes to an organization, and the level of application can vary due to context of the organization and the nature of its activities, products, and services (BSI, 2015).

As described by Goh (2014), construction-related organizations tend to employ environmental management systems such as ISO 14001 to assist the execution of sustainability at the organization level. ISO 14001 standard is used in place of sustainability policies to show the organizational commitment toward sustainability goals (Goh, 2014). ISO 14001 enables an organization to establish, implement, maintain, and improve the environmental management system and demonstrate conformity with the international standard in order to gain a competitive advantage and the trust of stakeholders.

ISO 14001 recorded the second highest number of ISO certificates in 2015, i.e., 319324 with an increase of 8% compared to 296736 certificates in 2014. According to ISO survey 2015, the sector with the highest number of ISO 14001:2004 certified companies is "Construction" (17.51%), followed by "Basic Metal and Fabricated Metal Products" (9.89%) and "Electrical and Optical Equipment" (9.59%). The ISO 14001 certificate is experiencing exponential growth in the past two decades. It gains soaring popularity following the growing attention on sustainable development. The certified organizations have grown 13 times compared to the year of 2000 (22847 certificates) and 1.8 times compared to the year of 2005. The number of certified organizations also increases at a pace of 33% compared to 2010 (Fig. 3.1).

A new edition of ISO 14001:2015 has also been published in September 2015 and this edition supersedes the previous edition ISO 14001:2004. Organizations are given a 3-year transition period to migrate their certification of environmental management system to the latest edition. In ISO 14001:2015, a high-level structure—a common structure and core text for all of its management systems—is introduced. The new structure allows easier integration with other management systems in view of the use of common terms, definitions, and structure in its management. In addition, the followings are the major revisions made in ISO 14001:2015 (ISO, n.d.):

- Increased prominence of environmental management within the organization's strategic planning processes;

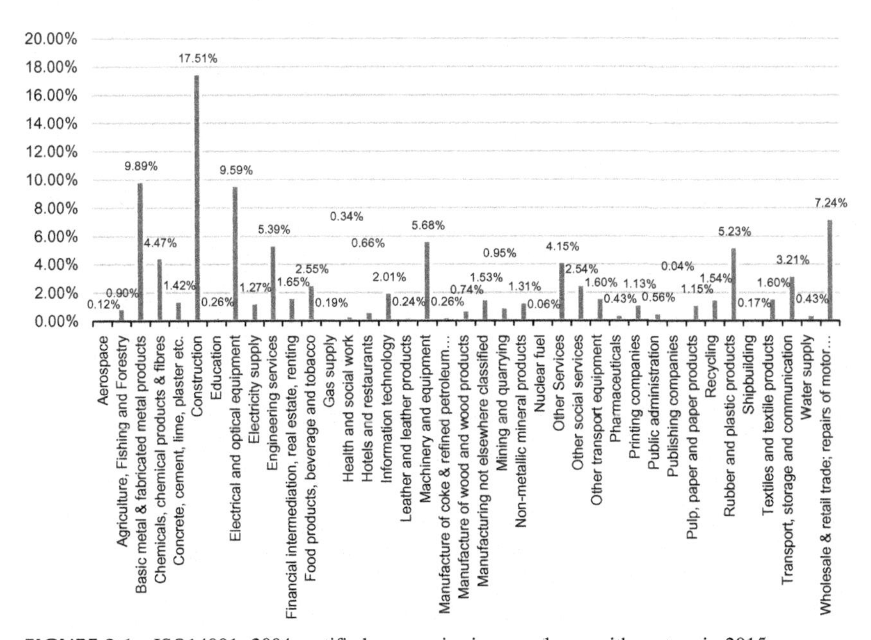

FIGURE 3.1 ISO14001: 2004 certified companies in accordance with sectors in 2015.

- Greater focus on leadership;
- Addition of proactive initiatives to protect the environment from harm and degradation, such as sustainable resource use and climate change mitigation;
- Improving environmental performance added;
- Lifecycle thinking when considering environmental aspects;
- Addition of a communications strategy.

Environmental management system is one of the systematic approaches for organizations to contribute toward environmental pillars of sustainability (BSI, 2015). However, it is pertinent to note that ISO 14001 needs to be implemented along with other sustainability tools since the focus of attention is paid mostly on the environmental pillar. Issues such as regionality and materiality, environment restoration, and cultural dimensions are not addressed in environmental management system (Ball, 2002).

3.5 BUILDING SUSTAINABILITY ASSESSMENT SYSTEMS

The concept of sustainable development is arguably vague and its meaning is often clouded by different interpretations. It is tough to envision and articulate a sustainable future and the assimilation of sustainable development in practice is often different from its principles in theory (Cole, 2000). In response to the rising demand for sustainable buildings, construction organizations have a tendency of using building sustainability assessment systems as part of their organizational commitment toward sustainable construction. The assessment systems include but not limited to Leadership in Energy and Environmental Design (LEED), BREAAM (Building Research Establishment Environment Assessment Method), SBTool (Sustainable Building Tool), Green Star, HK Building Environment Rating System Plus (BEAM Plus), Singapore Green Mark, and Malaysia Green Building Index.

The building assessment systems have a great role during the emergence of sustainable construction concept. In the early stage of sustainable practice, industry players have no much ideas to make their initiatives toward the sustainability goals. Sustainability appeared to be a complex issue in view of its broad coverage and multifaceted. There were no clear standards or policies to guide practitioners on how to implement sustainability and thus practitioners tend to adopt the building rating tools as a relevant means to demonstrate their contribution. Instead of developing sustainable policy, some organizations prefer to refer to building sustainability rating tools as a framework of their sustainable practice (Goh, 2014). In line with that, Ruparathna and Hewage (2015) also revealed that LEED certification is one of the main sustainability initiatives practiced by their studied organization.

Construction organizations are generally project based and the adoption of these building assessment tools offers indicators to construction decision makers in furthering sustainable construction practices. The role of these building sustainability rating systems is to provide a comprehensive assessment of the building sustainable performance by considering the resource use, ecological loadings, and health impacts associated with building production and operation (Cole, 2000). They set the standards and evaluate the extent to which building advance the sustainability goals. By awarding different score and rating levels to building projects, stakeholders could have better knowledge of the sustainability commitment of an organization. The rating tools are useful to manage sustainability issues at the project level (Whang and Kim, 2015).

Construction organizations employ the building sustainability assessment systems to help them to define and show their levels of achievement in sustainability. The assessment systems measure the building sustainability performance with a series of prescribed qualitative and quantitative criteria, ranging from energy and water to ecology and management (Goh and Rowlinson, 2013). They are designed under different certification packages to address a development at different stages such as new construction, in-use, refurbishment, as well as building with different design such as neighborhood development, interior design, and homes. The systems provide useful information on the good practices and measures to achieve sustainability goals (Goh, 2014; Wu and Low, 2010).

Building sustainability assessment systems, which have focused mostly on reducing environmental impacts, are currently central in moving their agenda toward a longer term and wider view. The whole project life cycle perspective and sustainability underpin their evaluation principles. They focus upon not only planning and design related criteria but also take into account other considerations in the cycles of occupancy, operation, and maintenance. In LEED v4, more social and economical sustainability dimensions are included to establish more comprehensive sustainability assessment tools for construction players. For instance, BREEAM and LEED v4 extend the evaluation criteria to regional context, neighborhood, and human experience.

The success of these assessment tools has brought sustainability into the mainstream of the construction industry and has increased the public awareness on sustainability in the built environment. This is evident when a sharp increase in the certification numbers is noticed. The registration of LEED increased from 5000 projects in 2006 to nearly 80,000 projects in 2016 and the number of BREEAM assessments certified also expanded from 34 before the year of 2000 to 425,690 in 2012 (Shutters and Tufts, 2016; BRE Global Ltd, 2014).

3.6 POST-OCCUPANCY MANAGEMENT

Sustainable buildings and structure comprises a great deal of interactive systems and facilities and sustainability performance is the outcomes between the environment, organization, and individual within them. Sustainability complexes generally push the envelope of practice and explore new disciplinary frontiers by introducing breakthrough, discoveries, and innovation in the construction applications. They also embrace high flexibility to adapt to unpredictable future changes of user needs, ownership, regulations, social norms, and technology advances. In view of that, continual learning and monitoring must be implemented to determine whether targets and expectations are met.

Nonetheless, there is a lack of studies on dynamic and life cycle thinking approach to integrate and manage different perspectives of sustainability in construction projects (Goh, 2014; Wang et al., 2014). Sustainable practice should be treated as "live systems" in which they change their behaviors continuously when interacting and harmonizing with the surrounding environment. A resurgence in post-occupancy management is thus driven following the uptake of sustainable development in the built environment.

Post-occupancy management (POM), also known as post-occupancy evaluation, is a systematic process of reviewing, examining, evaluating, and managing the effectiveness and efficiencies of occupied design environments. It considers three levels of performance evaluation, i.e. (1) health/safety/security level; (2) functionality/efficiency level; and (3) social, psychological, cultural, and esthetic level. It has a pivotal role to provide data for managing the performance of a development during the operation phases. Architects, engineers, and key participants in the design process have never or seldom been involved in a POM. It is vital to reinforce the relationship between design, construction, and operation of a development. There is a responsibility for design team to impart the knowledge of sustainable construction design and technologies to facilities management team and end users to manage sustainability practice.

POM consists of three intertwined strands (Cooper, 2001):

1. Design aid—a "feed-forward" mechanism to improve procurement, by using the POM outputs of previous development as the stating measurement point for briefing of the next project;
2. Management aid—a feedback loop approach to measure performance in relation to organizational efficiency and business productivity;
3. Benchmarking aid—a means for measuring progress toward sustainable development in the built environment, with two kinds of applications: one off and longitudinally.

As a management aid, the outputs of POM can be used to serve as a lesson learned to improve the functional fit of existing facilities and spaces. Real information on which to base the decisions is offered since most of the decisions made in the design stage are based on assumptions of development functions and user behaviors (Zimmerman and Martin, 2001). Demand response management system could be established from the POM results by identifying current use and any new demands for improving organizational effectiveness. Facilities managers and users can make adjustment to suit their needs for building spaces and systems better. For instance, energy use due to occupant behaviors are unregulated loads and they can only be determined after the building commissioning. The energy system can then be adjusted to operate more closely in matching the actual use, thus reducing owning and operating costs. To obtain the information, a knowledge of cultural, social, economical, and technical parameters within the building is essential.

Being a design aid, POM provides the basis to inform the appropriateness of the design for its intended functions. It identifies deficiencies of equipment performance and installation in which designers may not be aware of. Designers normally walk out after the project commissioning and they seldom go back and review the outcomes of their design decisions. Gaps are evident when designers often hold different perceptions and evaluation principles from clients, tenants, and occupants. Moreover, because of the long lead time of project delivery, the intended uses of a space could also change over time (Zimmerman and Martin, 2001). To avoid similar design deficiencies from happening, POM helps to inform design decisions of the next project by identifying inefficiencies and shortcomings associated with previous design solutions.

In relation to benchmarks, POM helps to review and track nonconformities of sustainable solutions as they intended. It identifies the gaps between inputs and outputs, design and operation, expectation and reality. It also makes provision of valuable information to support continuous improvement. Considering the importance of post-occupancy management, USGBC has also developed the LEED Dynamic Plaque as an award-winning tool to measure, monitor and score building performance.

3.7 CONCLUSION

Sustainable buildings should be treated as a process rather than a product in which their nature is continuously evolving, developing, and changing throughout their life cycles. A project will never be sustainable if it has not reviewed and monitored continuously throughout the life cycle. Selecting appropriate management approaches is essential to guarantee the success of sustainable applications. Irrespective of which approaches, establishing clear requirements and expectations to both internal and external stakeholders is a must to their full commitment and involvement toward the common sustainability goals.

REFERENCES

BRE Global Ltd. (2014). The Digest of BREEAM Assessment Statistics: Volume 01, 2014. Available at <http://www.breeam.com/filelibrary/Briefing%20Papers/BREEAM-Annual-Digest---August-2014.pdf>.

BSI (2010). BS 8093: 2010 Principles and Framework for Procuring Sustainably—Guide.

BSI (2015). BS EN ISO 14001: 2015 Environmental Management Systems—Requirements with Guidance for Use.

Ball, J., 2002. Can ISO 14000 and eco-labelling turn the construction industry green? Build. Environ. 37 (4), 421–428.

Cole, R.J., 2000. Building environmental assessment methods: assessing construction practices. Construct. Manage. Econom. 18 (8), 949–957.

Cooper, I., 2001. Post-occupancy evaluation—where are you? Build. Res. Inform. 29 (2), 158–163.

ECO-Buy, 2013. Sustainable Procurement Guide. Department of Sustainability, Environment, Water, Population and Communities, Australian Government, Melbourne.

Goh, C.S. (2014). Development of a capability maturity model for sustainable construction. HKU Theses Online (HKUTO).

Goh, C.S., Rowlinson, S., 2013. The roles of sustainability assessment systems in delivering sustainable construction. Proceedings 29th Annual ARCOM Conference 1363–1371.

Goodhew, S., 2016. Sustainable construction processes: A resource text. John Wiley & Sons, Chichester, UK.

Häkkinen, T., Belloni, K., 2011. Barriers and drivers for sustainable building. Build. Res. Inform. 39 (3), 239–255.

Hwang, B.-G., Tan, J.S., 2012. Green building project management: obstacles and solutions for sustainable development. Sustain. Develop. 20 (5), 335–349. Available from: https://doi.org/10.1002/sd.492.

ISO (n.d.). ISO 14001 Environmental Management Systems: Revision. Available at <http://www.iso.org/iso/home/standards/management-standards/iso14000/iso14001_revision.htm>.

Kubba, S., 2010. Green Construction Project Management and Cost Oversight. Butterworth-Heinemann, pp. 86–96.

Lewis, M., 2004. Integrated design for sustainable buildings. ASHRAE J. 46 (9), S22–S26, S28–S30.

McMurray, A.J., Islam, M.M., Siwar, C., Fien, J., 2014. Sustainable procurement in Malaysian organizations: practices, barriers and opportunities. J. Purch. Suppl. Manage. 20 (3), 195–207.

Ochieng, E.G., Wynn, T.S., Zuofa, T., Ruan, X., Price, A.D.F., et al., 2014. Integration of sustainability principles into construction project delivery. Arch. Eng.Technol. 3 (1), 1–5.

Ofori, G., Briffett, C., Gang, G., Ranasinghe, M., 2000. Impact of ISO 14000 on construction enterprises in Singapore. Constr. Manage. Econom. 18 (8), 935–947.

Rowlinson, S., McDermott, P. (Eds.), 2005. Procurement Systems: A Guide to Best Practice in Construction. Routledge, London, UK.

Ruparathna, R., Hewage, K., 2015. Sustainable procurement in the Canadian construction industry: current practices, drivers and opportunities. J. Clean. Product. 109, 305–314.

Shen, L.-y, Tam, V.W.Y., Tam, L., Ji, Y.-b, 2010. Project feasibility study: the key to successful implementation of sustainable and socially responsible construction management practice. J. Clean. Product. 18 (3), 254–259.

Shutters, C. & Tufts, R. (2016). LEED by the numbers; 16 years of steady growth. Available at <http://www.usgbc.org/articles/leed-numbers-16-years-steady-growth>.

UK DEFRA. (2013). London 2012 Olympic and Paralympic Games. The Legacy: Sustainable Procurement for Construction Projects. Available at <http://www.fsc-uk.org/preview.defra-the-legacy-sustainable-procurement-for-construction-projects.a-676.PDF>.

UK DOF (n.d.). Sustainable Construction in Procurement. Available at: <https://www.finance-ni.gov.uk/articles/sustainable-construction-procurement>.

UNEP, 2012. Sustainable Public Procurement Implementation Guidelines: Introducing UNEP's Approach. UNEP, Paris.

USGBC (2014). Green Building 101: What Is an Integrated Process? Available at <http://www.usgbc.org/articles/green-building-101-what-integrated-process>.

Wang, N., Wei, K., Sun, H., 2014. Whole life project management approach to sustainability. J. Manage. Eng. 30 (2), 246–255.

Whang, S.W., Kim, S., 2015. Balanced sustainable implementation in the construction industry: The perspective of Korean contractors. Energy and Buildings 96, 76–85.

Wu, P., Low, S.P., 2010. Project Management and Green Buildings: Lessons from the Rating Systems. J. Prof. Issue. Eng. Edu. Pract. 136 (2), 64–70.

Yao, R. (Ed.), 2013. Design and Management of Sustainable Built Environments. Springer, London.

Zimmerman, A., Martin, M., 2001. Post-occupancy evaluation: benefits and barriers. Build. Res. Inform. 29 (2), 168–174.

FURTHER READING

Preiser, W.F.E., 1995. Post-occupancy evaluation: how to make buildings work better. Facilities 13 (11).

Chapter 4

Indoor Environmental Quality

Oluyemi Toyinbo
Department of Environmental and Biological Sciences, University of Eastern Finland, Kuopio, Finland

4.1 INTRODUCTION

The term indoor environmental quality (IEQ) is broad and it encompasses the several conditions that make up the indoor environment (Steinemann et al., 2017). These conditions may influence the overall comfort and well-being of building occupants. Some of the many factors that contribute to IEQ include indoor thermal comfort, indoor air quality (IAQ), odor, sound quality, and lighting (Arif et al., 2016). Sometimes the term IEQ and IAQ are used interchangeably but there exist difference(s) between them, although the two concepts overlap. IEQ refers to the entire environmental condition inside a building that includes IAQ, thermal comfort, lighting, and so on, while IAQ is a part of IEQ that deals with the condition of air present indoor and its effect on occupants' health and comfort (Ole Fanger, 2006; Steinemann et al., 2017).

This chapter will discuss indoor environmental quality in sustainable construction technology and the importance of IEQ parameters meeting appropriate standard.

4.2 FACTORS AFFECTING INDOOR ENVIRONMENTAL QUALITY

In-depth studies have shown that there are interactions between the biological, physical, and chemical factors that are available indoors (Mitchell et al., 2007). The outdoor environment also affects IEQ especially in directly ventilated buildings where windows and doors are opened, making the filtering of air coming indoor difficult. For example, particulate matters from automobile emission and environmental tobacco smoke (ETS). Biological factors that affect IEQ are related to bacteria such as legionella and fungi such as mold. The build-up of these biological pollutants are mainly associated with

Sustainable Construction Technologies. DOI: https://doi.org/10.1016/B978-0-12-811749-1.00003-1

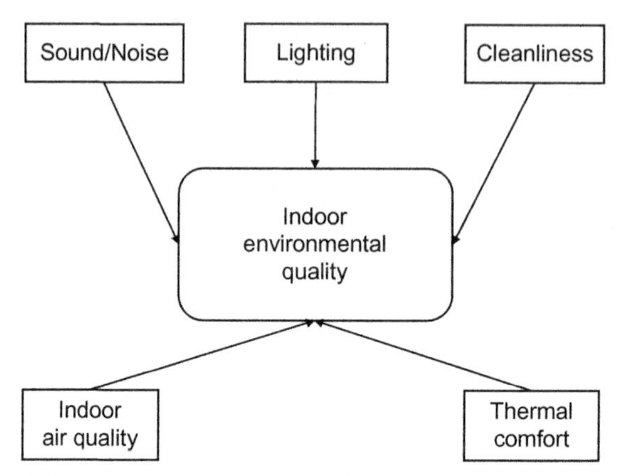

FIGURE 4.1 Different components affecting IEQ. *From Toyinbo, O., 2012. Indoor Environmental Quality in Finnish Elementary Schools and Its Effects on Students' Health and Learning (Master's degree dissertation). Department of Environmental Science, University of Eastern Finland. Available from: <http://epublications.uef.fi/pub/urn_nbn_fi_uef-20120654/urn_nbn_fi_uef-20120654.pdf> (accessed 01.08.18).*

moisture or water damage in the building that encourages their growth (Cho et al., 2016; Clausen et al., 2012). Constant and chronic exposure to these pollutants can cause or exacerbate an adverse health condition such as allergic reactions and respiratory diseases that include cough, eye irritation, fatigue, runny nose, and respiratory infections (Annesi-Maesano et al., 2012; Bidassey-Manilal et al., 2016; Katoto et al., 2018).

The physical factors are air movement, temperature, pressure, lighting as well as humidity. These factors according to Levin (1995) can directly affect occupants and their body response to their indoor environment.

The chemical factors affecting IEQ come from indoor or outdoor emissions such as ETS and emissions from indoor materials such as furniture, printers, and carpets that emit pollutants such as volatile organic compounds (VOCs) (Ho et al., 2011). Fig. 4.1 gives a better explanation of the different factors affecting the indoor environment.

4.3 COMPONENTS OF INDOOR ENVIRONMENTAL QUALITY

The components of IEQ includes thermal comfort, ventilation, noise, lighting as well as cleanliness as shown in Table 4.1.

Thermal comfort relates to the condition in which one is contented with the thermal environment. This sensation is affected by airspeed, radiant heat, humidity, surrounding temperature, and individual thermal resistance brought about by metabolic rate and clothing material as well as the number of clothes worn (Daghigh, 2015; Ekici and Atilgan, 2013). It is also affected by climate or season, location, age, and gender (Kim et al., 2013; Quang et al., 2014).

TABLE 4.1 Different Physical, Chemical, Biological, and Particle Factors That Affect IEQ

Physical factors	Indoor environmental quality		
	Chemical factors	**Biological factors**	**Particle factors**
Temperature	(Organic) VOCs, PAH, e.g., Benzo [a]pyrene, Formaldehyde (inorganic) CO_2, CO, SO_2, NO_x, O_3, NH_3, Radon (odors)	Molds (fungi)	Dust
Humidity		Bacteria	Tobacco smoke
Air pressure, air movement (draught)		Plant pollen	Fibers (e.g., asbestos)
Lighting		Dust mites	Combustion by-products
Noise		Animal dander	
Cleanliness			

Source: From Toyinbo, O., 2012. Indoor Environmental Quality in Finnish Elementary Schools and Its Effects on Students' Health and Learning (Master's degree dissertation). Department of Environmental Science, University of Eastern Finland. Available from: <http://epublications.uef.fi/pub/urn_nbn_fi_uef-20120654/urn_nbn_fi_uef-20120654.pdf> (accessed 01.08.18).

A 2015 study by Nam et al. on thermal comfort and clothing insulation in Korea shows children to prefer clothing with lower insulation and a lower temperature than adults due to their different metabolic activities. The study also shows that the preferred clothing insulation differs by gender. Thermal comfort is a function of air temperature, airspeed or movement, and season as shown in a study by Modeste et al. (2014). In the study, 69% of respondents were thermally comfortable in rainy season while 47% were thermally comfortable in the dry season. The study also shows that thermal comfort was related to air temperature and airspeed with 75% of respondents in favor of more air movement in their built environment. A review study by Andersen and Gyntelberg (2011) suggested an indoor temperature to be lower than 23°C for thermal comfort to be achieved in a building, but the minimum indoor temperature suggested by the World Health Organization is 18°C (WHO, 2007). According to ASHRAE standard 55, at least 80% of building occupants should be thermally comfortable for thermal comfort to be achieved in a building. The concept of adaptive thermal comfort that relates to the ability of humans to acclimatize to their surrounding environmental temperature sometimes makes comfort possible at a wider temperature range than that recommended (de Dear, 1998; Nicol and Humphreys, 2002). For example, although comfort temperature was found to be around 22°C in a study of the effect of urban green infrastructure (UGI) on thermal comfort in the Netherlands, the preferred temperature of respondents in the survey was 36°C (Wang et al., 2017). In another study by Schiavon et al. (2017), adaptive thermal comfort was better at temperature of 29°C rather than the approved thermal comfort set point of 23°C in Singaporean offices with occupants wearing cloth of clothing insulation of 0.7 clo. Occupants of green buildings were found to make more personal and psychological adjustments than environmental adjustments when compared to those in conventional building during thermal discomfort of too hot or too cold temperature (Azizi et al., 2015).

Ventilation on the other hand involves the exchange and circulation of air in a building. This includes the introduction of outdoor air supposedly rich in oxygen into the indoor environment, and the removal of stale indoor air with a high concentration of carbon dioxide (CO_2) due to metabolic activities and other indoor pollutants (Patton et al., 2016; Rim et al., 2015). Age, size, and physical activity affect the amount of oxygen used in a space and the amount of CO_2 produced (Persily, 1993). The amount of CO_2 exhaled is about 100 times that inhaled and this invariably increases the concentration of indoor CO_2, making it to exceed the outdoor concentration especially in densely populated buildings (Batterman, 2017). Ventilation helps in diluting this pollutants; it also helps in the cooling of the indoor environment especially when the introduced outdoor air is conditioned (MacNaughton et al., 2015; Sharpe et al., 2015). Research has shown that an increase in ventilation is related to a corresponding decrease in indoor temperature (Sekhar, 2016). This invariably means that an adequately ventilated building will provide a better thermal comfort for occupants of such building (Daghigh, 2015; Damiati et al., 2016; Jamaludin et al., 2015).

Lighting refers to illumination of a built environment either through natural means (e.g., daylighting) or artificially (e.g., lamps) to see objects clearly (Arif et al., 2016). The amount and source of lighting in a space is related to the visual comfort experienced by the occupants which research have shown can affect occupants well-being and performance (Aries et al., 2010; Arif et al., 2016; Veitch, 2001). The impact of visual comfort can also be felt after leaving a particular space such as workplace. For example, a study by Aries et al. (2010) shows that appropriate window lighting at work is associated with workers comfort and improve sleep quality at home after work.

Noise refers to unpleasant or unwanted sound that causes inconvenience in a built environment (Wong et al., 2008). This unwanted sound can irritate and distract occupants from effectively performing their tasks. The source of noise can be from faulty equipment indoor, heating ventilation, and air conditioning (HVAC) system or from the outdoor environment such as those from nearby company or from traffic (Kim, 2015).

Cleaning activities involves removing undesirable materials such as impurities from the environment (Huffaker and Phipatanakul, 2014; Toyinbo et al., 2018). The unwanted materials includes dirt, stains, and pathogenic agents in the environment. Hygienic activities such as hand washing improves cleanliness and reduce the transfer of infectious agents thereby curbing the spread of diseases. While cleaning is supposed to help improve IEQ, some cleaning activities can worsen the condition of the indoor environment. For example, sweeping of dry floor instead of mopping with water can make settled particles to be resuspended (Qian and Ferro, 2008; Toyinbo et al., 2018) and an ineffective cleaning and drying of wet mops after use can encourage bacterial growth and spread (Westwood et al., 1971).

4.4 INDOOR ENVIRONMENTAL QUALITY IN SUSTAINABLE CONSTRUCTION TECHNOLOGY (GREEN BUILDINGS VERSUS CONVENTIONAL BUILDINGS)

Sustainable construction technologies can be termed green construction or building since it involves the use of environmentally friendly and energy saving processes in the design (e.g., by using natural ventilating units) and construction of buildings while preserving building durability and occupants comfort and health (Braganca et al., 2014; Ding, 2008; Kibert, 2016; Steinemann et al., 2017). It is therefore necessary that processes needed to achieve an acceptable IEQ are energy saving and environmentally friendly, that is, they should be from renewable or biodegradable resources and have no or reduced emission of greenhouse gases (Breysse et al., 2011; Loftness et al., 2007). Sometimes in conventional buildings, the durability of the building structure and comfort of occupants is of more importance than energy saving processes during construction.

With the above, a sustainable and acceptable IEQ can be achieved by employing passive procedures or techniques in construction such as using green roof in construction, employing natural ventilation method, fitting energy-efficient windows that are treated (e.g., window shading) to conserve heat in the cold and warm seasons by reducing heat loss and heat gain, respectively (Taleb, 2014; Wang et al., 2016). This conservation technique in energy-efficient windows help to keep indoor temperature within an acceptable range in order to make thermal comfort achievable. Low-energy windows are not only cheap for constructions, but also the treatment given to them helps in filtering ultra-violet radiations of the sun that may cause excessive indoor temperature and damage home fittings such as carpets and furniture (Long and Ye, 2014; Wang et al., 2016; Zhou et al., 2013). The shading or coating also discourage heat loss in cold weather. In green roofing, the topmost part of a building is covered with vegetation either totally or partially (Saadatian et al., 2013). There is a waterproof membrane placed between the building and vegetation to support the vegetation as well as to discourage water leakage (Gaffin et al., 2009; Piro et al., 2017; Taylor, 2007). Although green roofing is not a recent discovery, its application in modern building coupled with membrane waterproofing technology makes its use safer for sustainable construction. Green roofing is ecofriendly, attractive (when flowers are planted on it), and energy efficient by protecting the building from severe temperatures (encouraging thermal comfort) since it provides a cooler and warmer indoor environment in warm and cold weather conditions, respectively (Ascione et al., 2013; Niachou et al., 2001). A 2011 study on the thermal performance of green roof by Becker and Wang (2011) shows green roofing to gain up to 74% more heat in the winter months and lose about 25% more heat in the summer months than conventional roofing. In another study by Parizotto and Lamberts (2011), green roof in temperate climate of Brazil was found to reduce heat gain by up to 97% and encourage heat gain by up to 50% during warm season than conventional roofing. In cold season, the green roofing reduces heat gain by up to 85% and reduces heat loss by about 50% thereby encouraging thermal comfort and energy efficiency of the buildings. Another important aspect of green roofing is the substrate thickness of the roof. Pianella et al. (2017) in their experimental study of substrate thickness and green roof performance found green roof with substrate thickness of 200 mm to provide better insulative effect than those with 150 and 100 mm thickness, respectively. The thickest of the three substrate investigated also provided the least temperature fluctuation but it is also the heaviest and this can be a disadvantage to smaller buildings. Although the cost of green roof is more than that of conventional roof, green roof is more durable, making it to be more of an economic advantage (Veisten et al., 2012).

A passive ventilation system (natural ventilation) for green building involves the opening of windows and doors for air circulation or the use of passive stack ventilation (PSV) system (Tong et al., 2017; Zhai et al., 2015). This fully depends on the outdoor wind speed, wind direction, and thermal buoyancy (Aflaki et al., 2015; Walker, 2016). This system operates on a low or no energy, saving about 30% and 80% of the building total energy used and conditioning energy, respectively (Tong et al., 2016; Walker, 2016). This helps to reduce carbon footprint that can affect the environment (Tong et al., 2016). Mechanical ventilation that uses mechanical force in distributing air is sometimes used in conventional building. This system is energy dependent and airflow rates depend on the efficiency of the system (WHO, 2009). Professionals do the operation and maintenance of this system. This as well as its use of energy makes it costly to run and it leaves carbon footprints (Ianniella, 2011; Rim et al., 2015). While natural ventilation is cheaper to run, its overreliance on wind speed and climatic conditions can encourage overventilation or underventilation, which may ultimately affect ventilation or airflow rates (Chu et al., 2015; Zhai et al., 2015). This problem can be catered for in mechanically ventilated buildings, where ventilation rates can be adjusted to conform to standard at any particular time. Naturally ventilated buildings that are near polluted areas such as processing industries and mining areas can be polluted with effluents from such places, since outdoor air are usually not filtered and conditioned as being done in some mechanically ventilated buildings before their introduction indoor (Fuoco et al., 2015). This can make raw unconditioned (with high or low temperature without dehumidification) air that may be contaminated to be introduced into the indoor environment (Fuoco et al., 2015). There is a constant exchange of air in green ventilation as opposed to some conventional buildings. This helps against sick building syndrome (SBS) associated with mechanically ventilated buildings (Joshi, 2008). This is because air recirculation is sometimes done in mechanically ventilated buildings by blocking outdoor air in order to quickly achieve indoor air conditioning and save energy (Joshi, 2008). This makes the concentration of indoor pollutants to increase.

According to ASHRAE standard 55, 80% of building occupants should always be thermally comfortable (Olesen and Brager, 2004). The green method of ventilation (natural ventilation) which is the PSV system as well as windows and doors opening for ventilation may not be enough to give occupants thermal comfort. This is because natural ventilation is difficult to adjust (ventilation rate mostly depends on outdoor airspeed and climatic condition) and ventilation is inversely related to temperature (Norbäck and Nordström, 2008; Toyinbo et al., 2016a). For example, all classrooms with natural ventilation has ventilation rate per student and per area that were below the Finnish building code recommendation of 6 *l/s per student* and 3 $l/s\ per\ m^2$, respectively, in a study by Toyinbo et al. (2016b). In the same study, ventilation rate was inversely associated with indoor temperature. An Austrian study that compared naturally and mechanically ventilated buildings found indoor air quality parameters in mechanically ventilated buildings to be better than those from naturally ventilated buildings (Wallner et al., 2015). A study by Yang et al. (2009) suggested mechanical ventilation for Korean schools in order to improve classroom ventilation rate. This shows that the comfort level of the naturally ventilated classrooms might be lower than the ones mechanically ventilated. In another study by Lu et al. (2015), 76% of occupants in naturally ventilated rooms want their room environment conditioned. When mechanically ventilated rooms that met ASHRAE standard 55 thermal comfort recommendations were converted to naturally ventilation rooms in a study by Daghigh et al. (2009), occupants became thermally uncomfortable.

Lighting, acoustics, thermal comfort as well as IAQ were better in green buildings as opposed to conventional buildings in a study by Liang et al. (2014). The concentration of indoor pollutants such as CO_2 and VOCs was also lower in the green buildings. It is important to know that people working in green certified buildings were found to have a higher sleep quality, cognitive performance, and lower health symptoms as compared to conventional buildings (MacNaughton et al., 2017, 2016). Nevertheless, a study by Paul and Taylor (2008) was unable to verify that green buildings are more comfortable than conventional buildings. IEQ parameters such as lighting, acoustics, and ventilation were found to be similar but thermal discomfort was noticed in the green buildings. However, Gou et al. (2013) and Holmgren et al. (2017) found evidence of bias toward green buildings by occupants with those believing in environmental sustainability showing more tolerance in order to keep the environment save. Other passive techniques that can be employed include making sure that sources of radiant heat like scanners and printers are removed during the warm season and introduced during the cold season so as to make thermal comfort achievable by discouraging extreme temperatures (too hot or too cold) indoor (Taleb, 2014). In all, green building seems to offer a better IEQ than conventional building as shown by Singh et al. (2010). In their study, office occupants that were moved from conventional buildings to green buildings had better work performance and reduced absenteeism due to stress and respiratory diseases.

4.5 CONSEQUENCES OF UNACCEPTABLE INDOOR ENVIRONMENTAL QUALITY

The consequences of an unacceptable IEQ relate to building occupants comfort and well-being while indoor. SBS, which is an acute health condition, related to time spent indoor and building-related illness (BRI), which are chronic health conditions, related to building conditions should not occur. While SBS is related to the time spent indoor in a particular building, BRI may persist even when an inhabitant has already left the building; meaning, there will be a prolonged recovery time of symptoms (Joshi, 2008; Lu et al., 2017; Salin et al., 2017). The causes of SBS are diverse and they include extreme temperatures, indoor pollution, as well as psychosocial condition of an individual. Green ventilation (natural ventilation) may help to relieve SBS symptoms since research has shown that occupants of mechanically ventilated buildings exhibit more SBS symptoms (Preziosi et al., 2004; USEPA 2007). SBS leads to more hospital visits, absenteeism, and reduced productivity (Preziosi et al., 2004; Wargocki et al., 2000).

The works by Dharmage et al. (2002) and Apostolakos et al. (2001) show the influence of moldy houses on asthma symptoms and the occurrence of hypersensitivity pneumonitis respectively, which affect occupants' performance. Some green applications in construction such as the use of recycled and waste based materials may negatively affect IEQ rather than improving them (Steinemann et al., 2017). These materials can radiate pollutants from materials utilized in their recycling processes; they can also reemit toxic substances already present in them from previous use (Raut et al., 2011; Steinemann et al., 2017). This will invariably affect the health and well-being of occupants. A review by Mendell (2007) shows evidence that suggests that emissions from indoor materials are detrimental to health by increasing the risk of respiratory and allergic health problems.

Thermal comfort must be achieved as well as adequate ventilation per person and per area. Extreme indoor temperature (too hot or too cold) is related to thermal discomfort, health symptoms, and reduced productivity (Toyinbo et al., 2016b). A similar association was found with inadequate ventilation (Bakó-Biró et al., 2012; Gaihre et al., 2014; Toftum et al., 2015).

A study by Preziosi et al. (2004) shows the adverse effect mechanical ventilation or air-conditioned building may have on occupants. The odds ratio for occupants visit to otorhinolaryngologist was 2.33 (95% CI = 1.35–4.04) in the mechanical ventilation group compared with the natural ventilation group, and 1.70 (1.13–2.58) for sickness leave. This shows the benefit of green ventilation that it is cheaper to construct and energy efficient, using lower energy than air-conditioned building (Nicol and Humphreys, 2002). However, in a more recent study by Wallner et al. (2015), mechanically ventilated buildings had overall better IEQ when compared to those using natural ventilation. This result is similar to that of Toftum et al. (2015) where naturally ventilated rooms were associated with ventilation inadequacy leading to high indoor CO_2 concentration, which is related to respiratory problems (Ferreira and Cardoso, 2014). A water, sanitation, and hygiene (WASH) study found unhygienic practices among students to increase their risk to contact *Escherichia coli* with relative risk (RR) of 1.36 (95% CI = 0.74–2.49) and 2.63 (95% CI = 1.29–5.34) for boys and girls, respectively (Greene et al., 2012).

A scientific review of 85 literatures by Cho et al. (2015) found the use of artificial light to be related to negative health outcomes in human. The review found exposure to artificial light to reduce melatonin secretion which affects circadian rhythm and ultimately sleep. Circadian rhythm disruption due to artificial light was also found to affect cardiovascular, psychological, and metabolic functions (Cho et al., 2015). In another study by Boubekri et al. (2014), workers in windowless environments with artificial lighting had lower achievements and sleep compared to those with windows and daylight. An experimental study that sampled 1000 individuals between the age of 19 and 80 in Stockholm, Sweden, found traffic-related noise to annoy 13% of respondents with 27% of the subjects having sleep disturbances (Bluhm et al., 2004). A review of environmental noise and health by Van Kamp and Davies (2013) show noise to be related to cardiovascular effects such as a raised blood pressure, sleep disturbances, annoyance, physiological effect such as fatigue, headache, and low quality of life, and cognitive effects such as impaired reading skills.

Each of the above or in connection with another has been linked with adverse health conditions, reduced efficiency, and absenteeism (Annesi-Maesano et al., 2013; MacNaughton et al., 2015; Mendell et al., 2013).

4.6 CONCLUSIONS

Although the aim of sustainable construction is to reduce energy consumption and use renewable materials in construction, the comfort of the building's occupants must also be achieved. It is therefore necessary that scheduled commissioning and maintenance of building systems should be done as recommended. Building components should be renovated to meet current specifications and ventilation systems upgraded or adjusted to meet the number of occupants at any particular time. As we move toward sustainable construction and look for greener ways to achieve IEQ, it is important that comfort is achieved. This shows the need to use the best sustainable method to achieve an acceptable IEQ and have other active methods on standby (e.g., hybrid ventilation) in case comfort is not achieved.

REFERENCES

Aflaki, A., Mahyuddin, N., Mahmoud, Z.A.-C., Baharum, M.R., 2015. A review on natural ventilation applications through building façade components and ventilation openings in tropical climates. Energy Build. 101, 153–162.

Andersen, I., Gyntelberg, F., 2011. Modern indoor climate research in Denmark from 1962 to the early 1990s: an eyewitness report. Indoor Air 21 (3), 182–190. Available from: https://doi.org/10.1111/j.1600-0668.2011.00716.x.

Annesi-Maesano, I., Hulin, M., Lavaud, F., Raherison, C., Kopferschmitt, C., de Blay, F., et al., 2012. Poor air quality in classrooms related to asthma and rhinitis in primary schoolchildren of the French 6 Cities Study. Thorax 67 (8), 682–688. Available from: http://dx.doi.org/10.1136/thoraxjnl-2011-200391.

Annesi-Maesano, I., Baiz, N., Banerjee, S., Rudnai, P., Rive, S., Group, S., 2013. Indoor air quality and sources in schools and related health effects. J. Toxicol. Environ. Health B 16 (8), 491–550.

Apostolakos, M.J., Rossmoore, H., Beckett, W.S., 2001. Hypersensitivity pneumonitis from ordinary residential exposures. Environ. Health Perspect. 109 (9), 979–981. Available from: https://doi.org/sc271_5_1835 [pii].

Aries, M.B.C., Veitch, J.A., Newsham, G.R., 2010. Windows, view, and office characteristics predict physical and psychological discomfort. J. Environ. Psychol. 30 (4), 533–541.

Arif, M., Katafygiotou, M., Mazroei, A., Kaushik, A., Elsarrag, E., 2016. Impact of indoor environmental quality on occupant well-being and comfort: a review of the literature. Int. J. Sustain. Built Environ. 5 (1), 1–11.

Ascione, F., Bianco, N., de' Rossi, F., Turni, G., Vanoli, G.P., 2013. Green roofs in European climates. Are effective solutions for the energy savings in air-conditioning? Appl. Energy. 104, 845–859. Available from: https://doi.org/10.1016/j.apenergy.2012.11.068.

Azizi, N.S.M., Wilkinson, S., Fassman, E., 2015. An analysis of occupants response to thermal discomfort in green and conventional buildings in New Zealand. Energy Build. 104, 191–198. Available from: https://doi.org/10.1016/j.enbuild.2015.07.012.

Bakó-Biró, Z., Clements-Croome, D.J., Kochhar, N., Awbi, H.B., Williams, M.J., 2012. Ventilation rates in schools and pupils' performance. Build. Environ. 48, 215–223.

Batterman, S., 2017. Review and extension of CO_2-based methods to determine ventilation rates with application to school classrooms. Int. J. Environ. Res. Public Health 14 (2), 145.

Becker, D., Wang, D., 2011. Green Roof Heat Transfer and Thermal Performance Analysis. Civil and Environmental Engineering Department, Carnegie Mellon University, Pittsburgh, PA.

Bidassey-Manilal, S., Wright, C.Y., Engelbrecht, J.C., Albers, P.N., Garland, R.M., Matooane, M., 2016. Students' perceived heat-health symptoms increased with warmer classroom temperatures. Int. J. Environ. Res. Public Health 13 (6), 566.

Bluhm, G., Nordling, E., Berglind, N., et al., 2004. Road traffic noise and annoyance—an increasing environmental health problem. Noise Health 6 (24), 43.

Boubekri, M., Cheung, I.N., Reid, K.J., Wang, C.-H., Zee, P.C., 2014. Impact of windows and daylight exposure on overall health and sleep quality of office workers: a case-control pilot study. J. Clin. Sleep Med. 10 (06), 603–611.

Braganca, L., Vieira, S.M., Andrade, J.B., 2014. Early stage design decisions: the way to achieve sustainable buildings at lower costs. Sci.World J. 2014, 365364. Available from: https://doi.org/10.1155/2014/365364 [doi].

Breysse, J., Jacobs, D.E., Weber, W., Dixon, S., Kawecki, C., Aceti, S., et al., 2011. Health outcomes and green renovation of affordable housing. Public Health Rep. 126 (Suppl. 1), 64–75.

Cho, S.J., Cox-Ganser, J.M., Park, J., 2016. Observational scores of dampness and mold associated with measurements of microbial agents and moisture in three public schools. Indoor Air 26 (2), 168–178.

Cho, Y., Ryu, S.-H., Lee, B.R., Kim, K.H., Lee, E., Choi, J., 2015. Effects of artificial light at night on human health: a literature review of observational and experimental studies applied to exposure assessment. Chronobiol. Int. 32 (9), 1294–1310.

Chu, C.-R., Chiu, Y.-H., Tsai, Y.-T., Wu, S.-L., 2015. Wind-driven natural ventilation for buildings with two openings on the same external wall. Energy Build. 108, 365–372.

Clausen, G., Høst, A., Toftum, J., Bekö, G., Weschler, C., Callesen, M., et al., 2012. Children's health and its association with indoor environments in Danish homes and daycare centres–methods. Indoor Air 22 (6), 467–475.

Daghigh, R., 2015. Assessing the thermal comfort and ventilation in Malaysia and the surrounding regions. Renew. Sustain. Ener. Rev. 48, 681–691.

Daghigh, R., Adam, N.M., Sahari, B.B., 2009. Ventilation parameters and thermal comfort of naturally and mechanically ventilated offices. Indoor Built Environ. 18 (2), 113–122.

Damiati, S.A., Zaki, S.A., Rijal, H.B., Wonorahardjo, S., 2016. Field study on adaptive thermal comfort in office buildings in Malaysia, Indonesia, Singapore, and Japan during hot and humid season. Build. Environ. 109, 208–223.

de Dear, R., 1998. Developing an adaptive model of thermal comfort and preference, field studies of thermal comfort and adaptation. ASHRAE Tech. Data Bull. 14 (1), 27–49.

Dharmage, S., Bailey, M., Raven, J., Abeyawickrama, K., Cao, D., Guest, D., et al., 2002. Mouldy houses influence symptoms of asthma among atopic individuals. Clin. Exp. Allergy 32 (5), 714–720.

Ding, G.K.C., 2008. Sustainable construction—the role of environmental assessment tools. J. Environ. Manage. 86 (3), 451–464.

Ekici, C., Atilgan, I., 2013. A comparison of suit dresses and summer clothes in the terms of thermal comfort. J. Environ. Health Sci. Eng. 11 (1), 32.

Ferreira, A.M.D.C., Cardoso, M., 2014. Indoor air quality and health in schools. Jornal Brasileiro de Pneumologia 40 (3), 259–268.

Fuoco, F.C., Stabile, L., Buonanno, G., Trassiera, C.V., Massimo, A., Russi, A., et al., 2015. Indoor air quality in naturally ventilated Italian classrooms. Atmosphere 6 (11), 1652–1675.

Gaffin, S.R., Khanbilvardi, R., Rosenzweig, C., 2009. Development of a green roof environmental monitoring and meteorological network in New York City. Sensors 9 (4), 2647–2660.

Gaihre, S., Semple, S., Miller, J., Fielding, S., Turner, S., 2014. Classroom carbon dioxide concentration, school attendance, and educational attainment. J. School Health 84 (9), 569–574.

Gou, Z., Prasad, D., Lau, S.S.-Y., 2013. Are green buildings more satisfactory and comfortable? Habitat. Int. 39, 156–161.

Greene, L.E., Freeman, M.C., Akoko, D., Saboori, S., Moe, C., Rheingans, R., 2012. Impact of a school-based hygiene promotion and sanitation intervention on pupil hand contamination in Western Kenya: a cluster randomized trial. Am. J. Trop. Med. Hyg. 87 (3), 385–393.

Ho, D.X., Kim, K.H., Sohn, J.R., Oh, Y.H., Ahn, J.W., 2011. Emission rates of volatile organic compounds released from newly produced household furniture products using a large-scale chamber testing method. Sci.World J. 11, 1597–1622. Available from: https://doi.org/10.1100/2011/650624 [doi].

Holmgren, M., Kabanshi, A., Sörqvist, P., 2017. Occupant perception of "green" buildings: distinguishing physical and psychological factors. Build. Environ. 114, 140–147. Available from: https://doi.org/10.1016/j.buildenv.2016.12.017.

Huffaker, M., Phipatanakul, W., 2014. Introducing an environmental assessment and intervention program in inner-city schools. J. Aller. Clin. Immunol. 134 (6), 1232–1237.

Ianniella, E., 2011. Ventilation systems and IAQ in school buildings. REHVA J. March, 26–29.

Jamaludin, N., Mohammed, N.I., Khamidi, M.F., Wahab, S.N.A., 2015. Thermal comfort of residential building in Malaysia at different micro-climates. Proc. Soc. Behav. Sci. 170, 613–623.

Joshi, S.M., 2008. The sick building syndrome. Indian J. Occup. Environ. Med. 12 (2), 61.

Katoto, C., Byamungu, N.L., Brand, A., Mokaya, J., Nawrot, T., Nemery, B., 2018. Ambient air pollution and health in sub-Saharan Africa: a systematic map of the evidence. A54. Indoor and Outdoor Air Pollution. American Thoracic Society, San Diego Convention Center, CA, USA, p. A1924.

Kibert, C.J., 2016. Sustainable Construction: Green Building Design and Delivery. John Wiley Sons. Inc., Hoboken, NJ .

Kim, J., de Dear, R., Candido, C., Zhang, H., Arens, E., 2013. Gender differences in office occupant perception of indoor environmental quality (IEQ). Build. Environ. 70, 245–256.

Kim, K., 2015. Sources, effects, and control of noise in indoor/outdoor living environments. J. Ergonom. Soc. Korea 34 (3), 265–278.

Levin, H., 1995. Physical factors in the indoor environment. Occup. Med. (Philadelphia, PA) 10 (1), 59–94.

Liang, H.-H., Chen, C.-P., Hwang, R.-L., Shih, W.-M., Lo, S.-C., Liao, H.-Y., 2014. Satisfaction of occupants toward indoor environment quality of certified green office buildings in Taiwan. Build. Environ. Available from: https://doi.org/10.1016/j.buildenv.2013.11.007.

Loftness, V., Hakkinen, B., Adan, O., Nevalainen, A., 2007. Elements that contribute to healthy building design. Environ. Health Perspect. 115 (6), 965–970. Available from: https://doi.org/10.1289/ehp.8988 [doi].

Long, L., Ye, H., 2014. How to be smart and energy efficient: a general discussion on thermochromic windows. Sci. Rep. 4, p. 6427.

Lu, C.-Y., Tsai, M.-C., Muo, C.-H., Kuo, Y.-H., Sung, F.-C., Wu, C.-C., 2017. Personal, psychosocial and environmental factors related to sick building syndrome in official employees of Taiwan. Int. J. Environ. Res. Public. Health 15 (1), 7.

Lu, S., Fang, K., Qi, Y., Wei, S., 2015. Influence of natural ventilation on thermal comfort in semi-open building under early summer climate in the area of tropical island. In: The 9th International Symposium on Heating, Ventilation and Air Conditioning (ISHVAC) Joint with the 3rd International Conference on Building Energy and Environment (COBEE), 12–15 July 2015, Tianjin, China. Available from: https://doi.org/https://doi.org/10.1016/j.proeng.2015.09.060.

MacNaughton, P., Pegues, J., Satish, U., Santanam, S., Spengler, J., Allen, J., 2015. Economic, environmental and health implications of enhanced ventilation in office buildings. Int. J. Environ. Res. Public. Health 12 (11), 14709–14722.

MacNaughton, P., Spengler, J., Vallarino, J., Santanam, S., Satish, U., Allen, J., 2016. Environmental perceptions and health before and after relocation to a green building. Build. Environ. 104, 138–144.

MacNaughton, P., Satish, U., Laurent, J.G.C., Flanigan, S., Vallarino, J., Coull, B., et al., 2017. The impact of working in a green certified building on cognitive function and health. Build. Environ. Available from: https://doi.org/10.1016/j.buildenv.2016.11.041.

Mendell, M.J., 2007. Indoor residential chemical emissions as risk factors for respiratory and allergic effects in children: a review. Indoor Air 17 (4), 259–277.

Mendell, M.J., Eliseeva, E.A., Davies, M.M., Spears, M., Lobscheid, A., Fisk, W.J., et al., 2013. Association of classroom ventilation with reduced illness absence: a prospective study in California elementary schools. Indoor Air 23 (6), 515–528.

Mitchell, C.S., Zhang, J.J., Sigsgaard, T., Jantunen, M., Lioy, P.J., Samson, R., et al., 2007. Current state of the science: health effects and indoor environmental quality. Environ. Health Perspect. 115 (6), 958–964. Available from: https://doi.org/10.1289/ehp.8987.

Modeste, K.N., Tchinda, R., Ricciardi, P., 2014. Thermal comfort and air movement preference in some classrooms in Cameroun. Revue Des Energies Renouvelables 17 (2), 263–278.

Nam, I., Yang, J., Lee, D., Park, E., Sohn, J.R., 2015. A study on the thermal comfort and clothing insulation characteristics of preschool children in Korea. Build. Environ. 92, 724–733.

Niachou, A., Papakonstantinou, K., Santamouris, M., Tsangrassoulis, A., Mihalakakou, G., 2001. Analysis of the green roof thermal properties and investigation of its energy performance. Energy Build. Available from: https://doi.org/10.1016/S0378-7788(01)00062-7.

Nicol, J.F., Humphreys, M.A., 2002. Adaptive thermal comfort and sustainable thermal standards for buildings. Spec. Issue Therm. Comfort Standards. Available from: https://doi.org/10.1016/S0378-7788(02)00006-3.

Norbäck, D., Nordström, K., 2008. An experimental study on effects of increased ventilation flow on students' perception of indoor environment in computer classrooms. Indoor Air 18 (4), 293–300.

Ole Fanger, P., 2006. What is IAQ? Indoor Air 16 (5), 328–334. https://doi.org/INA437 [pii].

Olesen, B.W., Brager, G.S., 2004. A better way to predict comfort. ASHRAE J. 46 (8), 20.

Parizotto, S., Lamberts, R., 2011. Investigation of green roof thermal performance in temperate climate: a case study of an experimental building in Florianópolis city, Southern Brazil. Energy Build. 43 (7), 1712–1722.

Patton, A.P., Calderon, L., Xiong, Y., Wang, Z., Senick, J., Sorensen Allacci, M., et al., 2016. Airborne particulate matter in two multi-family green buildings: concentrations and effect of ventilation and occupant behavior. Int. J. Environ. Res. Public Health 13 (1), 144.

Paul, W.L., Taylor, P.A., 2008. A comparison of occupant comfort and satisfaction between a green building and a conventional building. Build. Environ. 43 (11), 1858–1870. Available from: https://doi.org/10.1016/j.buildenv.2007.11.006.

Persily, A.K., 1993. Ventilation, carbon dioxide and ASHRAE Standard 62-1989. ASHRAE J. 35 (7), 42–44.

Pianella, A., Aye, L., Chen, Z., Williams, N.S.G., 2017. Substrate depth, vegetation and irrigation affect green roof thermal performance in a mediterranean type climate. Sustainability 9 (8), 1451.

Piro, P., Porti, M., Veltri, S., Lupo, E., Moroni, M., 2017. Hyperspectral monitoring of green roof vegetation health state in sub-mediterranean climate: preliminary results. Sensors 17 (4), 662.

Preziosi, P., Czernichow, S., Gehanno, P., Hercberg, S., 2004. Workplace air-conditioning and health services attendance among French middle-aged women: a prospective cohort study. Int. J. Epidemiol. 33 (5), 1120–1123.

Qian, J., Ferro, A.R., 2008. Resuspension of dust particles in a chamber and associated environmental factors. Aerosol. Sci. Technol. 42 (7), 566–578.

Quang, T.N., He, C., Knibbs, L.D., de Dear, R., Morawska, L., 2014. Co-optimisation of indoor environmental quality and energy consumption within urban office buildings. Energy Build. 85, 225–234.

Raut, S.P., Ralegaonkar, R.V., Mandavgane, S.A., 2011. Development of sustainable construction material using industrial and agricultural solid waste: a review of waste-create bricks. Constr. Build. Mater. 25 (10), 4037–4042.

Rim, D., Schiavon, S., Nazaroff, W.W., 2015. Energy and cost associated with ventilating office buildings in a tropical climate. PLoS ONE 10 (3), e0122310.

Saadatian, O., Sopian, K., Salleh, E., Lim, C.H., Riffat, S., Saadatian, E., et al., 2013. A review of energy aspects of green roofs. Renew. Sustain. Energy Rev. Available from: https://doi.org/10.1016/j.rser.2013.02.022.

Salin, J.T., Salkinoja-Salonen, M., Salin, P.J., Nelo, K., Holma, T., Ohtonen, P., et al., 2017. Building-related symptoms are linked to the in vitro toxicity of indoor dust and airborne microbial propagules in schools: a cross-sectional study. Environ. Res. 154, 234–239.

Schiavon, S., Yang, B., Donner, Y., Chang, V.W.-C., Nazaroff, W.W., 2017. Thermal comfort, perceived air quality, and cognitive performance when personally controlled air movement is used by tropically acclimatized persons. Indoor Air 27 (3), 690–702. Available from: https://doi.org/10.1111/ina.12352.

Sekhar, S.C., 2016. Thermal comfort in air-conditioned buildings in hot and humid climates—why are we not getting it right? Indoor Air 26 (1), 138–152. Available from: https://doi.org/10.1111/ina.12184.

Sharpe, T., Farren, P., Howieson, S., Tuohy, P., McQuillan, J., 2015. Occupant interactions and effectiveness of natural ventilation strategies in contemporary new housing in Scotland, UK. Int. J. Environ. Res. Public Health 12 (7), 8480–8497.

Singh, A., Syal, M., Grady, S.C., Korkmaz, S., 2010. Effects of green buildings on employee health and productivity. Am. J. Public Health 100 (9), 1665–1668.

Steinemann, A., Wargocki, P., Rismanchi, B., 2017. Ten questions concerning green buildings and indoor air quality. Build. Environ. 112, 351–358.

Taleb, H.M., 2014. Using passive cooling strategies to improve thermal performance and reduce energy consumption of residential buildings in UAE buildings. Front. Architect. Res. 3 (2), 154–165.

Taylor, D.A., 2007. Growing green roofs, city by city. Environ. Health Perspect. 115 (6), A306–A311.

Toftum, J., Kjeldsen, B.U., Wargocki, P., Menå, H.R., Hansen, E.M.N., Clausen, G., 2015. Association between classroom ventilation mode and learning outcome in Danish schools. Build. Environ. 92, 494–503.

Tong, Z., Chen, Y., Malkawi, A., Liu, Z., Freeman, R.B., 2016. Energy saving potential of natural ventilation in China: the impact of ambient air pollution. Appl. Energy 179, 660–668.

Tong, Z., Chen, Y., Malkawi, A., 2017. Estimating natural ventilation potential for high-rise buildings considering boundary layer meteorology. Appl. Energy 193, 276–286.

Toyinbo, O., Matilainen, M., Turunen, M., Putus, T., Shaughnessy, R., Haverinen-Shaughnessy, U., 2016a. Modeling associations between principals' reported indoor environmental quality and students' self-reported respiratory health outcomes using GLMM and ZIP models. Int. J. Environ. Res. Public. Health 13 (4), 385.

Toyinbo, O., Shaughnessy, R., Turunen, M., Putus, T., Metsämuuronen, J., Kurnitski, J., et al., 2016b. Building characteristics, indoor environmental quality, and mathematics achievement in Finnish elementary schools. Build. Environ. 104, 114–121.

Toyinbo, O., Obi, C., Shaughnessy, R., Haverinen-Shaughnessy, U., 2018. Ventilation, thermal comfort and cleanliness of high contact surface in Nigerian schools. In: Indoor Air 2018 Conference, 22–27 July 2018, Philadelphia, PA.

Toyinbo, O., 2012. Indoor Environmental Quality in Finnish Elementary Schools and Its Effects on Students' Health and Learning (Master's degree dissertation). Department of Environmental Science, University of Eastern Finland. Available from: <http://epublications.uef.fi/pub/urn_nbn_fi_uef-20120654/urn_nbn_fi_uef-20120654.pdf> (accessed 01.08.18).

USEPA (United States Environmental Protection Agency), 2007. The EPA Cost of Illness Handbook. U.S. Environmental Protection Agency, Washington, DC.

Van Kamp, I., Davies, H., 2013. Noise and health in vulnerable groups: a review. Noise Health 15 (64), 153.

Veisten, K., Smyrnova, Y., Klæboe, R., Hornikx, M., Mosslemi, M., Kang, J., 2012. Valuation of green walls and green roofs as soundscape measures: Including monetised amenity values together with noise-attenuation values in a cost-benefit analysis of a green wall affecting courtyards. Int. J. Environ. Res. Public Health 9 (11), 3770–3788.

Veitch, J.A., 2001. Psychological processes influencing lighting quality. J. Illumin. Eng. Soc. 30 (1), 124–140.

WHO (World Health Organization). (2007). Housing, energy and thermal comfort: a review of 10 countries within the WHO European Region. WHO Regional Office for Europe: Copenhagen, 2007; Available at <http://apps.who.int/iris/bitstream/handle/10665/107815/E89887.pdf?sequence=1&isAllowed=y> (assessed on 19.10.18).

WHO (World Health Organization). (2009). Guidelines for Indoor Air Quality: Dampness and Mold. WHO Regional Office for Europe, Copenhagen. Available from: <http://apps.who.int/iris/bitstream/handle/10665/164348/E92645.pdf;jsessionid = E8825FA47A0AA41D18B44F58405F3AA6?sequence = 1 > (accessed 24.09.18).

Walker, A., 2016. Natural Ventilation. Whole Building Design Guide (WBDG), A Program of the National Institute of Building Sciences. National Renewable Energy Laboratory, Washington, DC.

Wallner, P., Munoz, U., Tappler, P., Wanka, A., Kundi, M., Shelton, J.F., et al., 2015. Indoor environmental quality in mechanically ventilated, energy-efficient buildings vs. conventional buildings. Int. J. Environ. Res. Public Health 12 (11), 14132–14147. Available from: https://doi.org/10.3390/ijerph121114132 [doi].

Wang, Y., Runnerstrom, E.L., Milliron, D.J., 2016. Switchable materials for smart windows. Annu. Rev. Chem. Biomol. Eng. 7, 283–304.

Wang, Y., de Groot, R., Bakker, F., Wörtche, H., Leemans, R., 2017. Thermal comfort in urban green spaces: a survey on a Dutch university campus. Int. J. Biometeorol. 61 (1), 87–101.

Wargocki, P., Wyon, D.P., Sundell, J., Clausen, G., Fanger, P., 2000. The effects of outdoor air supply rate in an office on perceived air quality, sick building syndrome (SBS) symptoms and productivity. Indoor Air 10 (4), 222–236.

Westwood, J.C.N., Mitchell, M.A., Legacé, S., 1971. Hospital sanitation: the massive bacterial contamination of the wet mop. Appl. Microbiol. 21 (4), 693–697.

Wong, L.T., Mui, K.W., Hui, P.S., 2008. A multivariate-logistic model for acceptance of indoor environmental quality (IEQ) in offices. Build. Environ. 43 (1), 1–6.

Yang, W., Sohn, J., Kim, J., Son, B., Park, J., 2009. Indoor air quality investigation according to age of the school buildings in Korea. J. Environ. Manage. 90 (1), 348–354.

Zhai, Z.J., El Mankibi, M., Zoubir, A., 2015. Review of natural ventilation models. Energy Proced. 78, 2700–2705.

Zhou, J., Gao, Y., Zhang, Z., Luo, H., Cao, C., Chen, Z., et al., 2013. VO(2) thermochromic smart window for energy savings and generation. Sci. Rep. 3, 3029. Available from: https://doi.org/10.1038/srep03029.

Chapter 5

Life Cycle Energy Consumption of Buildings; *Embodied + Operational*

Rahman Azari
College of Architecture, Illinois Institute of Technology, Chicago, IL, United States

5.1 INTRODUCTION

The decade of 1970s was an important decade in the 20th century from an energy perspective. The Arab-Israeli war, which occurred in October 1973, caused a global oil crisis that began by the Arab members of the Organization of Petroleum Exporting Countries (OPEC) imposing an oil embargo on the United States and its allies. The global oil price nearly quadrupled in just about 6 months and the oil production declined. The oil crisis of 1973 prompted changes in the US policies with regard to energy and triggered implementation of standards and codes for energy efficiency. As an example, the first building energy code, ASHRAE Standard 90–75 Energy Conservation in New Building Design, was published in 1975 by the American Society of Heating, Refrigeration, and Air-conditioning Engineers (ASHRAE). The second world crisis of the decade occurred in 1979 because of which a shortage in national energy supply was proclaimed by the US government and temperature restrictions were established for nonresidential buildings.

Almost 39 years since 1979, the use of fossil-fuel-based energy is still growing which is a vivid reminder of its significance in modern society. An examination of the changes in the US society between 1979 and 2016 based on the US Census Bureau reveals 44% increase in the population, 62% increase in the number of households, and 50% increase in the square-footage of new single-family housing (Fig. 5.1). Looking at the trends of energy use, the past four decades have also experienced growth in two major end-use sectors including building stock and transportation (Fig. 5.2). The building stock's energy use has grown more than any other sector (at 45%)

Sustainable Construction Technologies. DOI: https://doi.org/10.1016/B978-0-12-811749-1.00004-3

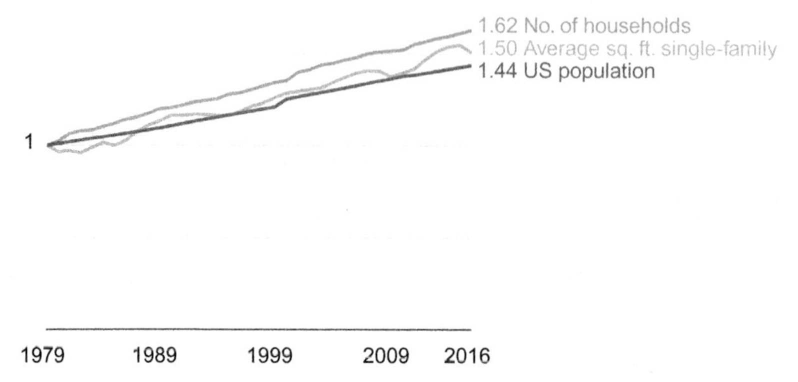

FIGURE 5.1 Trends of changes in population, number of households and new single-family housing units in the United States, based on the US Census Bureau data.

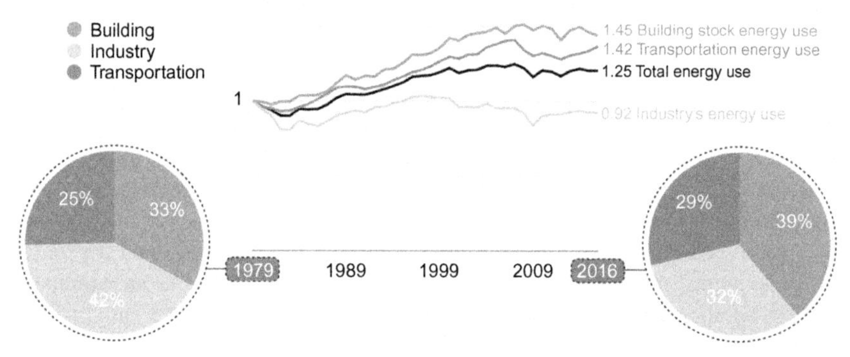

FIGURE 5.2 Trends of changes in energy use in the United States and contribution of end-use sectors to the US primary energy use, based on the US Energy Information Administration data (EIA, 2016).

since 1979, and its share in total primary energy of the United States has increased from 33% to 39% (Fig. 5.2). This energy that is used to operate the building is called operational energy. Adding the embodied energy, that is consumed by the industry sector to produce construction materials, can increase the current 39% energy use of the building stock to as high as 48%.[1] In transportation sector, the growth of energy use has been 42% and its contribution to total primary energy use has changed from 25% to 29% (EIA, 2016).

1. Architecture 2030 Challenge.

Since 1979, reducing the energy use of buildings has been a major concern of the professional and research community in the built environment (Pérez-Lombard et al., 2008). The focus has been on occupancy phase of building life cycle and operational energy use, and major successes have been achieved. Yet, a shift in paradigm seems to be underway with recent interests in buildings' life cycle energy use; i.e., the energy that is consumed over the entire life cycle of a building from raw material extraction to manufacturing of building components and systems, transportation, construction, operation, maintenance, and to demolition of buildings.

Life cycle energy has two significant components: operational and embodied energy. Much effort has been devoted in the past to increasing the efficiency of buildings with regard to operational energy; i.e., the energy that is consumed during the occupancy stage of building's life to heat, cool, illuminate, and run equipment and appliances in buildings. The professional community in architecture and engineering now have the design knowledge as well as technology at their disposal to develop buildings with potential to consume lesser operational energy. As design and technology upgrades alone have not been sufficient to guarantee operational energy efficiency, there is growing recognition that occupant behavior too needs to be accounted for in design and operation of low-energy buildings. In addition, energy-related codes, standards, and performance metrics and targets have been developed over the past decades in different countries in order to achieve low-energy buildings.

The second component of life cycle energy is embodied energy. Increasing efficiency for embodied energy has gained more attention only in recent years. Embodied energy is the energy that is used for extraction of materials, manufacturing of components, construction, maintenance, and demolition of building as well as all associated transportation. Despite recent developments, the embodied energy is comparatively a less developed field than operational energy research and there are serious limitations with regard to analytical methodologies, data availability, tools, and metrics that need to be addressed. Addressing these limitations will pave the way for more accurate estimations of embodied energy, and in turn building life cycle energy use.

This chapter provides an overview of information and knowledge in the field of life cycle energy use and reviews data, methodologies, design developments, and challenges to design buildings with lesser reliance on fossil-fuel energy and lower impacts on the environment.

5.2 EMBODIED AND OPERATIONAL; DEFINITIONS, DATA, AND DRIVERS

Operational energy is a well-established area of research and practice where rigorous data and figures about current levels of consumption for different building sectors exist, clear regionally sensitive design guidelines and code prescriptions have been established to inform and regulate building design, energy performance targets have been defined to serve as benchmarks for improvement purposes, and precise yet user-friendly energy simulation tools have been developed for the use of architects, engineers, and students.

The picture is less clear in embodied energy of buildings where data availability, quality (with regard to consistency, transparency, representativeness, etc.) and the lack of agreed-upon framework are still a challenge for estimation of embodied energy as well as other environmental impacts. Also, the design guidelines, performance targets, and tools in this area are less developed.

5.2.1 Embodied Energy

Embodied energy is used differently in the literature based on the stages of building life cycle that are included in its definition. Most literature approach embodied energy from a cradle-to-gate perspective and estimate it as the summation of the energy that is consumed directly or indirectly for production of construction materials used in a building (Dixit et al., 2010). This approach to embodied energy only incorporates the preconstruction stage (i.e., extraction of materials, manufacturing of products, components, and systems) of building life cycle. Some other literatures expand this definition to cradle-to-site, and incorporate both preconstruction and construction stages of the life cycle, and the associated transportation (Hammond and Jones, 2008). A more comprehensive definition of embodied energy is based on cradle-to-grave scope, which includes not only preconstruction and construction stages but also maintenance, demolition, and disposal stages of the building life cycle. This cradle-to-grave approach to embodied energy defines it as the total energy used in the entire life cycle of a building, excluding the energy that is used for the operation of building. Based on this approach, embodied energy is the summation of initial, recurring, and demolition embodied energies (Yohanis and Norton, 2002). Initial embodied energy is the total energy that is consumed to extract raw materials, manufacture and transport products and components, and construct a building. Recurring embodied energy is the energy that is required to maintain a building and repair or replace its materials and components.

The research and practice on embodied energy are gaining more interest in recent years, especially because the share of embodied energy in life cycle energy use is increasing as more high-performance energy efficient buildings are being built. However, there is no consensus yet amongst the scholarly community about the relative significance of embodied energy in life cycle energy use of buildings or the absolute embodied energy consumption levels per unit of floor area. Indeed, the significance of embodied energy can vary as a function of building's level of operational energy efficiency, as illustrated in

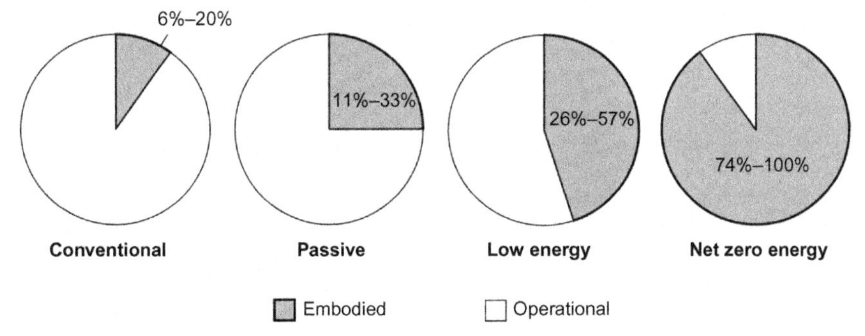

FIGURE 5.3 Share of embodied energy in life cycle energy use of residential buildings with various levels of operational energy efficiency. Source: *Data from Chastas, P., Theodosiou, Th., Bikas, D., 2016. Embodied energy in residential buildings—towards the nearly zero energy building: a literature review. Build. Environ. 105, 267–282.*

Fig. 5.3. Ramesh et al. (2010) report a share of 10%–20% for embodied energy, based on an article review effort that involves 73 conventional residential and office buildings. In another effort, Sartori and Hestnes (2007) examine the data on 60 conventional and low-energy buildings studied by the extant literature on energy use in buildings and conclude that embodied energy constitutes 2%–38% of total energy use in conventional buildings and 9%–46% in low-energy buildings. In a recent study, Chastas et al. (2016) study 90 cases of residential buildings and report embodied energy's share as being 6%–20% in conventional buildings, 11%–33% in passive buildings, 26%–57% in low-energy buildings, and 74–100% in net-zero energy buildings.

Based on a review of previous articles, Ding (2004) suggests that the embodied energy use in buildings ranges from 3.6 to 8.76 Giga Joule (GJ) per square meter of gross floor area (with a mean of 5.506 GJ/m^2) in residential buildings and from 3.4 to 19 GJ/m^2 of gross floor area (with a mean of 9.19 GJ/m^2) in commercial buildings. In another review study, Aktas and Bilec (2012) use more recent information and suggest an initial embodied energy (i.e., associated with preuse phase) range of 1.7–7.3 GJ/m^2 (with a mean of 4.0 GJ/m^2) for conventional residential buildings and 4.3–7.7 GJ/m^2 (with a mean of 6.2 GJ/m^2) for low-energy residential buildings. According to them, the higher initial embodied energy in low-energy buildings is due to thicker building skins and extensive use of insulation. Aktas and Bilec (2012) also show that the embodied energy of demolition phase ranges between 0.1% and 1% of total energy use in a residential building.

The variations in shares and absolute values of embodied energy use reported by the literature also occur due to differences in system boundaries, analytical methods, geographical locations, and data source and quality (age, completeness, representativeness, etc.) (Dixit et al., 2010).

At urban scale, compact cities with high-density downtown areas as compared with low-density urban sprawl offer lesser car dependency, better

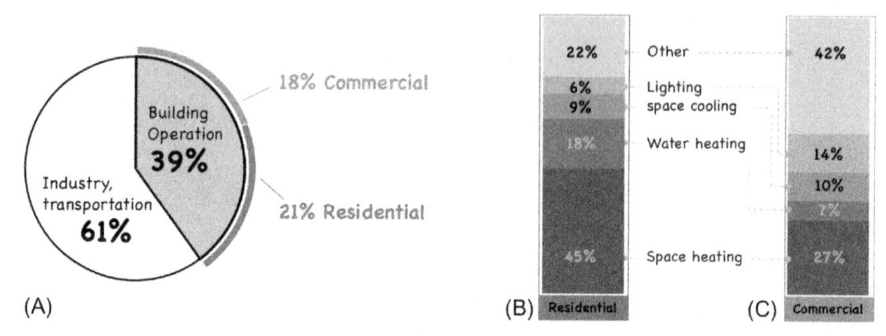

FIGURE 5.4 (A) Contribution of building operations to primary energy use, and (B, C) breakdown of buildings' site energy use. Source: *US Energy Information Administration.*

public transit services, lower building and transportation energy use, and lesser waste of electricity that is generated by power plants. Therefore, both embodied and operational energy of buildings can vary with the urban density too. Norman et al. (2006) conduct a life cycle assessment (LCA) analysis on two case-study residential buildings in Toronto; one being a high-density multistory compact condominium in the downtown area and the other one a low-density two story dwelling in suburban areas. Their findings suggest a 40% share of embodied energy in the low-density building's total energy use and 30% in the high-density building's. They also show that brick, windows, drywall, and concrete are the biggest contributors to embodied energy in both the cases, with a total of 60%–70% contribution (Norman et al., 2006).

Because embodied energy of buildings varies based on the choice as well as the quantity of construction materials, low embodied energy buildings are generally lightweight buildings constructed out of materials with lesser energy intensity. In addition, using locally produced materials reduces transportation and, therefore, lowers the embodied energy through lesser fuel quantity needed for transportation. Design for durability, reusing, and recycling too is critical in increasing efficiency for embodied energy.

5.2.2 Operational Energy

Operational energy is the energy that is used during the occupancy stage of building life cycle for space and water heating, space cooling, lighting, running the equipment and appliances, etc. According to the US Energy Information Administration, operation of buildings (commercial and residential) in the United States account for about 39% of primary energy consumption[2] and 40% of CO_2 emissions[3] annually (Fig. 5.4). Primary energy is the

2. http://www.eia.gov/tools/faqs/faq.cfm?id = 86&t = 1
3. http://buildingsdatabook.eren.doe.gov/ChapterIntro1.aspx

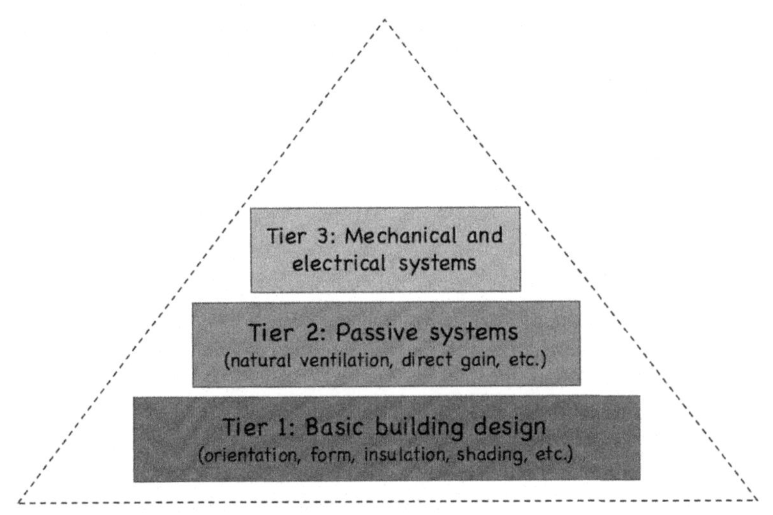

FIGURE 5.5 The three-tier approach to sustainable heating, cooling, and lighting. Source*: Adopted from Lechner, N., 2014. Heating, Cooling, Lighting: Sustainable Design Methods for Architects, fourth ed. John Wiley & Sons, New Jersey.*

energy contained in natural resources and raw materials (crude oil, for instance) that can be used in buildings or industry after being converted or transformed to other forms of energy. Residential and commercial buildings in the United States contribute 21% and 18% to primary energy use, respectively. About 62% of the total site energy in residential buildings is consumed for space heating (45%), water heating (18%), space cooling (9%), lighting (6%), and plug loads (22%).[4] Site energy is the energy that is delivered on site, without taking into account the losses that occur in conversion or transformation processes. In commercial buildings, more than 58% of site energy is consumed for space heating (27%), water heating (7%), space cooling (10%), and lighting (14%). The rest is used for ventilation (6%) and plug loads (36%).[5] Fig. 5.4 illustrates the breakdown of energy use in residential and commercial buildings.

Because heating, cooling, and lighting constitute a major part of operational energy consumption in buildings, there has been significant effort in the last decades to improve design and technology and reduce the energy demand of the building sector (Lechner, 2014). A popular approach in doing so is the three-tier approach to sustainable heating, cooling, and lighting, as illustrated in Fig. 5.5. In this approach, the basic building design, as the first step to high-performance buildings, is used to reduce the energy demand of buildings by avoiding unwanted solar gain in summer times and reducing

4. http://buildingsdatabook.eren.doe.gov/ChapterIntro2.aspx
5. http://buildingsdatabook.eren.doe.gov/TableView.aspx?table = 3.1.4

heat transfer through the building skin. In the second step, passive systems are used to achieve thermal and visual comfort of the occupants naturally. Mechanical and electrical systems are used only as the last step to achieve thermal comfort in parts of the year that passive systems alone cannot meet the needs of the occupants.

5.2.2.1 Drivers and Determinants of Operational Energy Consumption in Buildings

A primary purpose of buildings is to create a comfortable environment thermally, visually, and from indoor air quality (IAQ) perspective. Thermal comfort, as one of drivers of operational energy use in buildings, is defined by the American Society of Heating, Refrigeration, and Air-conditioning (ASHRAE) as "*the condition of mind that expresses satisfaction with the thermal environment*" and is affected by environmental (temperature, relative humidity, air velocity, and mean radiant temperature) and personal (clothing, activity) factors (ASHRAE, 2004). Thermally comfortable environment in buildings is created by the aid of natural (sun, wind) and mechanical means. Visual comfort is about the sufficiency and quality of the lighting environment and is achieved by the controlled use of natural and artificial light in buildings. Comfort with regard to IAQ is affected by concentration of contaminants (CO_2, VOC, etc.) and natural and mechanical introduction and conditioning of fresh air. Creating thermal, visual, and IAQ comfort in many types of buildings and in most climates is not possible without the use of fossil-fuel-based operational energy to run heating, ventilation, and air conditioning (HVAC) and electrical systems, as previously discussed.

The use of operational energy to create comfortable indoor environments varies significantly in buildings because of the effects of climatic, occupant behavioral, socioeconomic, and design and system-related energy use determinants. There is a wealth of literature in the field, which investigates how these factors determine the energy consumption in different types of buildings. An empirical study on Greek residential buildings revealed that poor envelope quality (with regard to insulation and window type) in buildings with low-income inhabitants results in higher-energy consumption (Santamouris et al., 2007). In another study on the new and old buildings in China, Chen et al. (2009) found out that the energy use in older buildings is lower per household than in newer ones. They also concluded that high-energy consumption of some households happened due to the use of aluminum-framed windows, large building area, and large household size (Chen et al., 2009). Yun and Steemers (2011) conducted a comprehensive study on the impact of behavior, physical, and socioeconomic factors on household energy consumption using the US Residential Energy Consumption Survey (RECS) 2001 dataset. They applied statistical methods and showed that the socioeconomic factors both directly and indirectly affect cooling energy use through building characteristics, user behavior, and equipment. They also found out that climate and user behavior, as related to equipment control, are the two most significant factors in cooling energy consumption, respectively (Yun and Steemers, 2011). Ye et al. (2011) empirically studied the effect of building characteristics on energy use in China. Their study revealed that window thermal insulation, among other building characteristics, has "*the greatest effect*" on generation of carbon emissions resulting from operational energy and that newer construction age and more energy-efficient envelopes (with respect to insulation of the walls and windows) will decrease the generation of these carbon emissions (Ye et al., 2011).

Yu et al. (2011) suggested that socioeconomic factors partly determine the building occupants' behavior and, therefore, are able to indirectly influence energy consumption in buildings. Ouyang and Hokao (2009) suggested that improving occupant behavior could result in 14% energy saving. Guerra Santin (2011) applied exploratory factor analysis as the statistical analysis technique on the Dutch households to investigate the user-related factors that impact heating energy consumption. Among the findings of the study was that the differences of energy consumption for various behavioral patterns or user profiles are not statistically significant (Guerra Santin, 2011). However, this study concluded that energy consumption tended to be higher in family households and lower in senior households, and that the couples with higher incomes are less concerned with saving energy (Guerra Santin, 2011). Vassileva et al. (2012) conducted an empirical research on 24 apartment buildings in Sweden to examine the effect of behavior of the occupants and their attitude toward energy on electricity consumption. In contrast to Guerra Santin (2011), they found a significant effect of user's behavior on energy consumption in Swedish households and that user's income also plays a major role with this respect, controlling for other variables.

5.3 CURRENT APPROACHES TO OPERATIONAL AND EMBODIED ENERGY PERFORMANCE

5.3.1 Beyond-Code Energy Requirements

Over the last two decades, green building rating systems have established tangible criteria for recognizing green buildings. These criteria are often beyond the requirements enforced by local codes. The criteria defined by the rating systems cover different aspects of sustainability, such as energy efficiency, water and resource conservation, IAQ. In energy-efficiency category, they often prescribe certain improvement percentages in operational energy performance of building and assign credit points accordingly. As an example, the "energy and atmosphere" category of the most recent version of LEED (Leadership in Energy and Environmental Design), LEED v4 that is widely applied as a standard rating system in the United States, requires a 5% improvement in operational energy performance of new construction projects, as compared with a baseline similar building that meets the energy codes (LEED, 2016). Based on LEED v4, a new building project can gain 1–18 points by demonstrating operational energy performance improvements of 6%–50%. Living Building Challenge, as one of the most ambitious rating systems, certifies only buildings that are net-zero energy or net-positive energy, as demonstrated by actual, rather than modeled, energy consumption. Net-zero energy buildings (NZEB) produce as much operational energy as they consume on a net annual basis using onsite renewable energy sources. In net-positive energy buildings (NPEB), based on Living Building Challenge, 105% of the building's operational energy use must be supplied by onsite renewable sources. NZEB and NPEB are achieved by significant energy demand reductions using building design and passive strategies first, and then implementation of efficient active technologies and renewable energy to meet the energy need of the building.

Green building rating systems now also recognize the importance of embodied energy and thus include prescriptions that require or encourage designers to recycle or reuse materials and buildings and specify materials with low embodied energy in construction of new buildings. Bullitt Center building in Seattle, Washington, for instance, is a Living Building Challenge certified net-zero energy building that also achieves a 3000-metric-ton embodied carbon footprint through the use of regionally produced materials, salvaged materials, and Forest Stewardship Council (FSC) certified wood (Bullitt Center, 2015).

5.3.2 Energy Simulation and Modeling

The first use of computers in estimating operational energy use of buildings started in the late 1960s by Tamami Kusuda, the American scientist who developed the National Bureau of Standards Load Determination (NBSLD) program for the estimation of thermal loads in dynamic conditions (Kusuda, 1976). Since then numerous operational energy simulation software packages have been developed which use environmental, building, and occupancy inputs as well as reliable methodologies and user-friendly graphical user interfaces (GUI) to help architects and engineers predict the energy consumption of their designs. These tools are also used for design optimization and benchmarking purposes. More than 140 tools for building performance assessment have been listed by IPBSA-USA, the United States regional affiliate of the International Building Performance Simulation Association, and can be found at its Building Energy Software Tools (BEST) directory website (BEST, 2016). The tools to estimate the embodied energy of buildings are limited and less sophisticated as compared with operational energy modeling tools; yet major developments have been made in recent years. Some embodied energy estimation tools are highlighted in Section 4.4.

Despite these modeling and estimation advancements for estimation of operational and embodied energy, there are still several challenges that need to be addressed. Here, three methodological challenges specific to life cycle energy use of buildings are listed.

5.3.2.1 Issue of Uncertainty in Operational and Embodied Energy Use Estimations

While operational energy simulation tools offer a strong means for the understanding of building energy performance, there is often discrepancy between their results versus the actual, i.e., measured, energy use. This uncertainty in energy simulation results occurs mainly due to the large number of variables in building operation that often times are not accurately accounted for in building energy models, difficulty in prediction of human behavior during building occupancy phase, and even inaccurate weather models used for energy use estimations. These factors can lead to inaccurate representation of the building under study, as well as the climate in which it is located, by the users of the energy simulation tools.

In general, the literature identifies three major types of uncertainty that affect building energy simulation results (de Wit, 2004):

- Specification uncertainty that occurs due to the lack of information on material properties, building geometry, etc.
- Modeling uncertainty that occurs due to simplification and assumptions made in the model.
- Scenario uncertainty, which is about uncertainty in outdoor climate conditions and occupant behavior.

One method to deal with the uncertainty in energy simulation is to use sensitivity analysis, which is about the assessment of the relationship between variations in input versus output parameters (Lam and Hui, 1996). The sensitivity analysis and energy audits can be used to calibrate the energy models. The energy model is considered to be calibrated and its results to be verified if the results fall within 5% of actual measured energy data.

The embodied energy estimations are subject to uncertainty too. Because LCA is a key method in embodied energy estimations, the LCA-related uncertainties classified by Baker and Lepech (2009a,b) are listed here. These include:

- Database uncertainty, which results from database unrepresentativeness (regional, temporal, technological, etc.).
- Model uncertainty, which results from the lack of knowledge about how the system under the study functions.

- Measurement error uncertainty, which is about the errors because of small sample sizes used in data collection or data collection errors.
- Uncertainty in researcher's preferences with regard to LCA goal and scope definitions and assumptions.
- Uncertainty in a future physical system, which is about uncertainty that occurs due to the inaccuracy of a model in representation of the physical system under study.

To deal with these LCA uncertainties, researchers may quantify and propagate uncertainty by the aid of statistical methods and sensitivity analysis.

5.3.2.2 *Analytical Methods and Tools*

A second methodological challenge in estimation of life cycle energy use has to do with the tools and software packages. Compared with the weather data- and geometry-based methods and software tools to predict operational energy use, the embodied energy is generally quantified through LCA methodology that tracks and aggregates the flows of energy as well as resources, emissions, and waste. While LCA is a well-established and strong methodology, current limitations with regard to availability of regionally representative inventory databases increases uncertainty of embodied energy estimations. In addition, the industry needs to develop architect friendly embodied energy and environmental impact modeling tools that could be used for analysis in earlier stages of design. Similar to many operational energy simulation tools, these tools also need to operate as embedded in Building Information Modeling (BIM) packages in order to facilitate convenient translation of building geometry into bill of materials and embodied energy estimations.

5.3.2.3 *Embodied Energy Performance Benchmarking*

Another challenge in design of low life cycle energy buildings has to do with embodied energy performance benchmarking. In operational energy field, performance metrics and targets have been defined and adopted that enable researchers and designers to communicate the performance of buildings and design alternatives in relation to established performance benchmarks. An example is the Architecture 2030 Challenge in the United States, which has established target energy use intensity (EUI) values for different types of buildings with the objective of achieving carbon neutrality by 2030[6]. Another example is the Energy Start Portfolio Manager program, which allows for comparison of building's energy consumption with that of similar typical buildings.

6. https://living-future.org/lbc/

There exists, however, a lack of widely accepted performance targets to be pursued for embodied energy of buildings. The industry needs to collect representative data on the embodied energy of typical buildings and define meaningful regionally sensitive performance targets for embodied energy.

5.4 LIFE CYCLE ASSESSMENT AND ENERGY ACCOUNTING METHODS

5.4.1 Process-Based Life Cycle Assessment

The main methodology that is used for life cycle energy analysis (LCEA) of buildings is the process-based LCA. Process-based LCA is a strong quantitative methodology that is able to capture the flows of energy, resources, emissions, and waste over the life cycle of a product, process, or a building. It can assess not only the embodied energy but also the environmental impacts in other categories such as global warming, smog formation, acidification, eutrophication. LCEA using process-based LCA is conducted in four stages, as suggested by the International Standard Organization (ISO 14040 and 14044, 2006a,b). As illustrated in Fig. 5.6, the four stages include: (1) goal and scope definition, (2) inventory modeling, (3) impact assessment, and (4) interpretation of results.

In the goal and scope definition stage of LCA-based LCEA, audience and applications of the study are defined, and the functional unit and system boundary are set. The functional unit, which provides a basis for comparison of alternatives, is often the unit floor area of a building during its life time. The system boundary is determined based on the stages in building life cycle that the researcher aims to investigate, and can vary from a cradle-to-gate and cradle-to-site to cradle-to-grave. In operational energy research, the system boundary is limited to occupancy stage of the building life cycle, and in life cycle energy analysis where both embodied and operational energy are

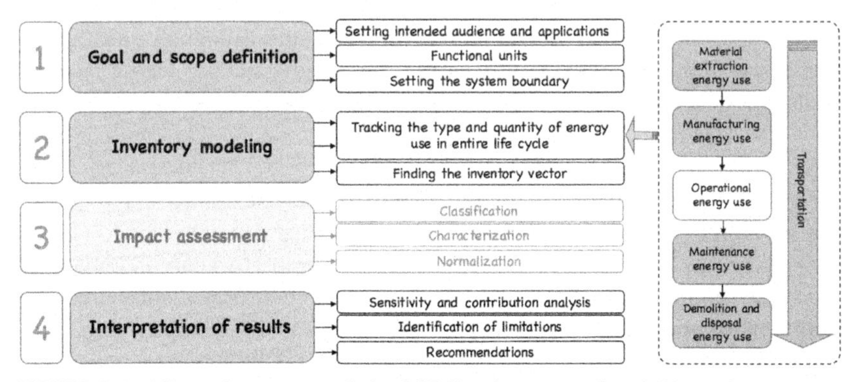

FIGURE 5.6 Life cycle energy analysis (LCEA) using process-based life cycle assessment (LCA).

of interest, a cradle-to-grave system boundary is ideal. However, demolition energy use is ignored in some LCEA studies because of data availability limitations as well as negligible share of demolition energy in life cycle energy use of a building.

In the second stage of the process-based LCA, the inventory modeling, the type and quantity of environmental inputs and outputs are identified, tracked, and quantified. In an LCEA type of LCA study, environmental inputs are limited to different types of energy that are consumed, and environmental emissions fall out of interest of researchers. The inventory modeling in an LCA-based LCEA study, therefore, tracks various types of energy (electricity, natural gas, etc.) that is consumed to develop, run, and demolish a building, for harvesting and mining, transportation of raw materials to manufacturing facilities, running the manufacturing machines and equipment, transportation of construction materials and products to construction site, and eventually demolition of building and transportation to waste facilities. Inventory modeling also accounts for the energy that is used in building for heating, cooling, lighting, and running the equipment in buildings when it is occupied; i.e., operational energy. In most LCEA studies, either an energy simulation software package (such as eQuest, EnergyPlus) is used to model and estimate the operational energy use or actual energy use from utility bills are used to feed the inventory modeling.

One computation method to quantify the environmental inputs/outputs in a process-based LCA study is defined by Heijungs and Suh (2002). Based on this method, a "technology matrix" (A) is constructed, which demonstrates all industrial inputs and outputs associated with the system boundary of a building. An "intervention matrix" (B) is also developed to demonstrate the type and quantity of environmental inputs (i.e., resources, electricity, and fossil-fuel energy) that are used for each and all industrial processes, as well as the type and quantity of environmental outputs (i.e., emissions and waste that are generated). The final demand vector (f) is then developed to represent the quantity of final product (i.e., a building of certain size and life span). Then, based on Heijungs and Suh (2002), the following equation is used to construct the inventory vector, which aggregates all environmental inputs and outputs throughout the complete life cycle of a building:

$$[g] = [B]^{*}[A]^{-1} \times [f]$$

where g is the inventory vector, B is the intervention matrix, A^{-1} is the inverse matrix of technology matrix, and f is the final demand vector.

Because environmental inputs and outputs in an LCEA study is limited to energy consumption, inventory vector therefore demonstrates all sources of energy used during the life of a building as well as their quantities.

Compared with other LCA studies that investigate different types of environmental impacts (such as global warming, acidification), an LCA-based LCEA analysis is focused on life cycle energy only. Therefore, the impact assessment stage, as the third stage in process-based LCA analysis, is more clear and less intensive in an LCEA study. In impact assessment of regular LCA studies, the environmental inputs/outputs that have been identified in inventory modeling stage are first assigned to their corresponding categories of environmental impact (classification) and then their magnitude is multiplied by certain impact-category-based characterization factors in order to achieve aggregate scores that would represent the contribution of building to impacts on the environment (global warming, for instance). In LCA-based LCEA analysis, however, only energy use-related data are collected; therefore, classification step would not be necessary because all relevant environmental inputs fall into the category of energy use, as the environmental impact category.

In LCEA, energy consumption at different stages and processes are translated into primary energy use and are aggregated into total primary energy consumption of buildings, often in megajoule (MJ).

5.4.2 Economic Input-Output-Based Life Cycle Assessment

Compared with process-based LCA, economic input-output (EIO)-based method of LCA is not used as widely for quantification of energy use and environmental impacts. EIO-based LCA uses annual input–output models of the US economy, that are reported by the US Department of Commerce, and links monetary values of the industry sector (such as building sector) to their environmental inputs/outputs (Hendrickson et al., 2005). The computation approach in this LCA method uses the following equation to quantify the energy use or environmental impacts (EIOLCA, 2008):

$$\mathrm{B} = \mathrm{R} \times \mathrm{X} = \mathrm{R} \times \left((I\text{-}A)^{-1} \times \mathrm{F} \right)$$

where B is the vector of total environmental inputs or outputs (such as emissions), R is the environmental input/output per dollar of output, I-A is the total requirement matrix, and F is the final demand vector.

EIO-based LCA takes into account wider system boundaries, compared with process-based LCA, and accounts for indirect as well as direct impacts of a sector on the environment. Therefore, this method tends to be more comprehensive in its environmental assessments. However, the key limitation with this method is that its results are sector specific, rather than product or process specific. In other words, the results of this LCA method would represent the energy use or environmental impacts associated with the average building produced by the US residential sector, for instance. Therefore, a wide variety of differences within a sector is not accounted for in EIO-based LCA. A variant of EIO-LCA method, hybrid EIO-LCA, aims to address this limitation by allowing for sector customization.

EIO-based LCA has been used in several studies. In an LCA study on the US residential sector, Ochoa et al. (2002) use the EIO-based LCA method and the US economy model of 1997 to assess residential electricity consumption and environmental impacts by accounting for all direct and indirect contributions of the system boundary. The study concludes that the construction phase in building life cycle is the largest contributor to economic activities, toxic air emissions, and waste, while electricity and energy are disproportionately consumed at operation and maintenance phase (Ochoa et al., 2002). In another study in the Swedish context, Nassen et al. (2007) compare primary energy use estimations using EIO-based LCA and process-based LCA and show that process-based LCA method used by other studies tends to underestimate the energy use of buildings, compared with the EIO-LCA method. According to Nassen et al. (2007), the higher estimation of energy consumption by 90% using EIO-LCA is partly because of "truncation errors due to the definition of system boundaries" in process-based LCA. Similarly, Säynäjoki et al. (2017) apply both EIO-based LCA and process-based LCA on a Finnish case-study building and achieve significantly different results; with EIO-based LCA results being almost two times greater than the process-based LCA results.

5.4.3 Other Life Cycle Energy Analysis Methods

In addition to LCA methodology, there exist other environmental accounting methods used to evaluate the environmental performance of buildings. Exergy analysis is a method that has gained more interest in recent years to investigate the life cycle environmental impacts of products and buildings. According to the first law of thermodynamics, energy can be converted from one form to another but cannot be created or destroyed in an isolated environment. Because energy, or mass, cannot disappear, they cannot be good indicators of resource depletion or consumption (Davidsson, 2011). Exergy, which is defined as "*... the amount of mechanical work that can be maximally extracted ...*" from a system (Wall, 1977), on the other hand, is a concept that can be used as a measure of quality and quantity of energy and materials and, therefore, exergy analysis can be used for accounting of energy and natural resources (Wall, 1977). In exergy analysis, energy carriers and natural resources used over the life cycle of a building are translated into their equivalent exergy in MJ and aggregated into total life cycle exergy consumption.

In another similar environmental accounting method, i.e., emergy analysis, all energy and environmental inputs in building life cycle are converted into their equivalent solar energy (expressed as *emjoule*) that is used directly or indirectly to make that building. This conversion happens through multiplication of energy and mass quantities by "*transformation coefficients.*" The results for different natural and energy resources are then summed up to make the life cycle emergy of a building. An example of emergy analysis in built environment is conducted by Pulselli et al. (2007), who use this method and investigate the emergy related to a residential-office building in Italy by accounting for all natural flows and energy over the building life cycle.

Another method that is used in some studies for environmental impact assessment is ecologically based LCA (Eco-LCA). Similar to EIO-LCA, this method also uses the US economy models to account for the impacts on the environment in different categories. However, Eco-LCA expands the system boundaries and examines the impacts on the ecosystem services that are not studied by other tools.

Table 5.1 summarizes different environmental accounting methods and their limitations and advantages.

5.4.4 Software, Tools and Databases to Estimate Embodied and Life-Cycle Energy Analysis

LCA researchers and practitioners can rely on a variety of public and private databases for their LCA studies. Some examples include: US Life Cycle Inventory (LCI) (2016) database by National Renewable Energy Laboratory (NREL), Chain Management Life Cycle Assessment (CMLCA) (2016) by Leiden University in the Netherlands, Ecoinvent (2016) by the Swiss Center for Life Cycle Inventories, and GREET (2016) by the US Department of Energy. More examples of LCA databases can be found in Table 5.2.

The industry has also developed software and tools that can be used for modeling and estimation of life-cycle environmental impacts of buildings. Most of these tools rely on environmental LCA framework, guidelines, and methodology as defined by ISO 14040 and ISO 14044. Athena Impact Estimator (Athena IE), SimaPro, and GaBi are examples of such tools. Athena Impact Estimator (Athena IE) is specific to construction industry in the North America while SimaPro and GaBi are broader in terms of coverage of products as well as geographical locations. A key limitation in many of these software as applied to LCA of buildings is the lack of capability to work with the digital model of buildings as generated in geometry modeling applications such as AutoCAD, SketchUp, Revit. Another related limitation is inability to simulate the operational energy use as part of life cycle energy use estimations. Therefore, the user would need to manually feed the LCA software with the operational energy use figure estimated independently by an energy modeling software package. To address the first limitation, Tally is a software and plugin to Autodesk Revit package that uses environmental

TABLE 5.1 Different Environmental Accounting Methods

	Process-based LCA	EIO-LCA	Eco-LCA	Exergy/emergy analysis
Description	• Tracks and quantifies environmental flows • Translates environmental flows into environmental impacts	• Translates the US economy models into environmental impacts	• Translates the US economy models into impacts on ecosystem services in addition to other environmental impacts	• Translates the energy and environmental flows into aggregate exergy in MJ, or aggregate solar energy in emjoule (emergy)
Limitations	• Time intensive • Data driven • Reliance on proprietary data • Data uncertainty	• Does not account for variations within sector • Does not allow for comparison of buildings within a sector • Data uncertainty	• Does not account for variations within sector • Does not allow for comparison of buildings within a sector • Data uncertainty	• Needs conversion factors for all energy sources and materials • Does not demonstrate environmental impacts
Advantages	• Produces product- and building-specific results • Allows for comparison of buildings	• Produces sector-specific results • Comprehensive results because of wider system boundaries • Allows for comparison of sectors • Reliance on public national data	• Does not allow for comparison of buildings within a sector • Allows for the study of impacts on ecosystem services	• Accounts for quality differences between different energy carriers or natural resources • Provides aggregate results

TABLE 5.2 Some Tools and Databases for Life Cycle Assessment and Embodied Energy Analysis

Tool/database	Developer	Main geographical coverage	Scope
Athena IE	Athena International	North America	Buildings and building components
AusLCI	Australian Life Cycle Assessment Society	Australia	Materials, products, and processes
BEES	U.S. National Institute of Standards and Technology	Unites States	Building products
Ecoinvent	Swiss Center for Life Cycle Inventories	Europe	Materials, products, and processes
ELCD	European Commission	Europe	Materials, products, and processes
GaBi	Thinkstep (PE International)	Germany, United States, Europe	Materials, products, and processes
GREET	DOE Argonne National Laboratory	United States	Transportation
Inventory of Carbon and Energy (ICE)	University of Bath	United Kingdom	Construction materials
Korean LCI	Korea Institute of Industrial Technology and Ministry of Environment	Korea	Materials, products, and processes
Okobaudat	German Federal Ministry of Transport, Building and Urban Development	Germany	Construction materials
SimaPro	PRe Sustainability	Europe, Australia	Materials, products, and processes
Tally	Autodesk	United States, Europe	Buildings and construction materials
US LCI	National Renewable Energy Laboratory	United States	Materials, products, and processes

LCA data provided by thinkstep, the German software company that has developed GaBi too, in order to estimate the environmental impacts of a building model generated in Revit.

5.5 CONCLUSION

Developments in design practices and available technology have made the goal of net-zero energy buildings possible, but the lack of a holistic life cycle energy perspective, methodological challenges in estimation of embodied energy, uncertainty of energy modeling results, performance benchmarking challenges, and behavioral barriers are some of the key obstacles in the path toward buildings that are truly carbon neutral. Also, the research community, policy-makers, and the industry need to increase the public awareness about the role of user behavior in achieving energy efficient buildings. Finally, the effects of operational energy efficiency strategies on changes in embodied energy and life cycle energy need to be clarified so that the architects would understand the embodied energy implications of achieving energy-efficient buildings that consume lesser operational energy.

REFERENCES

Aktas, C., Bilec, M., 2012. Impact of lifetime on US residential building LCA results. Int. J. Life Cycle Assess. 17 (3), 337–349.

ASHRAE (2004). Standard 55–2004: Thermal Environmental Conditions for Human Occupancy. American Society of Heating, Refrigerating and Air-Conditioning Engineers, Atlanta, GA.

Baker, J.W., Lepech, M.D. (2009a). Treatment of uncertainties in life cycle assessment. In: Proceedings of the ICOSSAR 2009—10th International Conference on Structural Safety And Reliability, Osaka, Japan, September 13–17, 2009.

Baker, J.W., Lepech, M.D., 2009b. Treatment of uncertainties in life cycle assessment. Proceedings of the ICOSSAR 2009—10th International Conference on Structural Safety and Reliability. Taylor and Francis., Osaka, Japan, September, 13–17, 2009.

BEST (2016). Building Energy Software Tools. International Building Performance Simulation Association (IBPSA), United States Retrieved from: <http://www.buildingenergysoftware-tools.com/software-listing>.

Bullit Center (2015). Living Building Challenge. Retrieved from: <http://living-future.org/bullitt-center-0>.

Chastas, P., Theodosiou, Th, Bikas, D., 2016. Embodied energy in residential buildings-towards the nearly zero energy building: a literature review. Build. Environ. 105, 267–282.

Chen, S., Yoshino, H., Levine, M.D., Li, Z., 2009. Contrastive analyses on annual energy consumption characteristics and the influence mechanism between new and old residential buildings in Shanghai, China, by the statistical methods. Ener. Build. 41 (12), 1347–1359.

CMLCA (2016). Chain Management Life Cycle Assessment. Leiden University, the Netherlands. Retrieved from: <http://www.cmlca.eu/> (accessed 11.12.16).

Davidsson, S., 2011. Life Cycle Exergy Analysis of Wind Energy Systems; Assessing and Improving Life Cycle Analysis Methodology. Uppsala University, Sweden.

De Wit, S., 2004. Uncertainty in building simulation. In: Malkawi, A., Augenbroe, G. (Eds.), Advanced Building Simulation. Taylor and Francis.

Ding, G. (2004). The Development of a Multi-Criteria Approach for the Measurement of Sustainable Performance for Built Projects and Facilities. PhD Thesis, University of Technology, Sydney, Australia.

Dixit, M., Fernandez-Solis, J., Lavy, S., Culp, C., 2010. Identification of parameters for embodied energy measurement: a literature review. Ener. Build. 42 (8), 1238–1247.

Ecoinvent (2016). Ecoinvent. Swiss Life Cycle Inventories. Retrieved from: <http://www.ecoinvent.org/> (accessed 11.12.16).

EIA. 2016. International Energy Outlook 2016. Available at: http://www.eia.gov/outlooks/ieo/world.cfm

EIOLCA, 2008. Economic Input-Output LCA, Green Design Institute. Carnegie Mellon University.

GREET (2016). Greenhouse gases, Regulated Emissions, and Energy use, in Transportation model. U.S. Department of Energy Argonne National Laboratory. Retrived from: <https://greet.es.anl.gov/ > (accessed 11.12.16)

Guerra Santin, O., 2011. Behavioural Patterns and User Profiles related to energy consumption for heating. Ener. Build. 43 (10), 2662–2672.

Hammond, G.P., Jones, C.I., 2008. Embodied energy and carbon in construction materials. Proceedings of the Institution of Civil Engineers - Energy 161 (2), 87–98.

Heijungs, R., Suh, S., 2002. Computational Structure of Life Cycle Assessment. Kluwer Academic Publishers, The Netherlands.

Hendrickson, C.T., Lave, L.B., Matthews, H.S., 2005. Environmental Life Cycle Assessment of Goods and Services: An Input-Output Approach. Resources for the Future Press, Washington DC.

ISO 14040 (2006a). Environmental Management, Life Cycle Assessment, Principles and Framework. International Standard Organization.

ISO 14044 (2006b). Environmental Management, Life Cycle Assessment, Requirements and Guidelines. International Standard Organization.

Kusuda, T. (1976). NBSLD, Computer Program for Heating and Cooling Loads in Buildings, National Bureau of Standards Science Series 69. National Bureau of Standards, Washington, DC.

Lam, J.C., Hui, S.C.M., 1996. Sensitivity analysis of energy performance of office buildings. Build. Environ. 31 (1), 27–39.

Lechner, N., 2014. Heating, Cooling, Lighting: Sustainable Design Methods for Architects, fourth ed. John Wiley & Sons, New Jersey.

LEED (2016). LEED v4 for Building Design and Construction. USGBC.

Lechner, N., 2014. Heating, Cooling, Lighting: Sustainable Design Methods for Architects, fourth ed. John Wiley & Sons, New Jersey.

Nassen, J., Holmberg, J., Wadeskog, A., Nyman, M., 2007. Direct and indirect energy use and carbon emissions in the production phase of buildings: an input–output analysis. Energy 32 (9), 1593–1602.

Norman, J., MacLean, H., Kennedy, C., 2006. Comparing high and low residential density: life-cycle analysis of energy use and greenhouse gas emissions. J. Urban Develop. Plan. 132 (1), 10–21.

Ochoa, L., Hendrickson, C., Matthews, H.S., 2002. Economic input-output life-cycle assessment of U.S. Residential Buildings. J. Infrastruct. Syst. 4 (132), 132–138.

Ouyang, J., Hokao, K., 2009. Energy-saving potential by improving occupants' behavior in urban residential sector in Hangzhou City, China. Ener. Build. 41 (7), 711–720.

Pérez-Lombard, L., Ortiz, J., Pout, C., 2008. A review on buildings energy consumption information. Energ. Buildings 40 (3), 394–398.

Pulselli, R.M., Simoncini, E., Pulselli, F.M., Bastianoni, S., 2007. Emergy analysis of building manufacturing, maintenance and use: Em-building indices to evaluate housing sustainability. Ener. Build. 39 (5), 620–628.

Ramesh, T., Prakash, R., Shukla, K.K., 2010. Life cycle energy analysis of buildings: An overview. Ener. Build. 42 (10), 1592–1600.

Santamouris, M., Kapsis, K., Korres, D., Livada, I., Pavlou, C., Assimakopoulos, M.N., 2007. On the relation between the energy and social characteristics of the residential sector. Ener. Build. 39 (8), 893–905.

Sartori, I., Hestnes, A.G., 2007. Energy use in the life cycle of conventional and low-energy buildings: a review article. Ener. Build. 39 (3), 249–257.

Säynäjoki, A., Heinonen, J., Junnonen, J.M., Junnila, S., 2017. Input–output and process LCAs in the building sector: are the results compatible with each other? Carbon Manag. 8 (2), 155–166.

U.S. Life Cycle Inventory Database (2016). National Renewable Energy Laboratory 2012. Retrieved from: <http://www.nrel.gov/lci/> (accessed 11.12.16).

Vassileva, I., Wallin, F., Dahlquist, E., 2012. Analytical comparison between electricity consumption and behavioral characteristics of Swedish households in rented apartments. Appl. Ener. 90 (1), 182–188.

Wall, G. (1977). Exergy – A Useful Concept Within Resource Accounting. Report No. 77-42. Institute of Theoretical Physics, Chalmers University.

Ye, H., Wang, K., Zhao, X., Chen, F., Li, X., Pan, L., 2011. Relationship between construction characteristics and carbon emissions from urban household operational energy usage. Energy Build. 43 (1), 147–152.

Yohanis, Y.G., Norton, B., 2002. Life cycle operational and embodied energy of a generic office building in the UK. Energy 27 (1), 77–92.

Yu, Z., Fung, B., Haghighat, F., Yoshino, H., Morofsky, E., 2011. A systematic procedure to study the influence of occupant behavior on building energy consumption. Ener. Build. 43 (6), 1409–1417.

Yun, G.Y., Steemers, K., 2011. Behavioural, physical and socio-economic factors in household cooling energy consumption. Appl. Ener. 88 (6), 2191–2200.

Chapter 6

Energy: Current Approach

Cheng Siew Goh
Department of Quantity Surveying and Construction Project Management, Heriot-Watt University Malaysia, Putrajaya, Malaysia

6.1 INTRODUCTION

The total world energy consumption grows from 546 quadrillion British thermal units (Btu) in 2012 to 815 quadrillion Btu in 2040 (EIA, 2016). Most of energy is produced using finite resources such as coal, natural gas, crude oil, peat and gas liquid. Global reliance on fossil fuels gives rise to over 50% projection of greenhouse gas emissions from 26.1GtCO_2 in 2014 to 37–40GtCO_2 by 2030 (IPCC, 2007). To prevent continuing extraction and combustion of the rich endowment of natural resources at current/increasing rates for the needs of growing populations and developing economics, it is essential to explore new energy sources and carriers which give energy services with long term security, affordable prices and minimal environmental impacts.

Buildings account for 40% of primary energy consumption worldwide and the projection of energy need in buildings is expected to increase 1.5% per year from 2012 to 2040 (EIA, 2016). The wave of sustainable development necessitates the needs for increased energy supply and efficiency. There is a trend to accelerate the uptake of renewable energy and energy efficiency measures in buildings and structures. A wiser use of energy in all aspects to decrease the energy loads and demands from users has also been highly promoted in both public and private policies. Energy saving has become one of the important agenda in carbon reduction policies as well as global sustainability practice, regardless of industries or sectors. For instance, UK Climate Change Act 2008 and UK Low Carbon Transition Plan 2009 envisaged to cut carbon emissions by using low-carbon sources in generating energy. German's Climate Action Plan 2050 has also considered restructuring the energy sectors as one of the key measures to achieve the climate targets. Deploying appropriate energy systems in buildings would promise great savings by reducing operating and maintenance costs in the long run.

Sustainable Construction Technologies. DOI: https://doi.org/10.1016/B978-0-12-811749-1.00018-3

6.2 ACTIVE STRATEGIES

Active strategies use advance technology solutions to create comfortable environment for users. Heating, ventilation and air conditioning (HVAC) systems and electrical lighting systems are the main focus of active strategies. Innovative supply technologies are also introduced to enhance the access to clean energy, improve energy security, and protect environment at all levels (local, regional and global).

6.2.1 Onsite Renewable Energy Generation

Renewable energy offers solutions for sustainable buildings and structures to be self-energy sufficient or positive energy – by contributing the surplus energy generated to the utility service providers. Renewable energy offsets the building energy needs by generating new supplies or sources. The common renewable energy resources for onsite generation include solar, wind, biomass, hydropower, and geothermal.

6.2.1.1 Solar Energy

The solar energy application relies on the availability of solar energy incident in a location. Irradiation that is measured in kWh/m^2 per duration is a key criterion to determine the solar energy and it is often classified into two major forms: Global Horizontal Irradiance (GHI) and Direct Normal Irradiance (DNI). GHI is the total amount of direct and diffuse solar radiation incident falling on a horizontal surface while DNI is the amount of solar irradiation received by a surface that is always held perpendicular to direct solar beam (Solargis, 2017; World Energy Council, 2016). Figs. 6.1 and 6.2 present the Solargis solar resource map in terms of GHI and DNI across different regions worldwide.

a. *Photovoltaic (PV) Systems*

Solar resource is an inexhaustible resource can be utilised actively to generate electricity using photovoltaic panels. Photovoltaic systems convert solar radiation to electricity by means of the photoelectric effect in materials such as silicon and selenium. Photovoltaic systems are installed on rooftops or building facades. They can be made from multiple materials that could generate different system efficiency. Major materials of PV panels include thin film, polycrystalline and monocrystalline. The relative efficiency of PV systems is 5%–11% for thin film, 13%–17% for polycrystalline, and 14%–19% for monocrystalline (Hayter & Kandt, 2011; HK EMSD, n.d.).

Photovoltaic systems can be either grid connected, off-grid or hybrid. With grid connected solar system, excess solar energy generated can be sold to the utility. The onsite production of solar energy is normally

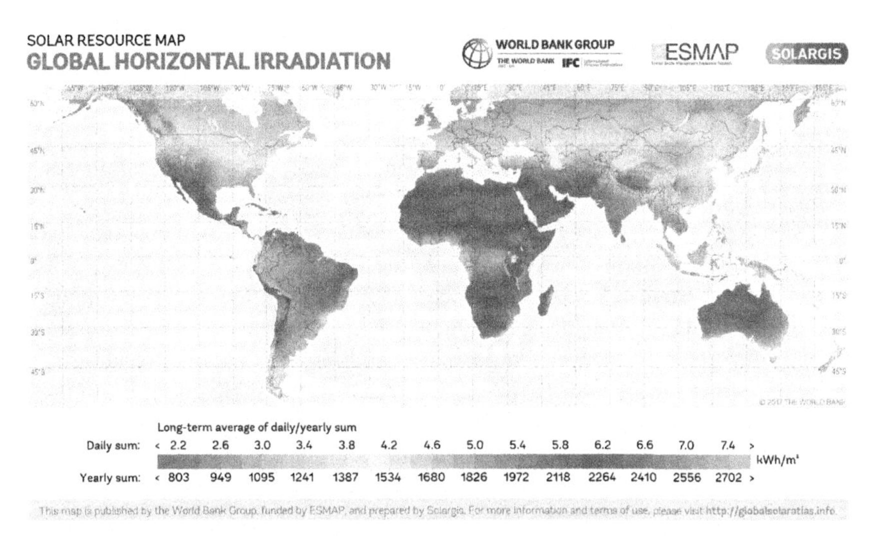

FIGURE 6.1 Global horizontal irradiation across regions worldwide.

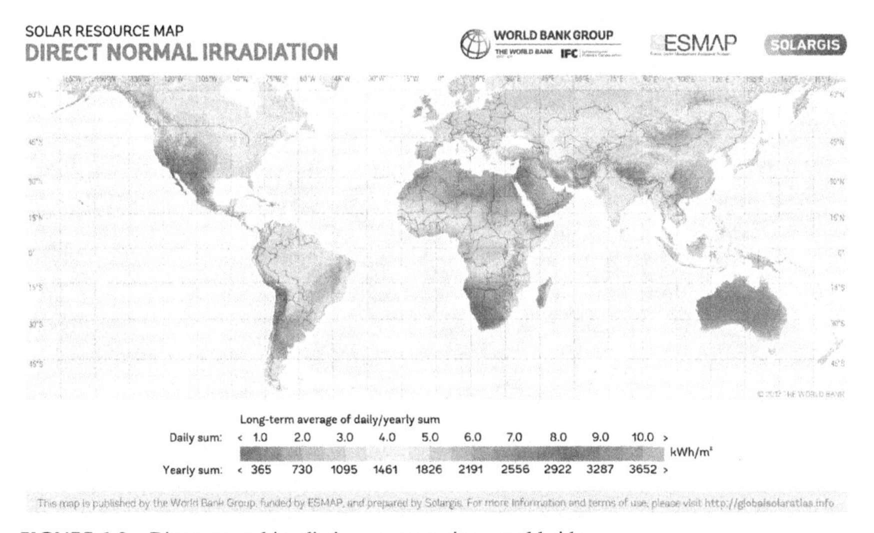

FIGURE 6.2 Direct normal irradiation across regions worldwide.

greatest at or near the time of building and utility peak loads, thereby reducing utility bills because of peak shaving (Strong, 2016). Off-grid solar system stores the electricity to a connected battery which serves as the main power supply. Inverters are required to convert the direct current (DC) power generated by solar cells to alternating current (AC) power.

Photovoltaic modules are normally available in different forms for various applications such as typical framed PV panels, PV glass units, flexible PV modules, and PV roof tiles (HK EMSD, n.d.). Building Integrated Photovoltaic (BIPV) is an architecture application that integrate photovoltaic technology into the building envelope to product electricity. By serving the dual function of building skin and power generators, BIPV can blend with architectural design with futuristic appearance.

However, the energy production of solar panels is not constant and fluctuates throughout the year, depending on the weather and seasons. In the United States, Energy Information Administration (EIA) estimates that about 50 million kilowatt hours (kWh) of electricity was generated by utility-scale PV power plants and 24 billion kWh by small-scale grid-connected PV systems in 2017 (U.S. EIA, 2018). It is therefore important to choose proper locations for the installation of solar systems. The installation needs to identify the optimum inclination for solar absorption. In the northern hemisphere like Hong Kong, PV arrays are usually installed facing south with a tilt angle near to the latitude to maximise the amount of electricity generation (HK EMSD, n.d.).

b. *Solar Hot Water/ Solar Thermal Systems*

Solar hot water systems convert solar energy to heat by transferring it to water running through a series of tubes/pipework behind the liner which absorb the heat. Solar hot water systems are categorised into low temperature (unglazed collectors), mid-temperature (flat-plate collectors), and high temperature (evacuated tube collectors). Factors affecting heat absorption are the angle and azimuth of the surface of the absorber, the insulation below the surface and glazing/ evacuation over the surface (Goodhew, 2016). In general, solar thermal systems can meet 20%–70% of domestic hot water needs, with efficiency of 35%–80%. They are comparatively cheaper, reliable and low maintenance, with lower roof space requirements.

6.2.1.2 *Wind Turbines*

Wind turbines convert the kinetic energy from wind into usable electrical energy. When the wind blows, propeller-like blades rotate to harness wind energy to spin the electric generator for electricity generation. Two major forms of wind turbines are available in the market: horizontal-axis turbines and vertical-axis turbines. Small wind turbine has an electricity production capacity of 10 kW while the large-scale wind turbines could generate electricity up to 5000–8000 kW (U.S. EIA, 2017). Wind turbines in buildings are often mounted on rooftops but these roof-mounted systems normally suffer from turbulence and low annual average wind speeds, making it less viable for majority urban sites (Teignbridge District Council, 2010). Besides that, there are also issues of building structural integrity and noise nuisance imposed on building users.

Similar to PV systems, the location for installing wind turbines must also be evaluated properly to determine the adequacy of wind flows for electricity generation. The evaluation also helps to identify a suitable wind turbine system for the specific location. The efficiency of wind energy varies greatly due to wide variations in quality of wind resources at the location of installation, with an efficiency of more than 40% if the location of wind energy system is carefully selected (Evans et al., 2009). It is important to underlie that wind turbines could not operate when wind speeds are either too high or too low. Besides that, wind energy systems also suffer from intermittency problems and an auxiliary energy system should be provided to prevent energy supply disruptions.

6.2.1.3 Biomass

Biomass is referred as organic matters derived from plants, animals, forestry and agriculture residues, and organic compounds of municipal and industrial wastes. Biomass technologies decompose organic matters to release their stored energy such as biofuels and bioenergy. Biomass energy is generated from either the combustion of biomass or the anaerobic gas from biomass. Biogas or biofuel such as ethanol and biodiesel is produced as an output from anaerobic digestions in vegetable or animal wastes. As shown by IPCC (2007), combustible biomass and wastes account for about 10% primary energy consumption with over 80% used for traditional fuels for cooking and heating in developing countries. Biomass is increasingly utilised to fuel electricity plant or combined heat and power (CHP) plant due to the low carbon emission (Teignbridge District Council, 2010).

6.2.1.4 Geothermal

Geothermal systems employ heat resources retained in shallow ground (3 meter upper the earth's surface), hot water, rock beneath the earth's surface and magma (molten rock located deep in the earth) (Hayter & Kandt, 2011). Geothermal heat pumps are the common application of geothermal energy by utilising the constant temperature of the earth as an exchange medium for heat, irrespective of seasonal change above the earth. They are capable of heating and cooling spaces by transferring heat from or into the heat exchanger. In winter, heat is removed from the heat exchanger and pumped into the indoor air delivery system; in summer, heat is removed from the indoor air stream into the heat exchanger (Hayter & Kandt, 2011). The removed heat can also be used as a source of hot water.

6.2.1.5 Hydropower

Hydropower uses the energy of water flowing from upstream to downstream to drive a turbine connected to an electrical generator, thereby generating electricity. Hydropower is a highly efficient and reliable energy source without much variations due to the constant flow. The energy produced is proportional to the product of volume flow rate and pressure head (Paish, 2002). The efficiency of hydroelectric plants is in the range of 80% to over 90% and this reduces with the relative sizes. Paish pointed out that micro hydropower systems tend to be 60%–80% efficient.

Hydropower is mostly used for large scale plants and it is quite limited on small or micro-hydropower schemes. There is no agreed definition of the size of small and microscale hydropower to date. The upper limit for small-scale hydropower ranges from 2.5 to 25 MW, though 10 MW is the most widely accepted value (Paish, 2002). Run-of-the-river systems are the typical micro-hydropower system used to power a single home, with no dam and water storage. Micro-hydropower can be either grid connected or standalone (off grid).

6.2.2 Energy Optimisation

Energy efficiency is a way of managing and restraining growth in energy consumption by delivering either the same service with less energy inputs or more services at the same energy inputs (IEA, 2013). As highlighted by Hayter and Kandt (2011) and Chwieduk (2017), energy efficiency measures should be considered prior to renewable energy strategies because the cost of efficiency measures is approximately half the cost of installing renewable energy which produce the same capacity as efficiency measures.

6.2.2.1 High Energy Efficient Products and Equipment

Improving the energy efficiency of heating and cooling equipment is critical to reduce energy consumption in the building sector. IEA (2013) found that most cooling systems today are often oversized and operate at less than maximum efficiencies. Variable speed drives/fans can be employed on chillers, pumps, cooling towers, fans and air handling units to lower electrical draw when electric motors are running at part load. They rectify the frequency of AC input voltage into DV and convert it back to variable frequency AC voltage to work at the required rate of partial load conditions. Adopting appropriate capacity and optimal design in accordance with the building cooling needs and design can reduce the overall energy demand of buildings. HK EMSD (2015) found that the introduction of variable speed drives has improved the performance and reliability of building air conditioning system while decreasing the equipment cost, hence achieving better economics of energy saving.

District heating is a high energy efficient measure of electricity production. It is a centralised energy system that produces steam, hot water or chiller water for individual buildings through a network of pipes. Heat can be produced and delivered more efficiently than individual systems.

The overall building energy efficiency can also be improved with the use of more efficient devices and appliances. Water heaters, refrigerators and clothes dryers are recognised as major energy consumers in buildings, with 18% of total energy use (DOE, 2015). Using high energy efficient appliances can increase productivity and bring dramatic energy savings to building users. Inefficient technologies shall be eliminated progressively to avoid energy waste.

6.2.2.2 Energy Recovery

As pointed out by IPCC (2007), approximately two-third of the input fuel's potential energy is lost as waste heat during the production of electricity, in addition to loss of transmission and distribution. To minimise this wasted energy, it should be recovered or reused to satisfy energy demand. Heat recovery is an energy recovery process of exchanging heat from a building's exhaust air stream to the make-up air stream. In summer, the outgoing exhaust air could be cooler than the incoming fresh air. An energy recovery system would transfer the heat energy from the incoming air supply to the outgoing exhaust air to reduce the incoming air temperature, thereby reducing energy necessary to processing the air for the required comfort level (RTM Engineering Consultants, 2016). Same principles apply in winter where the outdoor air temperature is lower than indoor air. Less energy is used to heat the incoming air stream. In addition to heat, humidity can also be recovered in the similar way.

A heat recovery system can be applied not only in systems with different indoor and outdoor climate but also variable refrigerant flow systems in buildings requiring simultaneous heating and cooling. serving for multiple zones. Heat recovery system circulates refrigerant between zones to transfer the reject heat from indoor units of zones being cooled to refrigerant of zones being heated.

a. *Heat Pumps*

Heat pumps move heat from a cool space to a warm space using electricity. Rather than generating heat, heat pumps provide equivalent space conditioning at as little as one quarter of the cost of operating conventional heating or cooling appliances (DOE, n.d.). Air source heat pumps which transfer heat between buildings and the outside air are more common than water-source and ground-source heat pumps (geothermal).

A range of innovations are introduced to heat pump applications for improved performance, including two-speed compressors, variable speed motor (on indoor fans, outdoor fans or both), desuperheater, scroll compressor and back-up burners. Two-speed compressors allow heat pumps operating close to the required heating or cooling capacity at any particular moment in order to save energy and reduce compressor wear (DOE, n.d.). Different spaces within buildings can be kept at different temperatures by using automatic dampers. As demonstrated by DOE, a desuperheater can recover waste heat from the heat pump cooling mode for water heating while scroll compressors can provide 5.6°–8.3 °C warmer air when in the heating mode.

b. *Combined Heat and Power (CHP) / Co-generation*

Combined Heat and Power (CHP), known as co-generation is a simultaneous production of electricity and heat (and potentially of cooling). Heat which is otherwise wasted in conventional power generation (separate heat and power generation) is recovered as the useful energy of electricity production in CHP. CHP is a new configuration of systems with advanced technologies (e.g. combustion turbines and reciprocating engines) that expands efficiency opportunities of energy generation systems. It generally comprises of a prime mover, an electricity generator, a heat recovery system and a control system.

As indicated by IEA (2013), CHP is operated at a generally high efficiency which ranges from 75%–90%. In IEA's appraisal, CHP is a reliable and cost effective technology that provides about 10% of total electricity generation worldwide, where more than 20% of electricity generated in Russia, Netherlands, Finland, Latvia, and Denmark. Although CHP is for industrial application, IEA also highlighted opportunities to deploy efficient and cost-effective CHP into smaller scale the building applications by using microturbines, Stirling engines and fuel cells.

c. *Regenerative Drives in Lifts/ Elevators*

Regenerative drives are a technology advancement that recycles energy rather than wasting it as heat. When the carriage travel downwards with heavy load or upwards with lighted load, the motor acts as a generator, transforming mechanical power into electrical power and pumping back into the electrical grid (Sniderman, 2012). The small amounts of energy generated from the elevator sporadic decelerations over time save energy and reduce excess heat in buildings (Sniderman, 2012). The energy saving arisen from lifts with regenerative drives differs according to the elevator traffic patterns. The application of elevators with regenerative power in Hong Kong shows energy efficiency of 20%–30% higher than conventional elevators (HK EMSD, 2015).

6.2.3 Energy Conservation

Apart from creating more sources and improving the efficiencies of energy solutions, it is also essential to reduce primary energy demands from the demand side of buildings. End-use efficiency could provide greater energy savings in building applications through leveraging information technology. A lack of systems coordinating load operation and controlling energy consumption could result in high energy wastage due to careless handling of energy appliances and systems by building users.

6.2.3.1 Demand Side Management System

Demand controlled ventilation employs advanced sensors to detect concentrations of carbon dioxide and other contaminants. The data collected is then sent to air handling controllers to determine the fresh air amount of the supply air. Air volume setpoints are shifted according to the air demand from room thermostat. Such system can adjust the ventilation needs upon occupancy, building functions and internal activities as well as other surrounding factors.

6.2.3.2 Demand Response

Demand response is a change in electricity use by demand-side resources to reduce energy consumption at times of high demands or when the system reliability is jeopardised. It is included by USGBC as one of the credits scores in LEED v4. Users are encouraged to reduce their usage or change their usage patterns during peak load times to allow utilities providers optimising their supply side energy generation and delivery systems. Service operators send an alert of curtailment events to participating users at specific period of time for a change in usage pattern or levels. Incentives (reactive policies) are rewarded to users participating in the programs when demand response events are called for. Tiered demand electricity pricing (proactive policies) is another alternative used in demand response strategies.

Demand response can reduce the overall energy needs and avoid additional loads on power generation facilities, transmission line and distribution systems, thereby ensuring electricity systems operate reliably and in equilibrium (Palensky & Dietrich, 2011). Service providers such as Johnson Controls and Siemens have developed some programmes and software tools to assist building operators shift their consumption out of peak periods such as cycling air conditioning, shifting production of heavy machinery off peak, and switching off/ deferring non-critical loads.

6.2.3.3 Smart Grid

Smart grid is introduced to offer interactive relationships between grid operators, utilities and users. Appliances and equipment in buildings are networked to give users access and operate through smart meter and computerised controls. Smart grids can respond to fluctuations in power generations.

6.2.3.4 Energy Management System

Energy use monitoring and measurement systems require efforts from both facilities management and property owners to evaluate the building energy usage in a periodical manner. A meter is required to facilitate the reporting of energy usage data as well as energy production data (if onsite energy generation systems are deployed). The results are analysed periodically to identify opportunities for improved building performance and energy cost saving. Anomaly within the building system can be detected and the peak load of energy demands can be identified. With energy management system, building performance can be monitored through a loop of commissioning, metering, reporting and re-commissioning. An interactive dashboard showing the building energy and water consumption is beneficial not only to track and assess the building performance but also to educate occupants of their energy use profiles.

6.2.3.5 Dynamic Energy Management System

Dynamic energy management is an innovative management approach to manage loads at the demand side by integrating energy use management principles, demand response, and distributed energy resource programs into a framework for energy savings (Parmenter et al., 2008). Smart devices and distributed energy resources with advanced control s and communication capabilities are integrated to develop a system operating based on user requirements, utility constraints, available incentives, and other variables such as weather and building occupancy (Parmenter et al., 2008).

6.2.3.6 Smart/ Intelligent Building Control System

Building automation system integrate technical features such as heating, cooling and ventilation systems, hot water, lighting, advanced facades and elevators. Time-and presence-dependent control systems are also used to switch on the electricity when the appliances and systems are not in use. Temperature of indoor environment can be controlled by operating fans, floor heating, chilled ceilings, lighting and blinds through the network infrastructure. Lighting can be adjusted through a presence and motion detector and integrated brightness sensors, without losing occupant comfort. The lighting sensors detect occupancy and measure the brightness required by the room. The lights are dimmed or turned off automatically when adequate daylight is available.

Smart controls merge building automation systems with information technology infrastructures. Interoperability and communications among different devices in the networks are enabled through leveraging information on open communication (Parmenter et al., 2008). Smart building control systems can learn from the past experience and apply that knowledge to future similar events. The building control system become adaptive based on user behaviours, external signals, and other surrounding factors. European Commission (2017) demonstrated two smart building projects—Flexom Apartment and The Edge Building in which smart solutions are integrated into the HVAC, lighting, alarm, shutter and space management system. It is claimed that smart building with Internet-of-Things–based sensor networks could reduce smart building cost from USD250,000 per square feet (using a building management system to USD50,000). The smart technologies become widely known when Facebook founder, Mark Zuckerberg launched an intelligent home automation system – Jarvis on 19 December 2016. As described in his facebook, Mark Zuckerberg used Jarvis to control the settings of lights, temperature, appliances, music and security of his building remotely after the interaction with his artificial intelligent systems in computers or phone devices.

6.3 PASSIVE STRATEGIES

Energy performance of buildings can only be optimised with careful consideration of passive measures such as materials, building forms, design and orientation, window location and sizing, and shading. Passive strategies are highly sensitive to meteorological factors and climatic factors for effective applications (Sadineni et al., Madala, & Boehm, 2011). Good passive design makes effective use of resources and surrounding environment conditions such as solar radiation, cool night air, temperature, humidity, air movement, and air pressure differences for improved comfort. The physical architectural configuration and interactions of building elements are normally analysed to maximise the site's potential for natural ventilation and daylights.

Unlike active design, passive design does not involve mechanical or electrical appliances. They are normally incorporated in the building design since the planning and design stage. There are two major forms of passive strategies employed in sustainable buildings: design and orientation and building envelope.

6.3.1 Design and Orientation

Proper orientation is fundamental to ensure buildings can reduce heat gain during summer and heat loss during winter. By taking the local climate and regional conditions into account, the energy need for heating, ventilation, and air conditioning can be reduced or even eliminated. Building windows should be always designed facing north or south. The west facing windows and walls should be minimised and shaded to avoid morning and evening sun that cause overheating and sun glare. Direct sunlight during the early afternoon (when solar angles are low and solar radiation is still high) should be always avoided. Apart from that, it is also important to consider summer and winter variations in the sun paths and direction as well as strength of wind to determine appropriate orientation for buildings. In Andersson et al. (1985)'s study, solar gain through east or west glazing is higher than south windows during summer (result in high cooling loads) but lower than south windows during winter (result in high heating loads). This has demonstrated disadvantages of east or west orientation in giving unnecessary energy needs.

In addition, a space called buffer zones can also be introduced to protect building interior against strong influence of the ambient environment (Chwieduk, 2017). Solar gains can be distributed better in buildings with buffer zones. Buffer spaces allow solar gains in winter for reducing heating requirements while delaying solar energy influence in time during summer. The size of a buffer space, the position and the transparent and opaque partitions need to be calculated properly by taking into account the astronomic relationship between position of the sun and surfaces under consideration at different times of the year (Chwieduk, 2017).

6.3.2 Building Envelope

A building envelope is a component separating the building indoor and outdoor environments including walls, fenestration (doors and windows), roofs, floors, ceilings and foundations. The insulating properties of building envelope and construction quality determine the flow of heat, moisture and air into and out of the buildings (DOE, 2015). As stated by DOE, foundation, walls and roofs account for 50% heating loads in residential buildings and 60% in commercial buildings, while window heat gain contributes greatly to both commercial and residential cooling loads. When selecting proper building envelopes for quality living and working environments, it is necessary to take into account the environmental impacts of temperature, humidity, air movement, rain, snow, solar radiation, wind forces, and other natural factors (Sadineni, Madala, & Boehm, 2011).

Thermal insulation is a material or combination of materials that retard the heat flow rate by conduction, convection, and radiation. It is a cost-effective measure to save energy because it reduces heat losses during heating and cooling. The insulation performance can be determined by either the thermal conductivity value (in W/mK) or the thermal transmittance coefficient (in W/m^2K). The common thermal insulation materials include glass wool (0.030 W/mK), extruded polystyrene (0.025 W/mK), fiberglass (0.041 W/mK), rockwool (0.041 W/mK), dry cellulose (0.078 W/mK), and polyurethane (0.027 W/mK) (Papadopoulos, 2005). Apart from heat resistance, thermal insulation also help to reduce mould and damp. Some materials can offer soundproofing functions. Thermal insulation can be either in forms of bulk insulation (batts, rolls, boards, forms, loose fill, or reflective insulation (aluminium foil or ceramic coating). Because spaces for additional insulation is often constrained, thin materials with high level insulation is valuable in the building applications, especially in retrofits (DOE, 2015).

Thermal mass is a material that stores warmth and coolness until it is needed within a space. The high heat capacity of thermal mass can absorb heat, store it over time, and release it when it is needed to alleviate heating and cooling loads. Phase change materials (PCM), which are a thermal mass are impregnated in walls as passive latent heat thermal energy storage (LHTES) systems to release heat when the outside temperature decreases (for instance, at night). PCMs could be benefited from peak load shifting and solar energy utilization and therefore could achieve maximum energy savings of about 30% (Khudhair and Farid, 2004; Sadineni et al., 2011).

Infiltration is the movement of air into the conditioned spaces through cracks, leaks or other openings while exfiltration is the air movement out of the building (Sadineni et al., 2011). Infiltration is caused by a pressure difference due to temperature difference between indoor and outdoor air (stack effect), wind movement and operation of mechanical ventilation and vented combustion devices. Since infiltration can affect the heating and cooling loads, moisture levels and temperature of indoor environments, uncontrolled infiltration into and out of buildings should always be minimised for energy savings.

Windows are effective parts of building climate control and lighting systems and their quality is determined by insulating value and transparency to the sun visible and infrared light (DOE, 205). As put forward by DOE, an ideal window would provide attractive illumination without glare, high levels of thermal insulation, infrared light transmission during heating mode, infrared light obstruction during cooling mode, and ultraviolet light hindrance. Technologies of window include solar control glazing, insulating glass units, low emissivity (Low-E) coatings, evacuated glazing, aerogels, gas cavity fills, vacuum glazing, multiple panes, switchable reflective glazing, suspended particles devices film, holographic optical elements, and low iron glass, in addition to improvements on window frame designs (Sadineni et al., 2011). The innovations are introduced to reduce solar heat gain, reduce air leakage rates, and allow sunlight transmittance. U-values (thermal conductivity), solar heat gain coefficient (SHGC) and visible transmittance (VT) are the main considerations for window performance.

Roofs and ceilings are critical because the highest heat transfer takes place at ceilings and roofs (DOE, 2015; Sadineni et al., 2011). They are greatly susceptible to solar radiation and other environmental changes and the control of air flow is important in roof and ceiling structures. Advances in energy efficient roofs include green roofs, photovoltaic roofs, domed and vaulted roofs, naturally or mechanically ventilated roofs, micro-ventilated roofs, high roofs, double roofs, solar reflective/cool roofs, and evaporative roof cooling (Sadineni et al., 2011).

Walls are also a predominant part of building envelope due to their high ratio in total building surface areas. Thermal insulation and thermal mass are often used as an energy efficiency measure in walls. High reflective walls with white and cool colour surfaces are also critical to reduce cooling loads, particularly in hot climate regions. Attention should be paid to moisture condensation issues in walls with thermal insulation since there is a formation of surface condensation when the relative humidity of ambient air is greater than 80% (Sadineni et al., 2011). R-value (thermal resistance) and moisture removal (mostly resulted from condensation) are major parameters to be considered in wall design and technologies.

6.4 CONCLUSIONS

There are numerous solutions and technologies of increasing energy supply and efficiency that could not be fully covered here. Buildings are an output of several complex, integrated and interrelated systems. Synergies between active strategies and passive strategies within buildings are fundamental to achieve energy optimisation. It is also pertinent to note that energy use of buildings also relies on appropriate and effective operations and maintenance during occupancy, in addition to the integration of good architecture and energy systems.

To facilitate the energy transition, the need of system thinking and the integration of energy efficiency and renewable energy technologies should be always emphasised in building design and construction.

REFERENCES

Andersson, B., Place, W., Kammerud, R., Scofield, M.P., 1985. The impact of building orientation on residential heating and cooling. Energy Build. 8 (3), 205–224.

Chwieduk, D.A., 2017. Towards modern options of energy conservation in buildings. Renewable Energy 101, 1194–1202.

DOE (2015). *Quadrennial Technology Review – Chapter 5 Increasing Efficiency of Building Systems and Technologies*. Available at https://energy.gov/sites/prod/files/2015/09/f26/QTR2015-05-Buildings.pdf.

DOE (n.d.). *Heat Pump Systems*. Available at https://energy.gov/energysaver/heat-pump-systems.

EIA. (2016). *International Energy Outlook 2016*. Available at http://www.eia.gov/outlooks/ieo/world.cfm.

European Commission. (2017). Smart Building: Energy Efficiency Application. Available at https://ec.europa.eu/growth/tools-databases/dem/monitor/sites/default/files/DTM_Smart%20building%20-%20energy%20efficiency%20v1.pdf.

Evans, A., Strezov, V., Evans, T.J., 2009. Assessment of sustainability indicators for renewable energy technologies. Renewable and Sustainable Energy Reviews 13 (5), 1082–1088.

Goodhew, S., 2016. Sustainable Construction Processes: A Resource Text. Wiley.

HK EMSD (2015). *Study Report on Application of Lift Regenerative Power*. Available at http://www.emsd.gov.hk/filemanager/en/content_764/applctn_lift_rgnrt_pwr.pdf.

HK EMSD (2015). Application Guide to Variable Speed Drives. Available at https://www.emsd.gov.hk/en/energy_efficiency/energy_management/publications/application_guide_to_variable_speed_drives_vsd/index.html.

HK EMSD. (n.d.). *Solar Photovoltaic*. Available at http://re.emsd.gov.hk/english/solar/solar_ph/solar_ph_to.html.

Hayter, S.J. & Kandt, A. (2011). *Renewable energy for existing buildings*. Paper presented at the 48th AiCARR International Conference Baveno-Lago Maggiore, Italy.

IEA, 2013. Transition to Sustainable Buildings: Strategies and Opportunities to 2050. IEA, France.

IPCC (Intergovernmental Panel on Climate Change), 2007. Working Group III Report: Mitigation of Climate Change, Fourth Assessment Report. Cambridge University Press, Cambridge.

Khudhair, A.M., Farid, M.M., 2004. A review on energy conservation in building applications with thermal storage by latent heat using phase change materials. Energy Convers. Manag. 45 (2), 263–275.

Paish, O., 2002. Small hydro power: technology and current status. Renewable and Sustainable Energy Reviews 6 (6), 537–556.

Palensky, P., Dietrich, D., 2011. Demand side management: Demand response, intelligent energy systems, and smart loads. IEEE Trans. Industr. Inform. 7 (3), 381–388.

Papadopoulos, A.M., 2005. State of the art in thermal insulation materials and aims for future developments. Energy Build. 37 (1), 77–86.

Parmenter, K.E., Hurtado, P., Wikler, G., & Gellings, C.W. (2008). Dynamic Energy Management, 2008 ACEEE Summer Study on Energy Efficiency in Buildings ACEEE, Washington, 10-118–10-132.

RTM Engineering Consultants. (2016). *Using energy-recovery systems to increase building efficiency*. Available at http://www.csemag.com/single-article/using-energy-recovery-systems-to-increase-building-efficiency/b8155b152d2cca8963e673b2162b3c48.html.

Sadineni, S.B., Madala, S., Boehm, R.F., 2011. Passive building energy savings: A review of building envelope components. Renewable and Sustainable Energy Reviews 15 (8), 3617–3631.

Solargis. 2017. Solar Resource Maps and GIS Data for 200 + Countries. Available at https://solargis.com/maps-and-gis-data

Sniderman, D. (2012). Energy Efficient Elevator Technologies. Available at https://www.asme.org/engineering-topics/articles/elevators/energy-efficient-elevator-technologies.

Strong, S. (2016). *Building Integrated Photovoltaics*. Available at https://www.wbdg.org/resources/building-integrated-photovoltaics-bipv.

Teignbridge District Council. (2010). *Renewable Energy and Sustainable Construction Study*. Available at https://www.cse.org.uk/downloads/reports-and-publications/policy/planning/renewables/teignbridge_energy_study_2010.pdf.

U.S. EIA. (2017). Wind Explained. Types of Wind Turbines. Available at https://www.eia.gov/energyexplained/index.php?page = wind_types_of_turbines.

U.S. EIA. (2018). Solar Explained. Photovoltaics and Electricity-Basics. Available at https://www.eia.gov/energyexplained/print.php?page = solar_photovoltaics.

World Energy Council. (2016). World Energy Resources – Solar.

Chapter 7

Sustainable Procurement and Transport of Construction Materials

Alireza Ahmadian Fard Fini[1] and Ali Akbarnezhad[2]
[1]*School of the Built Environment, University of Technology Sydney, Ultimo, NSW, Australia,* [2]*School of Civil and Environmental Engineering, University of New South Wales, Sydney, NSW, Australia*

7.1 INTRODUCTION

The construction industry is one of the top four sectors worldwide contributing to gross domestic product (GDP) in the economy and hence, plays a major role in national and international economic growth (Oxford Economics, 2015; IHS, 2015). According to Construction Intelligence Centre, the global construction industry has regained its growth rate in recent years from an average annual rate of 2.7% in 2012 to 3.8% in 2014 (CIC, 2016). A further rise to an average annual rate of 3.9% is forecasted over 2016–2020 for the global construction market, with a worth of US$10.3 trillion in 2020, while emerging markets, such as Middle East and Africa, will record a 5.3% growth over the same period (CIC, 2016; IHS, 2015).

The construction industry, however, faces enormous challenges in achieving and maintaining planned growths. The statistics show that the construction industry has experienced labor productivity stagnation, and even decline in some cases, over the last four decades, leading to considerable amount of waste in use of resources which can be prevented (CICA, 2002; Oxford Economics, 2015). One major cause of this poor performance is believed to be the outdated technologies that are in use in many construction organizations and the fact that the construction sector has not fully embraced the modern technologies and their benefits (Potocnik, 2013). Compared to other industries, the pace of adopting new technologies in the construction industry is considerably slow and there is a universal consensus on the need for a major technological transformation in the construction industry (Renz and Solas, 2016). Such a dramatic change is not only a need of the industry

Sustainable Construction Technologies. DOI: https://doi.org/10.1016/B978-0-12-811749-1.00005-5

itself, but also a moral obligation toward society. Adoption of more efficient technologies is considered as one of the fastest pathways to tackling some of the main economic, environmental, and social challenges faced by the construction industry through reducing the depletion of scarce resources, creating a more ecoefficient built environment, and boosting the global economic growth (Thormark, 2007; Zimmermann et al., 2005).

One of the main resources used by the construction industry, which can greatly benefit from improved productivity of new technologies is construction materials. Construction industry is the largest consumer of materials worldwide (Odijk and Bovene, 2014), consuming annually about two-fifths of the world's materials to produce building components (Augenbroe et al., 1998). The huge amount of materials used by the industry is a major contributor to environmental impacts of the industry including the increasing depletion rate of natural resources and carbon emissions.

The construction and building industry is responsible for up to 30% of worldwide greenhouse gas (GHG) emissions, with even higher figures reported in some countries (U.S. Environmental Protection Agency). A considerable portion of these emissions, referred to as embodied carbon, are associated with extraction, processing, transportation, and installation of materials. The recent advances in energy efficiency of building and enforcement of more stringent energy efficiency requirements by construction and building authorities have led to a growing share of embodied carbon in life cycle carbon of buildings in the new projects (Plank, 2008). The relative share of embodied energy, closely related to embodied carbon, in the total life cycle energy of buildings has been reported to increase from 20%, or lower, in conventional buildings to up to 40%–60% in low-energy buildings (Sartori and Hestnes, 2007; Thormark, 2007). The already high but yet growing impacts associated with construction materials used by the construction industry; highlight the importance of adopting more efficient and environmental friendly material procurement approaches. On the other hand, transport stage appears to be a hurdle for adoption of such approaches, particularly when offsite fabrication is regarded as a sustainable solution. The logistics parameters, mainly reflected in size and weight capacity of transport systems, may place constraints on potential of offsite fabrication which is inherently associated with bulky items. Moreover, transportation itself is the second contributor to global GHG emissions and a great deal of research has been performing to minimize its effects mainly through manipulating loading capacity in construction operation (Duffy and Crawford, 2013). With cost and time attributes of transport stage added to this, planning transport stage of construction materials requires as a hybrid approach in which long-term impacts of materials are considered.

Despite its effects on economic and environmental impacts of construction, little attention is generally paid to management of material procurement and transport based on sustainability requirements and whereby, the performance of construction industry in management of materials is far from ideal. A sustainable management approach should account for economic, environmental, and social impacts of the material flows in different phases of a building/infrastructure's life cycle. This is particularly important due to huge amount of solid waste produced by the construction sector. Construction and demolition waste constitute about 40% and 33% of total solid waste produced in the United States and Australia, respectively (Chiveralls et al., 2011; Odijk and Bovene, 2014). A great deal of effort has been made in the literature to develop strategies to reduce the impact of this waste through promoting the reuse and recycling of materials and estimating the economic and environmental impacts of different strategies (Akbarnezhad and Moussavi Nadoushani, 2014). However, maximizing the overall benefits of such strategies requires a life cycle approach where the overlapping effects of such strategies on different economic and environmental impacts incurred during life cycle of a material are evaluated during the material procurement stage (Akbarnezhad et al., 2014). Implementation of a sustainable material procurement approach necessitates a shift in paradigm from the early stage of material supply to reduce the impacts over the whole life cycle of a constructed facility. By starting with a proper choice of materials, the positive effects can even go beyond the life cycle of the facility through affecting the embodied energy and carbon and depletion of natural resources in the upstream (Bunz et al., 2006; Calkins, 2009; Kibert, 2008; Kim and Rigdon, 1998). With this in mind, sustainable procurement of construction materials, which requires a cradle-to-grave balance between economic, environmental, and social impacts, can concurrently serve to the needs of a sustainable life cycle for the constructed facilities (buildings or infrastructures). The construction industry's growth is dependent on the satisfaction of all its stakeholders and thereby, a holistic life cycle view that makes a balance between interests of different players along its value chain is vital.

Material procurement can considerably affect the economic, environmental, and social impacts of the construction projects. The material procurement decisions occur in early phases of design and planning when there is considerable opportunity to affect the life cycle impacts of the infrastructures/buildings. The materials may influence the impacts associated with all phases of a buildings life cycle including construction, operation, and end-of-life phases and thus provide a great opportunity for enhancing the life cycle sustainability of the buildings. A wide variety of material alternatives and supply chain structures are usually available for each project. Furthermore, new materials and innovative supply chain structures are made available commonly due to high rate of innovation in the materials field, adding to available options to choose from. Materials constitute enormous opportunities for innovation, as the European Commission estimates that 70% of product innovation across all industries is derived from new or improved materials (Renz and Solas,

2016). The construction industry, as the largest global consumer of materials, has a considerable potential for applying advanced construction components. The scope of applying innovations to material is truly wide, ranging from new multifunctional materials with radical property changes to incremental improvements through new combinations of traditional materials. The newly synthesized materials have the least chance to penetrate the market mainly because they lack a track record of success to gain acceptance from the customers (Potocnik, 2013; Renz and Solas, 2016). On the opposite edge of innovation spectrum, technology plays a central role to make a new product from combinations of existing materials (Renz and Solas, 2016). A major stream of such technology-embedded materials is prefabricated elements, which are growingly grabbing acceptance in the construction industry, yet are far inadequately widespread.

7.2 SCOPE OF THE CHAPTER

This chapter discusses the current state of the art and the new developments in the sustainable material procurement and transport strategies for construction industry. The focus is placed on a holistic approach for selection of sustainable material procurement and transport strategies based on their effect on the life cycle impacts of buildings/infrastructures. The rest of this chapter is structured as follows; Section 7.3 discusses "Current Approaches" toward adoption of sustainable construction technologies in procurement and transport of construction materials. Section 7.4 presents a "Literature Background" in which modern concepts of sustainability, life cycle assessment, supply chain structure, and transport of materials are reviewed. In Section 7.5, "New Approaches" toward procurement and transport of construction materials are presented through a hybrid framework, which addresses the weaknesses of Current Approaches. Section 7.6 demonstrates application of the alternative approaches in two case studies; one focusing on the entire process of procurement in a building project and the second one on transport stage in an industrial construction project. In Section 7.7, a summary of the discussions is presented with remarks on future of sustainable technology adoption in procurement and transport of construction materials.

7.3 CURRENT APPROACHES

Selection and adoption of sustainable procurement and transport strategies is a complex process. The construction industry is behind the universal adoption trend of sustainable technologies, mainly due to its traditional and temporary nature (Toole, 1998). The complexity associated with adopting sustainable construction technologies has roots in two major building blocks of such a milestone in the life of a construction firm. The first building block is related to the term "sustainability" and misperception about its goals and definitions, while the second feature is uncertainties and difficulties lying in construction technology adoption and changes to as-usual practice (Toole, 1998; Werkheiser and Piso, 2015). In the domain of material supply, on the other hand, the construction industry heavily relies on economic approaches, which oversimplify the procurement and supply chain process. To gain a better understanding of such issues and requirements of sustainable procurement strategies, the following sections elaborate on current approaches toward "sustainability," "technology adoption," and "material supply."

7.3.1 Objectives of Sustainable Procurement

Sustainability has been increasingly targeted as a goal in different construction projects from a single-story building to mega infrastructures. Since its introduction in 1987, it has, however, been diversely conceptualized and modeled under different circumstances (Altieri, 1987; Seghezzo, 2009). In line with this, the approaches to sustainability are different and are dependent on the context in which sustainable actions are defined. Therefore, achieving a common understanding about requirements of a sustainable approach is crucial prior to discussing sustainable procurement and transport strategies.

Sustainability was first conceptualized to address the increasing concerns about depletion of natural resources and disruptive impacts of human actions on the nature. Therefore, sustainability tends to commonly perceived to have purely environmental goals. In other words, environmental objectives have gained superiority in execution of construction projects (Robinson, 2004; Seghezzo, 2009). However, overemphasizing on environmental aspects, mainly advocated by environmentalists, has led to low adoption of sustainability practice. This fact has been acknowledged in the literature, through evaluating the low accomplishment rate of sustainability programs aimed purely at environmental assessment of construction materials (Werkheiser and Piso, 2015). While environmental concerns that have been overlooked for ages must be utterly placed in execution plans, disregarding interests of major project stakeholders, such as contractors, who share their capital with a hope of monetary benefits, is a radical change that cannot sustain. Thus, a balance between different interests is required to take a forward step to a sustainable world.

A renowned concept of sustainability reconciles economic, social, and environmental objectives in a three-pillar model where a Venn diagram is used to depict overlaps between these goals (Lozano, 2008). In this model, the overlap between economic and environmental objective produces "viable" alternatives; social objective nested in economic goals promotes "equitable" solutions while environmental dreams in conjunction with social targets come into "bearable" cure for the community. While a majority of national and international agencies tend to place their focus on one area at a time, a "sustainable" approach is achieved in the center where all three objectives overlap (Lozano, 2008). However, despite its enlightened conceptualization, this sustainability model does not explicitly suggest correlations between the three concentric components of sustainability (Werkheiser and Piso, 2015). This is despite the interrelated nature of different components of sustainability in the sense that any decision, such as selection of a procurement strategy, made to affect a particular impact, such as reducing the costs, tends to also affect the environmental and social impacts of the project.

Sustainability is argued to require adding a fourth dimension to the three-pillar model. In the fourth dimension, a future-oriented perspective is embedded in sustainability assessments in which the aim is to make a balance between current and future needs (Curran, 2006). By definition, a sustainable development is a "development that meets the needs of the present without compromising the ability of future generations to meet their own needs" (The World Commission on Environment and Development, 1987). Hence, time, as the fourth dimension, comes into trade-offs between different pillars of sustainability and directs how a specific pillar (such as the environment) may place constraints on other aspects (such as the economy). Under a time-dependent setting, sustainability analysis encounters with two new challenges. The first challenge is boundary of the system that must be considered and the second is the length of timeframe that may have sensible impacts on the assessments, and thus, is worthwhile to be taken into account.

In summary, a new approach to sustainability analysis of procurement and transport decisions made in construction projects should comprehensively determine the significance of economic, social, and environmental objectives. In addition, the assessment requires a timeframe specified for a system with predefined boundaries, which would facilitate tangible trade-offs between different objectives on the same par.

7.3.2 Current Approach to Adoption of Technologies and Procurement Strategies

Improving the sustainability of procurement and transport strategies commonly requires to adopt new technologies (e.g. prefabrication), new materials (e.g. structural composite materials), and new approaches (e.g. data-driven decision support systems), hitherto referred to all generally as technology. However, apart from a limited number of innovators and entrepreneurs, who may be early adopters, the dominant approach toward adoption of new technologies has been usually conservative in construction context (Toole, 1998). There has been a common root cause behind diverse range of factors contributing to such a conservative approach; that are uncertainties associated with adoption decisions. The uncertainties in technology adoption decisions can be classified into five groups.

7.3.2.1 Time Horizon

While the decision to adopt a construction technology may take place at a present time, consequences of this decision may last as long as technology's life, if not longer. Therefore, the assessment of technology involves forecasting consequences at least over the life cycle of a technology (Yu et al., 2013). This has been never easy particularly in a construction project environment where a series of interrelated incidents may affect intended benefits from a particular element or technology (Yu et al., 2012).

Two factors may be added to time horizon uncertainties in adoption of new technologies. First, when technology lifespan increases, predictability of its benefits in the remote future decreases (Bell, 1982). In fact, a new technology is presumed to bring an added value for the adopters. This benefit has never been a constant value over the technology lifespan (Yu et al., 2013). In the initial years, the earnings from technology may have a pure profit substance when the expenditures are limited to small regular costs, such as periodic maintenance cost and operation cost. As the technology gets older, then investors have to incur higher cost to keep the technology functioning. This includes cost of overhaul and upgrading which are usually huge values, compared to regular periodic cost items (Ortiz et al., 2009). Therefore, the longer the lifespan of a technology is, the higher uncertainty is involved in evaluating the adoption decision. Second, the technology adoption in a construction organization is not an isolated decision which would most affect the contractor in all cases. There are instances in which a technology's life is more spent with end-users (Odijk and Bovene, 2014). In other words, the lifespan of a technology begins with one party, who is usually the contractor, continues with a user, who is generally a client, and ends up with a third party, who may want to demolish or recycle the end-of-life products. An example of this is facilities or elements of a building that are initially selected by a contractor, handed over to buyers after completion of the project, and is transferred to another party who may demolish, refurbish, or recycle the facilities after the operation life (Odijk and Bovene, 2014, Potocnik, 2013). Therefore, the uncertainties associated with adoption of a technology increases when a technology has more than one user/owner over its life cycle.

Technology adopters have been predominantly encouraged to take a future-oriented approach in their assessment process. That is to avoid a quit decision based on incorrect comprehension of future benefits by performing a cost–benefit analysis mainly focused on the present time toward future usage (Curran, 2006). In addition to a forward look at the time horizon of a technology adoption, the adoption process may require a backward view over the stages behind the milestone a technology is owned. In the absence of such a holistic view, the adopters may underestimate the entire cost and benefit to acquire a specific technology. This is particularly important when sustainability concerns arise and analyses are not solely limited to financial trade-offs. In construction projects with inherently temporary nature, adopting a technology may be largely influenced by schedule pressure and its benefits in the construction site (Nepal et al., 2006). However, there are many stages before a product or machine is delivered at construction site, including manufacturing and transport stages, which if not accurately taken into accounts, may lead to big losses during execution phase (Ahmadian et al., 2015). Before-acquisition steps may have impact on succeeding stages through offsetting the benefits assumed for the use period. In addition, we should bear in mind that adopting a technology in construction projects is not always a yes or no decision to a sole technology. It is rather a comparison between close alternatives available to contractors, which demands for a thorough cradle to grave analysis (Chen et al., 2010).

7.3.2.2 Information

It has been proofed that information and knowledge about technology play a pivotal role in an adoption decision (Toole, 1998). Relevant literature suggests different theories that explain and model how individuals make decisions under different circumstances. In category of descriptive decision theories, such as status quo bias (Ritov and Baron, 1992) and prospect theory (Bern-Klug, 2010; Kahneman and Tversky, 1979), a common rule applied to analysis of decision mechanisms states that potential adopters may quit a technology when substantial information necessary for decision-making process is missing. On the other hand, it has been shown that early adopters of technology are those who have been able to collect effective information that can help to make informed decisions (Rogers, 1983). This is in agreement with these theories that model an "ideal decision maker" as a person who has gathered missing information, who is able to compute with perfect accuracy, and whose decision is fully rational (Toole, 1998).

Construction is a context with diverse types of data from pure technical to soft information, which are all needed in a decision-making framework. Access to information in construction industry is usually difficult due to a number of reasons. The first problem emerges when data is being collected. Due to dynamic nature of construction projects, data collection process is costly (Akhavian and Behzadan, 2012). Despite advances in automated data collection systems, the construction industry is still heavily reliant on manual methods of data collection such as log sheets and daily observations for the both technical and particularly soft form of data (Williams and Gong, 2014). Collection of soft data, however, has not been as regular and systematic as technical information. While technical information are usually properties, such as dimensions and temperature that palpably change over time and hence, can be easily distinguished and recorded, soft aspects are not tangible and difficult to be tracked (Kang et al., 2015). Therefore, construction firms are not equipped with techniques for regular tracking of soft statistics, which are usually costlier to be collected than technical data.

In addition to the internal sources of information that can be collected over the implementation phase, a construction project relies on information, which are supplied from external sources. A construction project involves long supply chains in which different parties provide services and materials. In parallel with supply chain of materials and services, there are supply chains of information that should flow to the project (O'Brien et al., 2004). This information may partly form the knowhow of service providers and material suppliers and hence, they may be reluctant to share. This reluctance particularly happens where there is no long-term partnership between project parties (Eriksson, 2010). When this information is required to assess a new technology, contractor organizations usually consider them as missing information to be replaced with guesswork-driven data (Frank, 2006). As such, the accuracy of decision-making may be compromised and lead to suboptimal technology adoption decisions.

Finally, construction organizations usually face a big amount of data stored in dispersed databases with unclassified formats. Currently, a key challenge is how to screen relevant information from redundant data (Bilal et al., 2016), as construction organizations need expertise to avoid getting lost in a pool of relevant and irrelevant data. Apart from relevance, the pattern of data plays a pivotal role in truly directing the decision-making process. Therefore, abundance of data will be effective in technology adoption if the scientifically sound methods are used to recognize likely correlations between different parameters affecting new technologies and their benefits (Bilal et al., 2016). For instance, it is essential to appreciate whether a specific parameter has a statistically significant effect on decision criteria based on statistical analysis of data or the significance of the effect is not proven by scientific methods and is just presumed by decision makers.

7.3.2.3 Evaluation Process

Understanding the current evaluation process for adoption of new technologies is required to understand the necessary characteristics of a sustainable procurement and transport strategy to increase its chances for adoption by the construction industry. The common practice for evaluation of construction technologies is rather qualitative, as shown in the relevant literature. Prioritizing the qualitative approaches over quantitative methods is attributed to psychological reasons. For instance, "word-of-mouth" has been reported as a top motivator of people in adopting a new technology or product (Cheung and Thadani, 2012). Therefore, potential adopters aggregate their information through communicating with individuals and agents who have already experienced that new technology and hence, their decision to adopt or quit the technology is influenced by those individuals and agents. The "theory of reasoned action" introduced by Ajzen and Fishbein (1980) explains how qualitative information received from others shape the decision to adopt or reject a technology. Under this model of decision-making, "perception" is the basis of technology assessments in which "perceived usefulness" and "perceived ease-of-use" are two main criteria of decision-making (Davis, 1989).

Another psychological motive behind tendency toward qualitative evaluations is the market that pushes people to make quick decisions. In other words, construction organizations believe that adapting to rapid changes in the market is essential to survive, protect the market share, and perhaps grow it (Rogers, 1983). Therefore, they have to rely on qualitative methods and expert judgment in adopting a new technology or product without knowing the costs and benefits of the adoption decision (Rogers, 2003). Relying on such models of decision-making may be justifiable in cases where the extent of technology's impacts is limited. Nevertheless, this should not be a central strategy in adopting a construction technology with long lifespan and consequences on other aspects of a construction project.

Apart from psychological reasons, issues of information and information management may contribute to taking qualitative assessment methods. The first issue is missing information and inabilities in gathering or replacing them in the decision-making framework (Bell, 1982). There are a considerable proportion of people who may think of a perfect technology analysis when all the necessary pieces of information are handy or they may get frustrated by waiting for a complete set of information to move forward (Rogers, 2003), leading to postponement of technology adoption. Another problem arises when the analysts are surrounded by big data, which are not usually classified, and are unscreened with many redundancies (Bilal et al., 2016). If there exists inadequate technical expertise in the contractor organization to manage these issues, they may tend toward qualitative assessments.

Qualitative assessments, particularly when performed under market/work pressure, can result into premature decisions (Butter, 1993). In the construction industry, diverse technological options are available for a specific activity of the business in which differences may look delicate in the first glance. Starting with qualitative assessments can lead to adopting an option that does not best suit the conditions of the organization. Over a long lifespan of a technology, the minor differences turn into huge losses that are hidden due to qualitative analysis. At the same time, a qualitative approach in the beginning is usually followed by intangible measures to assess the adopted technology during the operation stage. This does not provide a realistic understanding of effectiveness of technology in improving performance and operation of the business. Therefore, a vicious circle of qualitative decisions and qualitative performance assessments gradually reduces the profit margin that may end up in losing competitive advantage of the business, as evidenced by annual statistics of bankrupt companies which could not keep up with technological advancements (Sull, 1999; Thangavelu, 2015).

7.3.2.4 Decision Makers

While contractors are organizations that are in the center of a chain of supply and demand, the decision to adopt a construction technology has least gone beyond the center of that chain; it has remained in the control of contractor organization and is least influenced by other stakeholders (Slaughter, 1993). Nevertheless, there are a diverse range of external actors who can put an impact on contractors' goal setting and goal attainment. Similarly, they can influence both technology adoption decision and achieving the targeted outcomes from deployment of the technology. These groups include local and national authorities, architects and designers, suppliers, subcontractors, and clients/buyers (Moavenzadeh, 1991; Slaughter, 1993). The construction organization, however, believe that limiting the boundary of decision-making about adoption of a technology inside their own organization can decrease uncertainties associated with diversity of interests among these groups that may sometimes be conflictive (Toole, 1998). In other words, they may look at them as a source of uncertainty that can threaten a technology adoption decision through a reaction that may not comply with what the contractor may expect. This reaction is shaped not only by dissimilar interests, but also by different risk-taking attitude toward innovations (Toole, 1998; Toole and Tonyan, 1992). However, disregarding influential stakeholders of a construction company is just overlooking the main ingredients of success in adoption of new technologies. Instead, they should be looked at as counterparts that help in making an informed and mature decision, which increases the chance of success in the operation of technology.

7.3.2.5 Material Supply

Current approach toward material supply decisions in construction projects has two major drawbacks. First, construction contractors tend to oversimplify material categories by classifying the materials into two major groups of bulk and prefabricated materials (Rumane, 2011). According to this classification, bulk materials are usually standard off-the-shelf items/materials that can be instantaneously purchased, but undergo a relatively long fabrication/processing time in the construction site to become a part of the built facility. In contrast, prefabricated materials are engineered elements that may take a long time to be ordered and fabricated in the factory while they can usually become a major part of the built facility shortly after arrival in the construction site with the least on-the-job activities (Rumane, 2011; Ahmadian et al., 2015). This lookout rather uses prefabrication in compliance with the traditional construction process that allows a limited number of construction components to be factory made. It leaves little room to think of intermediate states where prefabrication can be a flexible strategy for fundamental changes in the construction process. Prefabrication strategy can be defined in various degrees for each element of a facility under different circumstances.

To overcome the challenge of widespread adoption and gain full potential of prefabrication, it is essential to primarily appreciate the rigorous manufacturing concept behind it and then integrate it into a life cycle assessment framework, which provides a long-term view at its pros and cons.

Second, construction contractors justify procurement of prefabricated materials mainly based on direct economic benefits to the construction stage. These benefits include efficiency and productivity improvements and reduction of weather and labor-related holdups (Potocnik, 2013; Zimmermann et al., 2005). While prefabrication has proven environmental and social benefits over the life cycle of the facility (Ortiz et al., 2009), contractors tend to overlook them in their decision-making process, presuming them as factors competing against project profit margin. Nevertheless, such judgments are not based systematic trade-offs between tangible effects of economic and noneconomic factors.

The purely economic view in adopting prefabricated components has also affected planning of transport stage in construction projects (Ahmadian et al., 2015). Offsite transportation of materials accounts for 10%–20% of the total project expenditure in a typical construction project (Shakantu et al., 2003). This has been a great stimulus to minimize the material procurement costs through optimization of the transportation stage (Irizarry et al., 2013; Said and El-Rayes, 2011). On the other hand, accounting for time in transport of prefabricated components has been a challenging task, especially in cases where the factory is far from the construction site, packaging features, such as oversized consignments, are limiting, and/or capacity of transport systems is constrained (Renz and Solas, 2016). Delayed shipment of construction materials can affect timely completion of projects and may result in considerable financial losses (Sambasivan and Soon, 2007). Moreover, transportation has been identified as the second top contributor to air pollution and global warming through carbon emissions (U.S. Department of Transportation). These necessitate a holistic view in which cost, time, and environmental factors are taken into account in adjusting transport parameters of prefabricated materials.

7.4 LITERATURE REVIEW

This section provides an overview of the concept behind material supply chain and its correspondence to procurement and transport strategies. Furthermore, the available literature on life cycle assessment, which is one of the key elements for supply of sustainable material, is succinctly reviewed. Finally, common techniques for multiattribute decision-making, which can correspondingly be employed in material supply process, is introduced.

7.4.1 Supply Chain Structure of Materials

The manufacturing industry takes supply chain structure as the basis of materials classification and thereby, classifies them into four nonoverlapping groups of Made-To-Stock (MTS), Assemble-To-Order (ATO), Made-To-Order (MTO), and Engineered-To-Order (ETO) (Babu, 1999; Olhager, 2003). This classification is made with reference to a point along the supply chain, known as "decoupling point," at which manufacturers take the first action in response to customers' need (Olhager, 2003). According to this conceptualization, while the stock is the first point where an MTS order is responded to, the first point of action taken on an ATO order is somewhere in the final stages of assembly line. For MTO materials, however, the decoupling point begins with procurement of raw components for the assembly abased on the purchase order. In ETO items, the decoupling point penetrates into the design stage where manufacturer takes the responsibility to design, procure, and fabricate based on a basic order (Olhager, 2003). Supply chain duration may be affected considerably by the type of the supply chain structure, changing from the longest to the shortest duration for ETO, MTO, ATO, and MTS, respectively (Tommelein et al., 2008).

In line with the aforementioned concept, the structure of a material's supply chain denotes the degree of prefabrication involved in its manufacturing and supply process. From this viewpoint, MTS materials are standard items or bulk materials readily available in the supplier's stock, which then will become an integral part of almost any component after they undergo a fabrication process in the construction site. The degree of prefabrication requested in this category of orders is nearly zero, while its onsite fabrication stage is the longest (Olhager, 2003). ATOs are components made from standard parts in the assembly line of the factory based on a standard design provided by a contractor. ATOs have a limited degree of prefabrication and are subject to various assembly operations on the construction site where different assemblies form a larger component of the built facility (Tommelein et al., 2008). However, degree of prefabrication involved in MTO products is higher than ATOs, as a manufacturer makes an assembly from diverse standard and nonstandard items based on a detailed order provided by a contractor (Babu, 1999). Correspondingly, this category of materials requires a short processing stage on the jobsite before final installation. Finally, ETO materials are customized factory-made assemblies that are transported to the construction site for installation with least on-the-job intervention and thus, have the highest degree of prefabrication.

7.4.2 Planning Offsite Transport of Prefabricated Materials

Importance of planning offsite transport of construction materials increases with an increase in degree of prefabrication (Renz and Solas, 2016).

Prefabricated components are commonly perceived to be large, and may reach oversized dimensions in fully prefabricated components or items with irregular shapes. Besides size, prefabricated elements may be characterized by weight increase particularly when made from high-density materials, such as steel or concrete. Delivery of oversized and/or overweight components has been problematic due to limitations of transport systems, including limited capacity of roads, bridges, and public dissatisfactions particularly in urban areas (Luskin and Walton, 2001). To protect safety, transport-planning authorities enforce traffic rules that place speed or time limits on the movement of construction materials in roads. However, such rules are usually inconsistent across different territories and add to the complexity of planning for timely arrival of materials (Ceuster et al., 2008; Luskin and Walton, 2001). Moreover, complexity of offsite transport planning increases further in cases where there is a large distance between factory and construction site. In timely delivery of prefabricated components, where procurement from overseas suppliers is quite common, mode choice is another important attribute of planning for transport stage. With increasing environmental concerns, mode choice is influenced by embodied energy and emissions of transportation, besides cost–time trade-offs (Zhang et al., 2015). Nonetheless, accounting for environmental impacts in planning the transport of construction materials should not result in overlooking the environmental benefits of prefabrication during the life cycle of the facility.

7.4.3 Life Cycle Analysis

Life cycle analysis (LCA) is a universal approach for quantifying the environmental impacts incurred in all stages of life cycle of a material to provide input into various decision-making processes. LCA may contribute considerably to promoting environmental sustainability goals through providing a measure to compare environmental impact of various alternatives available. Effective LCA is, however, reliant on availability and quality of information about manufacturing, use, maintenance, and disposal of materials. To cope with different levels of data accessibility and accuracy, ISO 14044 suggests three methodologies, namely, input–output, process-based, and hybrid approach, for systemic assessment of environmental impacts of a product throughout its life cycle (Curran, 2006; Treloar et al., 2000).

The process-based LCA involves identifying all materials and energy flows in different operations required for production of a material or provision of a service, followed by quantifying the corresponding environmental impacts of these flows. The collection of data required for process-based analysis may however be difficult, costly, and time consuming. With this in mind, subject to availability of such information, this approach will lead to the most reliable results (Cooper, 1999). Nonetheless, performing a detailed process-based LCA may be time consuming and costly, and thus, is generally performed by defining boundaries for processes to be evaluated and flows to be tracked. On the other hand, defining the systems boundaries for process-based LCA has been criticized for its subjectivity (Chang et al., 2012). In a different approach, the input–output analyses the aggregate industry-wide or national data to quantify the associated environmental impacts of different products and services by applying a top–down methodology (Su et al., 2010). The main advantage of input–output analysis is its ability to account not only for direct environmental impacts associated with production processes but also for the indirect impacts occurred in different stages of supply chain of the product. The disadvantages, however, include difficulties in linking monetary values with physical units and application of this method to an open economy with considerable noncomparable imports.

A well-known example of input–output method is the work done by the University of Bath, which provides an estimate on the embodied energy and carbon of a variety of construction materials (Hammond and Jones, 2008). The tools and formulations provided by the U.S. Environmental Protection Agency (EPA) for estimation of emissions and energy associated with on-road and nonroad vehicles, engines, and equipment can be considered as process-based methodologies. Hybrid method, however, is a combinatorial approach, which utilizes both of the aforementioned methods and may mainly rely on one and fill the gaps by the other method, if possible. For instance, whenever there is inadequate process data on a particular stage of a material's life cycle, hybrid analysis uses input–output data for this stage while process-based approach is used for the other stages (Treloar et al., 2000).

Sustainability of construction materials can alternatively be assessed using European standards EN 15804 and EN 15978 for "Sustainability of construction works." In EN 15804 and EN 15978, the emphasis is placed on detailed measurement of environmental impacts at a product level and building level, respectively. These two standards are interdependent and both necessary to perform a conclusive analysis. Therefore, users are guided through a set of rules to flexibly define system boundaries, set unit of measurements, and derive computational methods. This flexibility is mainly accommodated through modular division of life cycle stages. Compared to LCA approach suggested by ISO 14044, each stage of materials' life cycle is subdivided into its constituent phases, as shown in Fig. 7.1, whereby users are systematically allowed to establish a cradle-to-grave assessment approach if cradle-to-gate inputs are not adequately reliable. Furthermore, the comparability of different products with dissimilar unit of measurements is promoted in EN standards based on their functional use in the building. Nevertheless, the main challenge in carrying out EN-based assessments relates to incorporating impacts beyond the building's service life which are currently considered as a nonsplit module out of the system boundary (EN 15804: 2012; EN15978: 2011).

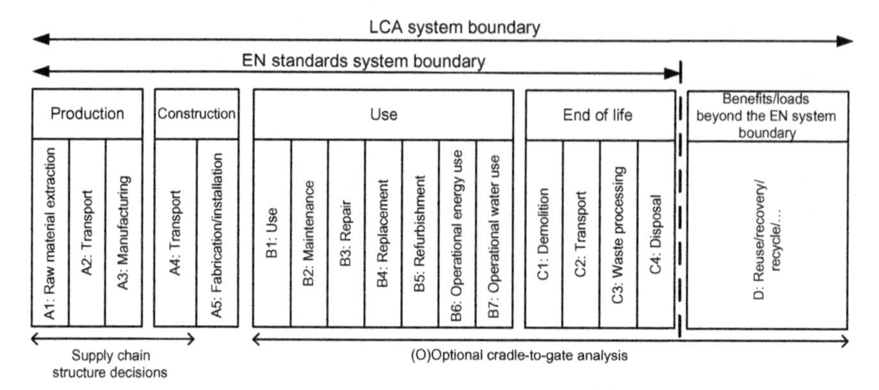

FIGURE 7.1 An overview of European standards for assessing sustainability of construction works.

7.4.4 Role of Life Cycle Thinking in Planning of Material Procurement

Life cycle thinking meets sustainability concerns at points where it provides decision makers with a robust tool in supply of materials that best makes a balance between current and future needs of the global community (ISO 14044). Such a balance is achieved through raising environmental awareness in overwhelmingly economic governance of material supply decisions. Concurrently, sustainability of material supply solutions is improved through accommodating diversity and long-term views in selection criteria. The relevant literature has identified a diverse range of cost, time, quality, social, and environmental factors affecting sustainable life cycle of materials, as summarized in Table 7.1.

Despite progresses made by life cycle thinking, it has been widely shaped on a linear chain of stages in the lifespan of materials in which the emphasis is placed on the use/operation phase while end-of-life practices are still suboptimal (Despeisse et al., 2015). Such practices are mainly limited to recycling of demolished materials, such as metals contaminated with concrete, by energy-intensive processing methods, which are not economically viable, compared to new materials readily available in the market (Odijk and Bovene, 2014). In the best cases, the recycled components are of low value, for instance used as gravel in roads, which are deemed as down-cycled materials (Akbarnezhad et al., 2013). To reshape the linear thinking, circular construction has been introduced to the industry that aims at revolutionizing current construction methods (Despeisse et al., 2015; Odijk and Bovene, 2014; Potocnik, 2013). The main difference between linear and circular construction lies in the replacement of material recycle with component reuse strategy. According to principles of circularity, buildings and infrastructures are designed in such a way that their components can be reused with no or

TABLE 7.1 Factors Affecting Sustainable Life Cycle of Materials Identified in the Literature

Criteria	Factors	Referred by
Cost	Manufacturing	Emmitt and Yeomans (2008)
	Transport	Irizarry et al. (2013)
	Construction	Said and El-Rayes (2011)
	Maintenance	Wong and Li (2008)
	Disposal	Asokan et al. (2009)
Time	Manufacturing	Tommelein et al. (2008)
	Transportation	Ahmadian et al. (2015)
	Construction	Dissanayaka and Kumaraswamy (1999)
Quality	Durability	Zhou et al. (2009)
	Load performance	Lo et al. (2008)
	Thermal performance	Zhou et al. (2009)
	Fire resistance	Lo et al. (2008)
	Constructability	Kibert (2008)
	Maintainability	Sirisalee et al. (2004)
	Aesthetic	Ashby and Johnson (2002)
Social and environmental sustainability	Health and safety	Spiegel and Meadows (2010)
	Indoor air quality (IAQ)	Nematchoua et al. (2015)
	Emission	Bank et al. (2011)
	Natural resource depletion	Kibert, 2008, Kim and Rigdon, 1998
	Water use	Bank et al. (2011)
	Energy use	Spiegel and Meadows (2010)
	Waste	Asokan et al. (2009)
	Recycle and reuse	

little processing at the end of facility's life (Potocnik, 2013). In other words, project elements are designed for disassembly, reuse, and adaptability to future functionalities with minimum aesthetical and technological modifications required at the facility's end-of-life (Peeters et al., 2017).

Prefabrication has been identified as a key step in removing the bottlenecks of shifting from linear to circular life cycle (Peeters et al., 2017).

It allows for disintegration and dismantling of project components at the end of lifespan and prevents loss of valuable and energy-intensive resources by demolition and low-value recycling (Odijk and Bovene, 2014; Peeters et al., 2017). In addition to environmental benefits, prefabrication brings up a new economic value through turning disposal cost into residual value for reusable components (Odijk and Bovene, 2014). On the other hand, prefabrication of project elements can affect operation phase through factory-sealed/insulated components of project that can improve acoustic and thermal performance of the facility (Peeters et al., 2017). If exactly prefabricated to a well-designed order, it allows easy assembly of different parts at the construction site that can improve aesthetic aspects of building/infrastructure. Easy assembly also improves productivity and minimizes waste generation during construction phase (Peeters et al., 2017). To precisely make to orders and achieve anticipated functionalities, however, the main challenge is to address fragmentation of project stakeholders (Odijk and Bovene, 2014; Potocnik, 2013).

7.4.5 Techniques for Multiattribute Decision-Making

Project stakeholders may have different, in some aspects conflictive, interests with regard to supply of materials. On the other hand, it is seldom the case that a particular material is the best option among available alternative with respect to all criteria. This is particularly the case, due to inherent trade-offs between economic and environmental impacts of materials in practice. Therefore, selection of the sustainable material in practice requires comparing the relative performance of different alternatives against a set of different criteria (Jato-Espino et al., 2014), rendering the selection process a multiattribute decision-making one. A number of techniques for multiattribute decision-making in which a transparent basis is provided for comparison and ranking of available alternatives against selection criteria have been proposed in available literature (Jato-Espino et al., 2014). Among these, Technique for Order of Preference by Similarity to Ideal Solution (TOPSIS) and Analytical Hierarchy Process (AHP) are the most commonly used techniques in different domains of decision-making such as bidding, material selection, structural analysis, and conflict resolution (Hwang and Yoon, 1981; Lourenzutti and Krohling, 2016; Reza et al., 2011; Wong and Li, 2008). TOPSIS and AHP also facilitate group decision-making through their collaborative platform for selection of criteria, weight assignment to each criterion, and assessment of alternatives (Jato-Espino et al., 2014). In doing so, AHP has been identified as a reliable tool for pairwise comparisons of criteria and alternatives with respect to a main goal set for the decision-making process (Reza et al., 2011; Wong and Li, 2008). TOPSIS, on the other hand, characterizes the ideal and the worst solution given the decision-making criteria and establishes principles of geometry to rank alternatives based on their distance from the ideal and the worst solution (Hwang and Yoon, 1981).

To acquire a robust decision-making framework, the literature suggests a combinatory approach where AHP assigns weights of criteria and TOPSIS determines ranking of alternatives (Reza et al., 2011). Weighting of criteria, however, can be consolidated through access to historical data about importance of each criterion. Therefore, modern decision support systems improve reliability of multiattribute decision-making methods using techniques, such as statistical analysis, for knowledge discovery from databases about historical effect of different criteria (Kumaraswamy and Dissanayaka, 2001).

7.5 A NEW COMBINATORIAL APPROACH TO PROCUREMENT OF CONSTRUCTION MATERIALS

In this section, a systematic framework for evaluating the material procurement strategies is introduced. The proposed framework relies on detailed sustainability assessment of various procurement options for different building components by considering the function of components and circumstances of the project. Due to its considerable effect on supply chain of materials, degree of prefabrication is considered as one of the main decision variables and is determined based on the notion of supply chain structure discussed in Section 7.4.

Reducing uncertainties and enhancing comprehensiveness are two central principles established to shape the framework, which are achieved by relying on:

1. a comprehensive set of measures which define economic, social, and environmental aspects of sustainability;
2. a set of databases that facilitate access to information required to quantify various performance measures;
3. life cycle thinking which accounts for life cycle sustainability of various procurement options;
4. a mathematical/numerical platform which quantifies the sustainability measures; and
5. a decision-making approach to rank various alternative procurement options by considering their relative performance in different criteria.

The main steps involved in proposed framework to decide on procurement of sustainable construction materials are presented in Fig. 7.2 and described in detail in the following sections.

7.5.1 Sequence of Decisions in Procurement of Materials

Procurement of construction materials involves a series of interrelated decisions that need to be made during construction phase. While such decisions are made before material's arrival in the construction site, i.e., manufacturing and transport stages, they may considerably affect the economic, environmental, and social impacts of other stages of the life cycle. With this in

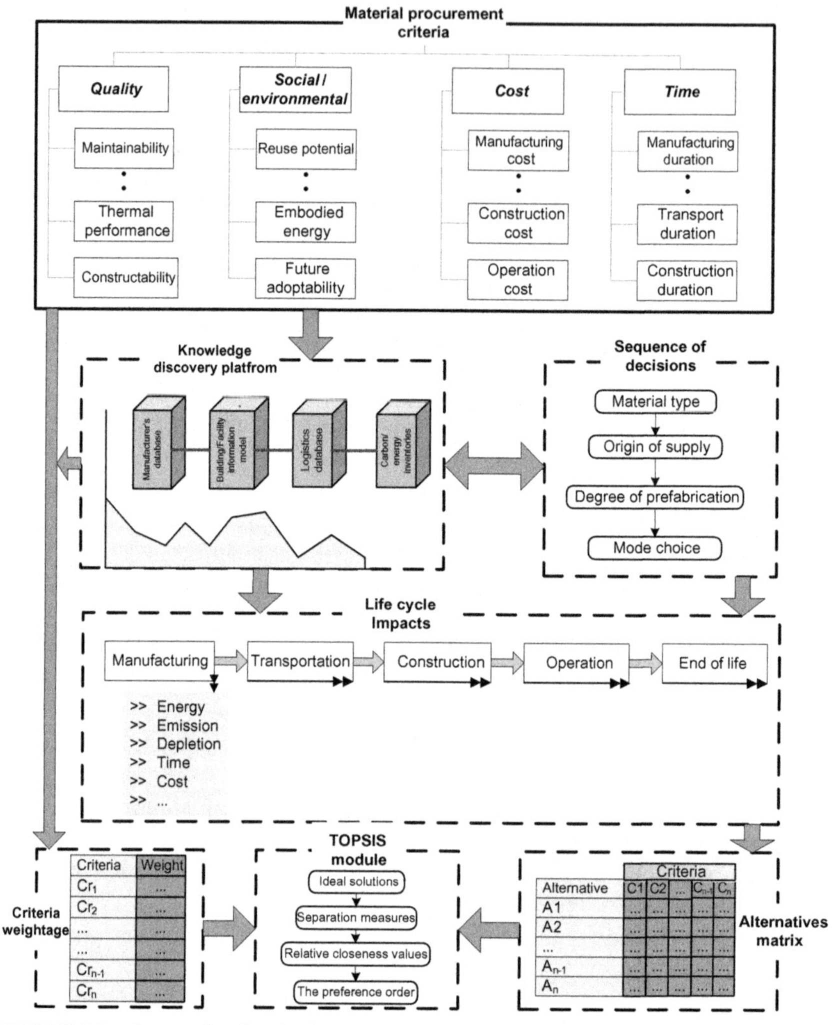

FIGURE 7.2 Steps undertaken by the proposed framework to rank supply decision alternatives.

mind, a life cycle approach is taken by proposed framework to accounts for consequence of such decisions on the whole life cycle of materials. The hierarchy of supply decisions consists of four levels:

1. *Type of materials*: Supply decisions begin with selection of materials from a list of technically feasible options for a specific component of the project. Usually, there exist several alternatives for type of materials used in fabrication of project components, which meet technical requirements. The decision to select a particular one is, hence, affected by sustainability criteria defined for the project and function of the

component. On the other hand, selection of materials may affect succeeding stages in procurement of materials, such as number and location of the suppliers available.
2. *Origin of supply*: The origin of supply decision takes into account the location of various suppliers, for both local and international suppliers, with respect to location of the project.
3. *Degree of prefabrication*: This is determined based on supply chain structures available in the market, namely MTS, ATO, MTO, and ETO, for a specific component.
4. *Mode choice*: Mode of transport is especially important when a degree of prefabrication is involved and/or materials are supplied from overseas. Choices can be any combination of road, air, sea, and rail.

7.5.2 Procurement Criteria

The framework establishes different sources to define criteria for sustainable procurement of materials. The sources used are twofold: (1) general documentations specifying sustainability objectives applicable to any new development/action, such as regulatory obligations imposed by public/governmental authorities, and (2) proprietary documentations stipulating objectives especially set for the current project, such as contractual clauses and conditions. While the first source of information tends to accommodate long-term public welfare in which socioeconomic and environmental concerns are dominant, the latter may preserve short-term benefits of those who are directly involved in the execution phase where economic features are privileged. Accordingly, as shown in Fig. 7.2, the main criteria for sustainable procurement, presenting different performance areas, are classified generally into four groups of cost, time, quality, and social/environmental criteria.

A diverse set of measures are defined for each criterion broad evaluation of the criteria which may result in imbalanced and unsustainable procurement decisions to ensure different aspects of each performance area are adequately evaluated. For fair comparison of all alternatives and to account for overlapping impacts of various decisions on performance in different phases of material's life cycle, the value of different measures is obtained by considering the entire life cycle of the materials, rather than a particular phase.

7.5.3 Life Cycle Assessment of Materials

The proposed framework includes a computational module in which life cycle impacts of material procurement decisions are assessed. This module is fed by the data extracted from different databases, including historical records of contractor on cost and time performance, products' databases of manufacturer, carbon and energy inventories, and logistics databases. Furthermore, facility information modeling platforms, commonly referred to as Building Information Modeling (BIM) in building projects and their equivalent industrial applications, may be utilized to structure a systematic analysis. The object-oriented features of BIM tools enable automatic classification of material life cycle data imported from different databases based on requirements and properties of the component. Besides, the use of BIM, when available, can expedite the LCA process by providing accurate quantity and spatial data in a considerably shorter time compared to conventional estimation methods. The development of such attributes and the use of BIM are not in the scope of this chapter and detailed information can be found in Fini et al. (2017). The following sections describe the methodology for quantifying the performance measures used in each criterion.

7.5.3.1 Cost

In the traditional approach to material procurement, the focus is placed mainly on the costs incurred in initial stages of material's life cycle up until the delivery of material to site, i.e., manufacturing and transport stages. In contrast, sustainable procurement also requires accounting for all the other costs incurred in other stages of life cycle of a material to estimate the life cycle costs. These include cost of design, order, supply/manufacture, transport, construction, use, and end-of-life processing of materials including the costs of demolition and landfilling, recycling or reuse, when applicable. Due to time value of money, the life cycle costing involves summing up the present value of all the costs incurred during life cycle of the material. In the proposed framework, the life cycle cost is used as the selected economic assessment measure for comparing various alternative material procurement options. Moreover, the present worth values are annualized to provide a fair basis for comparison of alternatives with nonidentical lifespans.

The value of the costs incurred in stages up to construction phase have present worth ($P_{g,s}$) which should be converted to their equivalent Annualized Present Value ($APV_{g,s}$) by:

$$APV_{g,s} = \frac{P_{g,s} i(1+i)^n}{(1+i)^n - 1} \tag{7.1}$$

where g, s, n, and i are alternative number, stage of lifespan (e.g., transport), lifespan, and interest rate, respectively (ref). Depending on the type of element to be purchased for the facility, the applicable cost of use may include annual cleaning ($APV_{g,a}$), annual energy saving ($APV_{g,e}$) obtained as a negative or positive value with reference to a base case material available in the contractor's database, and periodic maintenance ($F_{g,m}$). The last one is usually a future value that needs to be converted to its equivalent annual value ($APV_{g,m}$) by:

$$APV_{g,m} = \frac{F_{g,m}i}{(1+i)^n - 1} \tag{7.2}$$

Finally, the end-of-life value depends on the degree of prefabrication involved in the component, which may vary from a cost incurred by demolition of bulk materials to a residual benefit achieved through disassembly of prefabricated components for potential reuse. The construction industry currently lacks accurate data recorded on demolition and reuse, mainly due to its traditional approach toward procurement of materials. However, such costs/benefits are available based on rule of thumb estimates (Zahir, 2015). If available, these will be a single future expenditure/income ($F_{g,d}$) annualized by:

$$APV_{g,d} = \frac{F_{g,d}i}{(1+i)^n - 1} \tag{7.3}$$

The total cost associated with alternative g (AC_g) is then summation of all the cost/benefit items:

$$AC_g = \sum_{lifecycle} APV_{g,s} \tag{7.4}$$

According to principles of LCA, comparison of different alternatives is valid only if the "equivalent use" is the basis of cost estimations (Curran, 2006), and hence, the "equivalent use" should be defined depending on the type and function of the element in facility, beforehand.

7.5.3.2 Time

Time is a multifaceted decision variable in procurement of construction materials in which duration of manufacturing, transport, and construction are interrelated. One the one hand, there is usually a trade-off between the supply duration and onsite processing/fabrication duration. Nonprefabricated items with a shorter supply duration tend usually to require a longer onsite processing/fabrication than prefabricated items, which usually have a longer supply duration. Planning for on-time delivery of materials is important for both short and long lead items. Due to required onsite processing, late delivery of short lead items may delay other construction activities. Similarly, long lead items are generally critical components which their on-time delivery is a prerequisite for on-time start of many subsequent activities. Early delivery of long lead items is also generally not favored due to onsite space limitations for commonly large elements. With this in mind, it is of great importance to identify stage-specific parameters, which independently affect a stage, besides project-specific factors, using methods of knowledge extraction from relevant performance databases.

Fig. 7.3 depicts a flowchart developed to estimate the total delivery time associated with each arrangement of material procurement considering both project-specific and stage-specific parameters. As shown, the project specific

(A)

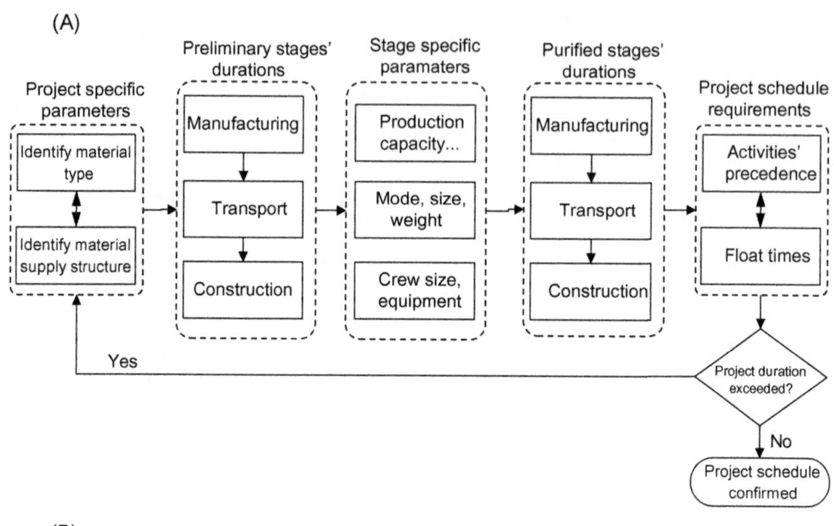

(B)

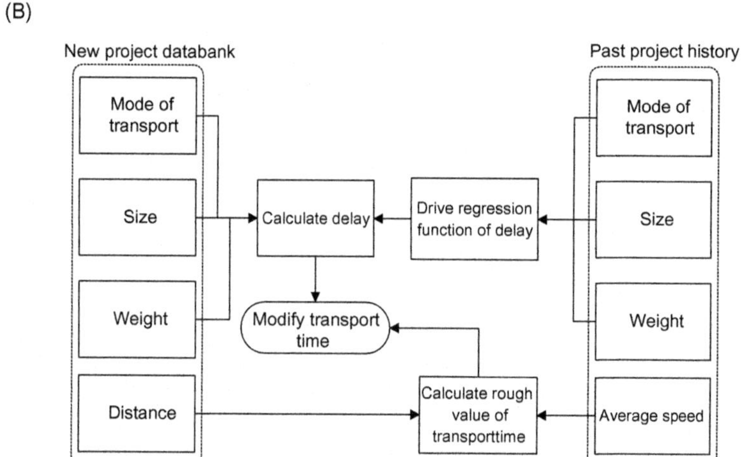

FIGURE 7.3 (A) The overall time estimation process. (B) Estimating transport time using statistical analysis.

attributes affecting the total time associated with all stages of ordering, manufacturing, transport, and construction include material type and supply chain structure. For instance, a great number of factors are considered in supply and use of hazardous materials rather than non-hazardous ones and a high number of people are involved in design and coordination for supplying custom prefabricated materials (i.e., ETO) and hence, more time is consumed along the supply process.

"Supplier capacity" in production stage and "crew size and construction machinery" in construction phase are, on the other hand, examples of stage-specific factors found to be correlated to duration of manufacturing and construction stages, respectively (Eng et al., 2014; Fini et al., 2016). The proposed methodology for establishing such correlations to estimate duration of a single stage of supply process, here "transport stage", is shown in Fig. 7.3B (Ahmadian et al., 2015). As seen, the estimation starts by identifying material supply chain structure for the component, based on the concept of decoupling point introduced by Olhager (2003). This is then followed by identifying the available modes of transport by considering the routes and distances between factory and construction site. Furthermore, an average speed corresponding to mode of transport in different routes is estimated and preliminary transport duration is estimated based on the distances to be traveled. In the next stage, descriptive clustering method is used to identify the size and weight limits from historical logistics data recorded for previous consignments. This is important because available literature highlights that size and weigh considerably affect transport time (Ahmadian et al., 2015). Therefore, it is required to perform a linear regression analysis to identify likely correlations between delayed transport as dependent variable and weight and size of consignments as well as mode of transport as independent variables. Such correlations are established to revise preliminary transport duration estimated based on average speed. The total duration from ordering to installation of an element in the facility is finalized by applying float time and activity precedence relationships, explained in detail in Ahmadian et al. (2015).

7.5.3.3 Quality

The proposed framework uses two categories of quality measures; namely, tangible and intangible indicators. Tangible indicators of material quality are generally physical properties with well-defined calculation procedures and include factors affecting the serviceability of the elements such as design/service life, load/pressure performance, thermal performance, fire resistance, leak tightness, and so forth (Zhou et al., 2009; Lo et al., 2008; Kibert, 2008). A tangible indicator may be characterized by one or more measures. In the latter case, a method is needed to define the combinatorial effect of different measures. This method can be as simple as product of different measures, regardless of compatibility of units. For instance, leak tightness of building components may be jointly defined by air/gas infiltration measured in "$L/s.m^2$" and water/liquid penetration in "minutes," which can be combined to define the overall leak tightness as follows:

$$Tightness = Tightness_{gas} \times Tightness_{liquid} \tag{7.5}$$

However, this method may only be valid for comparison purposes in procurement of material alternatives and may not be attributable to a real physical property. Moreover, the multiplication should be reversed (i.e., multiplied by reverse of opposites) in cases where measures defining an indicator have opposite effects (i.e., multiplied by reverse of an opposite measure).

Intangible quality indicators, on the other hand, include nonphysical measures constructability, maintainability, and aesthetic, as listed in Table 7.2. These indicators are usually assessed using linguistic terminologies, which define different levels of desirability for each indicator. The framework then converts such terminologies into a numerical scale, such as Likert scale, to provide quantitative inputs to multiattribute decision-making using TOPSIS.

TABLE 7.2 Intangible Measures for Quality, Environmental Sustainability and Social Sustainability of Material

Criteria	Indicator/ measure	Assessment scale
Quality	Aesthetic	1 = Extremely eye-catching
		2 = Very eye-catching
		3 = Moderately eye-catching
		4 = Slightly eye-catching
		5 = Ordinary
	Constructability	1 = Very hard to construct
		2 = Hard to construct
		3 = Fair to construct
		4 = Easy to construct
		5 = Very easy to construct
	Maintainability	1 = Nearly no component of the alternative is maintainable
		2 = A few components of the element are maintainable
		3 = Half of the components are maintainable
		4 = Many components of the alternative are maintainable
		5 = Almost all components of the alternative are maintainable

(Continued)

TABLE 7.2 (Continued)

Criteria	Indicator/ measure	Assessment scale
Environmental and social sustainability	Raw material abundance (R_Q)	1 = Scarce
		2 = Shortage
		3 = Poorly spread over the globe
		4 = Adequately available over the globe
		5 = Abundant
	Renewability (R_W)	1 = Nonrenewable
		2 = Centuries to be renewed
		3 = A century to be renewed
		4 = Decades to be renewed
		5 = A decade to be renewed
	Effect on indoor air quality (IAQ)	1 = No effect
		2 = Poor
		3 = Moderate
		4 = Strong
		5 = Very strong
	Safety grade	1 = A*
		2 = A
		3 = B

Source: Adapted from Fini, A.A.F., Akbarnezhad, A., Rashidi, T., Waller, S. 2017. BIM-enabled sustainability assessment of material supply decisions. Eng. Constr. Arch. Manage. 24(4), 668–695.

7.5.3.4 Environmental/Social

Similar to quality, measures defining environmental and social sustainability are classified into two groups of tangible and intangible. Table 7.2 lists four proposed intangible measures of environmental and social sustainability of material as well as the proposed performance assessment scale for each measure. As can be seen, safety grade has a definition stipulated in relevant standards for different materials/components and hence, its rating system may not be consistent with 1-to-5 Likert scale (Lo et al., 2008). Effect on indoor air quality (IAQ), raw material abundance (R_Q), and renewability (R_W) are rated using Likert scale. The last two intangible measures, i.e., R_Q and R_W, along with percentage (%) of recycled materials used in production of a component jointly define rate of material depletion (RMD_g) as:

$$RMD_g = \frac{R_Q \times R_W \times (100 - \%\ of\ Recycled\ materials)}{100} \tag{7.6}$$

while IAQ can be solely assessed as an intangible measure, it also be characterized by quantifiable measures such as indoor temperature, humidity, and air circulation during different seasons (Nematchoua et al., 2015).

The tangible measures are, however, diverse and spread over the life cycle of the building/facility. As a commonly used approach, environmental impacts associated with procurement of an alternative are primarily measured by embodied carbon (C_g) and embodied energy (E_g). Such impacts are computed for three stages: (1) cradle-to-gate emissions ($C_{g,CtoG}$ and $E_{g,CtoG}$) incurred during extraction, processing, and manufacturing substages; (2) transport emissions ($C_{g,t}$ and $E_{g,t}$) incurred during transport of material from supplier's site to construction site; and (3) onsite fabrication/installation ($C_{g,p}$ and $E_{g,p}$) emissions incurred due to operation of equipment, which are then summed up to estimate the total embodied energy and embodied carbon of the material:

$$E_g = E_{g,CtoG} + E_{g,t} + E_{g,p} \tag{7.7}$$

$$C_g = C_{g,CtoG} + C_{g,t} + C_{g,p} \tag{7.8}$$

where E_g and C_g are metered in MJ/m^2 and kgCO$_2$-e/m^2 for alternative g, respectively. The end-of-life emission of materials is, however, not included in the estimations, due to lack of access to reliable data. Fini et al. (2017) present a set of novel equations for computing the terms used in Eqs. (7.7) and (7.8) with detailed description of logics behind each. The novelty of equations mainly stems from three principles:

- first, definition of weight-to-area conversion factors which enable comparisons of dissimilar alternatives on the same par;
- second, taking the effect of supply chain structure (degree of prefabrication) on carbon emission and energy consumption into account; and
- third, decomposing the construction method through machinery type, age, and duration of use.

Furthermore, a diverse range of freely available databases, including Inventory of Carbon and Energy (ICE V2.0) and Australian National Inventory Report (AUS NIR) can be used to estimate the cradle to gate carbon emissions of various materials, which can then be modified to include transport and installation emissions.

In operation phase, water use, measured in L/m^2.year for cleaning purposes, is a prominent environmental factor established in the assessment of alternatives. The framework defines end-of-life environmental sustainability by relying on measures that characterize the extent of reuse, recycling, adaptability, and down cycle potential, all rated in percentage (%). Since no precise data are available in the existing databases, we use average industry data with regard to type of materials and degree of prefabrication, as suggested by ISO 14044 (Su et al., 2010).

The importance of social sustainability is also emphasized in the framework by accounting for the effect that material may have on acoustic performance (in dB of sound resistance/reduction) and light transmission (in %). The social sustainability indicators applicable to assessment of procurement decisions may vary considerably depending on the project specific conditions. Other social sustainability indicators including the effects of choice of materials on noise pollution, and community connectivity as well as job creation, local business development, etc., should be considered when applicable.

7.5.3.5 Group Decision-Making

7.5.3.5.1 Participants and Their Role

In this framework, procurement of materials is regarded as a group decision-making where influential stakeholders of a construction project, including client, contractor, and architect and/or engineer, participate in the process. Different areas of knowledge and experience brought by participants to the procurement evaluation process ensure alignment of decision with strengths of the organization and leads to a common understanding about a broad range of pros and cons of various decisions over the life cycle of materials. Different stakeholders in a project have diverse and, perhaps, conflictive interests particularly with respect to sustainability measures. For instance, while economy of the construction phase is of the highest importance from a contractor's point of view, architects may place their emphasis on noneconomic factors related to use/operation of the built facility. The inherent bias of different stakeholders toward different sustainability criteria may be used to obtain a balanced view of the relative importance of different criteria based on the organizational priorities, which will be reflected in the weights assigned to different criteria.

7.5.3.5.2 Assessment of Decision Criteria

The proposed method relies on stakeholder's opinion about suitability of each measure on a simple numerical basis, i.e., a 1-to-5 Likert scale, to rank measures from "least suitable" to "most suitable." The relative suitability of measure "*f*" is then computed by:

$$RI_f = \frac{\sum_{Respondents} w}{N \times V} \tag{7.9}$$

where w is the rank assigned to the measure by a participant, N is the total number of participants, and V is the highest scale which can be given in the assessment (here is 5). In the next stage, this index should be normalized and converted to a local weight (LW_f) within a corresponding criterion in comparison with the indexes calculated for the rest of measures:

$$LW_f = \frac{RI_f}{\sum_f RI_f} \tag{7.10}$$

To perform a thorough comparison between alternatives, a global weight is required for each measure. Therefore, a weight value should be assigned to different criteria of sustainability, i.e., cost, time, quality, social/environmental. The framework uses an analytical approach adopted from AHP where relative importance of the criteria can be computed through a set of pairwise comparisons (Golestanifar et al., 2011; Saaty, 1980). Such comparisons demand for a high level of experience and authority of participants who can affirm the overall significance of each criterion to a specific stakeholder. The framework, therefore, suggest that pairwise comparison of importance of different criteria should be done by project executives representing different parties. The pairwise comparison may be done using the same aforementioned Likert scale, however, with a definition tailored to pairwise comparisons, where 1 represents two equally important criteria and 5 evaluates one criterion five times more important than the other. Accordingly, a matrix consisting of all the pairwise comparisons made between different criteria is formed. Based on the procedure provided by Saaty (1980), a priority matrix and a priority vector of the criteria are computed. The priority vector assigns a weight to each criterion, denoted by Cr_0, so that:

$$\sum_{Criteria} Cr_0 = 1 \tag{7.11}$$

It is necessary to ensure the comparisons made are consistent and hence, the weights computed are valid. One way is to calculate a Consistency Rate (CR), as guided in the procedure, and test if $CR < 0.1$. Otherwise, the comparisons should be carefully readjusted (Saaty, 1980).

Upon acquiring the valid weight value of a criterion (Cr_0), the global weight of a measure (GW_f) is calculated by:

$$GW_f = Cr_O \times LW_f \tag{7.12}$$

The measures defining cost and time criteria are of the same dimension and can be summed up. Therefore, no local weight is required and the global weight, obtained using AHP, is used for these criteria:

$$GW_{cost} = Cr_{cost} \tag{7.13}$$

$$GW_{time} = Cr_{time} \tag{7.14}$$

7.5.3.6 Ranking the Supply Decision Alternatives

The procurement of construction materials is inherently a multiattribute decision analysis where different supply scenarios should be assessed against the measures of sustainability described/computed in Section 7.3, considering the global weight acquired in Section 7.3.5.1. This chapter uses TOPSIS to analyze the decisions and rank the alternatives. Selection of TOPSIS is due to (1) its rigor in providing an integrated computational platform for manipulating various measures with dissimilar dimensions through a straightforward normalization process (Lourenzutti and Krohling, 2016) and (2) its transparency in providing a tangible insight into potential areas of weaknesses for the selected alternative identified by comparison with the ideal solution (Hwang and Yoon, 1981; Jato-Espino et al., 2014).

TOPSIS module begins with creation of a $G \times F$ decision matrix of alternatives-criteria where each cell, denoted by x_{gf}, represents performance of an alternative, g, against a criterion, f. The starting decision matrix is then normalized to obtain the normalized decision matrix, $R = (r_{gf})_{G \times F}$. The entities of decision matrix, r_{gf}, are calculated using the following equation:

$$r_{gf} = \frac{x_{gf}}{\sqrt{\sum_G x_{gf}^2}} g = 1, 2, \ldots, G \text{ and } f = 1, 2, \ldots, F \tag{7.15}$$

The normalized matrix is then used to compute the weighted normalized decision matrix of H by applying the importance weights associated with each criterion to entities in its respective column in the matrix:

$$H = \left(h_{gf}\right)_{G \times F} = (GW_f \times r_{gf})_{G \times F} \therefore \text{subject to} \left(\sum_{f=1}^{F} GW_f = 1\right) \tag{7.16}$$

In the next stage, the positive and the negative ideal solutions, correspondingly denoted by A^+ and A^- are identified by:

$$A^+ = \left\{\left\langle \max\left(h_{gf} | g = 1, 2, \ldots, \right) | f \in F_+ \right\rangle, \left\langle \min\left(h_{gf} | g = 1, 2, \ldots, G\right) | f \in F_- \right\rangle\right\} = \left\{\alpha_f^+\right\} \tag{7.17}$$

$$A^- = \left\{\left\langle \min\left(h_{gf} | g = 1, 2, \ldots, G\right) | f \in F_+ \right\rangle, \left\langle \max\left(h_{gf} | g = 1, 2, \ldots, G\right) | f \in F_- \right\rangle\right\} = \left\{\alpha_f^-\right\} \tag{7.18}$$

where F_+ and F_- divide the criteria into sets of measures with desirable and undesirable impacts, respectively. This is followed by calculating the distances between each alternative and the positive ideal solution (d_g^+) and the negative ideal solution (d_g^-), as follows:

$$d_g^+ = \sqrt{\sum_{f=1}^{F} \left(h_{gf} - \alpha_f^+\right)^2} \tag{7.19}$$

$$d_g^- = \sqrt{\sum_{f=1}^{F} \left(h_{gf} - \alpha_f^-\right)^2} \tag{7.20}$$

Finally, an index representing the closeness to the positive ideal solution (S_g^*) is calculated by:

$$S_g^* = \frac{d_g^-}{d_g^+ + d_g^-} \tag{7.21}$$

The available decision alternatives are ranked using this index where the highest S_g^* denotes the best decision (Hwang and Yoon, 1981). The TOPSIS module also enables an analysis on sensitivity of rankings to the weights of criteria and interest rates used in the computations.

7.6 CASE STUDY

Two different cases are used to demonstrate the application of the new approach proposed for procurement of sustainable construction materials. The scope of first case is procurement of materials for a single element, a curtain wall, in a building project where hierarchy of procurement decisions is studied on LCA platform of the framework. Case one provides a practical example for implementation of sustainable material procurement strategies, discussed earlier in this chapter, by demonstrating the application of a comprehensive and quantitative framework used to promote consistency in the decision-making process. While the relevant literature places its focus on selecting more ecofriendly type of materials, this case study extends the scope of assessments to sustainability of a series of interrelated decisions connected with the supply chain structure. Furthermore, this case study aims to highlight the incremental nature of achieving sustainability in association with each stage of procurement. On the other hand, the importance of a paradigm shift from a one-sided environment-oriented way of thinking to comprehensive multifaceted sustainability thinking is demonstrated by a real-world case in which the main three pillars of sustainability, i.e., economic, social, and environmental pillars, are taken into considerations.

In the second case, however, the focus is placed on planning offsite transport stage of multiple elements in an industrial construction project where concurrency of construction activities may affect the choice of mode of transport and supply structure.

Through the second case study, a theory-based approach is demonstrated for knowledge discovery from databases in order to guide a decision-making process with escalated achievements. The case study utilizes a rigorous and pertinent theory/concept in designing the framework of data analysis. The advantages of this approach are twofold. First, this approach examines the current decision-making practice and identifies potential areas for improvement using a theory and/or concept related to the subject of decisions. Thus, any weaknesses associated with the existing practices are proactively identified and prevented from becoming the basis of analysis. Second, it facilitates interpretation of the results achieved from statistical analysis. This is enabled through providing a common point of reference by which the numerical findings are turned into a corresponding meaning with the aid of a utilized theory. The theory utilized in statistical analysis of shipment data is the concept of "decoupling point" by which supply chain structure of construction materials is identified. As described in Section 4.1, an insight is provided into diversity of construction materials with respect to dissimilarity of supply chain structures, which is particularly demonstrated through continual flexibility required for timely delivery of orders at construction site.

7.6.1 Case One: Procurement of Curtain Wall

The objective of this case study is to identify the most sustainable material procurement alternative for curtain wall of a six-story residential building with design life of 50 years, located in Melbourne, Australia. The total surface area of the curtain wall is estimated to be 1219.4 m^2. The curtain wall is planned to be erected by a construction crew of eight workers, working 8 hours and 5 days a week. We assume an average load factor of 75% for construction machinery in estimation of emissions and a discount rate of 8% in financial analysis.

Curtain wall is known as an element of the building envelop with significant effects on life cycle in sustainability of a building through separating the unconditioned and conditioned spaces (Memari, 2013). There is usually a wide variety of options for material of curtain wall, each with a different degree of prefabrication. In general, curtain wall systems may be divided into stick, semiunitized, unitized, and specialized custom wall fabrication systems, ranked in an ascending order according to required degree of prefabrication. Equivalent to MTS supply structure, in the stick system, standard items/materials are ordered and delivered to construction site, where a majority of processing, fabrication, and installation operations is performed. Semiunitized fabrication system, on the other hand, follows the ATO supply structure, where limited activities including final glazing and installation of standard prefabricated frames and infills are required on the construction site. In utilized fabrication, glazing process is transferred to the factory by providing further design data while onsite construction activities are limited only to fixing and placing of ready-to-install panels (MTO). Custom wall system is analogous to ETO supply structure where the manufacturer takes responsibility to customize design and production of a curtain wall based on basic architectural information (Kazmierczak, 2010; Martabid and Mourgues, 2015).

Table 7.3 lists 20 procurement alternatives considered in the case study, formed from different combinations of material type, supply structure, location of supply, and mode of transport. For the sake of simplicity, in this study the focus is placed only on selection of frame and infill materials, as two main components of a curtain wall. Despite the limited options for the frame materials, which include aluminum and steel, a wide range of material alternatives including glass, fabric veneer, brick veneer, stone veneer, and concrete is available for infills of the curtain wall considered in this study. Corresponding to each material alternative, there are different supply chain structures possibilities from Chinese or Australian manufacturers of curtain wall materials, as listed in Table 7.3. Further, available modes of transport for materials imported from China to the construction site include road/sea and road/air, while local suppliers use road transport to deliver materials.

The computational module of the framework performs an LCA with respect to each alternative and quantifies the measures defining the decision criteria. To identify the best supply decisions, a pairwise comparison of different criteria were performed by a team of three executives, representing contractor, architect, and client, as the main influential parties present in the time of execution. The resultant priority weights for cost, time, quality, and environmental/social sustainability are 0.47, 0.173, 0.08, and 0.277, respectively, with confirmed consistency (i.e., $CR = 0.076 < 0.1$). On the other hand, a team formed by 15 experts of the same parties, 6 from client, 3 from architect, and 6 from contractor, were used to assess the suitability of measures defining the functionality of a curtain wall and hence, the global weights of the measures were computed using the methodologies discussed earlier in Section 7.3. Moreover, to evaluate the sensitivity of the decision to selected importance weight, a second scenario where an equal weight of 0.25 is considered for all the criteria is studied and the results are compared with the base scenario.

Table 7.4A shows the ranking of alternatives for procurement of the curtain wall under the base scenario (i.e., before sensitivity analysis with respect to weight of measures). As shown, the results indicate that regardless of the supply chain structure, steel and concrete appear to be the most sustainable options for frame and infill materials based on the sustainability criteria considered (performance of a concreted curtain wall with steel frame is compared with the positive and negative ideal solutions against different sustainability criteria in Table 7.4B). On the contrary, glass, despite its popularity among architects (Kazmierczak, 2010), seems to be the last choice for infills due to its negative impacts on the life cycle of the building, mainly due to energy and cleaning aspects of operation phase. However, the results indicated that if ordered locally either as a stick (MTS) or semiunitized (ATO) system, glass may turn out to be a slightly more sustainable option compared to semiunitized (ATO) stone veneers.

TABLE 7.3 Curtain Wall Supply Decision Alternatives

Alternative ID	Frame	Infills	Supply chain structure				Source of supply	Transport mode		
			MTS	ATO	MTO	ETO		Road	Sea	Air
1	Aluminum	Glass	√				China	√	√	
2			√				China	√		√
3			√				Australia	√		
4				√			China	√	√	
5				√			China	√		√
6				√			Australia	√		
7						√	China	√	√	
8						√	Australia	√		
9		Fabric veneer			√		China	√	√	
10	Steel				√		China	√		√
11					√		Australia	√		
12		Stone veneer	√				China	√	√	
13			√				Australia	√		
14				√			China	√	√	
15				√			Australia	√		
16		Brick veneer	√				Australia	√		
17				√			Australia	√		
18		Concrete		√			China	√	√	
19				√			Australia	√		
20	Concrete		√				Australia	√		

TABLE 7.4 (A) Closeness of Alternatives to the Ideal Solution and Their Ranks Under Original and Equal Weight Scenarios. (B) Comparison of the Selected Alternative With the Best and the Worst Solution

(A) Assessment Scenario		S^*_1	S^*_2	S^*_3	S^*_4	S^*_5	S^*_6	S^*_7	S^*_8	S^*_9	S^*_{10}	S^*_{11}	S^*_{12}	S^*_{13}	S^*_{14}	S^*_{15}	S^*_{16}	S^*_{17}	S^*_{18}	S^*_{19}	S^*_{20}
Original	Closeness	0.268	0.232	0.295	.201	0.190	0.249	0.163	0.173	0.715	0.773	0.809	0.338	0.371	0.239	0.249	0.805	0.823	0.838	0.856	0.891
	Rank	12	16	11	17	18	13	20	19	8	7	5	10	9	15	14	6	4	3	2	1
Equal weight	Closeness	0.409	0.392	.464	0.332	0.365	0.438	0.243	0.288	0.659	0.173	0.763	0.380	0.493	0.322	0.418	0.677	0.783	0.731	0.812	0.787
	Rank	13	14	10	17	16	11	20	19	8	6	4	15	9	18	12	6	3	5	1	2

(B) Criteria	Measure	Unit of measurement	In situ concreted curtain wall	The best solution	The worst solution
Cost	Total	Annualized $/m^2$	10.075	10.075	85.716
Time	Total	Days	43.08	38.18	85.57
Quality	Design life	Years	40	40	12
	Load performance	Combinatorial	4	2	6
	Thermal performance	Combinatorial	0.0875	0.0021	0.1032
	Fire resistance	Minutes	150	150	45
	Aesthetic	Likert scale	5	1	5
	Air infiltration	L/s.m^2	0	0	2.1
	Water penetration	Minutes	60	60	17
	Constructability	Likert scale	2	4	1
	Maintainability	Likert scale	1	4	1
Environmental and social sustainability	Embodied energy	MJ/m^2	312.559	312.559	7858.232
	Embodied carbon	kgCO_2-e/m^2	34.547	18.361	572.418
	Raw material depletion	Combinatorial	4	45	2
	Reuse/recycle/down cycle potential	Percentage (%)	60	65	30
	Future adoptability	Percentage (%)	15	70	15
	Acoustic performance	dB of sound reduction	58	60	39
	Light transmission	Percentage (%)	0	75	0
	Effect on indoor air quality (IAQ)	Likert scale	4	4.25	2.25
	Waste in construction	Percentage (%)	16	5	16
	Water use in operation	L/m^2.year	0.4	0.2	14
	Safety grade	Scale	2	1	3

In comparison with stone panels, a glass curtain wall has more positive impacts on well-being of occupants through enhancing IAQ, light transmission, and safety. The second and third best choices of infills are brick and fabric veneer, respectively. When it comes to mode and structure of supply, however, local supply of unitized (MTO) fabric veneer is more favorable than bricklaying (MTS). Such mixed tendencies toward different degrees of prefabrication do not allow general advice on adoption of prefabrication.

The results of the sensitivity analysis performed indicate that the preferred type of material and degree of prefabrication may be considerably affected by the selected relative importance of different criteria. As shown in Table 7.4B, under original scenario, where a high importance weight of 0.47 is given to costs, in situ concreting using ready mix concrete appears to be the best procurement option for infills of the curtain wall. However, the results of sensitivity analysis showed that a further emphasis on noneconomic criteria through applying equal weights to all criteria results in a tendency toward a higher level of prefabrication as reflected by selection of locally precast concrete panels as the most sustainable option.

To investigate the environmental performance of alternatives ranked under original scenario, Fig. 7.4 shows the variations in the emission, as a global measure of environmental concerns, with ranking of the alternative procurement options. The rate of emissions has been computed using the equations listed in Section 7.3.4. Detailed derivations and computational methods have been provided by the authors in Fini et al. (2017).

It is observed that an improve in ranking of alternatives is associated with a reduction in carbon emission in two zones (from rank 1 to 4 and 9 to 13); yet fluctuating trends are seen after 14th. The existence and nonexistence of a relationship between emission and ranking appear to lie in manufacturing and construction technologies behind an alternative. When such technologies related to a specific type of curtain wall have been omnipresent and less complicated, as of those in concrete or veneer curtain walls,

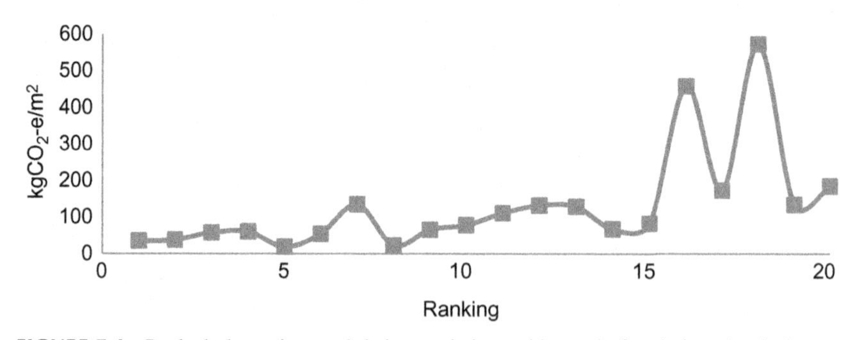

FIGURE 7.4 Ranked alternatives and their associations with trend of emissions (equivalent to $kgCO_2$ emitted per square meter of a curtain wall).

practices for reduction of embodied carbon are gradually aligned with other functional aspects of a curtain wall such as cost and quality.

On the other hand, local supply is usually preferred over international sourcing under different scenarios, particularly when a specific type of material and supply chain structure is available both locally and internationally. When materials are supplied internationally, mode choice is affected by the size and weight of orders. For instance, glass curtain walls supplied from China are usually bulky and heavy and hence, marine mode is prioritized over air mode. Light to medium weight consignments, such as fabric veneer panels, are, however, better to be transported by airplane.

Through the first case study, a valuable insight is gained about relativeness of sustainability in procurement of construction materials. Despite common interest, no dominant material, outweighing other alternatives in all the sustainability measures, can be found. Such an insight is provided by a comparative module in which superiority of an alternative over another is quantitatively examined with respect to every single measure. Therefore, through this approach, decision makers are made aware of the type and extent of weaknesses associated with a selected alternative and are enabled to clearly communicate those with all the stakeholders. Moreover, it is demonstrated that sustainability benefits achieved through a single decision-making step is marginal which can be either neutralized or enhanced in the next decision-making stage. As indicated by the results of this case study, a one-size-fits-all solution does not exist for a single stage of decision-making, and consideration of life cycle impacts and contextual conditions of the case project may bring a seemingly suboptimal choice, such as in situ concreted curtain wall, to the higher rank.

7.6.2 Case Two: Procurement of Materials in Industrial Construction Projects

To highlight the importance of accounting for offsite transport requirements of materials in selection of procurement strategy, the second case study presented in this chapter considers the procurement and transport of 149 orders in a real project involving construction of two oil refinery facilities, located in the Middle East. A majority of materials required in this project require a degree of offsite fabrication and have different characteristics in terms of weight and dimensions, which are listed in Table 7.5. The materials are divided in five groups including structural, mechanical, piping, instrumentation, and electrical. As shown, Table 7.5 highlights considerable variations in the characteristics of the materials to be procured in a typical industrial project.

Duration of offsite transport is an important factor to be considered in selection of sustainable materials. Historical records of companies can provide a reliable basis for estimating the duration of different types of materials by considering variations in material characteristics and project specific

TABLE 7.5 Characteristics of Purchase Orders Arrived at the Construction Sites (Weight and Size Categories)

Characteristics of orders	Material type				
	Structure	Mechanical	Piping	Instrument	Electrical
	(# of consignments)				
Weight (ton)					
Light (<1)	7	1	7	1	2
Medium (>1, <10)	6	5	22	4	5
Heavy (>10, <100)	9	14	15	3	4
Super heavy (>100)	21	12	11	0	0
Dimension (m^3)					
Small (<1)	9	3	6	1	0
Medium (>1, <10)	6	5	25	5	7
Large (>10, <100)	12	13	14	2	3
Super large (>100)	16	11	10	0	1

Source: Ahmadian et al., 2015

conditions. To investigate the correlation between transport duration and type of supply chain structure, the concept of supply chain structure was first applied to categorize order items into four groups of MTS, ATO, MTO, and ETO materials. This was then followed by performing linear regression analyses to identify correlations between deviations from estimated transport durations (i.e., delay in delivery), recorded by the project management team, and characteristics of the materials including size, weight, and mode of transport.

The identified correlations may be used as basis for prediction of deviations observed in previous projects for similar consignments and thus development of improved models to predict the offsite transport duration of future orders. The estimated supply duration can be in turn used as a decision-making criterion in selection of most sustainable procurement options. The results of the regression analysis performed on the data provided by the project management team in this case project for MTS materials, as a representative material type, is presented in the following:

$$d = 2.62 + 1.21 \times \mathrm{m} + 1.97 \times \mathrm{h} + 2.78 \times \mathrm{sh} + 0.71 \times \mathrm{s} - 0.88 \times \mathrm{w} - 1.42 \times \mathrm{v} \tag{7.22}$$

where d is delay in days per each 2500 km distance from factory/supplier to construction site; m, h, and sh, respectively, denote medium, heavy, and superheavy consignments; s is small size orders; and w and v represent railway and marine modes of transport, correspondingly. As shown, the regression model suggests that a growth in weight of MTS consignments increases the transport duration. The predicted delay due to selected material supply structure (i.e., MTS) can then be considered in decision-making, or be compensated through selection of an alternative mode of transport such as rail and marine modes. Selection of a different mode of transport to ensure faster delivery is, however, usually accompanied with changes in economic, environmental, and social impacts of the supply chain, which should be considered using appropriate measures discussed earlier. Similar conclusions are reached with regard to ATO materials in which weight is a limiting factor in timely delivery and air mode of transport is recommended to reduce transport time, where practical. In MTO and ETO materials, consignments of large and super large sizes are barriers of timely transport just in roads. However, the delays occurred in delivery of MTO and ATO are mainly attributable to before-transport stages, i.e., design and manufacturing.

The regression models derived were used to forecast delays in delivery of future orders from different material categories and a number of modifications were implemented by the project management team to minimize the predicted delays in procurement of materials. The implemented modifications and their observed effects on transport duration are reported in Table 7.6.

TABLE 7.6 Summary of Implemented Changes on Characteristics of Orders and Achieved Improvements

Item No.	Activity total float (TF) - days	Category of material	Forecasted delay before change-days/ 2500 km	Type of change	Mode of transport	Route	Source of origin	Supply chain structure	Expected transport delay improvement after change-days/ 2500 km	Achieved reduction in (%)
				Size and weight						
1	0≤ TF ≤2	MTO	2	√	√				0.1–1.36	71
2		MTO	1.9		√	√	√	√	0.18–1.12	100
3		ATO	3.1	√	√				0.12–1.4	83
4		ATO	2.3		√	√			0.73–1.4	56
5		MTS	3.2	√	√				0.76–1.64	75
6		MTS	4.4	√	√				0.54–3.39	100
7		MTS	1.9	√					0–0.71	91
8	3≤ TF ≤7	MTO	0.4	√	√				0-0.94	35
9		ATO	3.1		√				0.12–1.54	100
10		MTS	3.2	√		√	√		0–0.76	100
11		MTS	3.82		√				0.88–2.3	78
12	8≤ TF ≤13	MTO	1.72	√	√				−0.42–1.4	34
13		ATO	2.3	√					0.73–1.63	0
14		MTS	3	√					0.54–1.75	11

Source: Adapted from Ahmadian et al., 2015.

As can be seen, the type of changes range from modifying logistics parameters (i.e., size and weight, mode of transport, and route) to radical variations in sourcing and supply chain structures. In majority of the cases, such changes have been effective for achieving the expected improvements in timely delivery of materials at construction site, which can prevent likely stoppages due to shortage of materials. The results of this case study highlight the importance of differentiating various types and supply chain structure of materials in planning of procurement process as well as accounting for variations in transport requirements of different materials when comparing various procurement options. It also emphasizes on versatility of material procurement strategies needed for supply of thousands of diverse materials in mega projects, which need to be constructed in a limited duration, as reflected in dissimilar supply chain structures for different project components, flexible sourcing, and variable logistics parameters.

While the first case study demonstrates the application of the concepts discussed in this chapter to procurement of a single building element (i.e., curtain wall), the case study two applies the concepts to a considerably larger project where multiple project components have to be simultaneously supplied. Therefore, longitudinal planning for sustainable procurement and transport of a single construction product is turned into a matrix planning for supply of different components within a vortex construction environment, as depicted in Fig. 7.5. The logistics decisions are, however, influenced by narrow visibility of project conditions and needs. Concurrently, case study two provides a closer look inside knowledge discovery platform of the proposed framework. Therefore, it is demonstrated how project records may dynamically lead to adjusted decisions.

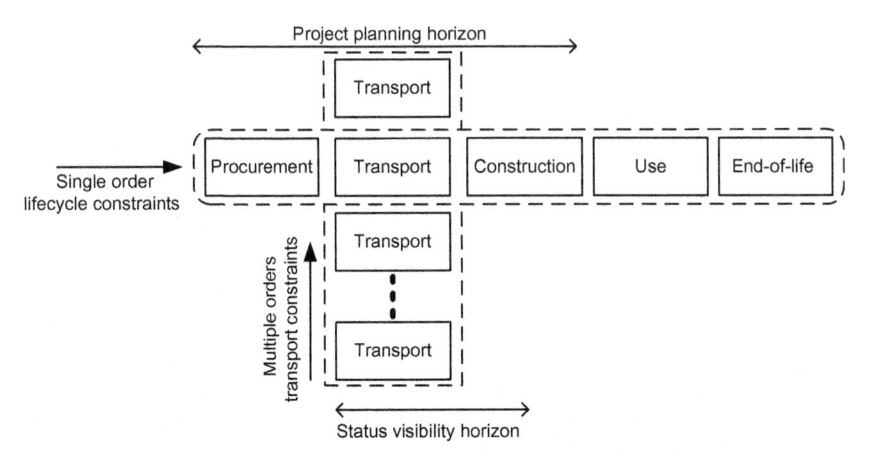

FIGURE 7.5 Longitudinal sustainability planning versus matrix logistics planning for construction materials.

7.7 CONCLUSIONS

An overview of current sustainability practices for procurement and transport of construction materials was presented by elaborating on existing approaches toward "sustainability," "technology adoption," and "material supply." It was found that objectives set for sustainable procurement are not adequately comprehensive and future oriented. Moreover, uncertainties associated with technology adoption related to time horizon, information, and evaluation process prevent steady movement toward sustainable materials technologies. Such uncertainties are intensified by incomplete knowledge of material supply strategies and unilateral decision-making in contractor organization for procurement and transport of materials.

The review of available literature highlighted the need for a comprehensive approach for identifying the most sustainable procurement strategy, among various available alternatives, by considering the life cycle economic, environmental, and social impacts of strategies.

To address this gap, a decision-making framework for systematic comparison and ranking of various procurement and transport options in a construction, project based on sustainability criteria, was proposed. The framework establishes a comprehensive set of measures defining economic, social, and environmental aspects of sustainability, stipulated in regulatory, contractual, and functional requirements for infrastructure/building. It then relies on a set of databases, which provide input information to novel formulations/methods developed for computation of life cycle impacts associated with various procurement options. The sequence of decisions in procurement of materials includes type of materials, origin of supply, degree of prefabrication, and transport mode choice. The pillars of sustainability are balanced in the decision-making process through inputs received from project parties with dissimilar interests, where weight of measures are assigned on a transparent and quantifiable basis. The framework determines the ranking of different procurement alternatives by means of TOPSIS in which areas of strengths and weaknesses for each alternative are easily trackable.

Two case studies were presented to (1) demonstrate the application of the proposed framework and (2) demonstrate the importance of accounting for variations in the offsite transport requirements of different procurement strategies as an important selection criterion. By demonstrating the process of selection of the most sustainable curtain wall system for a residential building, the first case study highlights that procurement of materials in practice involves a series of interrelated decisions where there is usually no dominant choice in the hierarchy of decisions under all circumstances. The type of the material, as classified based on degree of prefabrication, was found to considerably affect the sustainability of the material choices. Furthermore, for a particular material with a given supply chain structure, local supply was found to be prioritized over international sourcing. The preferred mode of transport was, on the other hand, found to be dependent on upstream characteristics of materials, i.e., type and supply chain structure, and their impacts on logistics attributes including size and weight. In the second case study, the importance of accounting for transport requirement of different procurement options was further highlighted through evaluating the delay in supply of various types of materials in an oil-refinery industrial construction project. It was shown that relatively accurate estimates of transport duration can be made by considering the characteristics of materials, which can then be used as input into comparative sustainability analysis of different procurement options.

Further research is needed to improve accuracy of evaluating sustainable performance of materials particularly over operation and end-of-life stages of buildings and infrastructures. In line with this, future studies should focus on application of smart material technologies integrated with advanced asset management tools in order to provide reliable and accurate input to LCA of material supply decisions.

REFERENCES

Ahmadian, F.F.A., Akbarnezhad, A., Rashidi, T., Waller, S., 2015. Accounting for transport times in planning off-site shipment of construction materials. J. Constr. Eng. Manage. 142 (1), 04015050.

Ajzen, I., Fishbein, M., 1980. Understanding Attitudes and Predicting Social Behavior. Prentice-Hall, Englewood Cliffs, NJ.

Akbarnezhad, A., Moussavi Nadoushani, Z.S., 2014. A computational method for selection of optimal concrete recycling strategy. Mag. Concr. Res. 67 (11), 543–558.

Akbarnezhad, A., Ong, K.C.G., Zhang, M.H., Tam, C.T., 2013. Acid treatment technique for determining the mortar content of recycled concrete aggregates. J. Test. Eval. 41, 441–450.

Akbarnezhad, A., Ong, K.C.G., Chandra, L.R., 2014. Economic and environmental assessment of deconstruction strategies using building information modeling. Automat. Constr. 37, 131–144.

Akhavian, R., Behzadan, A.H., 2012. An integrated data collection and analysis framework for remote monitoring and planning of construction operations. Adv. Eng. Inform. 26, 749–761.

Altieri, M.A., 1987. Agroecology: The Scientific Basis of Sustainable Agriculture. Westview Press, Boulder, CO.

Ashby, M.F., Johnson, K., 2002. Materials and design: the art and science of material selection in product design. Elsevier, Burlington, MA.

Asokan, P., Osmani, M., Price, A.D.F., 2009. Assessing the recycling potential of glass fibre reinforced plastic waste in concrete and cement composites. J. Clean. Prod. 17 (9), 821–829.

Augenbroe, G., Pearce, A.R., Kibert, C.J., 1998. Sustainable construction in the United States of America: a perspective to the year 2010. CIB-W82 Report. Georgia Institute of Technology, Atlanta, GA.

Babu, A.S., 1999. Strategies for enhancing agility of make-to-order manufacturing systems. Int. J. Agile Manage. Syst. 1, 23–29.

Bank, L.C., Thompson, B.P., Mccarthy, M., 2011. Decision-making tools for evaluating the impact of materials selection on the carbon footprint of buildings. Carbon Manage. 2 (4), 431–441.

Bell, D.E., 1982. Regret in decision-making under uncertainty. Operat. Res. 30, 961–981.

Bern-Klug, M., 2010. Prospect theory and end-of-life decision-making in the nursing home context. Gerontologist 50, 57–57.

Bilal, M., Oyedele, L.O., Qadir, J., Munir, K., Ajayi, S.O., Akinade, O.O., et al., 2016. Big Data in the construction industry: a review of present status, opportunities, and future trends. Adv. Eng. Inform. 30, 500–521.

Bunz, K.R., Henze, G.P., Tiller, D.K., 2006. Survey of sustainable building design practices in North America, Europe andAsia. Arch. Eng. 12, 33–62.

Butter, I.H., 1993. Premature adoption and routinization of medical technology—illustrations from childbirth technology. J. Social Issue. 49, 11–34.

Calkins, M., 2009. Materials for Sustainable Sites: A Complete Guide to the Evaluation, Selection, and Use of Sustainable Construction Materials. John Wiley & Sons, Hoboken, NJ.

Ceuster, G., et al., 2008. Effects of adapting the rules on weights and dimensions of heavy commercial vehicles as established within Directive 96/53/EC. European Commission, Belgium.

Chang, Y., Ries, R.J., Lei, S.H., 2012. The embodied energy and emissions of a high-rise education building: a quantification using process-based hybrid life cycle inventory model. Energy Build. 55, 790–798.

Chen, Y., Okudan, G.E., Riley, D.R., 2010. Sustainable performance criteria for construction method selection in concrete buildings. Automat. Constr. 19, 235–244.

Cheung, C.M.K., Thadani, D.R., 2012. The impact of electronic word-of-mouth communication: A literature analysis and integrative model. Decision Support Syst. 54, 461–470.

Chiveralls, K., Zillante, G., Palmer, J., Zuo, J., Wilson, L. & Pullen, S. 2011. Cleaning up the construction industry. The Conversation.

CIC 2016. Global Construction Outlook 2020. Construction Intelligence Center.

CICA 2002. Industry as a Partner for Sustainable Development.

Cooper, I., 1999. Which focus for building assessment methods environmental performance or sustainability? Build. Res. Informat. 27, 321–331.

Curran, M.A. 2006. Life Cycle Assessment: Principles and Practice. National Risk Management Research Laboratory, Reston, VA.

Davis, F.D., 1989. Perceived usefulness, perceived ease of use, and user acceptance of information technology. MIS Quarter. 13, 319–340.

Despeisse, M., Kishita, Y., Nakano, M. & Barwood, M. 2015. Towards a circular economy for end-of-life vehicles: a comparative study UK–Japan. In: Proceedings of the 22nd Cirp Conference on Life Cycle Engineering, vol. 29, pp. 668–673.

Dissanayaka, S.M., Kumaraswamy, M.M., 1999. Comparing contributors to time and cost performance in building projects. Build. Environ. 34 (1), 31–42.

Duffy, A., Crawford, R., 2013. The effects of physical activity on greenhouse gas emissions for common transport modes in European countries. Transport. Res. Part D-Transport Environment 19, 13–19.

Emmitt, S., Yeomans, D.T., 2008. Specifying Buildings: A Design Management Perspective. Elsevier, Burlington, MA.

EN 15804:2012. Sustainability of Construction Works: Environmental Product Declarations Core Rules for the Product Category of Construction Products.

EN 15978:2011. Sustainability of Construction Works, Assessment of Environmental Performance of Buildings-Calculation Method.

Eng, S.W.L., Chew, E.P., Lee, L.H., 2014. Impacts of supplier knowledge sharing competences and production capacities on radical innovative product sourcing. Eur. J. Operat. Res. 232, 41–51.

Eriksson, P.E., 2010. Improving construction supply chain collaboration and performance: a lean construction pilot project. Suppl. Chain Manage. Int. J. 15, 394–403.

Fini, A.A.F., Rashidi, T.H., Akbarnezhad, A., Waller, S.T., 2016. Incorporating multiskilling and learning in the optimization of crew composition. J. Constr. Eng. Manage. 142 (5), 04015106.

Fini, A.A.F., Akbarnezhad, A., Rashidi, T., Waller, S., 2017. BIM-enabled sustainability assessment of material supply decisions. Eng. Constr. Arch. Manage. 24 (4), 668–695.

Frank, S.J., 2006. Building an IP strategy. Intellectual Property for Managers and Investors. Cambridge University Press, New York.

Golestanifar, M., Goshtasbi, K., Jafarian, M., Adnani, S., 2011. A multi-dimensional approach to the assessment of tunnel excavation methods. Int. J. Rock Mechan. Mining Sci. 48, 1077–1085.

Hammond, G.P., Jones, C.I., 2008. Embodied energy and carbon in construction materials. Proc. Instn Civil. Engrs: Energy 161 (2), 87–98.

Hwang, C.L., Yoon, K., 1981. Multiple Attribute Decision Making: Methods and Applications. Springer-Verlag, New York.

IHS 2015. Construction Database.

Irizarry, J., Karan, E.P., Jalaei, F., 2013. Integrating BIM and GIS to improve the visual monitoring of construction supply chain management. Automat. Constr. 31, 241–254.

Jato-Espino, D., Castillo-Lopez, E., Rodriguez-Hernandez, J., Canteras-Jordana, J.C., 2014. A review of application of multi-criteria decision making methods in construction. Automat. Constr. 45, 151–162.

Kahneman, D., Tversky, A., 1979. Prospect theory: an analysis of decision under risk. Econometrica 47, 263–291.

Kang, Y., O'brien, W.J., O'connor, J.T., 2015. Information-integration maturity model for the capital projects industry. J. Manage. Eng. 31.

Kazmierczak, K. Review of Curtain Walls, Focusing on Design Problems and Solutions. Building Enclosure Science and Technology (BEST) Conference (BEST 2), 2010, Portland, OR, pp. 1–20.

Kibert, C.J., 2008. Sustainable Construction: Green Building Design and Delivery. John Wiley & Sons, Hoboken, NJ.

Kim, J.J. & Rigdon, B. 1998. Sustainable Architecture Module: Qualities, Use, and Examples of Sustainable Building Materials. National Pollution Prevention Center for Higher Education, Ann Arbor, MI.

Kumaraswamy, M.M., Dissanayaka, S.M., 2001. Developing a decision support system for building project procurement. Build. Environ. 36, 337–349.

Lo, S.M., Zhao, C.M., Liu, M., Coping, A., 2008. A simulation model for studying the implementation of performance-based fire safety design in buildings. Autom. Constr. 17 (7), 852–863.

Lourenzutti, R., Krohling, R.A., 2016. A generalized TOPSIS method for group decision making with heterogeneous information in a dynamic environment. Informat. Sci. 330, 1–18.

Lozano, R., 2008. Envisioning sustainability three-dimensionally. J. Cleaner Product. 16, 1838–1846.

Luskin, D.M., Walton, C.M., 2001. Effects of truck size and weights on highway infrastructure and operations. Project Summary Rep. 2122-S. Center for Transportation Research. University of Texas, Austin, TX.

Martabid, J., Mourgues, C., 2015. Criteria used for selecting envelope wall systems in Chilean residential projects. J. Constr. Eng. Manage. 141.

Memari, A., 2013. New developments in curtain wall and glazing systems. In: Memari, A. (Ed.), Curtain Wall Systems. American Society of Civil Engineers, Reston, VA, pp. 194–209.

Moavenzadeh, F., 1991. Innovation in housing. Constr. Business Rev. 1, 60–77.

Nematchoua, M.K., Tchinda, R., Orosa, J.A., Andreasi, W.A., 2015. Effect of wall construction materials over indoor air quality in humid and hot climate. J. Build. Eng. 3, 16–23.

Nepal, M.P., Park, M., Son, B., 2006. Effects of schedule pressure on construction performance. J. Constr. Eng. Manage.-ASCE 132, 182–188.

O'brien, W.J., London, K., Vrijhoef, R., 2004. Construction supply chain modeling: a research review and interdisciplinary research agenda. ICFAI J. Operat. Manage. 3, 64–84.

Odijk, S.V. & Bovene, F.V. 2014. Circular Construction: The Foundation under A Renewed Sector. Circle Economy.

Olhager, J., 2003. Strategic positioning of the order penetration point. Int. J. Product. Econom. 85, 319–329.

Ortiz, O., Castells, F., Sonnemann, G., 2009. Sustainability in the construction industry: A review of recent developments based on LCA. Constr. Build. Mater. 23, 28–39.

Oxford Economics, 2015. Construction and Engineering: Analysis and Forecasts for the Construction Sector. Oxford Economics Aggregates.

Peeters, J.R., Vanegas, P., Dewulf, W., Duflou, J.R., 2017. Economic and environmental evaluation of design for active disassembly. J. Clean. Product. 140, 1182–1193.

Plank, R., 2008. The principles of sustainable construction. The IES Journal Part A: Civ. Struct. Eng. 1 (4), 301–307.

Potocnik, J. 2013. Towards the Circular Economy: Economic and business rationale for an accelarated transition. Ellen MacArthur Foundation.

Renz, A., Solas, M.Z., 2016. Shaping the Future of Construction Industry Institute: A Breakthrough in Mindset and Technology. Industry Agenda: World Economic Forum. Geneva, Switzerland.

Reza, B., Sadiq, R., Hewage, K., 2011. Sustainability assessment of flooring systems in the city of Tehran: an AHP-based life cycle analysis. Constr. Build. Mater. 25, 2053–2066.

Ritov, I., Baron, J., 1992. Status-quo and omission biases. J. Risk Uncertain. 5, 49–61.

Robinson, J., 2004. Squaring the circle? Some thoughts on the idea of sustainable development. Ecol. Econom. 48, 369–384.

Rogers, E.M., 1983. The Diffusion of Innovations. The Free Press, New York, NY.

Rogers, E.M., 2003. Diffusion of Innovations. Simon and Schuster, New York, NY.

Rumane, A.R., 2011. Quality Management in Construction Projects; Monitoring and Control. CRC press, Taylor and Francis Group.

Saaty, T.L., 1980. The Analytic Hierarchy Process. McGraw-Hill, New York.

Said, H., El-Rayes, K., 2011. Optimizing material procurement and storage on construction sites. J. Constr. Eng. Manage. ASCE 137, 421–431.

Sambasivan, M., Soon, Y.W., 2007. Causes and effects of delays in Malaysian construction industry. Int. J. Project Manage. 25, 517–526.

Sartori, I., Hestnes, A.G., 2007. Energy use in the life cycle of conventional and low-energy buildings: a review article. Energy Build. 39, 249–257.

Seghezzo, L., 2009. The five dimensions of sustainability. Environ. Polit. 18, 539–556.

Shakantu, W., Tookey, J., Bowen, P., 2003. The hidden cost of transportation of construction materials: an overview. J. Eng. Design Technol. 1, 103–118.

Sirisalee, P., Ashby, M.F., Parks, G.T., Clarkson, P.J., 2004. Multi-criteria material selection in engineering design. Adv. Eng. Mater. 6 (1/2), 84–92.

Slaughter, E.S., 1993. Builders as sources of construction innovation. J. Constr. Eng. Manage.-ASCE 119, 532–549.

Spiegel, R., Meadows, D., 2010. A Guide to Product Selection and Specifications Green Building Materials. John Wiley & Sons, Hoboken, NJ.

Su, B., Huang, H.C., Ang, B.W., Zhou, P., 2010. Input-output analysis of CO_2 emissions embodied in trade: the effects of sector aggregation. Energy Econom. 32, 166–175.

Sull, D.N., 1999. Why good companies go bad. Harvard Busin. Rev. 77 (4), 42–57.

Thangavelu, P. 2015. Companies that went bankrupt from innovation lag [Online]. Investopedia. Available from: <http://www.investopedia.com/articles/investing/072115/companies-went-bankrupt-innovation-lag.asp> (accessed 16.10.16).

The World Commission on Environment and Development, 1987. Chapter 2: Towards Sustainable Development, Our Common Future. Oxford University Press, UK, p. 41.

Thormark, C. Energy and resources, material choice and recycling potential in low energy buildings. In: CIB Conference SB07 Sustainable Construction, Materials, and Practices, 2007 Lisbon, Portugal.

Tommelein, I., Ballard, G., Kaminsky, P., 2008. Chapter 6: Supply Chain Management for Lean Project Delivery in Construction Supply Chain Management Handbook. CRC Press, Boca Raton, pp. 6/1–6/22.

Toole, T.M., 1998. Uncertainty and home builders' adoption of technological innovations. J. Constr. Eng. Manage.-ASCE 124, 323–332.

Toole, T.M., Tonyan, T.D., 1992. The adoption of innovative building systems: a case study. Build. Res. J. 1, 21–26.

Treloar, G.J., Love, P.E.D., Faniran, O.O., Iyer-Raniga, U., 2000. A hybrid life cycle assessment method for construction. Constr. Manage. Econom. 18, 5–9.

Werkheiser, I., Piso, Z., 2015. People work to sustain systems: a framework for understanding sustainability. J. Water Resour. Plan. Manage. 141.

Williams, T.P., Gong, J., 2014. Predicting construction cost overruns using text mining, numerical data and ensemble classifiers. Automat. Constr. 43, 23–29.

Wong, J.K.W., Li, H., 2008. Application of the analytic hierarchy process (AHP) in multi-criteria analysis of the selection of intelligent building systems. Build. Environ. 43, 108–125.

Yu, W.D., Cheng, S.T., Wu, C.M., Lou, H.R., 2012. A self-evolutionary model for automated innovation of construction technologies. Automat. Constr. 27, 78–88.

Yu, W.D., Wu, C.M., Cheng, S.T., Chen, T.S., 2013. Enhanced function modeling for early assessment of conceptual innovative construction technologies. Automat. Constr. 36, 180–190.

Zahir, S., 2015. Approaches and Associated Costs of Building Demolition and Deconstruction. Masters' thesis. Michigan State University, East Lansing, Michigan.

Zhang, M., Janic, M., Tavasszy, L.A., 2015. A freight transport optimization model for integrated network, service, and policy design. Trans. Res.E Logist. Transport. Rev. 77, 61–76.

Zhou, C.C., Yin, G.F., Hu, X.B., 2009. Multi-objective optimization of material selection for sustainable products: artificial neural networks and genetic algorithm approach. Mater. Des. 30 (4), 1209–1215.

Zimmermann, M., Althaus, H.J., Haas, A., 2005. Benchmarks for sustainable construction–a contribution to develop a standard. Ener. Build. 37, 1147–1157.

Chapter 8

Sustainable Water Use in Construction

Muhammad Muhitur Rahman[1], M Ashiqur Rahman[2], Md Mahmudul Haque[3] and Ataur Rahman[2]

[1]*Department of Civil and Environmental Engineering, King Faisal University, Al Hofuf, Saudi Arabia,* [2]*School of Computing, Engineering and Mathematics, Western Sydney University, Sydney, NSW, Australia,* [3]*EnviroWater, Sydney, NSW, Australia*

8.1 INTRODUCTION

Water is the vital source of life on this planet. Human civilization is evolved around water. Our earliest ancestors used to live in cottages made of leaves, bamboos, and tree branches; these habitats had little water demand. In contrary, a modern building has much larger water footprint, e.g., construction materials used in making a modern building are generally highly processed that require significant quantity of water (Bardhan, 2011; Crawford and Treloar, 2005). Also, a modern building needs continuous water supply to meet numerous demands (e.g., drinking, washing, waste disposal, cooling, and swimming). At the same time, a modern building generates a large volume of stormwater runoff from the roof, driveway, and connecting roadways that can pollute our waterways (Landcom, 2004). Mining and construction activities need large volume of water for day-to-day operation. Furthermore, the wastewater, mine tailings, stormwater runoff and water from dewatering activities present notable water pollution problems. For example, arsenic from mine tailing can contaminate adjacent groundwater supply. Similarly, water runoff containing chromium from mine tailings can have serious consequence on the environment and public health due to its high solubility in water (Achal et al., 2013).

A typical building is associated with water use in direct and indirect forms. Direct form includes water consumed by workers, water used in washing aggregates, preparing raw concrete, curing concrete, dust suppression, and washing of hard surface and equipment. Indirect use is related to embodied water, which has been used in production of construction materials. A typical building also needs water during its operation, e.g., drinking

Sustainable Construction Technologies. DOI: https://doi.org/10.1016/B978-0-12-811749-1.00006-7

water, water for toilet flushing, gardening, washing, and recreational and cooling activities; in the planning phase, necessary measures should be adopted to reduce this type of water use (Ding, 2008). Wastewater is also being generated from the construction sites, which needs to be captured and treated before disposal to natural water bodies. Furthermore, water use in mining activities is a major component of water management as they need significant volume of water for operational purposes and also with respect to wastewater and stormwater runoff generated at construction sites (Dharmappa et al., 2000). Nevertheless, due to the call for sustainable use of water and regulatory requirements, various techniques and tools have been proposed in the last two decades to reduce water use in construction and mining sites and infrastructure operation.

Engineers, town planners, and architects should design buildings that are more environmentally sustainable, i.e., use minimum water and energy that can ultimately reduce greenhouse gas emissions. This may easily be achieved by a life cycle analysis that can identify the cost and impact of individual components of a building (Dewulf et al., 2009). A concept of zero energy buildings has drawn significant attention in recent time (Sartori et al., 2012). For example, in Europe it is expected that all the new buildings should be "nearly zero energy buildings" by 2012 (Recast, 2010). The United States is targeting marketable zero energy homes in 2020 and commercial zero energy buildings in 2025 (US DOE, 2010). These zero energy homes must be water efficient as water use is closely related to energy, e.g., pumping of water needs energy. Another concept known as "green buildings" has become quite popular. A green building targets at least 20% overall decrease in drinkable water use as compared to a traditional building (Kibert, 2004). For example, Cheng et al. (2016) found that green buildings in Taiwan can save on average 37% of water compared to the baseline water usage rate where water savings mainly arise from water-saving appliances, and building design that promotes water savings and rainwater utilization.

This chapter presents various forms of usage of water in construction activities and relevant sustainable management procedures for conserving and reducing wastage of water in buildings, industries, and mines. It also discusses recently developed sustainable tools and techniques that have been applied to save water usage in buildings.

8.2 SOURCES OF WATER USED IN CONSTRUCTION

Water for construction can be derived from a number of sources. One of the most common sources is groundwater, which is pumped from aquifers and used for construction activities. Groundwater is mainly used for mining construction activities in remote areas, where piped water and readily available surface water are not available. Use of groundwater involves pumping, which needs energy that is associated with greenhouse gas emissions. Another source of water is surface water (e.g., river and lake). Recycled water is also used in construction, e.g., preparing concrete mix using secondary and tertiary treated domestic wastewater (Silva and Naik, 2010).

ASTM C1602 (Standard Specification for Mixing Water Used in the Production of Hydraulic Cement Concrete) specifies waters, which are used for preparing raw concrete as follows:

Batch water: Water for preparing raw concrete is derived from city water, reclaimed city water, and water coming out of concrete production (e.g., washing of aggregates and curing of concrete).
Ice: It can be used in preparing raw concrete during hot weather. It should be noted that ice must be fully melted during the mixing process.
Water provided by truck operator: ASTM C94 (AASHTO M 157) allows water to be added on site if the sump is smaller than the specification by maintaining the allowable water–cement ratio.
Free moisture in aggregates and admixtures: Water contained in aggregates and admixtures must meet the criteria of allowable mixing water, e.g., water does not contain harmful ingredients.
Recycled water: Water that is unsuitable for drinking can be used in preparing raw concrete provided this water meets the criteria of mixing water (as per ASTM C1602).

Besides, "embodied water" associated with the preparation of construction materials needs to satisfy specific criteria. For example, water used in preparing cement and steel would have different properties. The use of such water in construction industry is discussed in the following section.

8.3 USAGE OF WATER IN CONSTRUCTION

Water is one of the essential components in construction, which is required in varying amount depending on the type, scale, and stages of a project. Examples include dust suppression during excavation phase of a civil engineering construction, compaction of soil, soaking of coarse aggregates, preparation of mortar, mixing of cement concrete, curing of concrete, washing of equipment, cleaning of construction sites, demolition (by water jet), landscaping, firefighting, and human consumption. Besides, a considerable volume of wastewater and stormwater is usually generated in a construction site, which often needs treatment before being reused or disposed to the natural environment.

Significant volume of water, termed as "embodied water," is used in the manufacturing process of numerous building materials such as steel, brick, cement, aluminum, glass, and carpet. In the manufacturing process of these materials, water may be used as a lubricant, cleaning agent, sealant, heat-transfer medium, solvent, air pollution control means, and range of other purposes depending on materials being manufactured (Thornback et al., 2015).

In the Australian perspective, the highest value of embodied water for a medium-rise office building construction was assessed as 20.1 kL/m^2 of gross floor area (McCormack et al., 2007). Such analysis also reveals that the water embodied in the actual construction process is considerably smaller (only 8% of the total embodied water of the materials together) compared to the water embodied in the major building materials (Bardhan, 2011). In another study (conducted in Melbourne, Australia) on a commercial building construction site with a gross floor area of 11,600 m^2, the water embodied in the building was found to be 54.1 kL/m^2 based on the input–output-based hybrid analysis technique (Crawford and Treloar, 2005). Besides, Bardhan (2011) estimated the embodied water to be about 27.6 kL/m^2 of built up area for a typical multistoried apartment building in India.

Moreover, for certain activities in construction such as mixing and curing of concrete, use of clean and contaminant free water is important. Potable water is often used in preparation of mortar and fresh concrete. Therefore, it is important to optimally use water in construction sites and to reduce any possible misuse through promoting effective water saving measures. For instance, the use of water reducing admixtures in fresh concrete and mortar preparation can reduce up to 10% water consumption, while the admixtures increase the workability of the concrete and mortar.

Although construction industry is considered to be one of the highest users of water along with energy and material resources, limited data are available on water use for specific purposes and little is known regarding how water use for a particular purpose varies during the life cycle of a construction site or a product (Guggemos and Horvath, 2006; Waylen et al., 2011). Bekker (1982) suggests that life cycle assessments (LCA) should be conducted considering the activities such as manufacturing of construction materials, construction, operation, maintenance, refurbishment, and demolition in order to quantify and analyze the embodied water content of commodities and products used in construction.

8.4 ASPECTS OF WATER MANAGEMENT STRATEGIES IN CIVIL CONSTRUCTION SITES

In order to minimize the impacts of construction industry on water resources in a given country/region, it is often recommended to select more environmental friendly and sustainable planning and design option during the project formulation and appraisal phases when the water conservation measures could be best incorporated, rather than at the later phase of the construction project (Ding, 2008). Hence, construction methods and building system design should be aligned well for effective use of water resources from the very planning stage of the project. This is particularly important in reducing the water use during the operation phase of the building, e.g., water to be used in toilet flushing.

Various provisions for water conservation such as use of water-saving fixtures, installation of rainwater harvesting system, and installation of a graywater system in a building to reduce potable water consumption need to be taken into consideration at the planning stage of the project. Moreover, special consideration should be given to finding measures for minimizing water consumption at the material production stage. Bardhan (2011) has indicated that the industries have to be more accountable regarding usage of fresh water for manufacturing building materials and should look for alternate ways to reduce the volume of embodied water through water reuse and recycling.

Thornback et al. (2015) identified four broad groups of construction products that can facilitate better water management in construction industry. These include plumbing fixtures that help reducing water use in buildings, the systems that enable home and building occupiers to recycle or use alternative sources to mains water, the product systems that channel or soak up rainwater runoff, and the products that help other product manufacturers reducing water use in the manufacturing processes. Construction industry can undertake appropriate strategies to adopt these products and technologies in building infrastructures and can attain the goal of water efficiency.

The issue of water pollution management in construction sites is another desired component of building industry. In order to minimize environmental impacts resulting from water pollution in construction activities, environmental assessment tools for construction must specifically be designed with due consideration to this aspect. In this regard, Tam et al. (2004) proposed the *Green Construction Assessment* model for Hong Kong, which included the issue of "water pollution control" as one of the 13 performance indicators for construction. Monitoring of water usage and promotion of water conservation, water reusing and recycling system, wastewater collection and treatment systems, and other measures related to water pollution are the proposed subindicators in this model.

The State of New South Wales (NSW) in Australia has provided a comprehensive guideline for construction water quality monitoring, which is often recommended for road works in NSW to control water pollution related environmental degradation during preconstruction, construction and postconstruction phases. Application of such management instruments in order to control water pollution in construction sites could also be effective for other parts of the world. Further, Houser and Pruess (2009) postulate that when construction projects apply appropriate best management practices, they exhibit minimal impact on overall water quality of adjoining water bodies.

In many construction sites, "dewatering" is often needed. This process is usually conducted by pumping construction wastewater from a hole, excavation, basement, trenches, footings, piers, or sediment trap (Lake Macquarie City Council, 2015). Certain water quality characteristics should be met before the wastewater is removed from a construction site to prevent any water pollution in the receiving water bodies According to Lake Macquarie City Council (2015), dewatered water from a construction site should meet the criteria that the pH is between 6.5 and 8.5, total suspended solid (TSS) is less than 50 mg/L and electrical conductivity is less than 200 μs/cm. Prior to dewatering, the water must be appropriately treated and filtered to eliminate major contaminants from water including sediments, oil, grease, paint, and acids. Generally, dewatering to the stormwater system at the construction sites is not allowed before acceptable treatment and filtration. Construction sites should also have appropriate flood protection measures so that the construction wastewater does not mix with local stormwater (Lake Macquarie City Council, 2015).

Several countries have put considerable efforts in developing methods for environmental building performance assessment such as BREEAM (United Kingdom); BEPAC (Canada); CEPAS (Hong Kong); GHEM (China); and GreenStar, BASIX, and AccuRate (Australia). Ding (2008) has conducted an analysis on the characteristics of many such methods; from this, it is found that only BASIX has put explicit focus on water efficiency issues in construction and building maintenance purposes. However, other methods may have indirectly addressed this issue under environmental performance in general.

Australia's web-based online planning assessment tool called the BASIX (Building Sustainability Index) was developed by the NSW Government for the construction industry. It aims to make all new residential houses in NSW energy and water efficient. BASIX is a mandatory part of all development application process and is evaluated through an online assessment tool, which checks the elements of a proposed design against sustainability index. The water section of BASIX targets reduction of potable water consumption of all the new residential houses in NSW. The benchmark was set as 247 L/person/day, which was the average potable water consumption of a pre-BASIX home and reduction target ranges from 40% and 0% subject to the climatic zone of NSW (NSW Government, 2016). In this system, the water consumption and savings in new residential developments are assessed on the basis of water requirements for landscaping, fittings and fixtures, pools and outdoor spas, central systems (such as cooling towers, fire sprinklers) and common areas, and provisions of alternative (non-potable) water supplies.

8.5 SUSTAINABLE MANAGEMENT OF RUNOFF FROM CONSTRUCTION SITES

Quality of "construction runoff" from erection sites are usually inferior to the stormwater that are generated from prevailing developed areas. The quality of construction runoff varies depending on type, extent, and complexity of construction activities. Studies have shown that sediments generated from an area under building and construction development could be 5–20 times greater than the sediment pollution in an existing developed area (Landcom, 2004). Hence, number of pollutants generated from the construction sites should not be discharged into the stormwater system or on the land where they may enter into the stormwater system. The pollutants which are most prevalent in the building and construction industry include (EPA-SA, 2015):

- Wastewater from brick, bitumen, and concrete cutting;
- Wash water from buildings;
- Wash-down water from cleaning equipment;
- Waste from high pressure water blasting;
- Waste from building construction or demolition;
- Waste from roof cleaning;
- Wash water from paintings, paint scraps, stripped paints, solvents, and varnish;
- Water and waste from plaster;
- Sawdust;
- Sand, gravel, and clay particles from soil erosion.

Some typical values of contaminants observed in urban stormwater are listed in Table 8.1. Moreover, several organic substances such as diuron, carbendazim, mecoprop, and phthalates are also observed in the construction runoff (Burkhardt et al., 2011). The organic substances in construction runoff mainly come from additives used in construction materials, such as bitumen for rooftop. These organic additives are generally used in the materials to control growth of unwanted microorganisms and to reduce deterioration of construction material through microbial activity.

Sustainable management of construction runoff is an integral part of the overall construction management practiced by the construction industry in developed world. If not properly managed, the runoff from a construction site may reach into stormwater system and increase the pollution load on stormwater treatment system. In addition, coarse sediments may deposit in stormwater drainage pipes and culverts reducing the design flow of stormwater in the network system. Deposition of coarse sediments may also occur in creeks and wetlands leading to increased flood risk causing damage to wetland aquatic life, and increasing the concentration of nutrients in the receiving water bodies. Finer sediments from construction runoff make the receiving water turbid and muddy, and undermine the esthetic appeal of constructed wetlands, especially in urban areas.

In Australia, construction runoff is managed by implementing different measures, which typically include (Landcom, 2004; EPA-ACT, 2011; Witheridge, 2012) (1) stockpile management, (2) drainage control, (3) erosion control, and (4) sediment control.

TABLE 8.1 Quality of Typical Untreated Urban Storm Water

Contaminants	Units	Values[a]
pH		6.35 ± 0.54
Turbidity	NTU[b]	50.93 ± 40.46
Suspended solids (SS)	mg/L	99.73 ± 83.60
Total dissolved solids (TDS)	mg/L	139.6 ± 17.30
Total organic carbon (TOC)	mg/L	16.90 ± 3.33
Alkalinity (as $CaCO_3$)	mg/L	35.21 ± 3.36
Oil and grease	mg/L	13.13 ± 8.11
Biochemical oxygen demand (BOD)	mg/L	54.28 ± 45.58
Chemical oxygen demand (COD)	mg/L	57.67 ± 17.22
Total nitrogen (TN)	mg/L	3.09 ± 2.33
Total Kjeldahl nitrogen (TKN)	mg/L	2.84 ± 4.14
Total phosphorus (TP)	mg/L	0.48 ± 0.413
Reactive phosphorus (DRP)	mg/L	0.664 ± 0.762
Aluminum (Al)	mg/L	1.19 ± 0.60
Iron (Fe)	mg/L	2.842 ± 1.246
Manganese (Mn)	mg/L	0.111 ± 0.046
Mercury (Hg)	mg/L	0.218 ± 0.105
Zinc (Zn)	mg/L	0.293 ± 0.153

[a] *Mean ± Standard deviation.*
[b] *Nephelometric turbidity unit.*
Source: NRMMC-EPHC-NHMRC, 2009. Australian Guidelines for Water Recycling: Managing Health and Environmental Risks (Phase 2)—Stormwater Harvesting and Reuse. ISBN 1921173459.

Stockpile management includes measures to prevent wind erosion and washing of soil and sand mainly from stockpiles of excavated earthwork and topsoil, and sand used for construction activities. Sand stockpiles should be covered with synthetic covers when wind erosion is an issue. For long-term (more than 28 days) stockpiles of soil, covering with mulch, chemical stabilizers, soil binders, or impervious blankets are recommended, provided runoff from the stockpile is directed to a sediment trap.

Temporary drainage systems are generally implemented in construction sites to control unauthorized diversion of construction runoff to stormwater system. Some common drainage control techniques include catch drain, diversion channel, level spreader, recessed rock check dam, sandbag check dam, chute, flow diversion bank, slope drain, and temporary downpipe

(Landcom, 2004). Sometimes, linings are used in channels and in chutes to reduce erosion. Examples of such lining techniques include geosynthetic lining, grass pavers, rock lining, erosion control mat, and grass lining.

Selection of erosion control measures at a given construction site depend son soil erodibility, steepness of land, existing soil cover, and the surface flow condition, e.g., velocity, duration, and frequency of the flow. Typical erosion control measures for a flat land (slope $<$ 1 in 10) may be achieved by using erosion control blankets, mulching, and revegetation. Measures for a mild sloped land (slope between 1 in 10 and 1 in 4) may include compost blankets, mats and mesh, bonded fiber matrix, and anchored mulching. For steeper land (slope $>$ 1 in 4) rock armoring, cellular confinement systems, and turf are generally used.

Coarse sediments in the construction runoff are usually trapped by sediment control techniques. However, some sediment traps can capture fine sediments and able to reduce the turbidity levels in the construction runoff. Examples of some sediment control techniques include coarse sediment trap, filter fence, sediment weir, sedimentation basin, and sediment trench. In some construction sites, sediments are also controlled in the entry and exit to the site. Some examples include rock pad, vibration grid, and wash bay.

In Australia, there is provision in the law to fine a construction company if the construction materials or excavated soils are washed into stormwater systems. For example, Brisbane City Council issues on-the-spot fines if a construction company does not take necessary measures in controlling erosion that can range from 20 to 50 penalty units (one unit is equivalent to AUD 121.90 as at July 1, 2016) (Brisbane City Council, 2016). In Australian Capital Territory (ACT), according to *Environmental Protection Act 1997*, such fine ranges between AUD 200 (for an individual) and AUD 1000 (for a company), and for more serious offenses the penalties may lead up to AUD 50,000, 6 months jail and a criminal record (EPA-ACT, 2011).

On-site construction runoff can be reused within the site for different activities including dust suppression, soil compaction, and for watering landscape. However, the quality of reused water should meet certain quality standard, which can be achieved by simple treatment strategies. According to Transport for NSW (2015), construction runoff can be reused for on-site activities if:

- the runoff shows no evidence of the presence of oil and grease;
- the pH level of the runoff is between 6.5 and 8.5; acidic or alkaline runoff can be adjusted to the mentioned range using appropriate base and acid, respectively;
- the total suspended solid (TSS) is less than 50 mg/L; increased TSS level in the runoff can be reduced using suitable flocculants (e.g., gypsum) or filtering through a suitable filter media (e.g., geofabric).

8.6 SUSTAINABLE WATER CONSERVATION MEASURES IN CONSTRUCTION INDUSTRY

In recent years, water usage in new buildings has been substantially reduced as a result of the application of a wide range of technologies and measures. Box 8.1 shows some measures to achieve sustainable water conservation. Now-a-days, installation of water-efficient fixtures instead of conventional high water-flow fixtures ensures conservation of potable water in buildings. Adoption of these measures have incurred some additional initial capital investment, though reports show that in most cases the payback period for this investment was less than a year, especially when the fixtures are frequently used (Spigarelli, 2012).

In the United States, the National Energy Policy Act-1992 has mandated the use of water-conserving plumbing fixtures. Table 8.2 summarizes the reduction of water use in the US construction industry since 1992 through the usage of low-flow plumbing fixtures. Further, it has been estimated that through moderate improvements (approximately 40%) in water efficiency, the federal government of the United States would save as much as US$240 million per year, which could be invested to supply enough water to a population of approximately 1.8 million (Bourg, 2016).

In order to increase water efficiency in residential buildings, Australia mandated WELS scheme in 2005, which has established a compulsory and consistent star labeling system for plumbing fixtures. As per the WELS all

BOX 8.1 Tips for Achieving Sustainable Water Conservation in Construction Industry

- Selecting the highest Water Efficiency Labelling Standards (WELS) rated fittings and fixtures available for the development, including efficient showerheads, dual flush toilets, and flow regulators in the taps.
- Installing an alternative water supply, such as a rainwater tank and connecting it to internal water systems (e.g., toilets and/or laundry).
- Installing an alternative water supply for gardening and using low water requirement plant species as part of the landscaping plan.
- Installing shades and a permanent covers for any pool or outdoor spa.
- Implementing additional options for multidwelling apartments (such as storm/rainwater collection for toilet and garden use and low water use landscaping).
- Conducting water audit, leak detection, and repair to avoid water loss.
- Using innovative ingredients (e.g., plasticizers) in concrete that reduces water requirement in raw concrete without reducing desired strength.

Source: After NSW Government, 2016. BASIX. Retrieved from: <https://www.basix.nsw.gov.au/basixcms/water-help/water-targets.html> (accessed 18.11.16).

TABLE 8.2 Reduction of Water Usage in the United States From Water-Efficient Plumbing Fixtures

Fixture type	Usage rate		
	Pre-1992 (conventional technology)	Post-1992 (high-efficiency technology)	% reduction since pre-1992
Water closet (lpf)	13.25	4.85	63.43
Urinal (lpf)	7.58	0.5	93.75
Shower (lpm)	20.82	6.6	68.18
Faucet (lpm)	11.36	1.89	83.33

lfp, Litre per flush; Lpm, Litre per minute.
Source: After Spigarelli, M., 2012. 10 Ways to Save Water in Commercial Buildings. CCJM Engineers Ltd., Chicago, IL. Retrieved from: <https://www.csemag.com/single-article/10-ways-to-save-water-in-commercial-buildings/8f74baabfcc8f672483b3b0353ccad16.html> (accessed 19.11.16).

toilets, showerheads, taps, clothes washers, and dishwashers are registered and rated in accordance with Australian and New Zealand Standard AS/NZS6400:2005, and fixtures are labeled to display their water efficiency before they are supplied for sale in the Australian market. The label exhibit zero to six-star rating that helps to do a quick comparative assessment of the water efficiency of the product and provides a figure relating to the water consumption or flow rate of the product, based on laboratory tests (WELS, 2016). Builders/owners can easily select the most water efficient fixtures for installation in buildings from such labels, which would reduce water usage rate (thus water and energy bill) significantly during the operation phase of a building.

Moreover, Australian state and local governments have introduced a series of regulations for reducing water consumption in buildings through mandating rainwater tanks (Eroksuz and Rahman, 2010) and a minimum of three-star water fittings in all new houses. The Australian National Housing Industry Association has also been promoting the installation of water efficient fixtures and fittings in existing houses.

Taiwan began implementing the "Green Building Label" for its construction industry since 1999. During the first 4 years' time, the country could award the label to over 20 building projects, while the number of applications for the award continued to grow (Cheng, 2003). This Green Building evaluation process put emphasis on water conservation issues and uses evaluation indices such as water usage (L/day) and adoption rate of water saving equipment. Cheng (2003) recommended the use of two-sectioned water saving toilets, water-saving showering devices without a bathtub, auto-sensor flushing device systems, and rainwater and intermediate graywater systems to be the effective ways of saving water in Taiwanese perspective.

In the United Kingdom, companies like Hanson that produces cement, concrete, aggregates, and asphalt for building industry has implemented a structured approach to water management. Thornback et al. (2015) report that in 2009 this company set business line targets to reduce water consumption for manufacturing building products. In order to achieve the targets, it employed specialist consultants and conducted activities such as water audit, leak detection and repair, bill validation and data logging. In 3 years, Hanson reduced overall water consumption from 10,192 to 9,266 ML. A rainwater harvesting system was also installed in its factory, which provided a double benefit in terms of reducing water use and in cooling the condensate steam (Thornback et al., 2015).

There are also examples of best management practices where products are being developed that allow other manufacturers and construction contractors to reduce water use in their activities. For instance, BASF-UK has produced a product (MasterGlenium superplasticizers) that enable concrete producers to significantly reduce (by up to 40%) water requirement to achieve a given consistency in fresh concrete whilst increasing strength, improving durability and reducing permeability (Thornback et al., 2015). Similarly, the Porotherm (a clay block walling) system invented by Wienberger-UK has also been offering water savings in the construction industry. Thornback et al. (2015) observed that the specially designed mortar used for Porotherm walls saves up to 95% water compared to traditional mortar.

8.7 MINING AND WATER USE

Water is an essential element in the mining industry. Water in mining industries is needed for different operational activities including transporting ore and waste in slurries and suspension, in cooling systems around power generation, for suppressing dust during mineral processing and around conveyors and roads, and for washing equipment (Prosser, 2011). Sustainable management of water in a mine is crucial for the safety of workers and surrounding environment. Attention is required for mines situated in remote arid areas, where water resource is limited and only proper management can ensure smooth operation of mining activities. Such management options include handling stormwater within the mining area, treatment of runoff water generated from mining activities, supplement to water intake through effluent reuse scheme, and sustainable disposal of effluents to environment. Prevention of groundwater contamination (e.g., due to seepage of mine tailings) at mine sites is also an important issue at many mine sites. Besides, accidents may occur from collapsing of dams in mine sites if they are not effectively managed. More recently, dam failure at a mine site in Brazil has caused significant financial and environmental damage including killing of 19 people (The Wall Street Journal, 2016).

8.7.1 Water Usage by Mining Activities

Developed countries like Australia explore conventional and nonconventional sources of water to use in the mining industries. In Australia, state-based water authorities extract water from rivers, lakes, dams, and groundwater, and supply to mining industries. Water authorities also supply treated wastewater and recycled water to mining industries. In addition to water supplied by the state water authorities, many mining companies extract ground water by themselves to use in the mining activities (ABS, 2014). Similar scenario was observed in the Unites States of America, where in 2010, total water withdrawal for mining purposes was 5.32 Billion gal/d. A major portion of this water (about 73%) was supplied by the groundwater (Maupin et al., 2014).

Amount of water used by mining industries depends on the nature and type of mining activity. For example, water usage in a coal mine is different to the usage in metallic and nonmetallic mines. Again, the quantity of water usage increases with the increased production of minerals. Fig. 8.1 shows yearly average water consumption by different mining industries in Australia. As shown in the figure, metal ore and coal mining in Australia are the highest consumers of water compared to other mining industries. Most of the mines in Australia are situated in Western Australia, which is the highest consumer of water among all the states of Australia, followed by Queensland and New South Wales.

8.7.2 Characteristics of Water Produced by Mining Activities

Quality of water generated from mining activities depends on the type of mine. According to Dharmappa et al. (2000) water from an underground coal mine can be classified as (1) mine water, (2) process water, (3) storm water, and (4) domestic wastewater.

Mine water is generated from seepage in the excavated areas of the mine and is generally collected in underground sumps. Mine water can be used for firefighting and dust suppression within the mining complex. The quantity and quality of mine water varies from mine to mine and depends on the depth to ground water table and local soil conditions. Generally, mine water contains higher total dissolved solids (500–2000 mg/L), considerable amount of hardness (500–2000 mg/L as $CaCO_3$), low BOD (5 mg/L) and COD (10–100 mg/L), high conductivity (0.6–10 dS/m), and moderate concentrations of trace elements (Dharmappa et al., 1995; Cohen, 2002).

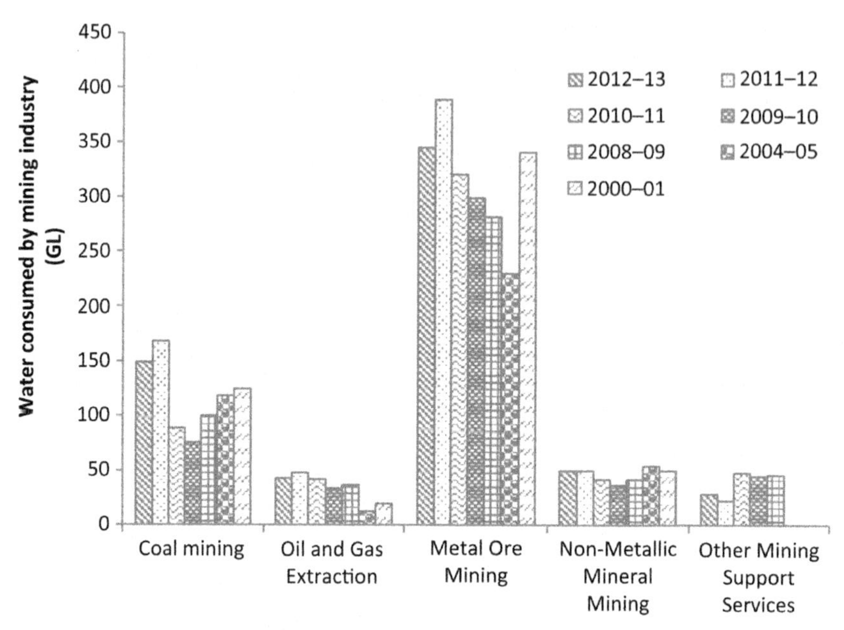

FIGURE 8.1 Water consumption by different mining industries in Australia. Sources: *ABS, 2006. Water Account, Australia 2004–2005. Canberra, Australia: Australian Bureau of Statistics; ABS, 2014. Water Account, Australia 2012–2013. Canberra, Australia: Australian Bureau of Statistics.*

The process water contains oil, coal particles, and other solid materials. Generally, sources of process water include coal conveyor, washing facilities in workshop, oil storage, diesel filling area, and on-site coal processing facility. The process wastewater contains close to neutral pH (8.24–8.87), moderate dissolved solid as TDS (140–801 mg/L), and high net filterable residues, NFR (400–13,650 mg/L) (Sivakumar et al., 1994; Wingrove, 1996).

The characteristics of stormwater runoff produced by moderate to heavy rainfall in a colliery largely depend on the type of the mining activities and on the land use of the exposed area at a mine site. According to Sivakumar et al. (1994), the suspended solids content in the storm water is considerably high (400–600 mg/L), whereas the dissolved solids are moderate (200–300 mg/L).

In a colliery, domestic wastewater constitutes mainly of graywater and blackwater. Sources of graywater and blackwater are bathhouse and kitchen, and toilet, respectively. The graywater generated from bathhouse is mainly contaminated by coal fine from the mine workers and soaps used in their showering. Graywater from bathhouse may contain moderate total solid (358–812 mg/L), net filterable residues (4–157 mg/L), and oil and gas (13–123 mg/L) (Wingrove, 1996; Dharmappa et al., 2000; Mishra et al., 2008).

8.7.3 Impact of Mining Activities on Water Resources and Remedial Options

Water that is not treated or reused within the mine, e.g., runoff and seepage may reach the surface water and seep into ground water from abandoned mine workings, waste rock, dump piles, tailings and flue dust pile (Street, 1993), and pollute water resources. In surface mining, runoff may get polluted due to the interaction between water and mineral ore itself. In addition, mining can expose previously unexposed and unweathered pyritic (iron sulfide, FeS_2) surfaces through blasting, and that exposure can result in acid mine drainage containing sulfuric acid (Pondja Jr et al., 2014).

Therefore, sustainable management options should be devised to avoid adverse effect of mining activities on water resources. Such management options are not straight forward and varies depending on mining activities (Rahman et al., 2016). For the purpose of on-site reuse of mine water, application of a combination of different treatment techniques brings better results than availing a single solution. Dharmappa et al. (1995) proposed a combination of conventional treatment process (e.g., coagulation, flocculation, and sedimentation) with a membrane process (e.g., microfiltration, ultrafiltration, or reverse osmosis) for coal mine water (Fig. 8.2). However, economic viability must be considered for implementing such treatment combination. In addition, biological treatment of mine tailings (bio-remediation) is also reported by some researchers (i.e., Johnson and Hallberg, 2005).

Runoff from mining operation should be managed sustainably. According to Singh et al. (1998), if not properly managed, runoff from mine area may overload water treatment dam, and pollute the already treated water. Management of runoff can be achieved by implementing catch drains, interceptor dykes, berms, open channels, and pipelines (Street, 1993). Increasing the vegetative ground cover will reduce quantity of soil particles to be picked up by runoff. Material storage areas (e.g., washery and workshop) can be enclosed with curbing barriers to divert runoff around the polluted areas. Trucks should be well positioned during loading and unloading operations to avoid spillage of materials, and should be kept covered in windy condition when opearted within the site.

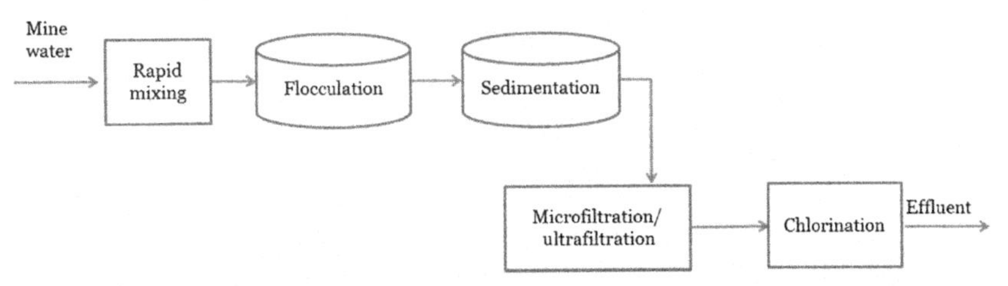

FIGURE 8.2 Typical treatment processes for treating coal mine water for reuse. Source: *After Dharmappa, H.B., Sivakumar, M., Singh, R.N., 1995. Wastewater minimization and reuse in mining industry in Illawarra region. In Proceedings of Water Resources at Risk, International Mine Water Association, Denver, Colorado, May 14–18, pp. 11–22.*

8.8 TOOLS AND TECHNIQUES TO REDUCE WATER USE IN BUILDINGS

8.8.1 Types of Water Use in Buildings

Amount of water use in buildings primarily depends on the sectors (e.g., residential, industrial, and commercial) and types of usage (e.g., indoor and outdoor). For example, in Central Coast in Australia, residential/household sector consumes around 68% of the annual urban water supply, whereas governmental, commercial, and industrial demand is around 22% as presented in Fig. 8.3 (Wyong Shire Council, 2016). In addition, amount of water usage is dependent on types of buildings (e.g., single story and multistory), number of occupants, landscapes, and water features. In a residential building, a notable volume of water is used in garden watering (Fig. 8.4) (AHSCA, 2014), whereas in office building a higher percentage of water is generally used in restroom, and cooling and heating purposes, and in a restaurant, most of the water is used in kitchen and dishwashing (Fig. 8.4) (EPA, 2016). Water usage is also varied significantly among the income groups. As indicated in Fig. 8.5, amount of domestic water use is similar among all the income groups but industrial water use is much higher in higher income countries than low- to moderate-income countries.

8.8.2 Strategies to Improve Water Sustainability in Buildings

One of the major aspects of new building construction is to create sustainable buildings that can reduce water consumption and improve water efficiency by water recycling or harvesting. Water sustainability in building is a different concept than water conservation, where water sustainability focuses

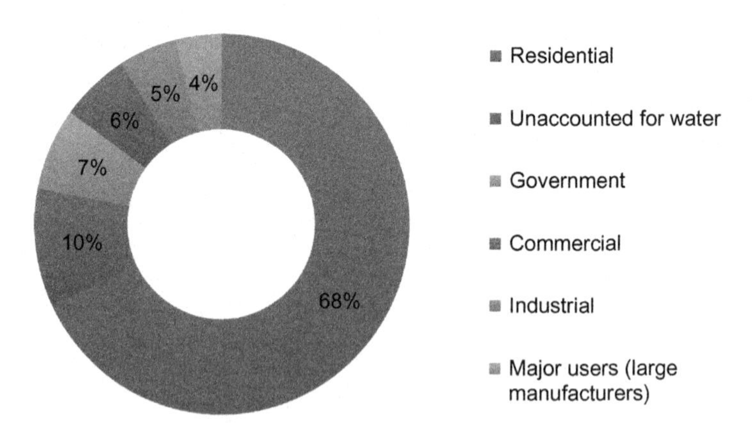

FIGURE 8.3 Water use by sectors in urban areas in Central Coast, Australia. Source: *Wyong Shire Council, 2016. Water Issues. Retrieved from: <https://www.blueplanet.nsw.edu.au/wi-water-consumption-on-the-central-coast/.aspx> (accessed 07.12.16).*

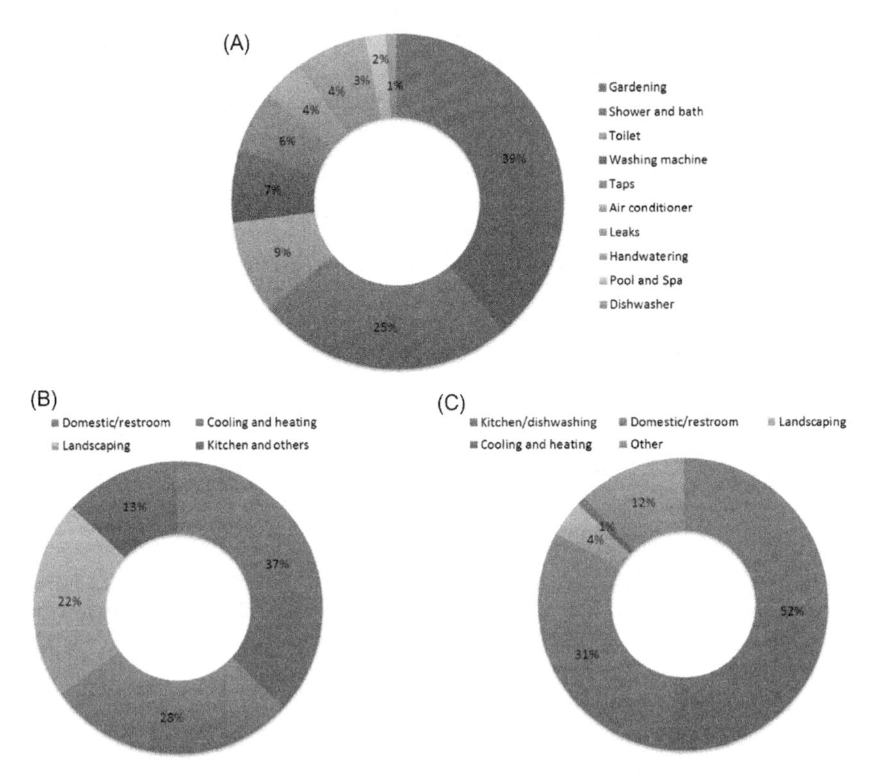

FIGURE 8.4 End use of water in (A) residential buildings, (B) office buildings, and (C) restaurants. Source*: EPA, 2016. Types of Facilities. Retrieved from: <https://www3.epa.gov/watersense/commercial/types.html> (accessed 07.12.16).*

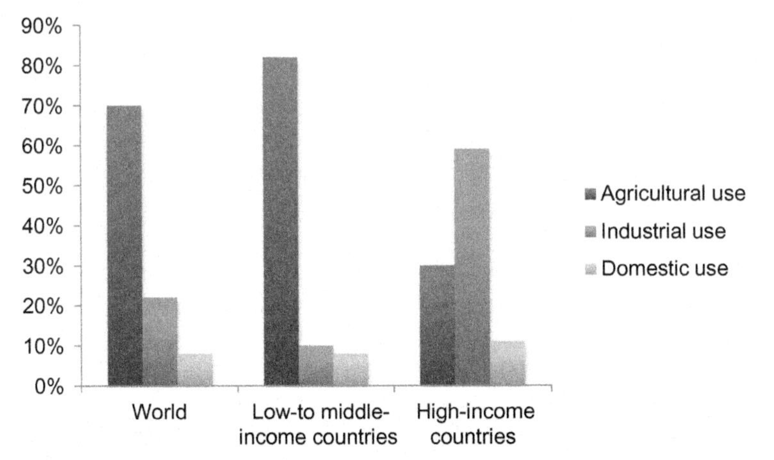

FIGURE 8.5 Water uses across income groups. Source: *EnviroSupply & Service, 2014. Water Usage by Industries Worldwide. Retrieved from: <http://envirosupply.net/blog/water-usage-by-industries-worldwide-interesting-facts-and-figures/> (accessed 07.12.16).*

more on reducing water "waste" without hampering daily needs and customer satisfaction. Water sustainability in buildings can be improved by ways, such as (1) rainwater harvesting, (2) graywater harvesting, (3) flush and flow fixtures in buildings, and (4) water wise landscaping.

8.8.2.1 Rainwater Harvesting

Rainwater is a relatively clean water source and with necessary caution it can be even used for potable consumptions. Importantly, it is a free source and can be collected in a considerable quantity from roof catchments and other pavement areas which can be used for various purposes (e.g., garden watering, toilet flushing, laundry, cooling and heating, hygienic use and drinking). Therefore, rainwater harvesting can play an important role in water sustainability by reducing the pressure on mains water supply. Many studies reported a good amount of water savings in a building by using rainwater harvesting system (Hajani and Rahman, 2014). For example, Muthukumaran et al. (2011) demonstrated that about of 40% potable water can be saved by using rainwater in a residential building in regional Victoria in Australia. Ward et al. (2012) found that an office-based rainwater harvesting system could save around 87% water (quantity of mains water saved) in a nondomestic building.

A rainwater harvesting system collects runoff from a catchment area (e.g., roof area and paved area) and generally consists of a storage tank, supply and distribution networks, and an overflow unit. Among the components of a rainwater harvesting system, the storage tank is normally the largest component of the total installation cost. Therefore, a proper economic analysis and design are necessary before implementing the rainwater harvesting systems in building in order to improve their performance and benefits, and to obtain short payback period. In order to evaluate the financial feasibility of rainwater harvesting system, life cycle cost assessment (LCCA) need to be incorporated during the planning of a building construction. LCCA is a method of evaluating the cost of a product over its life span where all the past, present, and future cash flows are converted to present values. Several studies have reported LCCA of a rainwater harvesting system and commented positive results on the financial feasibility of those systems, which could play an important factor in improving water sustainability in buildings. Zhang et al. (2009) evaluated the financial feasibility of rainwater harvesting system in high-rise buildings in four capital cities in Australia (e.g., Sydney, Perth, Darwin, and Melbourne) and demonstrated that all the cities had the potential to offer good financial return using rainwater harvesting system. They found shortest payback period (about 10 years) of rainwater harvesting system (about 10 years) in Sydney. Domenech and Sauri (2010) found the payback period in between 33 and 43 years in the metropolitan area of Barcelona in Spain for a single-family household depending on the rainwater tank size. Imteaz et al. (2011) reported that construction cost of a commercial rainwater tanks connected to large roofs in Melbourne could be recovered with 15–21-year time depending on the tank size, climatic conditions, and future water price.

8.8.2.2 Graywater Harvesting

Graywater is a kind of wastewater generated by certain uses in buildings such as wash hand basins, showers, and laundry rooms. This wastewater can be recycled on site for use in toilet systems, garden irrigation, and constructed wetlands. The reason behind the name "graywater" is due to its cloudy appearance and its status as being between potable water (generally referred as white water) and sewage water. Likewise, rainwater harvesting and graywater harvesting has also a great potential in water sustainability in buildings by minimizing water use from mains supply. In a study in residential buildings in Western Australia, Zhang et al. (2010) demonstrated that graywater reuse contributed to a notable amount of water savings of mains water supply by reducing the generation of wastewater by approximately 54.1% (88.1 m^3/lot/year).

Hence, graywater harvesting can be an important addition in achieving water sustainability in building. In order to achieve a properly functioning graywater harvesting system, early design and planning of the system including collection, treatment, storage, and distribution are important in building construction. A graywater harvesting system generally requires high investment cost; however, some studies reported positive financial feasibility of using graywater harvesting system in buildings. Godfrey et al. (2009) performed cost–benefit analysis of a graywater harvesting system installed in a school in Madhya Pradesh in India that was used for toilet flushing and irrigating the food crops. They reported monetary benefit of about 30,000 INR (1 USD = 42.5 INR) per year due to less water use from the water tanker. Stec and Kordana (2015) analyzed the financial feasibility of using graywater harvesting system in a multifamily residential building in Poland through LCCA and found that it was financially viable.

8.8.2.3 Flush and Flow Fixtures in Buildings

In order to improve water efficiency in buildings, installation of water efficient flush and flow fixtures in buildings can be a good option. The objective of installing water efficient flush and flow fixtures is to improve water sustainability by reducing water consumption in unit time or per flush without compromising the performance of the system. In a study done by Lee et al. (2011) in Miami-Dade County, United States, showed that installation of water efficient appliances saved around 28 (10.9%), 34.7 (13.3%), and 39.7 (14.5%) gallons of water per household per day for the showerhead, toilet, and clothes washer programs, respectively. Lee et al. (2011) also presented an excellent compilation of previous works on water savings through use of water efficient appliance where it can be found that a good quantity of water can be saved by using those appliances. Some of the water efficient flushes and fixtures include (1) dual flush toilets, (2) high-efficiency toilets, (3) ultra-low flow pressure assist toilets, (4) waterless urinals, (5) low flow and sensor-based sinks, (6) high-efficiency showerhead, (7) aerators for faucets, and (8) water-efficient dishwashers and washing machines. Integration of water-efficient flush and flow fixtures in building construction requires attention starting from the design phase of a building to its implementation phase.

8.8.2.4 Water Wise Landscaping

A considerable quantity of water is used in landscaping, and hence a saving here can make a notable difference in water use efficiency. Water wise landscaping can play an important role in achieving water sustainability. Water wise landscaping is a quality landscaping that conserves water and uses lesser quantity of water in growing and maintaining plants (Sevik and Cetin, 2015; Kjelgren et al., 2009). Some of the strategies in water wise landscaping are as follows:

1. Selection of less water hungry plants in garden.
2. Keeping turf grass to minimum.
3. Watering of plants only when needed, not by the clock.
4. Use of compost to improve soil's ability to resist evaporation.
5. Use of mulch on bare soil.

8.9 WATER USE AND GREENHOUSE GAS EMISSION

Water use within buildings requires energy mainly for heating the water and pumping of water from rainwater tanks to toilet flushing and other indoor use. For example, it was found that water use in residential buildings accounts for 5.4% of all the electricity and 15.1% of all the natural gas use in California, United States (CEC, 2005). This implies that a notable greenhouse gas emission is directly related to household water use.

Escriva-Bou et al. (2015) developed an end-use model using Monte Carlo simulation technique (Rahman et al., 2002) for water use for single-family households at 10 water utilities in California, United States. They noted that the water-related CO_2 emissions from single households accounts for 2% of overall per capita emissions. It was suggested that joint management of water and energy can reduce the future greenhouse gas emission.

Malinowski et al. (2015) noted that energy saving is directly related to water savings. They advocated for an integrated water management system in households based on rainwater harvesting and graywater reuse, which can make significant saving of potable water. They estimated that an integrated water management can save up to 3.8 billion kWh energy in the United States by adopting rainwater tanks to meet outdoor water demand (e.g., landscape irrigation and car washing).

8.10 CONCLUSION

Water is needed in direct construction activities and also for the production of construction materials as embodied water. Water use during life time of an infrastructure also needs to be considered in the planning and design phase of the infrastructure to save and conserve water in the long run. Water from construction sites can pollute the environment, e.g., inadequate erosion control measure can pollute waterways. Use of recycled water in construction has been increasing. Water use is directly related to greenhouse gas emission in many cases such as puming of rainwater to the in-house plumbing system and heating of water for domestic use. Concept of green buildings and zero energy buildings are becoming popular that aim to save water and energy. Rainwater harvesting and graywater reuse can save significant volume of water during the life time of a building, and hence these techniques should be adopted where possible. Use of water-efficient devices can save significant volume of water during the operation phase of a building. Loss of water due to leakage of water pipes and taps should be prevented to save water. It is expected that a greater environmental awareness, favorable government policy and continuing education will enhance the water use efficiency in the construction industry.

REFERENCES

ABS, 2006. Water Account, Australia 2004–2005. Australian Bureau of Statistics, Canberra, Australia.

ABS, 2014. Water Account, Australia 2012–2013. Australian Bureau of Statistics, Canberra, Australia.

AHSCA (2014). Water efficiency – Water Services Association of Australia. Retrieved from: <https://www.ahscansw.com.au/news/water-efficiency-water-services-association-australia> (accessed 07.12.16).

Achal, V., Pan, X.L., Lee, D.J., Kumari, D., Zhang, D.Y., 2013. Remediation of Cr (VI) from chromium slag by biocementation. Chemosphere 93, 1352–1358.

Bardhan, S., 2011. Assessment of water resource consumption in building construction in India. WIT Trans. Ecol. Environ. 144, 93–101.

Bekker, P.C.F., 1982. A life-cycle approach in building. Build. Environ. 17 (1), 55–61.

Bourg, J. (2016). Water Conservation. Millennium Energy LLC, National Institute of building Sciences, Washington, DC. Retrieved from: <https://www.wbdg.org/resources/water-conservation> (accessed 19.11.16).

Brisbane City Council (2016). Erosion and Sediment Control. Retrieved from: <https://www.brisbane.qld.gov.au/planning-building/applying-post-approval/after-approval/erosion-sediment-control-esc> (accessed 08.12.16).

Burkhardt, M., Zuleeg, S., Vonbank, R., Schmid, P., Hean, S., Lamani, X., et al., 2011. Leaching of additives from construction materials to urban storm water runoff. Water Sci. Technol. 63 (9), 1974–1982.

CEC (2005). California's Water–Energy Relationship. Prepared in Support of the 2005 Integrated Energy Policy Report Proceeding. California Energy Commission, CEC-700-2005-011-SF.

Cheng, C.L., 2003. Evaluating water conservation measures for green building in Taiwan. Build. Environ. 38 (2), 369–379.

Cheng, C.L., Peng, J.J., Ho, M.C., Liao, W.J., Chern, S.J., 2016. Evaluation of water efficiency in green building in Taiwan. Water 8 (6), 236. Available from: https://doi.org/10.3390/w8060236.

Cohen, D. (2002). Best practice mine water management at a coal mining operation in the Blue Mountains. MS thesis, University of Western Sydney, Sydney, Australia.

Crawford, R.H. and Treloar, G.J. (2005). An Assessment of the energy and water embodied in commercial building construction. In Proceedings of the 4th Australian LCA Conference, February, Sydney.

Dewulf, J., Vorst, G., Versele, N., Janssens, A., Langenhove, H., 2009. Quantification of the impact of the end-of-life scenario on the overall resource consumption for a dwelling house. Res. Conser. Recycl. 53 (4), 231–236.

Dharmappa, H.B., Sivakumar, M., Singh, R.N., 1995. Wastewater minimization and reuse in mining industry in Illawarra region. Proceedings of Water Resources at Risk, International Mine Water Association. Denver, Colorado, pp. 11–22, May 14–18.

Dharmappa, H.B., Wingrove, K., Sivakumar, M., Singh, R.N., 2000. Wastewater and stormwater minimisation in a coal mine. J. Clean. Product. 8 (1), 23–34.

Ding, G.K.C., 2008. Sustainable construction—the role of environmental assessment tools. J. Environ. Manage. 86, 451–464.

Domenech, L., Sauri, D.A., 2010. Comparative appraisal of the use of rainwater harvesting in single and multi-family buildings of the metropolitan area of Barcalona (Spain): social experience, drinking water savings and economic costs. J. Clean. Product. 11, 1–11.

EPA (2016). Types of Facilities. Retrieved from: <https://www3.epa.gov/watersense/commercial/types.html> (accessed 07.12.16).

EPA-ACT (2011), Environment Protection Guidelines for Construction and Land Development in the ACT, Australia. Retrieved from: <https://www.environment.act.gov.au>.

EPA-SA (2015), Stormwater Pollution Prevention for Building and Construction Activities. Environmental Protection Agency, South Australia, Australia. Retrieved from: <http://www.epa.sa.gov.au/environmental_info/water_quality/programs/stormwater/pollution_prevention_for_building_and_construction_activities>.

EnviroSupply & Service (2014). Water Usage by Industries Worldwide. Retrieved from: <http://envirosupply.net/blog/water-usage-by-industries-worldwide-interesting-facts-and-figures/> (accessed 07.12.16).

Eroksuz, E., Rahman, A., 2010. Rainwater tanks in multi-unit buildings: a case study for three australian cities. Resources. Conser. Recycl. 54 (12), 1449–1452.

Escriva-Bou, A., Lund, J.R., Pulido-Velazquez, M., 2015. Modeling residential water and related energy, carbon footprint and costs in California. Environ. Sci. Policy 30, 270–281.

Godfrey, S., Labhasetwar, P., Wate, S., 2009. Greywater reuse in residential schools in Madhya Pradesh, India—A case study of cost–benefit analysis. Resour. Conserv. Recycl. 53 (5), 287–293.

Guggemos, A.A., Horvath, A., 2006. Decision-support tool for assessing the environmental effects of constructing commercial buildings. J. Architect. Eng. 12 (4), 187–195.

Hajani, E., Rahman, A., 2014. Reliability and cost analysis of a rainwater harvesting system in peri-urban regions of Greater Sydney, Australia. Water 6, 945–960.

Houser, D.L., Pruess, H., 2009. The effects of construction on water quality: a case study of the culverting of Abram Creek. Environ. Monit. Assess. 155 (1), 431–442.

Imteaz, M.A., Shanableh, A., Rahman, A., Ahsan, A., 2011. Optimisation of rainwater tank design from large roofs: a case study in Melbourne, Australia. Resour. Conserv. Recycl. 55, 1022–1029.

Johnson, D.B., Hallberg, K.B., 2005. Acid mine drainage remediation options: a review. Sci. Total Environ. 338 (1–2), 3–14.

Kibert, C.J., 2004. Green buildings: an overview of progress. J. Land Use 19, 491–502.

Kjelgren, R., Wang, L., Joyce, D., 2009. Water deficit stress responses of three native Australian ornamental herbaceous wildflower species for water-wise landscapes. HortScience 44 (5), 1358–1365.

Lake Macquarie City Council (2015). Dewatering Operations on Construction Sites, Fact Sheet 8K, NSW, Australia. Retrieved from: <https://www.lakemac.com.au/downloads/A4763D2C815C268C7E872FC579F9B60C89343CCC.pdf> (accessed 19.11.16).

Landcom (2004). Soils and Construction – Managing Urban Stormwater: Volume 1, Fourth Edition, NSW. Retrieved from: <https://www.environment.nsw.gov.au/resources/water/BlueBookVol1.pdf>.

Lee, M., Tansel, B., Balbin, M., 2011. Influence of residential water use efficiency measures on household water demand: a four year longitudinal study. Resour. Conserv. Recycl. 56 (1), 1–6.

Malinowski, P., Stillwell, A., Wu, J., Schwarz, P., 2015. Energy-water nexus: potential energy savings and implications for sustainable integrated water management in urban areas from rainwater harvesting and grey-water reuse. J. Water Resour. Plan. Manage. 141, 12.

Maupin, M.A., Kenny, J.F., Hutson, S.S., Lovelace, J.K., Barber, N.L., Linsey, K.S., 2014. Estimated use of water in the United States in 2010: U.S. Geological Survey Circular 1405, 56 p. Available from: https://doi.org/10.3133/cir1405.

McCormack, M., treloar, G.J., Palmowski, L., Crawford, R., 2007. Modelling direct and indirect water requirements of construction. Build. Res. Inform. 35 (2), 156–162.

Mishra, V.K., Upadhyaya, A.R., Pandey, S.K., Tripathi, B., 2008. Heavy metal pollution induced due to coal mining effluent on surrounding aquatic ecosystem and its management through naturally occurring aquatic macrophytes. Bioresour. Technol. 99 (5), 930–936.

Muthukumaran, S., Baskaran, K., Sexton, N., 2011. Quantification of potable water savings by residential water conservation and reuse – a case study. Resour. Conserv. Recycl. 55, 945–952.

NRMMC-EPHC-NHMRC (2009), Australian Guidelines for Water Recycling: Managing Health and Environmental Risks (Phase 2)—Stormwater Harvesting and Reuse. ISBN 1921173459.

NSW Government (2016). BASIX. Retrieved from: <https://www.basix.nsw.gov.au/basixcms/water-help/water-targets.html> (accessed 18.11.16).

Pondja Jr, E.A., Persson, K.M., Matsinhe, N.P., 2014. A survey of experience gained from the treatment of coal mine wastewater. J. Water Resour. Protect. 6 (18), 1646.

Prosser, I.P., 2011. Water: Science and Solutions for Australia. CSIRO, Melbourne, Australia.

Rahman, A., Weinmann, P.E., Hoang, T.M.T., Laurenson, E.M., 2002. Monte Carlo Simulation of flood frequency curves from rainfall. J. Hydrol. 256 (3–4), 196–210.

Rahman, M.M., Hagare, D., Sivakumar, M., Singh, R.N., 2016. Sustainable management of process water and stormwater within a mine. In: Devasahayam, S., Dowling, K., Mahapatra, M.K. (Eds.), Sustainability in the Mineral and Energy Sectors. CRC Press-Taylor and Francis, New York, ISBN 978-1-4987-3302-1.

Recast, E.P.B.D., 2010. Directive 2010/31/EU of the European Parliament and of the Council of 19 May 2010 on the energy performance of buildings (recast). Off. J. Eur. Union 18 (06), 2010.

Sartori, I., Napolitano, A., Voss, K., 2012. Net zero energy buildings: A consistent definition framework. Ener. Build. 48, 220–232.

Sevik, H., Cetin, M., 2015. Effects of water stress on seed germination for select landscape plants. Pol. J. Environ. Stud. 24 (2), 689–693.

Silva, M., and Naik, T.R. (2010). Sustainable use of resources–recycling of sewage treatment plant water in concrete. In Proceedings of the Second International Conference on Sustainable Construction Materials and Technologies, Vol. 28, Ancona, Italy.

Singh, R.N., Dharmappa, H., Sivakumar, M., 1998. Study of waste water quality, management in Illawarra coal mines. In: Aziz, N. (Ed.), Coal 1998: Coal Operators' Conference. University of Wollongong and The Australasian Institute of Mining and Metallurgy, Wollongong, Australia, pp. 456–473.

Sivakumar, M., Morton, S., Singh, R.N., 1994. Case history analysis of mine water pollution in New South Wales. In Proceedings of the Fifth International Mine Water Congress. Nottingham 18–23 (September), 679–689.

Spigarelli, M. (2012). 10 Ways to Save Water in Commercial Buildings. CCJM Engineers Ltd., Chicago, IL. Retrieved from: <https://www.csemag.com/single-article/10-ways-to-save-water-in-commercial-buildings/8f74baabfcc8f672483b3b0353ccad16.html> (accessed 19.11.16).

Stec, A., Kordana, S., 2015. Analysis of profitability of rainwater harvesting, gray water recycling and drain water heat recovery systems. Resour. Conserv. Recycl. 105, 84–94.

Street, M., 1993. Water Runon/Runoff Management. Report. US Environmental Protection Agency, Office of Solid Waste, Washington, DC.

Tam, C.M., Tam, V.W.Y., Tsui, W.S., 2004. Green construction assessment for environmental management in the construction industry of Hong Kong. Int. J. Project Manage. 22, 563–571.

The Wall Street Journal (2016). Agence France-Presse/Getty Images. Retrieved from: <https://www.wsj.com/articles/brazil-police-say-bhp-vale-at-fault-for-dam-disaster-1465510198>.

Thornback, J., Snowdon, C., Anderson, J. and Foster, C. (2015). Water Efficiency: The Contribution of Construction Products. Construction Products Association, United Kingdom. Retrieved from: <https://www.constructionproducts.org.uk/media/87904/water_efficiency_report.pdf>.

Transport for NSW (2015). Water Discharge and Reuse Guideline. No. 7TP-SD-146/4.0. Retrieved from: <https://www.transport.nsw.gov.au/sites/default/files/b2b/projects/Water%20Discharge%20and%20Reuse%20Guideline%20-%207TP-SD-146.pdf>.

US DOE (2010). Building Technologies Program, Planned Program Activities for 2008–2012. Department of Energy, United States. Retrieved from: <http://www1.eere.energy.gov/buildings/mypp.html >.

WELS (Water Efficiency Labelling and Standards) Scheme, 2016. Water Rating. Australian Government. Retrieved from: <https://www.waterrating.gov.au/> (accessed 22.11.16).

Ward, S., Memon, F., Butler, D., 2012. Performance of a large building rainwater harvesting system. Water Res. 46 (16), 5127–5134.

Waylen, C., Thornback, J. and Garrett, J. (2011). Water: An Action Plan for Reducing Water Usage on Construction Sites, Report 009. Strategic Forum for Construction, Construction Product Association, United Kingdom. Retrieved from: <https://www.greenconstructionboard.org/otherdocs/SCTG09-WaterActionPlanFinalCopy.pdf>.

Wingrove, K. (1996). Wastewater Management in Illawarra Coal Mines. BE thesis, University of Wollongong, Wollongong, Australia.

Witheridge, G., 2012. Principles of Construction Site Erosion and Sediment Control. Catchment & Creeks Pty Ltd, Brisbane, QLD.

Wyong Shire Council (2016). Water Issues. Retrieved from: <https://www.blueplanet.nsw.edu.au/wi--water-consumption-on-the-central-coast/.aspx> (accessed 07.12.16).

Zhang, Y., Chen, D., Chen, L., Ashbolt, S., 2009. Potential for rainwater use in high-rise buildings in Australian cities. J. Environ. Manage. 91, 222–226.

Zhang, Y., Grant, A., Sharma, A., Chen, D., Chen, L., 2010. Alternative water resources for rural residential development in Western Australia. Water Resour. Manage. 24, 25.

FURTHER READING

De Gisi, S., Casella, P., Notarnicola, M., Farina, R., 2016. Grey water in buildings: a mini-review of guidelines, technologies and case studies. Civil Engineering and Environmental Systems 33 (1), 35–54.

Rahman, A., Keane, J., Imteaz, M.A., 2012. Rainwater harvesting in greater Sydney: water savings, reliability and economic benefits. Resour. Conserv. Recycl.g 61, 16–21.

<http://www.epa.sa.gov.au/environmental_info/water_quality/programs/stormwater/pollution_prevention_for_building_and_construction_activities>.

Chapter 9

Materials

Isidore C. Ezema
Department of Architecture, Covenant University, Ota, Ogun State, Nigeria

9.1 INTRODUCTION

The issue of sustainability has been in the forefront of environmental discourse for quite a while, especially within the last quarter of a century due mainly to the adoption, at the global level, of sustainable development as a preferred development paradigm (WCED, 1987). The built environment as represented by the building and construction industry plays a key role in sustainability of the environment in terms of resource consumption, energy use, greenhouse gas (GHG) emissions, among other indices. For example, it has been estimated that buildings consume about 50% of global resources (Edwards, 2002), 36% of energy (UNEP, 2007), and account for 33% of global GHG emissions (UNEP-SBCI, 2010). Buildings impact on the environment directly through its operation and use and indirectly through the materials and processes that bring them about. The direct impacts of building operation have been studied extensively with emphasis placed on low-impact buildings from the operational perspective. As a result, low-impact operational targets for buildings have been set and achieved to certain extents in many contexts especially in Europe (Koukkari and Braganca, 2011). The impacts of materials use and other associated building processes are also assuming considerable significance due to increased adoption of operational energy efficiency strategies. The major concerns of sustainable construction are resource efficiency, energy use efficiency, mitigation of harmful emissions to the environment, especially carbon emissions as well as occupants' well-being (Meriani, 2008).

Sustainable construction is a necessary condition for a sustainable built environment, which in turn contributes to overall sustainable development. The International Council for Research and Innovation in Building and Construction (CIB) defines sustainable construction as "the sustainable production, use, maintenance, demolition and reuse of buildings and constructions or their components" (CIB, 2004, p. 2). Hence, sustainable construction pervades the entire life cycle of a building. Sustainable construction has also been defined as "a holistic process aiming to restore and maintain harmony

Sustainable Construction Technologies. DOI: https://doi.org/10.1016/B978-0-12-811749-1.00007-9

between the natural and the built environments, and create settlements that affirm human dignity and encourage economic equity" (Du Plessis, 2002, p. 8). Sustainable construction practice aims to achieve efficiency in resource and energy use while at the same time promoting reduced emissions to the environment and enhancing well-being of the people. It demands the use of sustainable materials and the deployment of innovative processes of building construction (Du Plessis, 2002; Meriani, 2008).

From the materials perspective, the materials life cycle is critical. Building materials generate environmental impacts during their entire life cycle phases of manufacturing (raw material extraction, transportation, and processing), site construction, use and maintenance, and during the disposal or demolition phase. The construction industry is resource intensive. According to OECD (2015), the amount of materials extracted, harvested, and consumed worldwide increased by 60% between 1980 and 2008 and reached about 62 billion metric tons (Gt) per year in 2008 and is projected to reach 100 Gt by 2030. Interestingly, this growth has been mainly driven by increased global demand in three major material groups: (1) construction minerals, (2) biomass for food and feed, (3) and fossil energy carriers, which together constitute 80% of total global material extraction (OECD, 2012). Hence, the use of sustainable building materials and construction processes holds the key. Evidence from literature suggests that the major building materials in use are generally of high impact to the environment in their extraction, processing, use, and disposal (Bribian et al., 2011).

As a result, the use of sustainable building materials is advocated for building and construction projects as a way of contributing to sustainable construction (DuPlessis, 2007). Sustainable building materials have elicited different definitions from different quarters. Sustainable or green building materials are materials that promote resource efficiency, use low energy, and emit low carbon in their life cycle (Meriani, 2008; O'Brien, 2011). In addition, they have low environmental impact, are thermally efficient, energy efficient in production and use, and have low toxic effect on the environment (European Commission, 2011). Typically, sustainable materials are materials that can be sourced locally and are recyclable, renewable, and financially viable. Sustainable materials can also be high-performance lightweight and highly durable materials which are products of sophisticated high-tech production processes. In all, sustainable materials contribute minimally to environmental impact over their life cycle (Franzoni, 2011).

Advances have been recorded in recent years in the use of sustainable materials and sustainable construction methods due mainly to increased awareness regarding the advantages of sustainable construction. However, a lot remains to be done in both developed and developing countries. The sustainability of building materials can be better understood by examining their environmental impact. Relying on extant literature, this chapter examines the importance of materials in general and sustainable materials in particular to the attainment of sustainable construction. Also, the environmental impacts of building materials are examined from the perspective of life cycle assessment (LCA) with focus on embodied energy and carbon emissions. An overview of the LCA framework is examined and results from LCA studies used as a validation of the various approaches to sustainable construction subsequently outlined. The chapter further underscored the implications of the study findings especially with respect to the creation of an enabling institutional environment as main driver of sustainable construction. Also, the application of LCA results early in the project life cycle was recommended. In this respect, the need for environmental information on materials to be readily available so as to facilitate easy reference at design and specification stage of construction projects was stressed.

9.2 ENVIRONMENTAL IMPACT OF BUILDINGS

Buildings exert a lot of impact on the environment through material consumption, construction methods, use, and deconstruction. Building materials are almost entirely made from nonrenewable natural resources, which provide the major inputs to the production processes. There has been unprecedented growth in the demand for raw materials for the purpose of production due mainly to rapid urbanization and economic development (OECD, 2012). For example, mining of natural resources either for construction or as inputs for production of other materials has environmental, economic, and social consequences. One of such environmental consequences is pollution of both the air and waterways (Mensah et al., 2015). Also, it has been found that inland sand mining can contribute to economic loss through loss of agricultural land and depreciation of land value (Adedeji, Adebayo, & Sotayo, 2014). Similarly, Singh and Singh (2016) identified some social impacts of mining to include housing displacement, resettlement challenges, and sociopolitical conflicts. A number of indices are available to measure the environmental impact of buildings and building materials. Such indices include resource efficiency indicators (Science Communication Unit, University of West England Bristol, 2012), ecological footprint (Wackernagel & Rees, 1996; Bell et al., 2008), carbon footprint (Wiedmann & Minx, 2008), energy use intensity, and energy efficiency (Alyahya & Nawari, 2018; Yoon & Park, 2017). Building materials production from natural sources usually involve initial extraction and further processing prior to delivery to and use in construction sites. The processes involve energy use and carbon emissions, which constitute the hidden energy and carbon content of buildings. This hidden impact is also referred to as embodied energy and carbon emissions, and it is the energy used for building materials manufacture and building assembly processes and the carbon emissions associated with the processes.

Tools for assessing environmental impact and sustainability of buildings are varied and have developed rapidly in the attempt to empirically assess sustainability. However, according to Forsberg and Von Malmborg (2004), they can be broadly classified into three, namely, (1) environmental impact assessment (EIA), (2) qualitative building environmental rating systems, and (3) quantitative building LCA. EIA has well-developed applications, but has remained a site-specific environmental assessment tool with limited application to assessment of overall environmental sustainability. Building environmental rating system is a multicriteria method for evaluating the environmental performance of a building against an explicit set of criteria as enunciated by Cole (2003). The two most prominent building environmental rating systems are the Leadership in Energy and Environmental Design (LEED) and the Building Research Establishment Environmental Assessment Method (BREEAM). However, they remain relative measures of sustainability and their application is still largely voluntary in many countries. LCA is a product-based tool in environmental management which examines in quantitative terms, the burdens which a product or process imposes on the physical environment. Its purpose is to assess the environmental impact of a product or process over its life cycle with a view to identifying and evaluating opportunities for environmental improvements (USEPA, 2006). It is the life cycle approach that can be effectively used to assess the sustainability of building materials and building construction processes.

9.3 LIFE CYCLE ASSESSMENT

From its foundation in the 1960s to the pioneering work by the Society of Environmental Toxicology and Chemistry (SETAC), LCA has evolved into an international tool with well-developed methodologies as standardized by the International Organization for Standardization into ISO 14040 and ISO 14044 (UNEP, 2011). LCA application in products has engendered increased inclusion of environmental labeling on products thus promoting environmental products declarations. LCA measures the environmental impact of a construction product, component, or building under different categories. Hence, LCA can be conducted at both product and building level. Environmental impact categories refer to the specific impact on the environment, which is traceable to a building product or process. The United States Environmental Protection Agency (USEPA) identified 10 environmental impact categories commonly considered in LCA and they are: (1) global warming, (2) stratospheric ozone depletion, (3) acidification, (4) eutrophication, (5) photochemical smog, (6) terrestrial toxicity, (7) aquatic toxicity, (8) human health, (9) resource depletion, and (10) land use (USEPA, 2006). Environmental impact of buildings can also be assessed in terms of energy use and carbon emissions associated with the various life cycle phases. Energy use and carbon emissions are the environmental impacts mostly relied on in this chapter.

The LCA framework is shown in Fig. 9.1. It is a four-stage process comprising goal and scope definition, inventory analysis, impact assessment, and

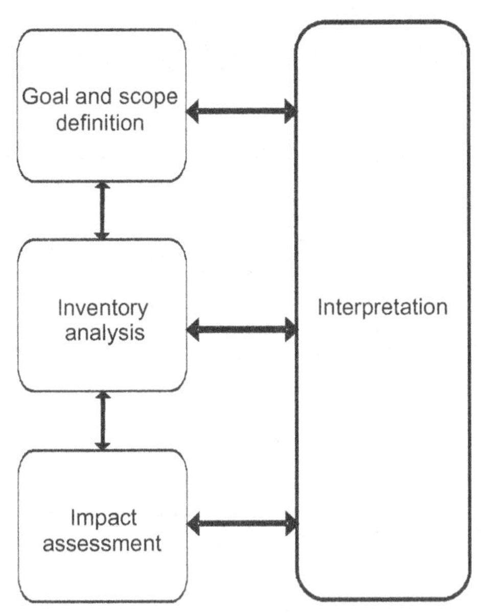

FIGURE 9.1 The life cycle assessment framework (ISO, 2006).

interpretation. The goal and scope definition outlines the purpose of the LCA and establishes the boundary conditions. Inventory analysis collects data pertaining to materials and energy flows while impact assessment evaluates the environmental impact in relation to the scope earlier defined. Interpretation combines the inventory stage and the life cycle impact assessment to arrive at conclusions that would lead to recommendations.

Of particular importance in LCA study is the life cycle boundaries which define the phase in the life cycle of product or building that is assessed. According to SETAC (1991), the life cycle phases of industrial products comprise of raw materials extraction; processing and manufacturing; distribution and transportation; use, reuse, and maintenance; recycling and waste management. With respect to the life cycle of buildings, Athena Institute (2009) identified the following life cycle phases: manufacturing of building materials; transportation of building materials; construction of the buildings; occupancy and renovation; and demolition and removal. In general, however, whole life cycle of buildings encompasses three main stages, namely, (1) embodied phase (resource extraction, manufacturing of building materials, building construction); (2) operational phase (building occupancy, maintenance); and (3) end-of-life phase (demolition, recycling, and disposal) as shown in Fig. 9.2.

All products impact on the environment at the various stages of the product life cycle which include:

1. Materials extraction and manufacture;
2. Transportation from factory to construction site;

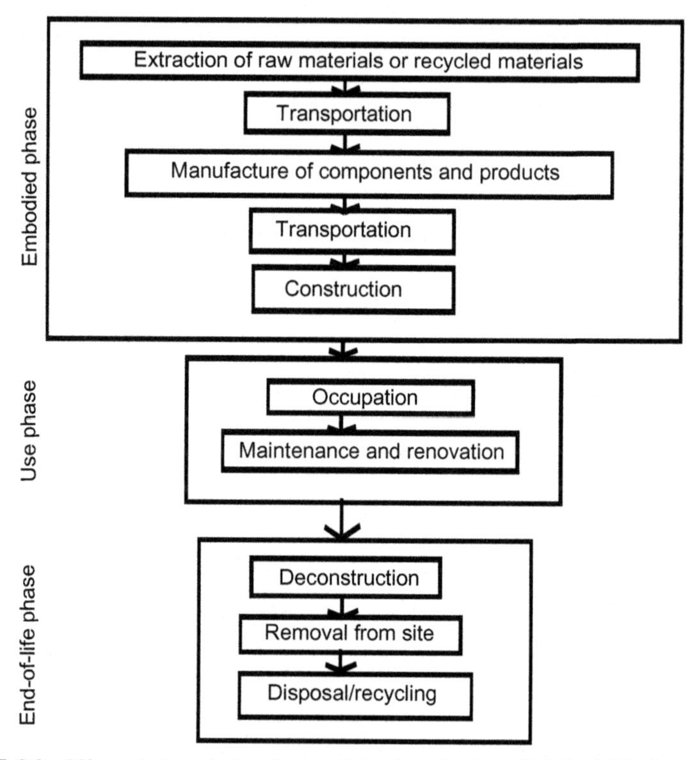

FIGURE 9.2 Life cycle boundaries. Source: *Monahan, J., Powell, J.C., 2011. An embodied carbon and energy analysis of modern methods of construction in housing: A case study using a life cycle assessment framework. Ener. Build. 43(1), 179–188.*

3. Site construction process involved in placement of the material at point of need;
4. Maintenance including replacement of failed components during the operational life of the building;
5. Deconstruction of the building including recycling and disposal of resultant waste.

All phases of a product life cycle are important in LCA but the material extraction and manufacture phase (cradle-to-gate phase) is usually the most critical.

LCA of buildings can also be conducted at the operational and embodied phases. The operational phase refers to the impact of building operation and use while the embodied phase refers to impact of building materials and construction processes. The product life cycle phases in Fig. 9.2 capture the embodied phase of a building life cycle. For conventional buildings the embodied phase has been estimated to be about 20% of total life cycle energy while operational phase accounted for 80% of total life cycle energy (Thormark, 2002, 2006). However, with increasing adoption of operational energy efficient measures, the embodied energy to operational energy ratio could be between 40% and 60%, respectively (Sartori and Hestnes, 2007; Thormark, 2002, 2006). This is an indication that the embodied aspect of building energy consumption could be higher that operational energy as buildings tend to achieve net zero operational energy. Hence, the choice of materials is crucial in determining the pattern of energy consumption and carbon emissions in buildings and research is increasing focusing in this area.

In a building, a number of energy and carbon intensive components have been identified from LCA studies. They include the substructure (foundations, retaining walls, and ground floor work), superstructure (frame, walls, roof, staircases), fixtures and fittings (doors, windows, cabinets), and finishes such as walls, floor, and ceiling (RICS, 2012). Building services also make substantial contribution to a building's energy and carbon profile (Chau et al., 2007). However, their contributions are generally complex to assess and offer limited opportunity for mitigation (RICS, 2012). Transportation can contribute substantially to the embodied energy and carbon of building materials if the materials are transported over long distances through importation. However, in terms of relative contributions to the embodied impact of buildings, the substructure, superstructure, the building envelope, roofing, and finishes account for the most environmental impact (Ezema, 2015). The major materials used in the aforementioned parts of the building are: Portland cement, Portland cement-based products, steel reinforcement bars, aluminum and other metal-based products, concrete aggregates, timber and timber products, ceramic tiles, PVC tiles, and paints (Ezema, 2015).

A number of other LCA studies reinforce the evidence that energy and carbon intensive materials such as cement and cement-based products, steel and nonferrous metals such as aluminum and copper dominate the construction industry. From a Hong Kong study of office buildings, Chau et al. (2007), found that the materials with the highest impact with respect to structure, envelope, and finishes are cement-based materials, metal-based materials, and tiles while copper is the most important material for building services. Some key building materials of considerable impact to the environment are explained further. Also, Bribian et al. (2011) identified cement-based materials, metal-based materials, insulation materials, ceramic tiles, and wall envelope materials as materials that have very high impact on the environment. The environmental implications of these materials are subsequently discussed.

9.3.1 Cement and Cement-Based Materials

The cement production process, for example, starts with mining of limestone, which is then crushed and ground to powder. It is then preheated to save energy before being transferred to the kiln, the heart of the process. The kiln is then heated to a high temperature of up to 1480 degrees to convert the material to a molten form called clinker. The clinker is then cooled and ground to a fine powder with other additives and transferred to storage silos for bagging or bulk transportation (Portland Cement Association, 2014). The production of cement is either through the wet or dry process with the dry process as the preferred option because of the lower energy intensity. Cement production accounts for about 5% of total anthropogenic emissions (IFC, 2017). Cement-based structures constitute the largest surface area of all man-made structures (Odigure, 2009). World cement demand was about 2.283 billion tons in 2005, 2035 million tons in 2007, and 2836 million tons in 2010 with an annual estimated increase of about 130 million tons (Madlool, Saidur, Hossaina & Rahim, 2011; Odigure, 2009). World total cement production for 2016 was about 4.2 billion tons with emerging markets playing a dominant role (IFC, 2017). The energy intensity of cement production ranges from 3.6 to 6.5 GJ/ton depending on production process and location of the production (Hammond and Jones, 2011; Ohunakin et al., 2013; Worrell et al., 2000).

9.3.2 Steel and Steel-Based Materials

Steel is the most important engineering and construction material in the modern world (Siemens, 2011). In terms of environmental impact, it contributes 4%–7% of anthropogenic carbon dioxide emissions (WSA, 2016a). The carbon intensity for primary steel production is 1.9 tons of carbon dioxide per ton of steel. Also, the iron and steel industry is considered the largest energy consuming industry in the world and one of the most important sources of carbon dioxide emissions (Hidalgo et al., 2003). However, the energy consumption of steel production is on the decrease in some countries due mainly to adoption of recycling and improved efficiency of machinery (WSA, 2016a). There are two routes for steel production, namely, (1) blast furnace and basic oxygen furnace (BF-BOF) and (2) electric arc furnace (EAF). The BF-BOF process accounts for 75% of world steel production while the EAF method is used to produce 25% of world total steel production (WSA, 2018). However, in terms of environmental impact, the BF-BOF method is about 2.5 times more carbon intensive than the EAF method (Li et al., 2012). Energy consumption in primary steel production has dropped by about 60% from 1960 value, but it is still high at about 20 GJ/ton in 2015 (WSA, 2016b). A lower value of 14.9 GJ/ton was reported for the US steel industry (Hassanbeigi et al., 2014). Hence, primary steel production is about three times more energy intensive than cement production.

9.3.3 Aluminum

Closely related to the steel industry is the aluminum industry. While the steel industry dominates the ferrous metals category, aluminum is the most important nonferrous metal used in building construction. The production process of aluminum involves mining of bauxite, the basic raw material, refining of bauxite to alumina (Bayer Process), conversion of alumina to aluminum ingots in a smelter plant, and semifabrication of the ingots through rolling and extrusion. The smelting of alumina into aluminum ingots is the most energy intensive metal industrial process in the world (Husband et al., 2009). The energy intensity of virgin aluminum production is about 10 times than that of steel and has been estimated at 218 GJ/ton (Hammond and Jones, 2011). Recycling is therefore encouraged in aluminum production as it can reduce energy intensity of production by up to 95% (Enerdata, 2013). As a result of high energy intensity of primary aluminum production, smelting plants are sited to take advantage of available raw materials as well as to benefit from relatively cheap sources of energy.

9.3.4 Insulation Materials

Insulation materials are indispensable in the drive to achieve operational energy efficiency in buildings. They are essential in the control of indoor environmental conditions thereby reducing the operational energy load of buildings. Insulation enhances the thermal performance of the building envelope by reducing heat gain in hot weather and heat loss in cold weather thereby limiting the need for cooling and heating. However, conventional insulation materials have high embodied energy values. In Europe, for example, the most used insulation materials are glass wool, stone wool, and expanded polystyrene (EPS) where they account for over 80% of insulation materials used (Papadopoulos et al. 2002). In embodied energy terms, EPS has an estimated intensity of 88.6 GJ/ton while glass wool and stone wool have intensities of 28 GJ/ton and 16.8 GJ/ton, respectively (Hammond and Jones, 2011). Additional properties of the insulation materials are presented in Table 9.1. Materials for insulation in buildings have been made mostly from organic and inorganic materials with high level of industrial processing. From the Inventory of Carbon and Energy (ICE) database, the production of EPS insulation material, for example, is about 4.4 times more energy intensive than the production of steel.

TABLE 9.1 Comparative Properties of Some Insulation Materials

Insulation material	Density (kg/m^3)	Thermal conductivity factor (W/Mk)	Thermal resistance —*R*-value (K^2/W)	Embodied energy coefficient (MJ/kg)
Rock wool	30–150	0.035–0.05	2.85	16.8
Glass wool	16–40	0.035–0.05	2.85	28
EPS	15–30	0.037–0.042	3.52	88.6

Source: Adapted from: Hammond, G., Jones, C., 2011. Inventory of Carbon and Energy (Version 2.0). Sustainable Energy Research Team, University of Bath, Bath, UK. and Utochikina (2014).

9.3.5 Bricks and Ceramic Tiles

Bricks are popular envelope materials for buildings while ceramic tiles are used for floor and wall finishes. The high primary energy demand in the production of brick and tiles is associated with the high consumption of fossil fuel during the manufacturing stage. It is estimated that firing of the bricks and tiles in the kiln consumes about 80% of the total energy of production (Bribian et al., 2011). In addition, some raw materials for brick and tiles manufacture are transported over long distances, which add to the transportation component of material production. Embodied energy of fired clay bricks and that of ceramic tiles have been estimated to be about 3 GJ/ton and 12 GJ/ton, respectively (Hammond and Jones, 2011).

9.4 STRATEGIES FOR MINIMIZING IMPACT OF CONSTRUCTION

From the foregoing scenario, reduction of environmental impact of buildings should target materials and construction processes. A number of approaches have been developed over time to facilitate sustainable construction from the perspective of building materials production and construction processes. They include improvements in the materials production process, minimizing waste through improved construction processes, conserving resources by reducing primary production and encouraging recycling, developing and using partial or full substitutes for high-impact materials, adoption of innovative construction methods, development and use of high-performance materials and the use of eco-friendly alternative and renewable materials. The above strategies can be further grouped into three: (1) production-based strategies, (2) design- and construction-based strategies, and (3) new and nonconventional materials. The strategies are presented in Table 9.2 and further explained.

9.4.1 Improved Materials Production Processes

The energy and carbon associated with materials production can be reduced by adopting improved production processes. LCA studies indicate that cement, ferrous metals, and nonferrous metals are the materials with the highest environmental impact in the built environment. For example, in the production of cement, according to (IFC, 2017), the most energy and carbon intensive processes take place during fuel combustion (38%) and clinker production (62%). Hence, huge energy savings and carbon mitigation can be achieved by switching from the predominantly used fossil fuels to alternative and renewable energy sources.

Also, considerable energy saving and carbon emissions reduction can also be achieved by improving existing production technology through the reuse of the waste heat from the furnace for electricity generation

TABLE 9.2 Strategies for Reducing Impact of Materials

Broad strategies	Specific strategies	Objectives
Production-based strategies	Improved materials production processes	Reduction of energy intensity
		Reduction of carbon intensity
		Reduced use of high impact production inputs
	Recycling	Resource efficiency
		Reduction of waste
	Materials substitution	Use of low impact substitutes for high impact materials
Design- and construction-based strategies	Innovative design and construction methods	Improvement of efficiency
		Reduction of waste
	Deconstruction and disassembly	Reuse and recycling of materials
New and nonconventional materials	Innovative materials	High performance(high-tech) materials
	Renewable materials	Eco friendly (natural) materials

Source: Author's adaptation.

(Legmann, 2002; Madlool et al., 2012). Waste heat recovery has become a growing trend in cement manufacturing. In a study of 11 emerging economies with substantial cement production capacity, only six have some installed heat recovery capacities with India leading the lot in terms of number of installations and installed capacity (IFC, 2014).

Another way of reducing the environmental impact of cement production is by reducing the clinker content in the final cement product. This can be done by substituting some percentage of clinker by some low-impact materials that have pozzolanic properties within the cement production process. In an LCA of Portland cement production, it was found that cement produced with a blend of clinker and pulverized fly ash (PFA) has lower environmental impact than the traditional process (Huntzinger and Eatmon, 2009). Also, in a study by Garcia-Segura et al. (2014), the GHG emissions of blended cement is more favorable than conventional process. In addition, concrete made from blended cement have been shown to be very durable (Oner and Akyuz, 2007; Rajdev et al., 2013). The production of blended cement by replacing a percentage of clinker with low-impact supplementary cementitious materials (SCM) can considerably reduce the environmental impact of cement production. Another method of reducing cement production impact is by continuous improvement in the efficiency of cement manufacturing equipment (Siemens, 2009).

9.4.2 Minimizing Environmental Impact Through Recycling

Recycling is the process of converting used materials, which would have been treated as waste into new products. It is an effective method of promoting energy efficiency and carbon dioxide reduction associated with materials production (Grimes, Donaldson & Gomez, 2008). Recycling decreases the need for the production of virgin materials thereby eliminating the energy and carbon emissions that would have been associated with virgin material production. The importance of recycling is particularly noteworthy in the production of ferrous metals, such as iron and steel, and nonferrous metals, such as aluminum and copper, where it can reduce the demand for natural resources.

According to the World Steel Association (WSA, 2016a), steel possess 100% recyclability. Also, energy cost in steel production is estimated to be between 20% and 40% of total production cost depending on the energy source (WSA, 2016a). However, steel has 100% recyclability with recycling accounting for up to 25% energy savings. For every tonne of virgin steel produced, 1.9 tonnes of carbon dioxide is emitted making steel one of the most carbon intensive materials for construction (WSA, 2015). The use of recycling helps to reduce the carbon intensity of steel production. An examination of the Inventory of Carbon and Energy (ICE), a database for embodied energy and embodied carbon coefficients of construction materials, indicates that carbon intensity of recycled steel is about 16% of that of virgin steel (Hammond and Jones, 2011).

The aluminum industry is another industry that relies on recycled materials to reduce its energy and carbon intensity. In the LCA study of aluminum shredder cables, Grimaud et al. (2016) found considerable environmental benefits through recycling. Also, Olivieri et al. (2006) demonstrated that aluminum recycling offers both environmental and economic advantages. The advantages include energy savings and reduction in bauxite mining. In addition, it offers economic advantage to countries involved solely in secondary production of aluminum.

9.4.3 Materials Substitution

Cement requires relatively high amounts of energy and associated carbon emissions in its production. Additionally, the chemical processes involved also generate carbon emissions. Reducing cement production impact by switching from fossil fuel to renewable and the use of waste heat recovery processes may be outside the control of most construction industry stakeholders. However, it is possible to reduce demand for cement by partially substituting cement with less carbon-intensive materials referred to as SCM in the production of cement-based building components in the construction site or offsite. By adopting substitutes that can reduce the demand for ordinary Portland cement (OPC), the construction industry is contributing to the reduction of the environmental impact of cement production (Huntzinger and Eatmon, 2009).

Concrete made with a blend of OPC and SCM has been shown to have comparable durability with concrete made with OPC. In this respect, the use of PFA, a by-product of coal-powered power stations, and Ground Granulated Blast-furnace Slag (GGBS), a by-product of steel production, can replace substantial quantities of OPC in typical construction and engineering applications. In an experimental study, Oner and Akyuz (2007) found that the compressive strength of concrete increased as the quantity of GGBS increased up to 55% substitution level before declining. The decline was attributed to the presence of unreacted GGBS in the paste acting as a filler material in the concrete paste. Similarly, a study showed that the optimum mix for PPC blended with PFA is 25% with compressive strength estimated at 23.5 N/mm^2 at 28 days (Rajdev et al., 2013). The result also showed that strength at 7 days was 70% of strength at 28 days thus indicating that the attainment of strength follows the normal pattern for concrete.

Other low-impact materials with pozzolanic properties include rice husk ash, silica fume, calcinated shale, calcinated clay, volcanic ash, and meta-kaolin. In addition, it has been shown that by adding SCM such as flyash and slag to ordinary Portland cement, improved workability, durability and long-term strength can be achieved (Wang and Ge, 2003). SCMs also reduce permeability and absorption in cement-based structures and also increase corrosion resistance (Dinakar et al., 2007). Hence, the use of SCMs as blend for OPC in cement-based structures reduces the demand for OPC, thus saving the energy and preventing the carbon emissions that would have been associated with the OPC production. However, technical barriers (quality consistency), market barriers (acceptability), and government barriers (appropriate regulations) tend to limit widespread use of SCMs in cement and cement-based constructions (Wescott, McNulty, VanGeem, & Gajda, 2010).

9.4.4 Innovative Construction Methods

The construction industry is dominated by materials with high environmental impact in terms of energy use and emissions. In addition, waste associated with inefficient construction processes is high and accounts for 10%–30% of landfill wastes (Begum et al., 2006). While the search for more durable low-impact materials continues, innovative ways of construction aimed at promoting sustainability through efficient use of resources are being undertaken. Key energy and carbon intensive building materials such as cement and steel reinforcement bars will continue to play prominent roles in building construction. The approach to be adopted is to find how to use these materials innovatively to limit their overall impact on the environment. Sustainable approaches to construction such as lean construction, offsite manufacturing construction, and prefabrication to limit wastes associated with in situ construction are recommended.

The lean concept or lean thinking was originally developed and applied in the auto manufacturing industry where it helped to maximize value while limiting waste (Aziz and Hafez, 2013). Lean thinking now pervades the production and assembly process where it has been adjudged as revolutionary. Lean construction is the application of the principles of lean manufacturing to the construction industry. The purpose is to improve efficiency and reduce wastes often associated with the construction industry. The lean philosophy was introduced into construction by Koskela (1992) with construction production conceptualized as transformation, flow, and value generation. The lean approach can be streamlined into five, namely, (1) identification of value from client's perspective, (2) identification of processes to deliver the value stream, (3) achieving continuous flow of work processes, (4) use of pull planning and scheduling, and (5) perfection of the process through continuous improvement (Aziz and Hafez, 2013). The main purpose of lean construction is to improve construction productivity and minimize waste. Lean construction is often associated with integrated project delivery, which has been described as a project delivery process that collaboratively harnesses the knowledge, talents, and insights of all participants in order to improve project value, reduce waste, and optimize efficiency. Hence, lean has been associated with sustainability and has come to be associated with any innovative way to make building design and construction more efficient.

One of the direct fallouts from the lean principles is the concept of offsite manufacturing (OSM) of components and site assembly of the components. OSM is the process of incorporating prefabrication and preassembly to produce units and or modules that are then transported to the site to form a permanent work. It is often used synonymously with prefabrication, preassembly and preconstruction (Nadim and Goulding, 2010). Sometimes the term "modern methods of construction" (MMC) is used to describe the process of factory producing building components and assembling them on site (National Audit Office, 2005). In a UK study of energy and carbon analysis of MMC, it was found that the use of MMC resulted in 34% reduction in embodied carbon when compared to traditional methods (Monahan and Powell, 2011).

9.4.5 Building for Deconstruction and Disassembly

The conventional material end-of-life processes such as demolition, transportation, and disposal to landfill have environmental implications. The concept of deconstruction addresses the desire of the construction industry to shift the accepted end-of-life destination for most building materials from landfill to reuse or recycling (Olson, 2010). Hence, deconstruction refers to careful removal of all building materials for recycling and/or reuse. It is therefore a sustainable way of carrying out demolition and reducing pollution. Deconstruction and disassembly are closely related. Deconstruction practice is facilitated if building construction processes emphasize preassembly at initial construction and disassembly rather than demolition at the end of useful life. This method of construction likens the building construction process to the manufacturing process where each component is differentiated and can easily be detached from the whole. In an LCA study involving four deconstruction scenarios, O'Brien et al. (2012) found that the most favorable environmental impact was achieved with 100% and 44% manual deconstruction with the salvage material reused within 20 m radius of the deconstruction to reduce transportation impact. In general, major structural materials such as wood, steel, concrete, and masonry can be designed and constructed with deconstruction in mind. The following strategies are usually recommended: use of panelized construction (wooden structures), use of precast members (concrete structures), use of interlocking bricks and blocks (masonry structures), use of fasteners and bolted connections (steel structures), and the separation of services from building structure (Shumaker, 2009).

9.4.6 Use of New and Innovative Materials

One of the ways of reducing resource consumption is by the use of high-performance materials such as building integrated photovoltaic systems (BIPVs) and nanomaterials. Photovoltaic panels are transiting from being mere appendages on completed buildings to full integration as a building component. BIPVs in addition to their primary function of providing renewable electricity also perform the functions of weather protection, noise reduction, and glare protection. Even though the use of BIPVs may lead to increase in indoor temperature especially in the tropical regions (Ekoe-a-Akata et al., 2015), its overall effect is positive to the environment.

Nanomaterials are lightweight high performance and durable products of nanotechnology. Nanotechnology is a specialized scientific discipline based on the discovery that when materials are broken down into subatomic scale known as nanoscale, they can exhibit properties significantly different from the properties of the same material at large scale (Wong, 2014). Through manipulation of particles at nanoscale, there are limitless possibilities for improving the performance of materials, structures, and devices. Nanotechnology holds special prospects for the construction industry through the use of manufactured nanotechnology products (MNPs). A number of possibilities for improvement in materials through nanotechnology were outlined by Arora et al. (2014). Nanopolymer coating has been shown to inhibit corrosion in steel where acid and alkaline induced corrosion is very common (Abdel-Hameed et al., 2013). Nanotechnology has also facilitated the production and use of nano self-cleaning and antimolding surfaces which have potential to facilitate efficient building maintenance. Nanocoating such as by using Titanium dioxide film applied to glass surfaces, solar panels, wall surfaces, and wall cladding materials can facilitate self-cleaning of the surfaces (Benedix et al., 2000; Quagliarini et al., 2012).

In another study, Tsang et al. (2016) showed that the use of organic photovoltaic charging units achieved 39%–98% reduction in life cycle impacts when compared to the conventional silicon material. Also, energy payback period for the alternative material indicated shorter periods of between 220 and 118 days.

9.4.7 Use of Eco-friendly Renewable Materials

The use of high-performance, high-tech sustainable materials may not always meet the needs of technologically less developed societies. In this respect, materials coming from renewable sources, where available, present a very good opportunity. There is growing evidence especially in literature that the use of locally available materials is an effective strategy to achieve energy efficiency and carbon mitigation at the embodied phase of buildings (Reddy and Kumar, 2010; Utama and Gheewala, 2009). Also, in a study by Myers et al. (2012) with the study context as Australia, it was found that by using renewable materials to replace some conventional building materials and components, it was possible to reduce embodied energy by 28% from 7.5 to 5.4 GJ/m^2.

The renewable materials worthy of note are usually naturally occurring fibers mainly from plant sources. Plant fibers are classified into three, depending on the part of the plant that provides the fiber (Mwaikambo, 2006): (1) seed/fruit fibers are extracted from seeds and fruits and examples include palm kernel fibers, coir or cocoanut fibers and cotton fibers; (2) bast or stem fibers are extracted from stems of plants such as bamboo, hemp, banana, and plantain; (3) leaf fibers are extracted from leaves of plants such as sisal and abaca.

Also, hemp, a plant often seen in negative term, is gradually coming into good acceptance as a renewable building material especially in Europe and the United States where laws now favor its cultivation. Hemp concrete has been shown to be more eco-friendly than conventional concrete panels for nonload bearing walls (Pretot et al., 2014). In South Africa, a hemp house has been built and it is being acclaimed as Africa's most sustainable building. Pretot et al. (2014) conducted an LCA of hemp concrete wall. It was found that while the highest impact was attributed to the production phase, it was also found that when compared to traditional construction materials, hemp concrete has a low impact on the environment. In addition, the hemp plant itself helped to reduce climate change as a result of its effect on carbon sequestration through photosynthesis. In an earlier study, Ip and Miller (2012) found that with a functional unit of 1 m^2 wall area, 300 mm thick, 82.7 kg of CO_2 could be sequestered with a net life cycle greenhouse emission reduction of 36.08 $kgCO_{2\text{-}e}$.

Natural plant fibers such as kenaf have shown considerable potentials as eco-friendly building materials. Studies by Ardente et al. (2008) and further studies by Batouli and Zhu (2013) indicate that kenaf fiber-based insulation materials have lower environmental impact than synthetic insulation materials. Also, in a study that compared the environmental impacts of earthen and conventional plaster, it was found that the environmental impact of earthen plaster based on clay is substantially less than the impact of conventional plaster based on Portland cement or hydraulic lime (Melia et al., 2014).

9.5 IMPLICATIONS

The strategies outlined in the foregoing can be grouped into four broad categories, namely, (1) improved material production processes for conventional building materials, (2) production and use of innovative building materials, (3) development of innovative construction and deconstruction methods, and (4) use of low-impact building materials. The roles of knowledge and technology are central in driving the aforementioned strategies. Also, different levels of implementation of sustainable construction strategies have been achieved in different contexts, thus suggesting varying degrees of challenges to, and opportunities for the adoption of sustainable construction practices. Hence, this section examines the drivers and barriers of sustainable construction as well as the role of enabling environment for the adoption of sustainable construction practices. For most contexts, sustainable construction remains work in progress.

9.5.1 Barriers and Drivers

The barriers to the adoption of sustainable material strategies are similar in nature but occur in varying degrees across geographical boundaries. They include level of skill and knowledge (Ahn et al., 2013; Alsanad, 2015), difficulty associated with adoption of new methods and technology (Hakkanin and Belloni, 2011), challenges of initial cost and long payback period (Ahn et al., 2013; Smith, 2012), import dependency (Smith, 2012), cultural change (Ametepey et al., 2015; Kasai and Jabbour, 2014) and lack of legislation (Ametepey et al., 2015). In order to overcome the barriers, there is a need to continuously improve on knowledge and skills on sustainable construction by all stakeholders. On the other hand, the drivers have been identified as rising energy cost (Ahn et al., 2013; Smith, 2012; Windapo, 2014), resource conservation (Ahn et al., 2013; Smith, 2012), legislation (Smith, 2012; Windapo, 2014), and self-legislation (Smith, 2012). The barriers and drivers are moderated by additional factors such as appropriate institutional support including legislation, promotion of environmental product declarations (EPDs), and adoption of building environmental assessment methods as outlined subsequently.

9.5.2 Institutional Support and Legislation

A critical evaluation of strategies to facilitate the drive for sustainability in the built environment underscored the central role governments at different levels can play (Raynsford, 2000). Government at the national level may commence the intervention by subscribing and domesticating international and regional protocols and agreements that encourage the reduction of the environmental impact of building and construction. A good example is the Energy Performance of Buildings Directive (EPBD) of the European Union, which in turn was inspired by the Kyoto Protocol on the reduction of carbon emissions. The EPBD which deals mainly with operational energy in buildings was first issued in 2002 and reissued in 2010 and requires all new buildings to be energy efficient by 2020 and all new public buildings to be energy efficient by 2018 (European Parliament, 2010). Individual countries in the European Union subsequently enacted legislations that would lead to the achievement of the target. Periodic evaluations are carried out to monitor progress made toward achieving the targets. Appropriate legislation to back the directives is subsequently enacted at national levels. Other roles of government include regulations made pursuant to the legislations and the use of incentives where necessary.

In the area of building materials and construction, the European Committee for Standardization (CEN) through CEN/TC350 (Sustainability of Construction Works) had set guidelines for the development of standardized methods for the assessment of the sustainability of new and existing construction works. These guidelines include core rules for the development of EPDs of construction products. EPDs with respect to construction products is guided by ISO 21930 (Sustainability in Building Constructions—Environmental Declaration of Building Products) and by CEN/TC350 for European countries. EPDs disclose quantified life cycle environmental information of building materials and products using well-established guidelines with the overall aim of encouraging demand and supply of products with favorable environmental impact (Tritthart et al., 2011).

9.5.3 Promoting Environmental Products Declarations

Closely related to institutional support is the promotion of EPDs. Sustainable material use is facilitated by the availability of accurate environmental information for building materials and products. EFDs are useful in this regard and are usually derived from product LCAs. They are based on applicable product category rules (PCRs), as defined in ISO 21930 for building products and are a set of specific rules, requirements and guidelines used for developing environmental declarations for products that can fulfill equivalent functions. PCRs determine what information should be gathered and evaluated for EPDs. The PCRs ensure that all products categories are harmonized (Tritthart et al., 2011). The EPD platform ensures that there is free flow of reliable environmental information about building products generated through LCA methodologies to all stakeholders to help in the selection and use of materials with reduced environmental impact. The role of government and government agencies is very crucial in this respect.

The above scenario of available material environmental information and procedures for achieving environmental responsibility characterize developed countries. The developing countries do not yet have systems for collecting, storing, and using data on environmental impact of materials on a continuous and widespread basis. In addition, many developing countries import large quantities of building materials from developed countries, which add to the environmental impact. The few LCA studies are mostly dependent on foreign databases for life cycle inventory as local inventories are in early stages of development (Harding et al., 2015). More LCA studies on materials in developing countries using local data are needed in order to provide background knowledge to inform decision-making. The volume of LCA studies in developing countries is indicative of the lack of adequate supporting framework for such studies (Maepa et al., 2017).

9.5.4 Adoption of Environmental Assessment Tools

Self-regulation through voluntary acceptance of building sustainability rating systems by building industry stakeholders has also been shown to be useful in promoting sustainable construction practices. The two leading building sustainability assessment tools, namely, (1) Leadership in Energy and Environmental Design (LEED) in the United States and (2) Building Research Establishment Environmental Assessment Method (BREEAM) in the United Kingdom are still largely voluntary building sustainability assessment methods. However, some level of certification is prescribed for public buildings. As a result, their influence and acceptance is growing in their countries of origin and internationally. Also, LEED Version 4 provides credits for using materials that have published LCAs and EPDs. Similarly, BREEAM materials credit takes care of extraction, production, transport, installation, and end-of-life using LCA methodologies compliant with ISO 14044. Hence, through EPDs, LCA become fully embedded in building sustainability assessment methods. By incorporating material credits into building sustainability assessment, LEED and BREEAM are stimulating building products manufacturers to improve the environmental rating of their products thereby encouraging designers to specify them for buildings. However, the score allocated to materials in the assessments have been estimated at 13% and 9%, respectively, which are considered too low to encourage designers to pay attention to the selection of sustainable materials (Ding, 2014).

9.6 CONCLUSION

The importance of sustainable construction to overall sustainable development has been underscored. It was also stressed that the use of high-impact materials and conventional construction practices still dominate the building and construction industry. The central role of materials in the attainment of sustainable construction was highlighted. In this respect, the route to sustainable construction was identified to be through improved material production processes, recycling, materials substitution, innovative construction methods, building for disassembly and deconstruction, use of innovative materials, and use of eco-friendly materials.

LCA was presented as a standardized method of quantifying a material's environmental impact over its life cycle in order to facilitate the availability of transparent material information in addition to facilitating the production and use of low-impact materials. The processes of collecting and disseminating environmental information regarding building materials are fairly well defined in developed countries where protocols, legislations, regulations, and voluntary adoption of building sustainability assessment methods have combined to drive the uptake of sustainable construction practices. In developing countries, building environmental assessments are still not well developed. LCA studies are few and far between are mostly dependent on external databases for life cycle inventory. Appropriate measures to encourage its use in developing countries are necessary. The role of policy interventions in encouraging attainment of sustainable construction is considered pivotal.

Sustainable construction cannot be achieved without effective collaboration of stakeholders such as government, manufacturers, building industry practitioners, and the public at different levels. In driving an innovation, awareness is fundamental. Also, effective information dissemination in a transparent manner to all stakeholders is important in order to facilitate uptake. The selection of materials for building and construction should be based on sound environmental information. Also, early integration of materials selection in the project planning stage is necessary. In all, the role of appropriate policy framework, driven by government in collaboration with building construction stakeholders, can engender the collective action needed to deliver sustainable construction. In this respect, the central role of legislation and the creation of enabling institutional environment to promote sustainable construction were underscored. Also, where it is found to be necessary, appropriate incentives may be introduced.

REFERENCES

Abdel-Hameed, R.S., Abu-Nawwas, A.H., Shehata, H.A., 2013. Nano-composites as corrosion inhibitors for steel alloys in different corrosive media. Adv. Appl. Sci. Res. 4 (3), 126–129.

Adedeji, O.H., Adebayo, H.O., Sotayo, E.I., 2014. Assessing the environmental impacts of inland sand mining in parts of Ogun State, Nigeria, Ethiop. J. Environ. Stud. Manage 7 (5), 478–487.

Ahn, Y.H., Pearce, A.R., Wang, Y., Wang, G., 2013. Drivers and barriers of sustainable design and construction: the perception of green building experience. Int. J. Sust. Build. Urban Develop. 3 (1), 35–45.

Alsanad, S., 2015. Awareness, drivers, actions, and barriers of sustainable construction in Kuwait. Proced. Eng. 118, 969–983.

Alyahya, A.A., Nawari, N.O., 2018. Towards net-zero energy in the hot-dry regions: building envelope design strategies for single family house. J. Sus Dev. 11 (3), 45–70.

Ametepey, O., Aigbavboa, C., Ansah, K., 2015. Barriers to successful implementation of sustainable construction in the Ghanaian construction industry. Proced. Manuf. 3, 1682–1689.

Ardente, F., Beccali, M., Cellura, M., Mistretta, M., 2008. Building energy performance: a LCA case study of kenaf-fibres insulation board. Ener. Build. 40, 1–10.

Arora, S.K., Foley, R.W., Youtie, J., Shapira, P., Wiek, A., 2014. Drivers of technology adoption — the case of nanomaterials in building construction. Technol. Forecast. Social Change. Available from: https://doi.org/10.1016/j.techfore.2013.12.017. (accessed 15.11.16).

Athena Institute (2009). Life Cycle Impact Estimator. Athena Sustainable Materials Institute and Morrison Hershfield Consulting Engineers, Canada.

Aziz, R.F., Hafez, S.M., 2013. Applying lean thinking in construction and performance improvements. Alexandria En. J. 52, 679–695.

Batouli, S.M. & Zhu, Y. (2013). Comparative life cycle assessment study of kenaf fibre-based and glass fibre-based structural insulation panels. In: International Conference on Construction and Real Estate Management. Karlsruhe, Germany, 10–11 October, 2013. Available from: https://doi.org/10.1061/9780784413135.036.

Begum, R.A., Siwar, C., Pereira, J.J., Jaafar, A.H., 2006. A benefit-cost analysis on the economic feasibility of waste minimization: The case of Malaysia. Resour. Conserv. Recycl. 48, 86–98.

Bell, I., Carry, V., Kuperus, S., Myers, L., Walsh, A., Walton, S., 2008. An ecological footprint analysis of the Department of Zoology, University of Otago. Otago Manage. Grad. Rev. 6, 1–20.

Benedix, R., Dehn, F., Quaas, J., Orgass, M., 2000. Application of titanium dioxide photocatalysis to create self-cleaning building materials. LACER 5, 157–168.

Bribian, I.Z., Capilla, A.V., Usón, A.A., 2011. Life cycle assessment of building materials: Comparative analysis of energy and environmental impacts and evaluation of the eco-efficiency improvement potential. Build. Environ. 46, 1133–1140.

CIB, 2004. 50 Years of International Cooperation to Build a Better World. CIB, Rotterdam, Netherlands.

Chau, Y.K., Yik, F.W.H., Hui, W.K., Liu, H.C., Yu, H.K., 2007. Environmental impacts of building materials and building services components for commercial buildings in Hong Kong. J. Clean. Product. 15, 1840–1851.

Cole, R.J., 2003. Building environmental assessment methods: a measure of success. Fut. Sustain. Constr. (Special Issue) 1–8.

Dinakar, P., Babu, K.G., Santhanam, M., 2007. Corrosion behavior of blended cements in low and medium strength concretes. Cement Concr. Composit. 29, 136–145.

Ding, G.K.C., 2014. Life cycle assessment (LCA) of sustainable building materials: an overview. In: Eco-efficient Construction and Building Materials. Life Cycle Assessment (LCA), Eco-labelling and Case Studies. Woodhead Publishing Limited, Cambridge, UK, pp. 38–62.

Du Plessis, C., 2002. Agenda 21 for Sustainable Construction in Developing Countries: A Discussion Document. CIB and UNEP-IETC Project. CSIR Building and Construction Technology, Pretoria, South Africa.

Du Plessis, C., 2007. A strategic framework for sustainable construction in developing countries. Constr. Manage. Econom. 25, 67–76.

Edwards, B., 2002. Rough Guide to Sustainability, first ed. RIBA Publications Ltd, London, UK.

Ekoe-a-Akata, M.A., Njomo, D., Mempouo, B., 2015. The effect of building integrated photovoltaic system (BIPVs) on indoor temperature and humidity (IATH) in the tropical region of Cameroon. Fut. Cities Environ. 1 (1). Available from: https://doi.org/10.1186/s40984-015-0002-y.

Enerdata (2013). The State of Global Energy Efficiency: Global and Sectoral Energy Efficiency Trends. ABB Ltd, Zurich, Switzerland. Retrieved from: <www.abb.com/energyefficiency> (accessed 13.10.14).

European Commission, 2011. Reducing the environmental impact of building materials. Retrieved from: ec.europa.eu/environment/integration/research/newsletter/pdf/232na1_enpdf. (accessed 21.09.18).

European Parliament, 2010. Directive 2010/31/EU on the Energy Performance of Buildings (Recast). Official Journal of the European Union, Brussels, Belgium.

Ezema, I.C. (2015). Profiling the Environmental Sustainability of Residential Buildings in Lagos, Nigeria Using Life Cycle Assessment. Ph.D. thesis, Covenant University, Ota, Ogun, Nigeria.

Forsberg, A., Von Malmborg, F., 2004. Tools for environmental assessment of the built environment. Build. Environ. 39, 223–228.

Franzoni, E., 2011. Materials selection for green buildings: which tools for engineers and architects? Proced. Eng. 21, 883–890.

Garcia-Segura, T., Yepes, V., Alcala, J., 2014. Life cycle greenhouse gas emissions of blended cement including carbonation and durability. Int. J. Life Cycle Assess. 19 (1), 3–12.

Grimaud, G., Perry, N., Laratte, B., 2016. Life cycle assessment of aluminium recycling process: case of shredder cables. Proced. CIRP 48, 212–218.

Grimes, S., Donaldson, J., Gomez, G.C. 2008. Report on the environmental benefits of recycling, Bureau of International Recycling. Retrieved from: http://www.mgg-recycling.com/wp-content/uploads/2013/06/BIR_CO2_report.pdf. (accessed 23.09.18).

Hakkanin, T., Belloni, K., 2011. Barriers and drivers for sustainable building. Build. Res. Informat. 39 (3), 239–255.

Hammond, G., Jones, C., 2011. Inventory of Carbon and Energy (Version2.0). Sustainable Energy Research Team, University of Bath, Bath, UK.

Harding, K.G., Basson, L., Brent, A., Friedrich, E., van Vuuren, P.J., Mbohwa, C., et al. (2015). Status and prospects of life cycle assessment in South Africa. In: Life Cycle Management Conference, August 2015, Bordeaux France. Retrieved from: <www.researchgate.net/publication/281292750>.

Hassanbeigi, A., Price, L., Chunxia, Z., Aden, N., Xiuping, L., Fangqin, S., 2014. Comparison of iron and steel production energy use and energy intensity in China and USA. J. Clean. Product. 65, 108–119.

Hidalgo, I., Szabo, L., Calleja, I., Ciscar, J.C., Russ, P., Soria, A., 2003. Energy consumption and CO_2 emissions from the World Iron and Steel Industry. Institute for Prospective Technological Studies. Joint Research Centre, European Commission.

Huntzinger, D.N., Eatmon, T.D., 2009. A life cycle assessment of Portland cement manufacturing: comparing the traditional process with alternative technologies. J. Clean. Product. 17, 668–675.

Husband, C., McMahon, G., van der Veen, P., 2009. The aluminium industry in West and Central Africa: lessons learnt and prospects for the future. Extractive Industries and Development Series No. 13. The World Bank, Washington DC.

International Finance Corporation (IFC), 2014. Waste Heat Recovery for the Cement Sector: Market and Supplier Analysis. International Finance Corporation, Washington DC.

International Finance Corporation (IFC), 2017. Improving Thermal and Electric Energy Efficiency at Cement Plants: International Best Practice. International Finance Corporation, Washington, DC.

International Organization for Standardization (ISO), 2006. Life Cycle Assessment (ISO14040). ISO, Geneva, Switzerland.

Ip, K., Miller, A., 2012. Life cycle greenhouse gas emissions of hemp-lime wall constructions in the UK. Resour. Conserv. Recycl. 69, 1–9.

Kasai, N., Jabbour, C.J.C., 2014. Barriers to green building at two Brazilian Engineering Schools. Int. J. Sustain. Built Environ. 3, 87–95.

Koskela, L., 1992. Application of the New Production Philosophy to Construction. Technical Report No. 72, CIFE. Stanford University, CA.

Koukkari, H., Braganca, L., 2011. Review on the European strategies for energy-efficient buildings. Int. J. Sustain. Develop. Urban Develop. 2 (1), 87–99.

Legmann, H. (2002). Recovery of industrial heat in the cement industry by means of the ORC process. In: Proceedings Cement Industry Technical Conference, Jacksonville. Florida. USA, 5-9 May, 2002, pp. 29–35. Available from: https://doi.org/10.1109/CITCON.1006482.

Li, Y., Xu, W., Zhu, T., Qi, F., Xu, T., Wang, Z., 2012. CO2 emissions from BF-BOF and EAF steelmaking based on material flow analysis. Adv. Mat. Res. 518–523, 5012–5015. Available from: https://doi.org/10.4028/www.scientific.net/AMR.518-523.5012.

Madlool, N.A., Saidur, R., Hossaina, M.S., Rahim, N.A., 2011. A critical review on energy use and savings in the cement industries. Renew. Sustain. Ener. Rev. 15 (4), 2042–2060.

Madlool, N.A., Saidur, R., Rahim, N.A., 2012. Investigation of waste heat recovery in cement industry: a case study. IACSIT Int. J. Eng. Technol. 4 (5), 665–667.

Maepa, M., Bodunrin, M.O., Burman, N.W., Croft, J., Engelbrecht, S., Ladenika, A.O., et al., 2017. Life cycle assessment in Nigeria, Ghana and Ivory Coast. Int. J Life Cycle Assess. (7).

Melia, P., Ruggeri, G., Sabbadini, S., Doltelli, G., 2014. Environmental impacts of natural and conventional building materials: a case study on earth plasters. J. Clean. Product. 80, 179–186.

Meriani, S., 2008. Available technologies for local building materials. UNIDO & International Centre for Science and High Technology, Trieste, Italy. Retrieved from: https://institute.unido.org/wp-content/uploads/2014/11/12. (accessed 20.09.18).

Mensah, A.K., Mahiri, I.O., Owusu, O., Mireku, O.D., Wireko, I., Kissi, E.A., 2015. Environmental impacts of mining: a study of mining communities in Ghana. Appl. Ecol. Environ. Sci. 3 (3), 81–94.

Monahan, J., Powell, J.C., 2011. An embodied carbon and energy analysis of modern methods of construction in housing: A case study using a life cycle assessment framework. Ener. Build. 43 (1), 179–188.

Mwaikambo, L.Y., 2006. Review of the history, properties and application of plant fibers. Afr. J. Sci. Tech. 7 (2), 120–133.

Myers, F., Fuller, R. & Crawford, R.H. (2012). The potential to reduce the embodied energy in construction through the use of renewable materials. In: Proceedings of the 46th Annual Conference of the Architectural Science Association, Queensland, Australia. Retrieved from: <http://dro.deakin.edu.au/view/DU: 30051745>.

Nadim, W., Goulding, J.S., 2010. Offsite production in the UK: The way forward? A UK construction industry perspective. Constr. Innov. 10 (2), 181–202.

National Audit Office, 2005. Using Modern Methods of Construction to Build Homes More Quickly and Efficiently. Report by the National Audit Office. National Audit Office, London, UK.

O'Brien, M., 2011. Resource efficient construction: A systematic approach to sustainable construction. Eco-innovation Brief No. 4, Wuppertal Institute. Retrieved from: http://ec.europa.eu/environment/archives/ecoinnovation2012/1st_forum/pdf. (accessed 21.09.18).

OECD (2012), Sustainable Materials Management – Making Better Use of Resources. OECD, Paris, France. Retrieved from: <https://doi.org/10.1787/9789264174269-en> (accessed 18.09.17).

OECD (2015), Material Resources, Productivity and the Environment. OECD, Paris, France. Retrieved from: <https://doi.org/10.1787/9789264190504-en> (accessed 18.09.17).

Odigure, J.O. (2009). Duality of Cement-Based Structures: Mitigating Global Warming and Building Collapse. Inaugural Lecture Series 13. Federal University of Technology, Minna, Nigeria.

Ohunakin, O.S., Leramo, O.R., Abidakun, O.A., Odunfa, M.K., Bafuwa, O.B., 2013. Energy and cost analysis of cement production using the wet and dry processes in Nigeria. Ener. Power Eng. 5, 537–550.

Olivieri, G., Romani, A., Neri, P., 2006. Environmental and economic analysis of aluminium recycling through life cycle assessment. Int. J. Sustainable Develop. World Ecol. 13 (4), 269–276.

Olson, T.P. (2010). Design for deconstruction and modularity in a sustainable built environment. Retrieved from: <https://cmec.wsu.edu> (accessed 21.09.17).

Oner, A., Akyuz, S., 2007. An experimental study on optimum usage of GGBS for the strength of concrete. Cement Concr. Composit. 29, 505–514.

O'Brien, E., Guy, B., Lindner, A.S., 2012. Life cycle assessment of the deconstruction of military barrack: Ft McClellan, Anniston, AL. J. Green Build. 1 (4), 166–183.

Papadopoulos, A., Karamanos, A. & Avgelis, A. (2002). Environmental impact of insulatiom materials at the end of their useful life. Retrieved from: <https://www.researchgate.net/publication/237448672> (accessed 19.09.17).

Portland Cement Association (2014). How Cement is Made. Portland Cement Association, Washington, DC. Retrieved from: <www.cement.org/cement-concrete-basics/how-cement-is-made> (assessed 02.10.14).

Pretot, S., Collet, F., Garnier, C., 2014. Life cycle assessment of a hemp concrete wall: Impact of thickness and coating. Build. Environ. 72 (C), 223–231.

Quagliarini, E., Bondioli, F., Goffredo, G.B., Cordoni, C., Munafo, P., 2012. Self-cleaning and de-polluting stone surfaces: TiO_2 nanoparticles for limestone. Constr. Build. Mater. 37, 51–57.

RICS. (2012). Methodology to Calculate Embodied Carbon of Materials. RICS Information Paper IP32/2012. Royal Institute of Chartered Surveyors, UK. Retrieved from: <www.rics-books.com> (accessed 29.10.14).

Rajdev, R., Yadav, S., Sakale, R., 2013. Comparison between Portland pozzolana cement and processed fly ash blended ordinary Portland cement. Civil Environ. Res. 3 (6), 24–30.

Raynsford, N., 2000. Sustainable construction: the role of government. Proc. Inst. Civil Eng. 136 (6), 16–22.

Reddy, B.V.V., Kumar, P.P., 2010. Embodied energy in cement stabilised rammed earth walls. Ener. Build. 42, 380–385.

SETAC (Society of Environmental Toxicology and Chemistry), 1991. A Technical Framework for Life Cycle Assessment. Society of Environmental Toxicology and Chemistry, Washington DC.

Sartori, I., Hestnes, A.G., 2007. Energy in the life cycle of conventional and low-energy buildings: a review article. Ener. Build. 39, 249–257.

Shumaker, D., 2009. Design for deconstruction. Seminar on Sustainable Architecture, School of Architecture, The University of Texas at Austin. Retrieved from: https://soa.utexas.edu/sites/default/disk/.../09_0_su_shumaker_daniel_paper_ml.pdf. (accessed 21.09.18).

Siemens, 2009. How does cement production become energy efficient and environmentally friendly? Sustainable solutions optimize production processes and reduce CO2 emissions. Siemens AG Industry Sector (Cement), Erlangen, Germany.

Siemens. (2011). Process Analytics in the Iron and Steel Industry – Case Study. Siemens Industry Inc., Houston, TX. Retrieved from: <www.usa.siemens.com/processanalytics> (accessed 25.09.14).

Singh, P.K., Singh, R.S., 2016. Environmental and social impacts of mining and their mitigation, National seminar ESIMM, Kalkota, India, September 20. Retrieved from: https://www.researchgate.net/profile_singh35/publications/...308937912. (accessed 29.10.18).

Smith, S. (2012). Sustainable Construction: Key Drivers & Barriers. Retrieved from: <www.smithandwallwork.com/wp.../SaW_UoC_sustainability_panel_Sept_2012.pdf> (accessed 21.09.17).

Thormark, C., 2002. A low energy building in a life cycle—Its embodied energy, energy need for operation and recycling potential. Build. Environ. 37, 429–435.

Thormark, C., 2006. The effect of material choice on the total energy need and recycling potential of a building. Build. Environ. 41, 1019–1026.

Tritthart, W., Staller, H., Peuportier, B., Cai, X., Wetzel, C., Zabalza, I., et al. (2011). Use and Availability of Environmental Product Declarations (EPD), LoRe – LCA-WP2-D2.3-IFZ-2011-12-15.

Tsang, M.P., Sonnemann, G.M., Bassani, D.M., 2016. Life cycle assessment of cradle-to-grave opportunities and environmental impacts of organic photovoltaic solar panels compared to conventional technologies. Solar Ener. Mater. Solar Cells 156, 37–48.

UNEP. (2007). Buildings and Climate Change: Status, Challenges and Opportunities. United Nations Environment Programme. Retrieved from: http://www.hdl.handle.net/20.500-11822/7783. (accessed 25.01.15).

UNEP. (2011). Global Guidance Principles for Life Cycle Assessment Databases: A Basis for Green Processes and Products. United Nations Environment Programme. Retrieved from: <www.unep.org/pdf/Global-Guidance-Principles-for-LCA.pdf> (accessed 23.07.13).

UNEP-SBCI. (2010). Common Carbon Metric for Measuring Energy Use and Reporting Greenhouse Gas Emissions from Building Operations. Retrieved from: <http://www.unep.org/sbci/pdfs/UNEPSBCIcarbonmetric.pdf> (accessed 13.06.13).

USEPA. (2006). Life Cycle Assessment: Principles and Practice. Retrieved from: <http://www.epa.gov./ORD/NRMRL/lcaccess> (accessed 12.12.12).

Utama, A., Gheewala, S.H., 2009. Influence of material selection on energy demand in residential houses. Mater. Design 30, 2173–2180.

Utochikina, E., 2014. Heat insulation materials from environmental aspect. Retrieved from: https://www.theseus.fi/bitstream/handle/10024/81138/Utochkina_Evgeniia.pdf;sequence = 1. (accessed 24.09.18).

Wackernagel, M., Rees, W.E., 1996. Our ecological footprint: Reducing human impact on earth. New Society Publishers, British Columbia.

Wang, K. & Ge, Z. (2003). Evaluating Properties of Blended Cement for Concrete Pavements. Final Report, Centre for Portland Cement Concrete Pavement Technology in Association with Iowa State University.

WCED, 1987. Report of the World Commission on Environment and Development: Our Common Future. Retrieved from: http://www.un-documents.net/our-common-future.pdf"www.un-documents.net/our-common-future.pdf. (accessed 21.09.18).

Wescott, R.F., McNulty, M.W., VanGeem, M.G., Gajda, J., 2010. Prospects for expanding the use of supplementary cementitious materials in California. Coalition for Sustainable Cement Manufacturing and Environment (Keybridge Research). Retrieved from: http://keybridgedc.com/wp-content/uploads/2013/11/Keybridge-CSCME-SCM-Feb-2010.pdf. (accessed 24.09.18).

Wiedmann, T., Minx, J., 2008. A definition of carbon footprint. In: Pertsova, C.C. (Ed.), Ecological Economics Research Trends, first edition, Nova Science Publishers, New York, pp. 1–11. Retrieved from: https://www.novapublishers. (accessed 24.09.18).

Windapo, A.O., 2014. Examination of green building drivers in the South African construction industry: economics versus ecology. Sustainability 6, 6088–6106.

Wong, S., 2014. An overview of Nanotechnology in building materials. Can. Young Sci. J. 14 (2), 18–21.

World Steel Association (WSA). (2015). Steel's Contribution to a Low Carbon Future and Climate Resilient Societies. Worldsteel Position Paper. Retrieved from: "http://www.worldsteel.org/". (accessed 12.12.16).

World Steel Association (WSA). (2016a). Fact Sheet: Energy Use in the Steel Industry. Retrieved from: <www.worldsteel.org> (assessed 14.09.17).

World Steel Association (WSA). (2016b). Sustainable Steel: Policy and Indicators. Retrieved from: <www.worldsteel.org> (accessed 20.09.17).

World Steel Association (WSA). (2018). Fact Sheet: Energy Use in the Steel Industry. Retrieved from: <www.worldsteel.org>. (accessed 20.09.18).

Worrell, E., Martin, N., Price, L., 2000. Potentials for energy efficiency improvements in the US cement industry. Energy 25 (12), 1189–1214.

Yoon, S.H., Park, C.S., 2017. Objective building energy performance benchmarking using data envelopment analysis and Monte Carlo sampling. Sustainability 9 (5), 780. Available from: https://doi.org/10.3390/su9050780. 1-12.

Chapter 10

Emissions

G.K.C. Ding
School of Built Environment, University of Technology Sydney, NSW, Australia

10.1 INTRODUCTION

Buildings have great impacts on the environment and their construction consume approximately 40% of the stone, sand, and gravel; 25% of the timber; and 16% of the water used annually in the world (Arena and de Rosa 2003). In addition, the construction industry has been regarded as a major consumer of energy and emitter of greenhouse gases (GHGs). It is claimed that buildings are responsible for approximately 19% of GHGs emitted globally (IPCC, 2014) or approximately 29% in Canada (Finnveden and Moberg, 2005), 30% in the United States (Levermore, 2008), and 50% in the European Union (Yan et al., 2010). Hence buildings offer the greatest potential for reducing global GHG emissions (Giesekam et al., 2016) and much attention has been given to ways of improving its efficiency (Ali et al., 2013; Hong et al., 2015).

In the past, research studies focused on improving operating efficiency as it was regarded as the stage that consumes most of the energy and generates most of the GHGs during the building life cycle. However, GHG emissions due to buildings operation have been reduced progressively through shifting to the use of renewable energy, and the advancement and incorporation of technologies into the design and construction of green buildings (Meggers et al., 2012).

With the reduction in operating emissions, embodied emissions (EEs) are now attracting more attention in building life cycle as EEs are permanent and take immediate effect from when the raw material is extracted (Dimoudi and Tompa, 2008; Yan et al., 2010; Oregi et al., 2015; Chen and Ng, 2016; de Wolf et al., 2016). According to Omar et al. (2014), more than 90% of GHG emissions are generated from the upstream boundary of the supply chain in product manufacturing. Hong et al. (2015) and Huisingh et al. (2015) state that the EEs in the cradle-to-gate stage that includes manufacturing of materials, transforming them into products/components for the construction, and transporting them to site can account for between 88% and

Sustainable Construction Technologies. DOI: https://doi.org/10.1016/B978-0-12-811749-1.00008-0

96% of the total emissions. The increased attention on reducing EEs in the material is challenging and can be met by lower emission materials and by changing the perspective people have on building materials (Chau et al., 2012; Meggers et al., 2012).

This chapter begins with a discussion of the various policies, standards, and programs related to GHG emissions in an international perspective, followed by the calculation methods. This chapter also reviews the life cycle assessment (LCA) approach in the assessment of emissions of buildings in various stages and the development of sustainable building products and technologies to reducing emissions in buildings. The chapter ends with a discussion on emission strategies and two international case studies.

10.2 INTERNATIONAL POLICIES, STANDARDS, AND PROGRAMS ON REDUCING GREENHOUSE GAS EMISSIONS

GHG emissions are considered to be the main contributor to global climate change. GHGs act as a blanket of gases that trap heat and make the planet warmer, and as we have seen, it is closely related to the energy consumption at various stages in a building life cycle. The largest source of GHG emissions is fossil fuels for electricity, heat, and transport. According to the Kyoto Protocol, GHG includes various types of gases, of which carbon dioxide (CO_2), methane (CH_4), and nitrous oxide (N_2O) are closely related to the construction industry (Cole, 1999; Acquaye and Duffy, 2010; Mao et al., 2013) and account for more than 97% of the total global warming potential (GWP) (Nadoushani and Akbarnezhad, 2015).

GHG emissions in the building have different impacts on global warming and are mainly measured in the impact category of GWP in LCA. CO_2 is usually used as a reference standard for GHG effects (Cole, 1999). All other gases are expressed as a ratio of the warming caused by an equivalent mass of CO_2 and is termed as CO_2 equivalent ($CO_{2\text{-}e}$). The GWP of the energy-related GHGs is regulated over a 100-year time frame and the GWPs for each of the above emissions were added to give total $CO_{2\text{-}e}$ intensities for the construction sector (Acquaye and Duffy, 2010). Table 10.1 presents the GWP values of CH_4 and N_2O relative to CO_2 over a 100-year time frame.

Over the years, governments and international organizations, such as the US EPA, International Standard Organisation, around the world have developed policies, regulations, and standards to reduce GHG emissions. Policy makers have introduced regulations in the construction industry requiring improvements in the building fabric and performance for energy efficiency. This has prompted the issue of the European Union Energy Performance of Building Directive (Giesekam et al., 2016) and the Australian National Construction Code Section J (ABCB, 2010). However, these policies and regulations principally focused on the operating GHG emissions associated with energy consumption for heating and cooling, lighting, and ventilation

TABLE 10.1 Global Warming Potential Values Relative to CO_2

Greenhouse gases	Chemical formula	Global warming potential for 100 years
Carbon dioxide	CO_2	1
Methane	CH_4	21
Nitrous oxide	N_2O	265

Source: IPCC, 2014. Climate change 2014: impacts, adaptation, and vulnerability. Part A: Global and sectoral aspects. In: Field, C.B., Barros, V.R., Dokken, D.J., Mach, K.J., Mastrandrea, M.D., Bilir, T.E., et al. (Eds.), Contribution of Working Group II to the Fifth Assessment Report of the Intergovernmental Panel on Climate Change. Cambridge University Press, Cambridge, UK/New York.

but have not been extended to the embodied emission associated with the initial production of building materials and products.

Table 10.2 summarizes some of the key international policies, standards, and initiatives to address the gap. The most important one is the International Organization for Standardization (ISO) standards. In the 1990s, an environmental management standard was adopted by the ISO as part of its 14000 standard series. The purpose was to provide consistency, transparency, and credibility in the assessment, monitoring, reporting, and verification of GHG emissions (ISO, 2016a,b).

In this series ISO 14040 Standards focus on establishing principles and framework, and 14044 Standards provide requirements and guidelines for the LCA study (ISO, 2006a,b). According to the ISO 14040 and 14044 Standards the LCA framework includes four steps: (1) goal and scope definition, (2) inventory analysis, (3) life cycle impact assessment, and (4) interpretation. The ISO standard contained in these Standards is fundamental to the standardization and the ability to generalize findings pertaining to LCA and life cycle inventory (LCI) studies. Finkbeiner (2014) states that ISO 14040 and 14044 are the foundation for all the quantification standards.

In addition to the ISO 14000 environmental management standard series, the ISO 14064 was developed and launched in 2006 (ISO, 2006c). The ISO 14064 contains three parts to define how to quantify, monitor, report, and verify GHG emissions at the organizational and project levels (Wintergreen and Delaney, 2007; Baumann and Kollmuss, 2010; Chen and Ng, 2016). Parts 1 and 2 are the guidance of GHG quantification at organization and project level, respectively, whilst Part 3 deals with the validation and verification of GHG assertion. The other ISO standards related to GHG emissions include ISO 14065 provides requirements for GHG validation and verification bodies for use in accreditation or other form of recognition, ISO/TS 14067 is the technical specifications of carbon footprints of products, and ISO/TR 14069 is the guidance for the application of ISO 14064-1.

TABLE 10.2 International Policies, Standards, and Programs

Policies/Standards/ Programs	Countries	Year	Aims
ISO 14040 Environmental management—life cycle assessment—Principles & framework	International	2006	• To provide consistency in the principles and framework to undertake LCA study
ISO 14044 Environmental management—life cycle assessment—Requirements & guidelines	International	2006	• To provide requirements and guidelines for the LCA study
ISO 14064 Greenhouse gases accounting and verification	International	2006	Contains three parts to define how to quantify, monitor, report and verify GHG emissions at organization and project levels • ISO 14064-1: specifies principles and requirements for the quantification and reporting of GHG emissions and removals • ISO 14064-2: specifies principles and requirements and provides guidance for the quantification, monitoring and reporting of activities intended to cause GHG emission reductions or removal enhancements • ISO 14064-3: specifies principles and requirements and provides guidance for the validation and/or verification of GHG assertions in accordance with ISO 14064-1 or ISO 14064-2
ISO 14065 Greenhouse gases—Requirements for GHG validation and verification bodies for use in accreditation or other forms of recognitions	International	2007 2013	• To provide requirements for bodies that undertake GHG validation or verification using ISO14064

(*Continued*)

TABLE 10.2 (Continued)

Policies/Standards/ Programs	Countries	Year	Aims
ISO/TS 14067 Greenhouse gases—Carbon footprint of products—Requirements and guidelines for quantification and communication	International	2013	• Includes principles, requirements, and guidelines for the quantification and communication of carbon footprint of products • Uses principles of ISO 14044 but it adds some specificity
ISO/TR 14069 Greenhouse gases—Quantification and reporting of greenhouse gas emissions for organizations—Guidance for the application of ISO 14064-1	International	2013	• Describes the principles, concepts, and methods of quantification and reporting GHG emissions in terms of direct emissions, energy indirect emissions and other indirect emissions for all organizations
PAS 2050 Carbon footprint	United Kingdom	2008	• Provides guidelines for the assessment of life cycle GHG emissions of goods and services
PAS 2080 Carbon management in infrastructure	United Kingdom	2016	• Set out the general principle and guidelines for managing carbon for the transport, energy, water, waste and communication sectors
GHG Protocol Product Standard	International	2011	• To provide requirements for measuring, managing and reporting on GHG emissions • To enable meaningful comparison between products
Carbon Footprint	International	2003	• To measure the amount of carbon in the atmosphere contributing to global warming and climate change at product, project and city scales • To account for both direct and indirect emission caused by activities of individual, organization or nations

The GHG Protocol Product Standard is another well-known international standard, which was jointly developed by the World Resources Institute (WRI) and the World Business Council for Sustainable Development (WBCSD) in 2011. The GHG Protocol aims at providing requirements for measuring, managing, and reporting GHG emissions of products and services. It also sets requirements and a framework for public reporting on GHG emissions. According to Garcia and Freire (2014), the GHG Protocol provides an important platform to compare performance between products and for the reduction of GHG emissions in the manufacturing processes.

Carbon footprint (CF) is another approach that has been widely used to measure GHG emissions at product, project, and city scales. CF is an account of GHG produced to support directly and indirectly human activities, usually expressed in equivalent tons of CO_2. The term CF was used to refer to the GWP and was first defined by Høgevold (2003, cited in Cucek et al., 2012) that relates the amount of carbon in the atmosphere contributing to global warming and climate change. However, CF is not expressed in terms of the area though it has "footprint" in its name. The total amount of GHG is simply measured in mass units (e.g., kg, tonnage) and no conversion to an area unit is required (Galli et al., 2012).

In the assessment of CF it is important to allocate the in-boundary and transboundary GHG emissions in order to support policy making strategies in its reduction. WRI has developed a concept of "scope" to describe the in-boundary and transboundary emissions helping to define the boundaries of CF assessment (Ranganathan et al., 2004). The three scopes are:

- Scope 1 Direct Emissions—This scope encompasses the direct emission from combustion of in-boundary fossil fuels.
- Scope 2 Indirect Emissions—This scope includes indirect emissions from energy generated outside the established in-boundary limits as a consequence of the consumption of grid electricity.
- Scope 3 Indirect Emissions—This scope includes the whole indirect emission from processes of the supply chain life cycle for materials and energy carriers which are produced outside of the boundary. It should be noted that in a consumption-based approach, the life cycle GHGs emitted from products produced within the boundary that are exported for consumption outside should be subtracted.

According to the Environmental Protection Agency (EPA, 2007) and WRI (Ranganathan et al., 2004), the first and second scopes are imperative for corporate accounting along with the consideration of applicable items of Scope 3 that formulate a supply chain GHG mitigation strategies. As there is an increase in the number of CF product specifications and there is a growing uncertainty about which method to use and how to report, the ISO/TS 14067 (2013) was therefore developed in 2008 to fill the gap. ISO/TS 14067 was developed as a consensual international standard of the CF of products to ensure consistency in the way it is determined and reported. It is largely built on the ISO 14040, 14044, and 14025 (Garcia and Freire, 2014; ISO/TS, 2013; Chen and Ng, 2016).

Nationally, the UK PAS (Publicly Available Specification) 2050 was first developed in 2008 by the British Standard Institution (BSI) and further revised in 2011 to align with the GHG Protocol Product Standard (Garcia and Freire, 2014). PAS 2050 is the first consensus based and internationally applicable standard that specifies requirements for the assessment of the life cycle GHG emissions of goods and services. It is built upon the LCA guidelines and requirements articulated in the ISO 14040 and 14044 by adopting a life cycle approach to GHG emissions assessment (Sinden, 2009; BSI, 2011). This methodology accounts for emissions of all GHGs and most importantly CO_2, N_2O, and CH_4, and each gas is converted to a CO_2 equivalent. Furthermore, the PAS 2050 also deals with other relevant methods and approaches in the field of GHG assessment such as ISO 14064. In 2016, the PAS 2080 was developed as the world's first industry standard on managing and reducing whole life carbon emissions in infrastructure that includes transport, energy, water, waste, and communication. The PAS 2080 was developed by BSI and sponsored by The Green Construction Board (2016).

10.3 CALCULATION METHODS

10.3.1 Traditional Approaches

There are different ways to calculate environmental impacts and GHG emissions and these include process-based, economic input–output (EIO), and hybrid methods. These methods are used to develop an inventory analysis of activities in relation to the life cycle of a building. According to EN 15978 Standard (CEN, 2011) the building life cycle is divided into stages of production, construction, use, and end-of-life (see Fig. 10.2). A building life cycle consists of initial raw material extraction and production, transportation, construction, operation, and eventual demolition on site at the end of the life cycle (Yan et al., 2010). Emission-related activities are assessed and collected. Once emission-related data are collected, emission factors are used to convert these data into GHG emissions. An emission factor is a ratio of the amount of a GHG emitted as a result of a given unit of activity (Wei et al., 2013). Then GWP is used to convert individual GHG into an equivalent value of CO_2. GWP is a measure of all gases set in equivalent to CO_2 and has a GWP of 1 kg as indicated in Table 10.1. Other gases are expressed as a multiple of CO_2 such as CH_4 is expressed as 21 kg $CO_{2\text{-}e}$ for a 100-year time frame (Table 10.1). The source of emission data can be obtained first hand from primary data onsite but this would be time consuming and costly. In addition, data can also be obtained from official statistical data from government bodies, scientific peer-reviewed literature, and emission factors from manufacturers.

The process-based method is known as a bottom-up approach that focuses on the different activities associated with a product or service around process flow diagrams (Guggemos and Horvath, 2005; Omar et al., 2014). For every process, all materials and energy used in the process are identified and then converted into GHG emissions by multiplying it with the corresponding GHG emission coefficient (Yan et al., 2010). However, this method may be restricted by the confidentiality and unavailability of certain data (Chang et al., 2012) and fails to include the upstream boundary of material production (Omar et al., 2014).

The EIO analysis-based method is known as a top-down approach that considers both direct and indirect impacts of a product or service of the entire economic supply chain (Gerilla et al., 2007; Yan et al., 2010; Omar et al., 2014). This method bases the estimation on the input–output table to assess GHG emission in the construction industry that includes both direct and indirect impacts such as those that may arise from upstream supply chains (Yan et al., 2010). A geographical region, such as a national economy, is used as a boundary of analysis which incorporates economy-wide interdependencies. This method is widely used and based on EIO technique and publicly accessible data on sectoral economic and energy performance (Chang et al., 2012). The main problems of this method include uncertainty of data quality, data aggregation, the homogeneity assumption, age of data, and capital equipment (Omar et al., 2014).

The hybrid method combines the strength of the process based and the EIO method to overcome the weaknesses inherited in both calculation methods (Chang et al., 2012). This hybrid method uses the process-based method to calculate GHG emissions of different activities during the production and the EIO method to calculate GHG emissions from upstream production. The total GHG emission is the sum of both methods. In this method, the input–output data were used to extract the most important pathways and then replaced with by the process-based data (Omar et al., 2014). According to Chang et al. (2012), the hybrid method is considered a nearly perfect tool for the assessment of life cycle GHG emissions.

10.3.2 Digital Approaches

In addition to the traditional calculation methods as discussed above, GHG emissions can also be calculated using computer software on an LCA approach (de Wolf et al., 2016):

- Proprietary computer software such as GaBi from Germany and SimaPro from the Netherland are commonly used to calculate GHG emissions on a life cycle approach;

- OpenLCA developed since 2006 by GreenDelta 2 from Germany is an open source software that helps users perform LCAs of buildings;
- Athena Institute is a nonprofit organization based in Canada and has developed a tool called the Athena Eco Calculator;
- TALLY tool is the first LCA app to calculate environmental impact of buildings that extracts data from Revit models;
- eTool is an Australian web-based LCA tool for environmental assessment of materials, products, processes, and whole building.

With the advent of digital technologies such as Building Information Modeling (BIM), virtual reality (VR), and augmented reality (AR), the assessment of GHG emission of buildings has now been taken into a new era and has increasingly attracted attention in the construction industry for the design and construction of buildings. The concept of digital technology is to build a three-dimensional (3D) virtual representation of buildings, so that building components can be properly planned before tangibly build the building. The 3D virtual model helps to improve the quality of documentation, increase productivity, and improve visibility while minimizing negative impact on the environment (Wang et al., 2013; Shafiq et al., 2015).

BIM model contains geometric and functional properties of smart objects for visualization and simulation to facilitate the interdisciplinary flow of information and data for building projects over the life cycle. While BIM has the ability to provide visualization of buildings, VR and AR can excel design to real-time experience to give the sense of scale, depth, and spatial awareness by using 3D augmented modes, images, voiceovers, and animation (Ding et al., 2014). Nowadays, when BIM model works alongside with VR and AR, it creates a platform to explore different design options to permit 3D visualization to be manipulated in a real environment (Whyte et al., 2000; Wang et al., 2013). The integration of BIM and VR/AR provides a direct and dual-directional link so that changes can be transferred automatically in order that more sustainable built environment outcomes through better planning and analysis of design options can be achieved.

There is an emerging trend to consider sustainability in BIM to support an early incorporation of LCA to achieve the optimal design option to minimize environmental loads (Stadel et al., 2011; Memarzadeh and Golparvar-Fard, 2012). The incorporation of sustainability in BIM models can turn traditional BIM models into Green BIM models. The Green BIM model is a model-based process of generating and managing data to analyze the performance of buildings in order that direct feedback can be applied and tested to improve environmental performance of buildings (Motawa and Carter, 2013). One such approach is the integrated model developed by Memarzadeh and Golparvar-Fard (2012) that combines BIM with *n*-dimensional AR to enable benchmarking and monitoring of embodied carbon to be undertaken for both preconstruction and construction phases. The integrated model helps to test the conceptual design to identify the lowest carbon option that has the potential to reduce carbon emission during design and construction phases and carbon savings during operating phase.

The data-rich BIM model provides an ideal platform for the assessment of GHG emissions, which can be undertaken in two ways. The design data in a BIM model can be extracted to create an inventory of building materials and carbon intensity can then be applied for each material quantity to calculate GHG emissions at various stages of a building's life cycle (Alwan and Jones, 2014). BIM tools have the ability to export data to other computer-modeling platforms for energy analysis to enhance building's performance and efficiency. Therefore, alternatively plug-in programs to BIM models can be used to analyze energy consumption and GHG emissions (Stadel et al., 2011; Motawa and Carter, 2013).

10.4 LIFE CYCLE ASSESSMENT

In the past studies, GHG emissions have focused on the operating stage as this stage represents the major portion of emission during a building's life cycle. However, attention has also been given to the GHG emissions embodied in other stages such as material manufacturing, transportation, construction, and disassembly at the end of the building life cycle (de Wolf et al., 2016). At each stage, the scale and magnitude of GHG emissions are different and the focus of operating GHGs will not give a full picture of emissions of a building. Therefore, a holistic approach will be needed to establish an accurate assessment of total GHG emissions.

The assessment of GHG emissions of a building life cycle is often equivalent to an LCA study of a building. LCA is a systematic framework that focuses on dealing with the inflows and outflows and offers a methodology for assessing the environmental impacts associated with products and processes spanning from the extraction of raw materials to the ultimate disposal (Klöpffer, 2006).

The idea of LCA was conceived in Europe and the United States in the late 1960s and early 1970s. It was not until the late 1980s and early 1990s that LCA received wider attention in response to increased environmental awareness and concern for energy usage and associated emissions (Azapagic, 1999). The principal objectives of LCA include quantifying and evaluating the environmental performance of a product/process, so that decision makers can optimize their selections among alternatives. Additionally, LCA provides a basis for assessing potential improvements in the environmental performance of a product/system so as to decrease its overall environmental impacts. This can be done in an overall sense or targeted to improve specific stages during the life cycle (Wei et al., 2008; Ahn et al., 2010). The relationship between these four steps is presented in Fig. 10.1. As detailed in the

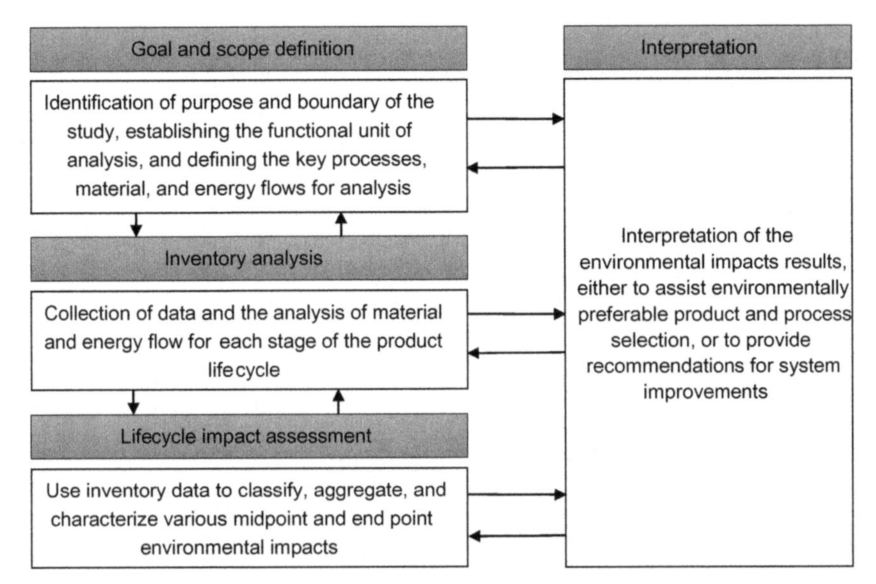

FIGURE 10.1 The life cycle assessment (LCA) framework. Sources: *From: ISO, 2006a. ISO 14040: Environmental Management—Life Cycle Assessment—Principles and Framework. International Standards Organization, Geneva; ISO, 2006b. ISO 14044: Environmental Management—Life Cycle Assessment—Requirements and Guidelines. International Standards Organization, Geneva.*

ISO 14040 the LCA framework includes steps of goal and scope definition, inventory analysis, life cycle impact assessment, and interpretation.

Over the years, LCA has been used to conduct detail analysis of environmental impacts. Cole (1999) has examined LCA on a selection of alternative wood, steel, and concrete structural assemblies; Wei et al. (2008) used LCA to investigate energy and emissions for different building components. Verbeeck and Hens (2010) applied an LCA methodology to analyze the whole building while Li et al. (2010) focused on using LCA to study the environmental impact of building processes. Goggins et al. (2010) studied the use of blast furnace slag to reduce emissions from concrete. Others have considered the prospect of limiting emissions via the use of industrialized methods of construction and found that prefabrication causes lower emissions than traditional methods of construction (Mao et al., 2013). Similarly, Monahan and Powell (2011) compared a novel off-site panelized timber frame system with two traditional alternatives and found that EEs for a semi-detached house was 34% less than the comparative traditional methods.

10.4.1 The Process

The life cycle emissions of a building according to LCA are the summation of total GHG emissions attributed to the production, construction, maintenance, end-of-life demolition, and transportation (de Wolf et al., 2016) and Fig. 10.2 shows the LCA process of a building life cycle as defined by EN 15978 (CEN, 2011):

- Initial EEs for all the processes of materials/products extraction and manufacturing;
- Emissions of fuel consumption for plants and equipment used for construction activities onsite;
- Recurrent EEs of materials/products used and emission of fuel in relation to maintenance and replacement during the operating stage;
- Emissions of fuel for plants and equipment used in the demolition of buildings at the end-of-life stage; and
- Emissions of fuel used for the transport of materials/products from distribution centers to building sites, or for disposal of construction and demolition waste to landfill or recycling facilities.

LCA is a term used fairly loosely. The common types of LCA or partial LCA include cradle-to-gate (from the stage of extraction to product production but excluding distribution), gate-to-gate (the manufacturing process and

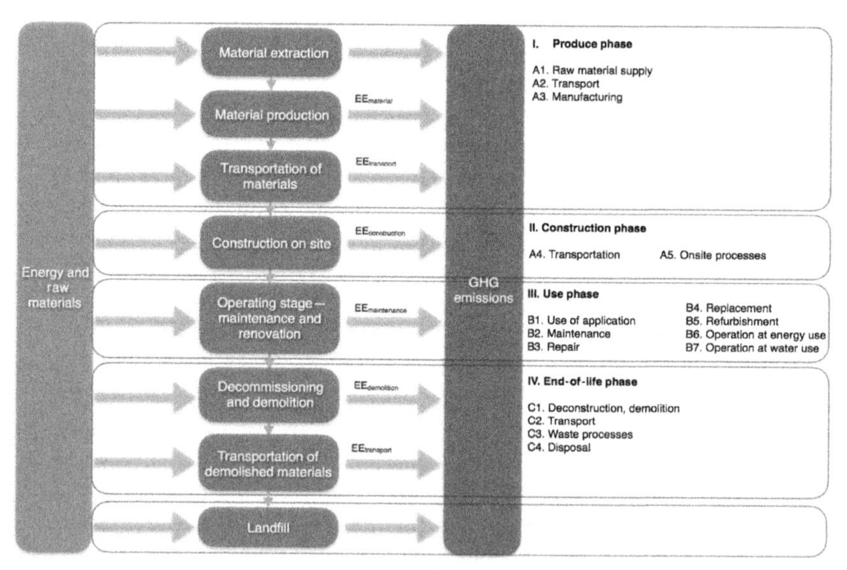

FIGURE 10.2 Life cycle assessment (LCA) process and building life cycle. Sources: *From CEN, 2011. Sustainability of Construction Works—Assessment of Environmental Performance of Buildings—Calculation Method. British Standard Institute, UK; Chau, C.K., Hui, W.K., Ng, W.Y., Powell, G., 2012. Assessment of CO2 emissions reduction in high-rise concrete office buildings using different materials use options. Resour. Conserv. Recycl. 61, 22–34; Oregi, X., Hernandez, P., Gazulla, C., Isasa, M., 2015. Integrating simplified and full life cycle approaches in decision making for building energy refurbishment: benefits and barriers. Buildings 5, 354–380.*

distribution to site), cradle-to-grave (includes the whole lifespan from extraction through use to disposal), and cradle-to-cradle (involves all the processes from extraction to end-of-life of a product with the addition of recycling) (Puettmann and Wilson, 2005; Mitchell and McFallen, 2008). The system boundary will be determined by the objective of each study and the availability of data. The LCA study in the construction industry is commonly referred as cradle-to-grave meaning that emissions after the end-of-life stage such as emissions related to reuse and recycling are not included.

10.4.2 Emissions at Product Phase

EEs account for the energy-related emissions in the production of building materials and components. These emissions include raw material extracting, transporting raw materials to factories, producing building materials and components, final transporting the finished products to building sites. The amount of GHG emissions associated with building materials/components production is highly dependent on the type and the amount of energy used such as oil, diesel, electricity in the extraction and manufacturing process and the associated quantities involved (Chau et al., 2012). Therefore, the calculation of EE of individual building materials requires the quantities per functional unit (e.g., kg, m^2 or m^3) and the coefficient of individual GHG expressed in $kgCO_{2\text{-}e}$. Hence, the EE is derived by multiplying the quantity of materials with corresponding $CO_{2\text{-}e}$ coefficient using the following formula (Yan et al., 2010; Oregi et al., 2015; de Wolf et al., 2016):

$$EE_{material} = \sum_{i=1}^{n} Q_i \times MEC_i \tag{10.1}$$

where

$EE_{material}$ = EE for material extraction and production ($kgCO_{2\text{-}e}$)
Q_i = quantity of material$_i$ (kg, m^2 or m^3)
MEC_i = emission coefficient per unit of material$_i$ ($kgCO_{2\text{-}e}$ per kg, m^2 or m^3)

10.4.3 Emission of Transportation

Transportation emissions include the energy used for the transportation of building materials and components from regional distribution centers to building sites, and the transportation of disposing construction and demolition waste from building sites to landfill or recycling facilities. Materials from factories or distribution centers will be delivered to building sites via trucks covering the outward trip (usually full load) and return trip (empty). The emissions in relation to transportation is directly related to the fuel type, number of truckloads, total distance travel (outward and return trip), fuel consumption, and emission coefficient per fuel type using the following formula (Yan et al., 2010; Zhang et al., 2013; de Wolf et al., 2016):

$$EE_{transport} = \sum_{i=1}^{n} D_i \times T_i \times TF_i \times TEC_i \tag{10.2}$$

where

$EE_{transport}$ = EEs from fuel combustion in the transportation of material from factory to sites ($kgCO_{2\text{-}e}$)
D_i = total distance (km)
T_i = no of truck loads for transporting building material$_i$ (No)
TF_i = fuel consumption per truck load (liter per km)
TEC_i = emission coefficient per unit consumption of fuel$_i$ ($kgCO_{2\text{-}e}$ per liter)

10.4.4 Emissions at Construction Phase

The construction stage of a building refers to a series of construction activities, including the erection of the structure and installation of building materials and components of a building. The emissions at this stage will relate to energy consumed by plants and equipment and the associated transport to dispose the construction waste to landfill or recycling facilities. According to Ortiz et al. (2009), the construction stage accounts for a total of 8%–20% of energy consumption in a building life cycle. Although the environmental impacts in this stage are small compared to the operating stage of the life cycle (Zhang et al., 2013), the excessive consumption of energy and generation of emissions during site operation can be substantial since the effect is immediate (Ortiz et al., 2009). Li et al. (2010) and Huisingh et al. (2015) further state that although the construction stage in a building life cycle is relatively short, the density of the GHG emissions is higher than that in the operating and maintenance stages.

Emissions at the construction phase are calculated by summing up the duration of plants and equipment used onsite multiplied it by the fuel consumption per type and the associated emission coefficient per liter of fuel consumed using the following formula (Yan et al., 2010; Kua and Wong, 2012; Zhang, et al., 2013; de Wolf et al., 2016):

$$EE_{construction} = \sum_{i=1}^{n} P_i \times PF_i \times PEC_i \tag{10.3}$$

where

$EE_{construction}$ = EEs from fuel combustion of construction plant or equipment ($kgCO_{2\text{-}e}$)
P_i = duration of plant or equipment used on site for construction activity$_i$ (hours)

PF_i = fuel consumption per plant or equipment used for construction activity$_i$ (kWh or liter per hour)
PEC_i = emission coefficient per unit consumption of fuel$_i$ (kg$CO_{2\text{-}e}$ per liter)

10.4.5 Emissions at Operating Stage

During the operating stage, the building is maintained and refurbished throughout its life cycle in order to maintain the functional capacity. In the past research studies of GHG emissions have focused on this stage as this stage has a long lifespan that generates most of the GHGs during the building's life cycle. Emissions at this stage are referred to as recurrent EE when they relate to the use of materials and components to refurbish and maintain the building over its lifetime (Zhang et al., 2013). In addition, emissions may also be associated with the energy use in running the facility such as heating, cooling, hot water, lighting, ventilation, and power for any appliances that allow the building to serve its intended function (Yan et al., 2010; Ibn-Mohammed et al., 2013).

Therefore, the service life and durability of materials/components are among the most important factors affecting the recurrent EE calculation (Zhang et al., 2013). This is particularly important for the material selection in the design and construction of building envelope as it will impact on the demand for heating and cooling loads to provide indoor comfort to users. Over a building's service life, one or multiple replacements may require. The shorter the service life of a material, the greater the quantity of ongoing maintenance and repair and the greater will be the recurrent EE associated with manufacturing and installing replacement materials throughout a building's life cycle. The recurrent EE is calculated using the following formula (excluding emissions from running the facility) (Yan et al., 2010; Kua and Wong, 2012; Zhang et al., 2013):

$$EE_{maintenance} = \sum_{i=1}^{n} \left[Q_i \times \left(\frac{BSL}{ESL_i} - 1 \right) \right] \times OEC_i \qquad (10.4)$$

where

$EE_{maintenance}$ = EEs for material used at the operating stage for maintenance or refurbishment (kg$CO_{2\text{-}e}$)
Q_i = quantity of material$_i$ used for maintenance or refurbishment (kg, m^2 or m^3)
BSL = reference service life of the building
ESL_i = estimated service life of material$_i$
OEC_i = emission coefficient per unit of material$_i$ used for maintenance or refurbishment (kg$CO_{2\text{-}e}$ per kg, m^2 or m^3)

The building materials and components used for maintenance or refurbishment are to be transported from factory or distribution center to the site. The emission associated with transportation of these materials and components are established as above.

10.4.6 Emissions at the End-of-Life Stage

When a building reaches the end of its useful life, demolition is inevitable. The most common disposal methods are landfilling, recycling, or incineration. The wastes produced by demolition as well as the waste disposal to the landfill have negative impacts on the environment and is some of the major waste flows in the world (Gao et al., 2015; Scheuer et al., 2003; Huang et al., 2013). The demolition activities represent about 2%–5% of resource consumption and 0.4% of total energy consumption (Scheuer et al., 2003). In Australia, 30%–40% of the solid waste comes from construction and demolition, which is nearly 1 ton per person annually (Pun et al., 2006).

Broadly speaking, recycling and disposal of demolition waste impact on the conservation of resources. With the recycling and reuse of demolition waste can be reduced and hence the associated emissions. According to Gong et al. (2012), the demolition stage of a building life cycle is important as it provides an opportunity for the recovery, and utilization of the dismantled and abandoned building materials. Nonrecycled waste causes the loss of raw materials and takes up land space for final disposal (Moffatt and Kohler, 2008). The emissions at this stage will include the estimated running time of the plants and equipment used in demolishing the building multiplied by the average consumption of electricity and/or fuel per unit of time and the associated emission coefficient per liter of fuel consumed using the following formula (Kua and Wong, 2012; Zhang et al., 2013):

$$EE_{Demolition} = \sum_{i=1}^{n} P_i \times PF_i \times PEC_i \tag{10.5}$$

where

$EE_{Demolition}$ = EEs associated to the end-of-life stage ($kgCO_{2\text{-}e}$)
P_i = duration of plant/equipment$_i$ used for demolition (hours)
PF_i = fuel consumption per plant/equipment$_i$ used (kWh or liter per hour)
PEC_i = emission coefficient per unit consumption of fuel$_i$ ($kgCO_{2\text{-}e}$ per liter)

At this stage transportation of demolition waste to either the landfill or recycling facilities for sorting and recycling will be required. The emission will therefore be associated with the truckloads, distance of travel, and fuel consumption, and it will be estimated as shown in Section 10.4.3.

10.5 SUSTAINABLE BUILDING PRODUCTS AND TECHNOLOGIES

Traditionally, buildings are built of heavy materials such as structural steel, reinforced concrete, and masonry. The use of concrete columns, beams, and floor plates are often mixed with brickwork wall infill panels and other façade materials to create appealing architectural features. The building industry has adapted to the use these materials in buildings and this has helped to reduce time and cost. However, sustainability has not been given a high priority in industry practices, either in design or construction. The emphasis on carbon reduction in the construction industry is only just gaining momentum and this is the reason why heavy materials have continuously been dominated in building designs.

In the late 1960s and early 1970s, people started to worry about the ability of ecosystems to support the ever-increasing economic activities (Bateson, 1973; Barrett & Odum, 2000). Throughout the world, the building industry is responsible for high levels of pollution as revealed in many research studies. Industrialized building methods, based on the widespread use of high-energy materials such as aluminum, cement, concrete, and steel, must now comply with new directives for the protection of the environment. The construction industry, whilst important for every society, is also responsible for environmental protection.

10.5.1 Innovative Products From Renewable Sources

As a means of achieving the goal of sustainable construction, modern innovations such as engineered wood products (EWPs) have attracted much attention because they are renewable and have made it possible to overcome the legislative hurdles to structural uses of timber in construction. The innovation factors along with sustainability benefits are contributing motivations for greater interest in using timber in buildings. The last 6–7 years have seen the construction of innovative timber structural buildings exceeding the usual height limits of traditional timber buildings (Thomas, 2015).

The advancement in understanding of timber building behavior, along with the refinement of EWP design and performance, has demonstrated the applicability of wood as a primary structural support system instead of as architectural features only (Robertson et al., 2012). The use of EWP has provided consistency and reliability in structural performance through the negation of typical natural defects found in timber. A wide variety of EWPs is now in use.

Studies of various countries have shown that buildings with wood structures require less energy and emit less GHG during their life cycle than buildings with other types of structures (Buchanna and Levine, 1999; Bribian et al., 2011). Cole (1999) compared the energy and GHG emissions of wood, steel, and concrete structural assemblies and found that structural timber frames generate GHG emissions 2.3–2.5 $kgCO_{2\text{-}e}/m^2$ compared with reinforced concrete (RC) structure of 13–20 $kgCO_{2\text{-}e}/m^2$. Petersen and Solberg (2005) in a quantitative analysis of materials in Norway and Sweden found that the quantity of GHGs avoided by replacing steel with wood in buildings is 0.06–0.88 $kgCO_{2\text{-}e}$ per kg input of timber while replacing concrete with wood reaches 0.16–1.77 $kgCO_{2\text{-}e}$ per kg.

EWPs have the additional ability to contribute to climate change mitigation as they withdraw CO_2 from its natural cycle and carbon stored was accounted for as a negative value in the GWP category (Robertson et al., 2012). Robertson et al. (2012) undertook a research study where they compared traditional RC structure with a laminated timber hybrid design for a five-story office building in Canada. Results reveal that GWP of timber design was 71% lower than the RC design, but it also stored 3131 ton of $CO_{2\text{-}e}$ within the building structures and envelopes. This clearly demonstrates the advantages of EWP over traditional RC structure.

EWPs such as cross laminated timber (CLT) and laminated veneer lumber (LVL) are more environmentally friendly materials, in particular, the end-of-life stage. EWPs, on the one hand, could essentially be disassembled as individual components for reuse in other buildings. On the other hand, if they could be chipped and reprocessed to lower grade wood products, e.g., particleboard, oriented strand board. Finally, they can be converted to usable energy in the form of heat or electricity generations as a biofuel energy source (Robertson et al., 2012). The EWPs are manufactured from wood which is a renewable resource. These products perform better and cost less compared with solid wood of the same dimensions for structural building elements. The use of adhesives and resins in the production of EWPs has health issues. Moreover, research studies demonstrate that emission levels can be reduced by adding polyvinyl acetate in the production of resins or using lignin-based adhesive in EWPs (Lei et al., 2007; Kim and Kim, 2005).

10.5.2 Advancement of Technologies in Material Production

As discussed previously conventional design and construction of buildings use steel, concrete, and masonry but these materials have high-energy intensity as well as high emission content. There is ongoing research into alternative materials, technologies, and practices that would help limit the amount of carbon-intensive materials. Research by Xing et al. (2008) on the environmental impact of steel and concrete construction of office buildings found that RC structure requires approximately 25% more energy than did steel in the production of materials with the associated GHG emissions. The production processes of building materials seem to offer potential opportunities to mitigate GHG emissions.

Over the years, research studies have been done to substitute conventional heavy materials with lower environmental impact and natural materials such as straw, hemp, and earth (Giesekam et al., 2014). They further state that instead of just replacing conventional materials with natural materials, natural materials can be combined to form low-impact products such as panelized prefabricated timber and straw bale systems, lime, and gypsum to replace traditional cement plasters, sheep's wool for insulation materials, timber, and steel hybrid structures in buildings.

Concrete is the second most consumed materials in the world and cement production is one of the key contributors to GHG emissions. According to Gao et al. (2015), the manufacturing process of cement has made it the most significant carbon emitter, accounting for about 7% of total CO_2 emission globally. Ishak and Hashim (2015) further state that the clinker production generates about 90% of the CO_2 emissions in the production of cement. Therefore, minimizing the need for carbon-intensive cement products is an essential part of reducing the EEs of construction materials (Giesekam et al., 2014; Gao et al., 2015; Huisingh et al., 2015).

Strategies have been developed to reduce its CO_2 emissions in cement production and these include waste heat recovery, substitute of fossil fuel with renewable energy sources, improve energy efficiency of manufacturing processes, and substitution with low carbon cement by using supplementary cementitious materials such as fly ash, silica fume, copper slag, sewage, ground-granulated blast furnace slag (Liu et al., 2015; Crossin, 2015; Yang et al., 2015). Gao et al. (2015) further state that GHG emissions can be reduced by replacing carbonate-containing materials with noncarbonate materials and changing the clinker ratio in cement production.

The iron and steel industry is the world's biggest energy consumer with the largest share in the world's economy. According to Remus et al. (2013), world steel production has increased considerably since 2000 and exceeded 1000 tons for the first time in 2004 and it has increased by a further 20% in 2006. The main reason why world steel production increased by 20% from 2004 to 2006 was that China's steel demand increased by 232% from 2004 to 2006. Globally, China ranks first among the world's iron and steel producers, followed by Japan and the United States. China's iron and steel production accounts for approximately 50% of the 2015 total, and it generates approximately 28% of the world's total CO_2 emissions (Olmez et al., 2016; Xu and Lin, 2016).

Iron and steel production is highly intensive in both energy and materials. It is crucial to the world's resource conservation, energy efficiency, and emission reduction. The production of iron and steel has high impacts on GWP as it directly involves the burning of fossil fuels. The various industrial processes of iron, coke, sinter, and steel production contribute to more than 60% of the total energy-related GHG emissions, of which iron production is the largest (74%) (Tian et al., 2013; Olmez et al., 2016).

According to Quader et al. (2016), 1 ton of steel manufacturing emits approximately 1.8 tons of CO_2. Therefore, over the years, iron and steel production has attracted much attention to improving its manufacturing process in order to reduce emissions. Tian et al. (2013) suggest a 3 R approach to reduce emissions from the iron and steel sector, which includes gas reuse and energy cascading program, waste iron and steel collection and recycling programs to reduce the use of virgin materials, and to improve the quality of iron and steel products through better quality control programs to extend lifespan of these products. Remus et al. (2013) and Quader et al. (2016) suggest the use of energy-saving equipment in the steel manufacturing process and improving the efficiency of energy conversion facilities.

Moya and Pardo (2013) suggest a two-level technological improvement in the production process: Best Available Technologies (BATs) and Innovative Technologies (ITs). BATs refer to the most effective and advanced technologies, methods or operations used in iron and steel production for improving performance and limiting discharges (Remus et al., 2013; Moya and Pardo, 2013). BATs cover environmental management systems, process-integrated and end-of-pipe measures, waste prevention and management in the production of iron and steel, e.g., top-pressure recovery turbine, coke dry quenching, high temperature air combustion (Tian et al., 2013).

ITs refer to emerging technologies to reduce GHG emissions in the iron and steel production process. One of the most important ones is the Ultra-Low Carbon Dioxide Steelmaking (ULCOS) project, which is one of the CO_2 mitigation strategies in Europe. ULCOS is a consortium of 48 European companies and organizations from 15 European countries. ULCOS aims at reducing CO_2 emissions from steel production by using the most advanced techniques to reduce CO_2 emissions by at least 50% by 2030 (Remus et al., 2013; Moya and Pardo, 2013). The main focus of the project includes using top gas recycling blast furnace in capturing and sequestering CO_2; use of carbon-lean fuels and reducing agents by replacing carbon with hydrogen or electricity; and the use of sustainable biomass (Remus et al., 2013).

10.5.3 Advancement of Technologies in Reducing Operating Emissions

Energy consumption and associated GHG emissions during the operating stage of a building is the main source of impact on the environment. The operating energy constitutes approximately 40%–50% of global demand and is mainly utilized to provide heating, ventilation, and air-conditioning (HVAC) in buildings (Prasartkaew and Kumar, 2010; Vakiloroaya, 2012; Kalnaes and Jelle, 2015). The main goal of HVAC is to provide comfort to building occupants as people spend more than 90% of time indoor (Steinemann et al., 2017) and its high energy consumption has initiated a great deal of research aiming at energy savings due to related GHG emissions.

In Australia, buildings accounted for approximately 25% of the total energy consumed and 10% of total GHG emitted in 2009 (Higgins et al., 2014). As stated in the report of the IPCC (2014), the commercial sectors can save approximately 1.4 billion tons of CO_2 by 2020 and these GHG emissions can be reduced by approximately 29% or even at zero-net with further commitment. Prasartkaew and Kumar (2010) propose the use of fully renewable energy based hybrid air-conditioning system using lithium bromide-water as the absorbent-refrigerant, whereby solar energy is used to heat the water at the collector field and is pumped to the tank by a pump, activated by the controller. This system can save electricity energy in the range of 25%–40% compared to the conventional refrigeration system.

The development of phase change materials (PCMs) is another approach of providing heating and cooling in buildings. It is considered as a latent heat storage unit to provide a solution for thermal management. PCM is a substance with a high heat of fusion that melts or freezes within a specific temperature range, leading to absorbing or releasing large amounts of energy from the immediate environment as it changes from solid to liquid or vice versa (Kalnaes and Jelle, 2015). de Gracia et al. (2010) state that the use of PCM increases the thermal inertia of the building and therefore improves the thermal comfort and reduces the use of energy for heating and cooling. PCM absorbs heat during the day through the melting process and releases the heat by coagulating at night.

de Gracia et al. (2010) investigated the environmental impact of using PCM in a typical Mediterranean construction on an LCA approach. Results reveal that using PCM in building materials can reduce energy consumption during the operating of buildings. Kalnaes and Jelle (2015) present an overview of PCMs used in building materials stating that large parts of energy usage are directly related to the heating and cooling of the building. The development of PCMs can reduce peak heating and cooling loads and still maintain a comfortable indoor environment due to smaller temperature fluctuations.

PCM can also be applied to any porous building materials to function in the same way. Baetens et al. (2010) discuss various ways PCM can be added in the manufacturing process of building materials to enhance their thermal performance such as adding PCM in wallboards with low thermal mass in conventional framed designs, adding PCM in concrete to form thermcrete to store and distribute thermal with structural properties. PCM can also be incorporated ventilation systems, passive heating and cooling systems to enhance their functionality (Kalnaes and Jelle, 2015). These PCM enhanced materials can provide thermostatic properties that can reduce heating and cooling loads in buildings.

10.5.4 Advancement of Technologies in Carbon Capture and Storage

CO_2 can be removed from the atmosphere by various processes and commonly used approaches include sequestration in timber, soil, and ocean. Recently, the use of CO_2 capture and storage (CCS) has also attracted much attention in power production and in CO_2 intensive industrial plants such as steel production (Huisingh et al., 2015; Quader et al., 2016). CCS is regarded as a bridging technology with the potential to stabilize CO_2 emissions in the atmosphere (Remus et al., 2013; Quader et al., 2016). CCS includes CO_2 separation from industrial processes. The separated CO_2 is then compressed to a liquid or supercritical state and transported to a storage site via pipelines or ships. The compressed CO_2 is stored either in on-shore or off-shore geological formations (Remus et al., 2013). One of the downsides of CCS is the CO_2 separation and compression is an energy intensive process, which may increase CO_2 emissions and other air pollutants.

10.6 EMISSION REDUCTION STRATEGIES

10.6.1 Use Low Impact Materials

Construction materials take in different types, qualities, shapes, and forms. They are important for the survival of the industry. However, if the manufacturing processes of these materials are not improved the construction industry will remain unsustainable and continue to be responsible for the degradation of the environment. Sustainable construction is considered as a way for the building industry to contribute positively to achieving the goal of sustainable development (Ding, 2008; Vatalisa et al., 2011). Low-impact materials play an important role in the delivery of sustainable buildings and their selection at the early design stage will provide an opportunity to bring together all the sustainability strategies to be realized for a project as it will influence all downstream processes in achieving sustainability in buildings (Shen et al., 2010). Low impact materials refer to materials that are non-toxic, sustainably produced or recycled and require little energy to process.

In addition, materials should be harvested from renewable sources such as timber, straw bale, and earth. These new material have become more competitive with mainstream conventional materials due to their low-impact properties. However, there are negative perceptions on the use of these materials and the concerns about durability, structural capacity, fire resistance, acoustic and thermal performance, and insect resistance. Nonetheless, advancements in technology have improved their properties in these respects. For example, the advancement in timber technologies has improved the performance of timber so that in many applications it can now compete with conventional materials. Research has shown that fire, acoustic, and durability issues can be overcome (Frangi et al., 2009; Gagnon and Kouyoumji, 2011). However, since these renewable materials are not mainstream building materials, the legislative challenges in building codes and standards have been main barriers for an increased uptake of their uses.

Bribian et al. (2011) suggest that there is a need to modify the current legal framework in order to promote the design of buildings with wooden structures rather than structures based on conventional materials such as RC as wood is a low-impact and low emissions material that require less industrial processing, better resistance to fire, ability to capture and confinement of CO_2.

10.6.2 Extend Building Lifespan

The designed lifespan of buildings can also play a significant role in sustainability and hence GHG emissions (Tian et al., 2013). Conventional heavy materials tend to have a long lifespan but the change of fashion or taste may shorten the lifespan of these buildings well before they are physically obsolescent (Tian et al., 2013; Giesekam et al., 2014). Meggers et al. (2012) suggest that the lifespan of buildings can be maximized so that the use of virgin materials can be minimized. Meggers et al. (2012) go on to state that any use of materials or systems that may be rapidly outdated should be avoided during the design development and building components should be designed for reuse or raw material recovery as early in the process as possible.

It is, however, also important to extend the lifespan of existing buildings to reduce the demand for resources to construct new buildings (Kohler, 1999). Energy embodied in existing buildings will be retained in the building fabric and equipment, and thus not contribute to emissions (Chau et al., 2015). Reducing the use of virgin materials should be a key target when aiming to reduce energy and CO_2 emissions from existing buildings. However, current practices frequently result in buildings designed with a very limited life expectancy because the natural capital used is not considered to have any significant value leading to the abuse of these resources. Chau et al. (2012) state that a reduction of about 17% of CO_2 emissions could be achieved if 15%–30% of the existing structural and nonstructural building elements were retained.

10.6.3 Maximize Design of Building Structures

With regard to the design of building structures, Meggers et al. (2012) suggest an adoption of lightweight building designs that minimize the consumption of raw materials and also the use of concrete with expanded aggregates in building structures to improve insulation. Giesekam et al. (2014) further suggest to use lightweight design in conjunction with structural member optimization to minimize the excess use of materials, but this will require changes in design practices or innovative manufacturing processes.

10.6.4 Improve Project Delivery Onsite

According to a study by Acquaye and Duffy (2010), structural activities account for the majority (76%) of direct emissions followed by groundworks (10%). They therefore suggest that the onsite activities are to be targeted in order to reduce direct emission. Prefabrication that reduces onsite activities should be encouraged as replacements (Robertson et al., 2012). This is supported by Guggemos and Horvath (2006), who conducted research into energy use and emissions of construction processes and found that prefabrication should be favored over onsite fabrication due to higher productivity, reduced equipment use, and reduced waste generation.

Mao et al. (2013) compared the GHG emissions between prefabrication and conventional construction project in China with similar results. Their results show that prefabrication produces less GHG emissions per square meter compared to conventional construction—336 and 368 $kgCO_{2\text{-}e}/m^2$, respectively—approximately 10% less. They go on to state that with the total GHG emission approximately 85% being the EEs of building materials. Therefore, reducing onsite activities and shifting from conventional materials to low emission materials through good design and construction will reduce the overall emission intensity in the building industry.

10.6.5 Increase Reuse and Recycling of Materials

The construction industry is an industry slow to change. Significant concept changes in the practices of the construction industry to facilitate the reuse and recycling of building materials can be made possible if they can be considered as early as possible in the design development. Such changes could be made to favor the disassembly of the construction materials at the end of their service life and by selecting materials for the recycled sources and assembly techniques are significant in the reduction of GHG emissions (Bribian et al., 2011). Giesekam et al. (2014) state that increased reuse and recycling will require design for disassembly and an increased focus on the end-of-life project stage during design. Good early planning in the design and construction of buildings is essential.

10.7 INTERNATIONAL CASE STUDIES

10.7.1 Pines Calyx, England

Pines Calyx is an innovative low-energy building located in St Margaret's Bay, Kent, England (Bothwell and Watkins, 2015). The building was

FIGURE 10.3 The Pines Calyx, United Kingdom. Sources: *From https://www.kentvenues.co.uk/wp-content/uploads/2015/04/Pines_Calyx_Headline.jpg.*

completed in 2006 with a gross floor area of 400 m^2. Pines Calyx was the first conference venue in Europe designed and built using cradle-to-cradle principles and is the smallest CF building in the United Kingdom (de Wolf et al., 2016) (Fig. 10.3).

The building has attracted much attention and won numerous awards for its innovative design and sustainability features. The energy and emission-related features include (Carbon Free Group, 2016):

- Reduced building footprint to minimize site disturbance and soil erosion;
- Maximized materials use onsite to reduce energy-related transportation and construction activities;
- Salvaged and recycled local waste to manufacture building materials onsite such as timber removed from site or immediate vicinity has been utilized in the construction and excavated waste clay and chalk have been used for the construction of walls and roofs;
- Used earth tube in conjunction with a heat exchange unit to provide heating in winter and cooling in summer to reduce operating energy and emissions;
- Adopted orientation to maximize natural lighting and solar gain;
- Used efficient double- and triple-glazed wooden framed windows to reduce heat loss during winter months;
- Improved thermal mass of walls and roofs by using waste chalk and clay to reduce heating and cooling loads.

The building was built using salvaged waste materials onsite as opposed to conventional construction methods using masonry or reinforced concrete (de Wolf et al., 2016). The walls were built utilizing chalk from excavated materials onsite using rammed earth techniques. This reduced emission related to transport of materials from factories and waste to landfills. The salvaged chalk was placed in 100–150 mm layers to form the wall using pneumatic rammers (Jones, 2007). The rammed chalk walls were a maximum of 650 mm thick and 7 m high, providing good thermal massing properties. According to de Wolf et al. (2016), the use of rammed chalk in walls has an environmental impact of four times lower than that of historical techniques of existing typical buildings.

The roofs/ceilings were constructed using the technique of Guastavino style vaults as opposed to conventional construction methods. The use of this technique was the first time in the United Kingdom (Jones, 2007). The Guastavino style Timbrel vaults form a self-supporting arch that has resulted in the minimum use of concrete and formwork that are high in both energy intensity and emissions (Jones, 2007; Ramage, 2007). The tiles used to build the Timbrel vaults were manufactured from waste clay sourced from a local quarry.

The use of chalk to form the rammed earth walls and clay for the dome has provided a good thermal mass to maintain a more constant temperature inside the building. The use of chalk and clay as building materials helps to maintain an indoor relative humidity between 40% and 60% to provide comfort to users. Both chalk and clay are hydroscopic, releasing and absorbing atmospheric moisture in the environment to improve indoor comfort and thus reducing the dependency on heating and cooling. The natural finishes of chalk and clay also require no plastering and painting and thus further minimizing consumption of building materials (Carbon Free Group, 2016).

The design of the building was grounded on the passive design principles of Passivhaus standard (Bothwell and Watkins, 2015). The building features high levels of insulation, air-tightness construction, mechanical ventilation heat recovery, and an earth tube to improve energy efficiency in the design approach. The earth tube design was used to preheat fresh air in winter and provide cool air in summer so that heating and cooling loads can be reduced to minimize emissions. The tube is approximately 16 m long, constructed with a concrete pipe with an internal diameter of 0.6 m and a wall thickness of 100 mm. The tube is located approximately 1.5 m below ground level. According to the research by Bothwell and Watkins (2015), the earth tube design has demonstrated its efficiency in providing heating and cooling with very low carbon emissions.

The building has achieved 299 MJ/m^2 and 33 kg CO_2/m^2, respectively, for embodied energy and carbon compared to 1393 MJ/m^2 and 147 $kgCO_2/m^2$ for a conventional construction. The savings are 79% and 78%, respectively, for embodied energy and carbon. With the use of chalk to construct the wall and clay for the dome the building is very energy efficient in reducing operating energy and CO_2 emission. The building consumes approximately 35 $kWh/m^2/year$ and emits 11.3 $kgCO_2/m^2/year$, compared to best practice similar design of 75 $kWh/m^2/year$ and 33.6 $kgCO_2/m^2/year$, respectively.

The savings are 66% and 67%, respectively, for operating energy and CO_2 emissions (Carbon Free Group, 2016).

10.7.2 Chau Chak Wing Building, Australia

The Business School at the University of Technology Sydney is situated in the Dr Chau Chak Wing Building in the Sydney CBD completed in November 2014. This building is the first Australian building designed by Frank Gehry and was awarded 5 Star Green Star Education Design v1 Rating certified by the Green Building Council of Australia. The building contains 14 levels (12 levels above ground and 2 levels below ground) with a gross floor area of 18,413 m^2 and total project cost of $180 million.

This university building has utilized sustainable building technology to reduce energy consumption and hence GHG emissions. As discussed previously approximately 40%–50% of the energy consumption is with the HVAC system to provide comfort to users, in particular for university building to enhance students' learning environment. The building constructed with high-performance glazing system and insulated double-glazed curtain wall with solar control and low emissivity coatings. The high-performance double-glazed curtain efficiently reduces the overall HVAC consumption in the building (Shameri et al., 2011). Chan et al. (2009) studied the energy saving of cooling in office buildings in Hong Kong and their results reveal that approximately 26% savings of energy for cooling can be achieved with high-performance double-glazed curtain system. The building is also insulated with low emissive and higher R-values of foamular extruded polystyrene for the building plaza decks to save energy.

Smart air-conditioning systems using variable speed drives are switched off automatically when spaces are empty for an extended time. The air-conditioning systems also use an energy efficient air cooled chilled water and condensing gas boiler plant instead of a conventional water cooled plant to reduce GHG emissions.

One of the largest energy consumers in buildings is lighting. It accounts for approximately 15% of total energy demand and the utilization of natural lighting can save approximately 20%–40% of that energy (Ryckaert et al., 2010; Yun et al., 2010). The Chau Chak Wing Building has been designed to maximize natural daylighting through its glass paneled curtain wall. Each larger floor is divided into six floor segments and the building façade folds in between these elements in order to bring natural daylight deep into the center of the floors. Throughout the building LED and T5 lightings with zoning and dimming control enhance energy efficiency. Adjustable blinds control glare and provide comfort indoor. Externally lightings are fitted with daylight sensors to automatically adjust artificial lighting requirements.

The interior design with open space environment and visible internal staircases encourage walking up and down via staircases to reduce lift energy use and improve the health and well-being of occupants. The building is also linked with real-time sustainability performance data on a publicly accessible viewing screen to help raise awareness of energy consumption. Users can use the screen to tap into the displays of the actual building meters and see the impact of energy consumption. The sustainable technologies that have been incorporated in the design and construction of the building are estimated to save 30%–50% energy over benchmark tertiary educational buildings with similar functional spaces (Anonymous, 2015).

With regard to materials the building used EWP for two oval classrooms and collaborative theater. The EWP includes 8–12 m floor span glulam beams connected to concrete by coach bolts combined with bird's mouth notches for the floor structure. This is the world's longest timber-concrete composite floor construction and the first in Australia (Basaglia et al., 2015). A total of 150 large laminated timber beams each weight up to 2 tons with spans ranging from 8 to 12 m timbers are either reused, recycled, or sourced from certified sustainable forests, e.g., Radiata Pine glulam beams from New Zealand, Victorian ash stairways and Hoop Pine for joinery. Other low emission materials include fly ash (a waste product from power station) replacing a proportion of cement in order to reduce EEs in the manufacturing process (Basaglia et al., 2015).

10.8 CONCLUSION

Buildings are major energy consumers and GHG emitters throughout their life cycles. To reduce the production of emission throughout the building's life cycle, it is necessary to promote the use of best techniques available, support innovations in production plants and processes, and replace the use of finite natural resources with waste products generated in different production processes. Over the years, international collaboration has grown in the development of policies, standards, and programs with an aim to reduce GHG emissions and hence mitigate global warming. This international collaboration has proved to be successful in raising the awareness among governments and professionals in the construction industry. The development of the ISO series underpins the approach to using LCA in assessing GHG emissions in materials, products, processes, and buildings. LCA has now become the framework for the assessment of life cycle GHG emissions and is used widely for developing strategies for improved efficiency in buildings.

Emission efficiency in buildings has been improved greatly with the use of sustainable building technologies. HVAC system has been targeted for improvement as it has been considered the major consumer of operating energy (approximately 40%–50%). Advancement in technologies has resulted in the improvement in efficiency from shifting to fully renewable energy-based system, use of latent heat storage units such as PCMs have demonstrated reductions in energy-related emissions. Emission reductions can also be achieved through advances in the manufacturing processes of building materials, replacing conventional materials with more environmentally friendly materials and shifting to renewable sources. The uses of wood and EWP have attracted much attention due to their natural and low emission features. Research studies have demonstrated that wood and wood products have achieved approximately 71% lower emissions than conventional RC structures. Emission reductions have also been achieved through improvements in the manufacturing processes of energy intensive materials such as cement, iron, and steel. These include substitute fossil fuel with renewable energy sources, producing green cement by using supplementary cementitious materials such as fly ash, and reuse and recycling of materials to reduce emissions. The construction industry has a great potential to contributing to mitigate the global challenge of climate changes. With the advancement of energy efficient technologies and improvement in the manufacturing of building materials and products, the construction industry can become more efficient and environmentally friendly.

REFERENCES

ABCB, 2010. BCA Section J—Assessment and verification of an alternative solution, Australian Building Codes Board. Australian Government and States and Territories of Australia, Canberra, ACT.

Acquaye, A.A., Duffy, A.P., 2010. Input-output analysis of Irish construction sector greenhouse gas emissions. Build. Environ. 45, 784–791.

Ahn, C., Lee, S., Peña-Mora, F., Abourizk, S., 2010. Toward environmentally sustainable construction processes: the US and Canada's perspective on energy consumption and GHG/ CAP emissions. Sustainability 2 (1), 354–370.

Ali, A.A.M., Hagishima, A., Abdel-Kader, M. & Hammad, H. (2013). Vernacular and modern building: estimating the CO2 emission from the building materials in Egypt. In: Proceedings: Building Simulation Cairo—Towards Sustainable & Green Life, June 23–24, Cairo.

Alwan, Z., Jones, P., 2014. The importance of embodied energy in carbon footprint assessment. Struct. Surv. 32 (1), 49–60.

Anonymous (2015). Celebrating the Unconventional. Retrieved from: <www.claybrick.org.za/ news/celebrating-unconventional logon> (accessed 01.10.16).

Arena, A.P., de Rosa, C., 2003. LCA of energy and environmental implications of the implementation of conservation technologies in school buildings in Mendoza-Argentina. Build. Environ. 38, 359–368.

Azapagic, A., 1999. Life cycle assessment and its application to process selection, design and optimization. Chem. Eng. J. 37, 1–21.

Baetens, R., Jelle, B.P., Gustavsen, A., 2010. Phase change materials for building applications: A state-of-the-art review. Ener Build. 42, 1361–1368.

Barrett, G.W., Odum, E.P., 2000. The twenty-first century: the world at carrying capacity. BioSci. 50 (4), 363–368.

Bateson, G., 1973. Steps on an Ecology Mind. Jason Aronson Inc. Northvale, US.

Basaglia, B., Lewis, K., Shrestha, R. & Crews, K. (2015). A comparative life cycle assessment approach of two innovative long span timber floors with its reinforced concrete equivalent in an Australian context. In: International Conference on Performance-based and Life-cycle Structural Engineering, Brisbane, Queensland, December 9–11, pp. 1433–1442.

Baumann, T., Kollmuss, A., 2010. GHG Schemes Addressing Climate Change: How Iso Standards Help? International Standards Organization, Geneva.

Bothwell, K. & Watkins, R. (2015). Pines Calyx earth tube performance. In: Proceedings: PLEA 31st International Conference, September 9–11, Bologna. Retrieved from: <https://kar.kent.ac.uk/53216/1/P_618pdf logon> (accessed 17.08.16).

Bribian, I.Z., Capilla, A.V., Uson, A.A., 2011. Life cycle assessment of building materials: Comparative analysis of energy and environmental impacts and evaluation of the eco-efficiency improvement potential. Build. Environ. 46, 1133–1140.

BSI, 2011. Specification for the Assessment of the Life Cycle Greenhouse Gas Emissions of Goods and Services PAS 2050:2011. British Standards Institution, London, UK.

Buchanan, A.H., Levine, S.B., 1999. Wood-based building materials and atmospheric carbon emissions. Environ. Sci. Policy. 2 (6), 427–437.

Carbon Free Group (2016). An Outstanding Carbon Neutral Conference Centre in Kent. Retrieved from: <www.carbonfreegroup.com/case-study/pines-calyx Logon> (accessed 15.09.16).

CEN, 2011. Sustainability of Construction Works—Assessment of Environmental Performance of Buildings—Calculation Method. British Standard Institute, UK.

Chan, A.L.S., Chow, T.T., Fong, K.F., Lin, Z., 2009. Investigation on energy performance of double skin façade in Hong Kong. Energy Build. 41 (11), 1135–1142.

Chang, Y., Ries, R.J., Lei, S., 2012. The embodied energy and emissions of a high-rise education building: A quantification using process-based hybrid life cycle inventory model. Ener. Build. 55, 790–798.

Chau, C.K., Hui, W.K., Ng, W.Y., Powell, G., 2012. Assessment of CO_2 emissions reduction in high-rise concrete office buildings using different materials use options. Resour. Conserv. Recycl. 61, 22–34.

Chau, C.K., Leung, T.M., Ng, W.Y., 2015. A review on life cycle assessment, life cycle energy assessment and life cycle carbon emissions assessment on buildings. Appl. Ener. 143 (0), 395–413.

Chen, Y., Ng, S.T., 2016. Factoring in the embodied GHG emissions when assessing the environmental performance of building. Sustain. City Soc. 27, 244–252.

Crossin, E.C., 2015. The greenhouse gas implications of using ground granulated blast furnace slag as a cement substitute. J. Clean. Prod. 95, 101–108.

Cole, R.J., 1999. Energy and greenhouse gas emissions associated with the construction of alternative structural systems. Build. Environ. 34 (3), 335–348.

Cucek, L., Klemes, J.J., Kravanja, Z., 2012. A review of footprint analysis tools for monitoring impacts on sustainability. J. Clean. Prod. 34, 9–20.

de Gracia, A., Rincon, L., Castell, A., Jimenez, M., Boer, D., Medrano, M., et al., 2010. Life cycle assessment of the inclusion of phase change materials (PCM) in experimental buildings. Ener. Build. 42 (9), 1517–1523.

de Wolf, C., Bird, K. & Ochsendorf, J. (2016). Material quantities and embodied carbon in exemplary low carbon case studies. In: Habert, G., Schlueter, A. (Eds.), Proceedings: Sustainable Built Environment (SBE) Regional Conference, Expand Boundaries: System Thinking for theBuilt Environment, Zurich, June 15–17, pp. 726–733.

Dimoudi, A., Tompa, C., 2008. Energy and environmental indicators related to construction of office buildings. Resour. Conserv. Recycl. 53 (1–2), 86–95.

Ding, G.K.C., 2008. Sustainable construction—The role of environmental assessment tools. J. Environ. Manage. 86 (3), 451–464.

Ding, L., Zhou, Y., Akinci, B., 2014. Building information modelling (BIM) application framework: the process of expanding from 3D to nD. Automat. Constr. 46, 82–93.

EPA, 2007. Climate Leaders Program. US EPA, Washington, DC.

Finkbeiner, M., 2014. Product environmental footprint – breakthrough or breakdown for policy implementation of life cycle assessment? Int. J. Life Cycle Assess. 19, 266–271.

Finnveden, G., Moberg, A., 2005. Environmental systems analysis tools—an overview. J. Clean. Prod. 13, 1165–1193.

Frangi, A., Knobloch, M., Fontana, M., 2009. Fire design of timber slabs made of hollow core elements. Eng. Struct. 31 (1), 150–157.

Gagnon, S., Kouyoumji, J., 2011. Acoustic performance of cross-laminated timber assemblies. In: Innovations, F. (Ed.), CLT Handbook. FP Innovations, British Columbia.

Galli, A., Wiedmann, T., Ercin, E., Knoblauch, D., Ewing, B., Giljum, S., 2012. Integrating ecological, carbon and water footprint into a "Footprint Family" of indicators: Definition and role in tracking human pressure on the planet. Ecol. Indicat. 16, 100–112.

Gao, T., Shen, L., Shen, M., Chen, F., Liu, L., Gao, L., 2015. Analysis on differences of carbon dioxide emission from cement production and their major determinants. J. Clean. Prod. 103, 160–170.

Garcia, R., Freire, F., 2014. Carbon footprint of particleboard: A comparison between ISO/TS 14067, GHG Protocol, PAS 2050 and Climate Declaration. J. Clean. Prod. 66, 199–209.

Gerilla, G.P., Teknomo, K., Hokao, K., 2007. An environmental assessment of wood and steel reinforced concrete housing construction. Build. Environ. 42 (7), 2778–2784.

Giesekam, J., Barrett, J., Taylor, P., Owen, A., 2014. The greenhouse gas emissions and mitigation options for materials used in UK construction. Ener. Build. 78, 202–214.

Giesekam, J., Barrett, J.R., Taylor, P., 2016. Construction sector views on low carbon building materials. Build. Res. Inform. 44 (4), 423–444.

Goggins, J., Keane, T., Kelly, A., 2010. The assessment of embodied energy in typical reinforced concrete building structures in Ireland. Ener. Build. 42 (5), 735–744.

Gong, X., Nie, Z., Wang, Z., Cui, S., Gao, F., Zuo, T., 2012. Life Cycle Energy Consumption and Carbon Dioxide Emission of Residential Building Designs in Beijing. J. Industr. Ecol. 16 (4), 576–587.

Guggemos, A.A., Horvath, A., 2005. Decision support tool for environmental analysis of commercial building structures. Construction Research Congress 2005 San Diego, California, April 5–7: Broadening Perspectives 92–97.

Guggemos, A.A., Horvath, A., 2006. A decision-Support tool for assessing the environmental effects of constructing commercial buildings. J. Architect. Eng. 12, 187–195.

Higgins, A., Syme, M., McGregor, J., Marquez, L., Seo, S., 2014. Forecasting uptake of retrofit packages in office building stock under government incentives. Ener. Policy 65, 501–511.

Hong, J., Shen, G.Q., Feng, Y., Lau, W.S., Mao, C., 2015. Greenhouse gas emissions during construction phase of a building: a case study in China. J. Clean. Prod. 103, 249–259.

Huang, T., Shi, F., Tanikawa, H., Fei, J., Han, J., 2013. Materials demand and environmental impact of buildings construction and demolition in China based on dynamic material flow analysis. Resour. Conserv. Recycl. 72, 91–101.

Huisingh, D., Zhang, Z., Moore, J.C., Qiao, Q., Li, Q., 2015. Recent advances in carbon emissions reduction: policies, technologies, monitoring, assessment and modelling. J. Clean. Prod. 103, 1–12.

Ibn-Mohammed, T., Greenough, R., Taylor, S., Ozawa-Meida, L., Acquaye, A., 2013. Operational vs. embodied emissions in buildings—a review of current trends. Ener. Build. 66, 232–245.

IPCC, 2014. Climate change 2014: impacts, adaptation, and vulnerability. Part A: Global and sectoral aspects. In: Field, C.B., Barros, V.R., Dokken, D.J., Mach, K.J., Mastrandrea, M.D., Bilir, T.E., et al.,Contribution of Working Group II to the Fifth Assessment Report of the Intergovernmental Panel on Climate Change. Cambridge University Press, Cambridge, UK/ New York.

Ishak, S.A., Hashim, H., 2015. Low carbon measures for cement plant—a review. J. Clean. Prod. 103, 260–274.

International Standard Organisation (ISO), 2006a. ISO 14040: Environmental Management—Life Cycle Assessment—Principles and Framework. International Standards Organization, Geneva.

International Standard Organisation (ISO), 2006b. ISO 14044: Environmental Management—Life Cycle Assessment—Requirements and Guidelines. International Standards Organization, Geneva.

International Standard Organisation (ISO), 2006c. ISO 14064: Greenhouse Gases Accounting and Verification. International Standards Organization, Geneva.

ISO/TS, 2013. ISO/TS 14067: Carbon Footprint of Products—Requirements and Guidelines for Quantification and Communication. International Standards Organization, Geneva.

Jones, W. (2007). Pines Calyx – Setting the Sustainable Standard. Retrieved from: <www.building.co.uk/pines-calyx-setting-the-sustainable-standard/3100028.article Logon> (accessed 15.09.16).

Kalnaes, S.E., Jelle, B.P., 2015. Phase change materials and products for building applications: a state-of-the-art review and future research opportunities. Ener. Build. 94, 150–176.

Kim, S., Kim, H., 2005. Effect of addition of polyvinyl acetate to melamine-formaldehyde resin on the adhesion and formaldehyde emissions in engineered flooring. Int. J. Adhes. Adhes. 25 (5), 456–461.

Klöpffer, W., 2006. The role of SETAC in the development of LCA. Int. J. Life Cycle Assess. 11 (1), 116–122.

Kohler, N., 1999. The relevance of green building challenge: an observe's perspective. Build. Res. Inform. 27 (4/5), 309–320.

Kua, H.W., Wong, C.L., 2012. Analysing the life cycle greenhouse gas emission and energy consumption of a multi-storied commercial building in Singapore from an extended system boundary perspective. Ener. Build. 51, 6–14.

Lei, H., Pizzi, A., Du, G., 2007. Environmentally friendly mixed tannin/lignin wood resins. J. Appl. Polym. Sci. 107 (1), 203–209.

Levermore, G.J., 2008. A review of the IPCC assessment report four, Part 1: The IPCC process and greenhouse gas emission trends from building worldwide. Build. Serv. Eng. Res. Technol. 29 (4), 349–361.

Li, X., Zhu, Y., Zhang, S., 2010. An LCA-based environmental impact assessment model for construction processes. Build. Environ. 45, 766–775.

Liu, G., Yang, Z., Chen, B., Zhang, J., Liu, X., Zhang, Y., et al., 2015. Scenarios for sewage sludge synergic reduction and reuse in clinker production towards regional eco-industrial development: a comparative energy-based assessment. J. Clean. Prod. 103, 371–383.

Mao, C., Shen, Q., Shen, L., Tang, L., 2013. Comparative study of greenhouse gas emissions between off-site prefabrication and conventional construction methods: two case studies of residential projects. Ener. Build. 66, 165–176.

Meggers, F., Leibundgut, H., Kennedy, S., Qin, M., Schlaich, M., Sobet, W., et al., 2012. Reduce CO_2 from buildings with technology to zero emissions. Sustain. Cities Soc. 2, 29–36.

Memarzadeh, M., Golparvar-Fard, M., 2012. Monitoring and visualisation of building construction embodied carbon footprint using DnAR-N-dimensional augmented reality models. Constr. Res. Congr. 1330–1339.

Mitchell, P. & McFallen, S. (2008). Timber LCI case study: the practicalities of representing Australian veneer, plywood and LVL. In: Proceeding: World Sustainable Building Conference, September, 22–26, Melbourne, Australia.

Moffatt, S., Kohler, N., 2008. Conceptualizing the built environment as a social-ecological system. Build. Res. Inform. 36 (3), 248–268.

Monahan, J., Powell, J., 2011. An embodied carbon and energy analysis of modern methods of construction in housing: a case study using a lifecycle assessment framework. Ener. Build. 43 (1), 179–188.

Motawa, I., Carter, K., 2013. Sustainable BIM-based evaluation of buildings. Proced. Social Behav. Sci. 74, 419–428.

Moya, J.A., Pardo, N., 2013. The potential for improvements in energy efficiency and CO2 emissions in the EU27 iron and steel industry under different payback periods. J. Clean. Prod. 52, 71–83.

Nadoushani, Z.M., Akbarnezhad, A., 2015. Computational method for estimation of life cycle carbon footprint of buildings. In: Blackman, C. (Ed.), Carbon footprinting: New developments, reduction methods and ecological impacts, USA. Nova Science Publishers.

Olmez, G.M., Dilek, F.B., Karanfil, T., Yetis, U., 2016. The environmental impacts of iron and steel industry: a life cycle assessment study. J. Clean. Prod. 130, 195–201.

Omar, W.W., Doh, J., Panuwatwanich, K., 2014. Variations in embodied energy and carbon emission intensities of construction materials. Environ. Impact Assess. Rev. 49, 31–48.

Oregi, X., Hernandez, P., Gazulla, C., Isasa, M., 2015. Integrating simplified and full life cycle approaches in decision making for building energy refurbishment: benefits and barriers. Buildings 5, 354–380.

Ortiz, O., Bonnet, C., Bruno, J.C., Castells, F., 2009. Sustainability based on LCM of residential dwellings: a case study in Catalonia, Spain. Build. Environ. 44 (3), 584–594.

Petersen, A.K., Solberg, B., 2005. Environmental and economic impacts of substitution between wood products and alternative materials: A review of micro-level analysis from Norway and Sweden. Forest Policy Econom. 7, 249–259.

Prasartkaew, B., Kumar, S., 2010. A low carbon cooling system using renewable energy resources and technologies. Energy Build. 42, 1453–1462.

Puettmann, M.E., Wilson, J.B., 2005. Life-cycle analysis of wood products: cradle-to-gate lci of residential wood building materials. Wood Fibre Sci. 37, 18–29.

Pun, S.K., Liu, C., Langston, C., 2006. Case study of demolition costs of residential buildings. Constr. Manage. Econom. 24 (9), 967–976.

Quader, M.A., Ahmed, S., Ghazilla, R.A.R., Ahmed, S., Dahari, M., 2016. Evaluation of criteria for CO_2 capture and storage in the iron and steel industry using the 2-tuple DEMANTEL technique. J. Clean. Prod. 120, 207–220.

Ramage, M., 2007. Guastavino's vault construction revisited. Constr. Hist. 22, 47–60.

Ranganathan, J., Corbier, L., Bhatia, P., Schmitz, S., Gage, P., Oren, K., 2004. The Greenhouse Gas Protocol: A Corporate Accounting and Reporting Standard, revised ed. World Resources Institute and World Business Council for Sustainable Development, Washington, DC.

Remus, R., Monsonet, M.A.A., Roudier, S., Sancho, L.D., 2013. Best Available Techniques (Bats) Reference Document for Iron and Steel Production, Industrial Emission Directive 2010/75/EU. Joint Research Centre of the European Commission, Seville, Spain.

Robertson, A.B., Lam, F.C.F., Cole, R.J., 2012. A comparative cradle-to-gate LCA of mid-rise office building construction alternatives: laminated timber or reinforced concrete. Buildings 2, 245–270.

Ryckaert, W.R., Lootens, C., Geldof, J., Hanselaer, P., 2010. Criteria for energy efficient lighting in buildings. Ener. Build. 42, 341–347.

Scheuer, C., Keoleian, G.A., Reppe, P., 2003. Life cycle energy and environmental performance of a new university building: modelling challenges and design implications. Ener. Build. 35 (10), 1049–1064.

Shafiq, N., Nurrundin, M.F., Gardezi, S., Kamaruzzaman, A.B., 2015. Carbon footprint assessment of a typical low rise office building in Malaysia using building information modelling (BIM). Int. J. Sustain. Build. Technol. Urban Develop. 6 (3), 157–172.

Shameri, M.A., Alghoul, M.A., Sopian, K., Zain, M.F.M., Elayeb, O., 2011. Perspectives of double skin façade systems in buildings and energy saving. Renew. Sustain. Ener. Rev. 15 (3), 1468–1475.

Shen, L.Y., Tam, V.W.Y., Tam, L., Ji, Y.B., 2010. Project feasibility study: the key to successful implementation of sustainable and socially responsible construction management practice. J. Clean. Prod. 18 (3), 254–259.

Sinden, G., 2009. The contribution of PAS 2050 to the evolution of international greenhouse gas emissions standards. Int. J. Life Cycle Assess. 14, 195–203.

Stadel, A., Eboli, J., Ryberg, A., Mitchell, J., Spatari, S., 2011. Intelligent sustainable design: Integration of carbon accounting and building information modelling. J. Profess. Issues Eng. Educat. Pract. 137 (2), 51–54.

Steinemann, A., Wargocki, P., Rismanchi, B., 2017. Ten questions concerning green buildings and indoor air quality. Build. Environ. 112, 351–358.

The Green Construction Board, 2016. PAS 2080: 2016 Carbon Management in Infrastructure. The British Standards Institution, London, UK.

Thomas, D., 2015. The Increase of Timber Use in Residential Construction in Australia: Towards a Sustainable Residential Development Model. PhD Thesis. University of Technology Sydney, Australia.

Tian, Y., Zhu, Q., Geng, Y., 2013. An analysis of energy-related greenhouse gas emissions in the Chinese iron and steel industry. Ener. Policy 56, 352–361.

Vakiloroaya, V., 2012. Design optimisation of the cooling coil for HVAC energy saving and comfort enhancement. Sustainability 32 (4), 1209–1216.

Vatalisa, K.I., Manoliadis, O.G., Charalampides, G., 2011. Assessment of the economic benefits from sustainable construction in Greece. Int. J. Sustain. Develop. World Ecol. 18 (5), 377–383.

Verbeeck, G., Hens, H., 2010. Life cycle inventory of buildings: a calculation method. Build. Environ. 45, 1037–1041.

Wang, X., Love, P.E.E., Kim, M.J., Park, C.S., Sing, C.P., Hou, L., 2013. A conceptual framework for integrating building information modelling with augmented reality. Automat. Constr. 34, 37–44.

Wei, H.L., Ni, J.R., Xu, N., 2008. Energy, material and pollutant intensity analysis in the life cycle of walling materials. Ener. Sour. t A (30), 1367–1381.

Wei, X., Lai, J., Zhang, J., 2013. Carbon-emission calculation of electromechanical energy consumption of different structures during the construction phase. J. Chongqing Univ. 12 (2), 67–74.

Whyte, J., Bouchlaghem, N., Thorpe, A., McCaffer, R., 2000. From CAD to virtual reality: modelling approaches, data exchange and interactive 3D building design tools. Automat. Constr. 10 (1), 43–55.

Wintergreen, J. & Delaney, T. (2007). ISO 14064: International standard for GHG emissions inventories & verification. In: Proceedings of the 16th Annual International Emissions Inventory Conference, May 14–17, Raleigh, NC. Retrieved from: <https://www3.epa.gov/ttnchie1/conference/ei16/session13/wintergreen.pdf; logon> (accessed 30.09.16).

Xu, B., Lin, B., 2016. Regional differences in the CO2 emissions of China's iron and steel industry: regional heterogeneity. Ener. Policy 88, 422–434.

Xing, S., Xu, Z., Jun, G., 2008. Inventory analysis of LCA on steel- and concrete-construction of office buildings. Ener. Build. 40, 1188–1193.

Yan, H., Shen, Q., Fan, L.C., Wang, Y., Zhang, L., 2010. Greenhouse gas emissions in building construction: a case study of one Peking in Hong Kong. Build. Environ. 45 (4), 949–955.

Yang, K., Jung, Y., Cho, M., Tae, S., 2015. Effect of supplementary cementitious materials on reduction of CO2 emissions from concrete. J. Clean. Prod. 103, 774–783.

Yun, G.Y., Hwang, T., Kim, J.T., 2010. Performance prediction by modelling of a light-pipe system used under the climate conditions of Korea. Indoor Built Environ. 19, 137–144.

Zhang, X., Shen, L., Zhang, L., 2013. Life cycle assessment of the air emissions during building construction process: a case study in Hong Kong. Renew. Sustain. Ener. Rev. 17, 160–169.

Chapter 11

Sustainable Construction Technology Adoption

Loosemore Martin[1] and Forsythe Perry[2]
[1]*Faculty of the Built Environment, University of New South Wales, Sydney, Australia,*
[2]*University of Technology Sydney, Sydney, Australia*

11.1 INTRODUCTION

Recent influential reports like the World Economic Forum's "Global Risks 2017" have identified the failure of climate-change mitigation and adaptation to be among the greatest risks to humanity in the foreseeable future (WEF, 2017). The construction industry is at the center of this debate. Greenhouse gas emissions from the building sector have more than doubled since 1970 and the industry currently consumes nearly one-third of global energy consumption, making it responsible for about one-third of total direct and indirect energy-related carbon dioxide (CO_2) emissions (IEA, 2013, IPCC, 2014). Energy demand in buildings is also predicted to rise by almost 50% between 2010 and 2050 and a 25% reduction in total energy use would represent energy savings of more than 40 exajoules (EJ), equivalent to total annual current energy use in India and Russia combined (IEA, 2013; IPCC, 2014). Historically, the construction industry has also contributed hugely to environmental pollution, waste, and degradation. For example, in the United Kingdom, construction waste consumes more than 50% of overall landfill volume with 70 million tons of construction waste being discarded annually. Similarly, the US construction industry generates over 100 million tons of construction waste per annum and contributes approximately 29% of the solid waste going to landfill. In Australia, construction activities generate 20%–30% of all the waste entering landfills and in Hong Kong the figure is around 25% which amounts to about 3850 tons per day of construction waste (Lu and Yuan, 2013). The construction industry is also a huge consumer of natural resources with approximately 3 billion tons of natural raw materials (40%–50% of the total flow in the global economy) being used each year in the manufacturing of building products and components worldwide. Buildings in use have also been estimated to be responsible for 12% of global water use (UNEP-SBCI, 2014).

Sustainable Construction Technologies. DOI: https://doi.org/10.1016/B978-0-12-811749-1.00009-2

Many of the technological advances we are likely to see in the future construction industry will be driven by the need to address the above challenges, with new sustainable technologies set to revolutionize the built environments we live in and the way we interact with them. For example, new technological innovations will be needed to reduce the reliance of our buildings on carbon-based energy production. To this end, an increasing body of research produced by researchers (Morris, 2013) and organizations like the Australian Sustainable Built Environment Council (2016) has argued for the greater use of new sustainable building technologies such as roof-top solar panels; photovoltaic cells and nanoparticles integrated into building materials and components such as concrete, roof tiles, glass, and paints; wind and geothermal energy microgenerators; heat pumps and new high-efficiency air conditioning and ventilation systems; LED lighting; and intelligent building management systems and sensors which collect and analyze data and adjust building settings in real time to optimize energy usage, etc. In more recent applications of emerging nanotechnologies, new piezoelectric fabrics are being developed which generate energy from movement and from light with obvious but as yet untested applications to the skins of buildings which could generate energy from solar and wind movement, making buildings net generators of renewable energy in the future rather than energy sinks (Rushton, 2010). Advocates argue that not only do such technologies lead to lower carbon emissions but they also lead to increased jobs and productivity and as much as $20 billion in energy savings across Australia (a return on investment of around 2:1). Given the current lack of guidelines and standards as to what constitutes a carbon neutral building, the Australian Government is currently developing a draft National Carbon Offset Standard for Buildings and the National Carbon Offset Standard for Precincts, which will provide a voluntary framework, standard and guideline for property owners and developers to measure, reduce, offset and report emissions, enabling them to legitimately make carbon neutral claims for building and precinct operations.

Building designs will also need to change, as will the technologies we use to build them. For example, Armstrong (2010) argues that as we move to a more sustainable future we will see taller buildings in our built environments reconceived as vertical urban garden cities which use resources more efficiently and increase urban density to reduce car dependency and energy consumption. For example, London only had two skyscrapers in 1999 but 26 skyscrapers were constructed last year alone, with 28 tall buildings forecasted for completion this year, 40 in 2018 and 455 more in the pipeline, according to the 2017 New London Architecture Tall Building Survey (New London Architecture, 2017).

Despite continuous advances in technology, the construction industry has generally lagged other industries in this area (Barlow, 2000; CII, 2008). The Australian Bureau of Statistics reports that only 31% of the construction companies in Australia undertook any form of innovation activity during

2012–2013 compared with the average of 42% for all Australian businesses (ABS, 2014). New technologies fail (i.e., are not adopted by enough customers to make a sustainable business) at a rate of 40%–90%, depending on the technology type (Cierpicki et al., 2000; Gourville, 2005) and there are numerous barriers to their adoption in construction, e.g., see Skibniewski (2014), Alkalbani et al. (2013), Samuelson and Björk (2013), Hinkka and Tätilä (2013). More recently, Kibert (2016) shows that while many advances have been made in making buildings and the building process more sustainable, there is a lot more than could be done in encouraging the adoption of sustainable development strategies and technologies.

The aim of this chapter is to address this challenge by exploring the process of sustainable technology adoption in construction using a case study of massive timber construction adoption in Australia. As well as helping to mitigate climate change risks, the adoption of new sustainability technologies like massive timber construction can hold other benefits for firms. For example, Hottenrott et al. (2016) found that firms that introduce new environmental technologies can achieve significantly higher levels of productivity if they also adapt their organizational structures to suit. Other potential benefits for construction firms showing that they care about the environment by investing in these technologies include leaving a lasting positive legacy in the communities in which they build; competitive advantage (with socially responsible clients); demonstrable corporate citizenship and social responsibility; improved employee recruitment, engagement, and retention; compliance with growing environmental regulations; improved community engagement and public relations; widening markets and customer base; positive reputation (communities, clients, shareholders, employees, and other stakeholders); investment opportunities through growing socially responsible investment markets (Loosemore and Lim, 2016). Today, the public is better informed, educated, and empowered than at any time in history to understand and influence the impact of business activities on their lives and are willing to change their consumer habits away from firms, which do not align with their sustainability values.

11.2 SUSTAINABLE CONSTRUCTION TECHNOLOGY DIFFUSION

According to Widen et al. (2008) and Brandon and Shu-Ling (2008), innovation diffusion is the process by which an innovation's acceptance spreads through the construction industry social system bringing about a change in its structure and functioning for the benefits of its stakeholders. According to Rogers (1962), whose theories of innovation diffusion are as relevant today as they were 50 years ago, the diffusion process follows an *s*-curve. This starts with the slow uptake of the new technology by a group called "first movers" (about 2.5% of the business population) who are then followed by successive groups of consumers adopting the new technology: early adopters which typically represent about 13.5% of the population; The early majority which typically represent about 34% of the population; the late majority or followers typically represent 34% of the population; and then laggards, resisters, and rejecters which represent around 16% of the population and are the last to do anything.

The diffusion of a Sustainable Construction Technology does not happen in one step and in order to understand the process in more detail, Rogers (2003) proposes a model which comprises of the following five stages.

1. Knowledge stage—when an individual is exposed to an innovation's existence and gains some understanding of how it works.
2. Persuasion stage—when an individual forms a favorable or unfavorable perception or attitude toward the innovation.
3. Decision-making stage—when an individual decides to adopt or reject the innovation.
4. Implementation stage—when an individual puts the innovation into use.
5. Confirmation stage—when an individual seeks reinforcement of the decision to adopt and continues with or rejects or reverses that decision.

Many researchers have explored the factors that determine the speed of technological innovation diffusion through innovation networks and many of these studies have focused on green technology adoption. For example, Walker and Lloyd-Walker (2011) found the four key obstacles to small firms adopting green production practices were:

1. Perceptions of minimum environmental impact of their activities.
2. A lack of convincing business case for change.
3. The targeting of government policies toward large firms.
4. High resistant to change.

Bollinger (2011) found that the adoption of environmentally friendly technologies is slower if there is uncertainty about the performance implications of adopting such technologies, if firms have long equipment replacement cycles or if firms do not have sufficient information to evaluate whether the switch is in their interests. Jumadi and Zailani (2011) found that organizations with knowledge and experience of green technologies, with structures and cultures which allow it to be adopted and with a business environment that encourages it, will be more likely to adopt a sustainable technology than organizations which do not have these attributes. Allan et al. (2014) reviewed the existing literature on the diffusion of green technologies and concluded while it was very heterogeneous and unorganized. They showed that the flow of information is a critical factor in the diffusion process as are networks and framing issues, values, and inherent cognitive biases, which influence decision-making behavior. While policy instruments do have some impact, it also appears to be heavily context dependent and varies across different approaches. For example, Arvanitis et al. (2016) looked at policy-induced green technology adoption is Australia, Germany, and Switzerland, and found that government policies to encourage the uptake of such technologies were largely ineffective compared to voluntary agreements and demand-related factors. They also found that different types of financial incentives worked in different countries—taxes in Germany and subsidies in Austria.

Outside the generic sustainable technology adoption literature discussed above, there has also been a considerable amount of research into technology adoption in construction. For example, Goulding et al. (2007) found that the uptake of hybrid concrete technologies as a viable solution to traditional frame design has been inhibited by a general lack of information and by designers, who tend to stay loyal to "traditional," tried and tested technologies. Gambatese and Hallowell (2011) found that technological innovations diffused more readily when the amount of training required to understand them and the perceived risk of failure was low. This was also the case when the perceived benefits of innovations were clearly pointed out (particularly, cost savings, quality improvement, and increased productivity). Other studies such as Loosemore (2014) have found the timing of the innovation to be important to effective diffusion as well as other factors such as the existence of supporting innovations, the complexity of the innovation, the trialability of the innovation, and the amount of supervision and coordination needed to support it. Sargent et al. (2012) found that the adoption of new information technologies requires changes to existing working practices and processes, which are often viewed as disruptive by members of the organization. They argue that understanding the factors that can influence individuals' intention to utilize technology can assist managers to implement strategies to increase and improve the uptake of technologies and improve the innovation adoption process. Murphy et al. (2015) expressed concern that the construction industry lacks the mechanisms to effectively implement new innovative technologies and products. They propose an innovation management model (IMM) which establishes the prioritized sequence of stakeholder activities required to implement a new technology as a risk-based approach to instill confidence in construction industry stakeholders to adopt new technologies.

There has also been some important research, which has focused on sustainable technology adoption in construction, although this has largely focused on the house building industry. For example, Nelms et al. (2007) developed a framework based on green roof technology to help project stakeholders and policy-makers systematically identify and evaluate the implications and relative merits of a range of alternative sustainable technologies. Christie et al. (2011) explored why homeowners were not adopting energy-efficiency technologies when these not only reduce environmental problems, but also improve health, comfort, economic, and social well-being. They found that although most homeowners declare an apparent preference for energy efficiency, there were inconsistencies in their decisions and actions, as demonstrated by the lack of success of numerous intervention schemes. The motivations behind this apparent reluctance to adopt energy-efficiency technologies were investigated and it was found that a large proportion of homeowners had exaggerated perception of risk associated with these technologies caused by numerous social and cognitive biases that affect their decision to opt for the "safest" option. Christie et al. (2011) conclude that to encourage the uptake of new sustainable technologies, it is vital to better understand the apparently irrational motivations behind homeowners' decisions and how the perception of change can be minimized. Lees and Sexton (2012) also looked at the role that the end users (or occupants) play in the use of new sustainable solar thermal technologies in domestic construction and found that households had little knowledge of, or interest in, the technology prior to moving into the houses. Lees and Sexton (2012) therefore concluded that the "market pull" of consumer design or imagination for such technology is weak. Akadiri and Fadiya (2013) produced empirical evidence to indicate a positive effect of government regulations, top management commitment, and construction stakeholder pressures on sustainable construction practices and technologies. Nahmens and Reichel (2013) explored the adoption process of high-performance building technologies, including alternative wall systems, in the housing sector in hot-humid climates and found that builders identified cost as the top constraint to adoption, followed by a slow learning curve and the lack of proper marketing channels to transition from construction to sale. Lees and Sexton (2014) investigated the adoption of low- and zero-carbon technologies by house builders, and found that house builders tend to select a narrow range of technologies and that these choices are made to minimize the disruption to their standard design and production templates. Their results challenge the dominant technical rationality that assumes technical efficiency and cost benefits are the primary drivers for technology selection. Instead, they found that while these drivers play an important role, it is mediated by the logic of maintaining the standard design and production templates, emphasizing the need for construction diffusion of innovation theory to be problematized and developed within the context of business and market regimes constrained and reproduced by resilient technological trajectories. Berry et al. (2014) investigated the role of eco open home events in imparting motivation and accelerating behavioral change in the adoption of new sustainable technologies and solutions. Drawing on a wide range of visitor survey datasets, and a social learning perspective, the results point to a consistent positive experience for attendees and high rates of perceived learning which encourage and support engaged attendees to undertake low-energy renovations. More recently, Abbott et al. (2015) throw light on the role of regulation on the adoption and diffusion of microgeneration technologies in the United Kingdom and found that these technologies are not normally selected on their technical merit but on their cost effectiveness in complying with rules and regulations. They conclude that although regulation does trigger innovation and the adoption of new products and technologies, the intent that underpins the regulations (such as reduction in carbon emissions) are often only partially achieved because it cannot control people's behavior in the use of those new technologies. Thorpe and Wright's (2015) evaluation of the use of sustainable materials found that experience of using the material previously was the leading issue restricting their adoption while other factors included cost of material and the availability of standards or codes of practice.

11.3 THE IMPORTANCE OF SITUATIONAL CONTEXT

What is clear from the previous discussion is the diversity of factors that impact on technology adoption and the way this apparently changes according to situational contexts. For instance, the case presented later in this chapter must contend with identifying the situational context needed to elevate massive timber construction to being a strongly demanded technology in construction markets.

Here, sustainable construction occupies an interesting and somewhat different space to many other areas of technological innovation. For instance, it sustainability is fundamentally acknowledged as being necessary for global existence, but at a more corporate level it does not necessarily meet the normal business motives of production-based income streams and profitability. Instead, its uptake often relies on the likes of government intervention, special market niches, social responsibility, and altruism.

Construction is also well known as being both complex and fragmented with many client and supplier relationships (Loosemore, 2014). The complexity and interplay between them can make it difficult to determine who the main decision-making parties are in driving the adoption of new technologies. For instance, with regard to the previously mentioned massive timber construction, this must include attention to what timber is worth as sustainability technology. There are also market behavior factors such as how is it dealt with in the supply chain; who is prepared to pay for it; and are the intended advantages project or organizationally oriented? For full impact, the value proposition for adoption must present sustainability as a "need" more so than a peripheral "want." It must penetrate beyond technology for technology's sake (Forsythe, 2014). It must make sense and be understood within the situational context of the construction market.

11.4 CASE STUDY A: THE SUSTAINABILITY VALUE OF MASSIVE TIMBER CONSTRUCTION TECHNOLOGY

In acting upon the previous discussion, the selection of massive timber construction as an area of study is predicated by the level of interest it is generating in global construction markets. It is currently evolving from dominance in traditional low-rise residential housing construction to adoption in larger scale multistory building markets—see for instance the recently completed

17 story Brock Commons building in Vancouver, Canada. Ongoing momentum in the context of larger scale mass timber buildings has also been dealt with in terms of increasing adoption by authors such as Jones et al. (2016) and Kremer and Symmons (2015, 2018).

It is worth noting that while previously mentioned technologies such as ink jet, valve, biomimicry, and nanotechnologies are enticing, the realization of these in terms of immediately measurable carbon savings are still many years away. Their high-tech nature also makes them initially expensive and risky for main flow adoption. On the other hand, massive timber construction is an adaptation of a traditional material in a way that can immediately meet the construction industry's need for new sustainable solutions see for instance a comparison of steel, concrete and mass timber buildings by Rajagopalan et al. 2017 and Ding and Forsythe (2018). The main vehicle for delivering such buildings, at least in terms of new timber technology adoption, is broadly referred to as massive timber construction. It involves large and solid panel, beam, and column elements such as cross-laminated timber (CLT), glue laminated timber (GLT), laminated strand lumber (LSL), and laminated veneer lumber (LVL). All represent engineered timbers with much improved structural capabilities relative to basic sawn timber. Authors such as M. Kuittinen et al. (2013) indicate that such materials are resource efficient and are necessary for ensuring sustainable timber construction.

The solid mass associated with such construction can provide a large-scale impact on reducing carbon emissions. For instance, 1 m^3 of wood is said to store approximately 1 ton of CO_2 —no other materials can positively impact on the environment to the same extent (Lehmann, 2012). On the same theme, many sources assert that wood can significantly reduce CO_2 emissions relative to other materials. Reasons include the way that dried wood products (as used in construction) consist of 50% carbon which has been drawn from CO_2 removed from the atmosphere by growing trees; wood is also said to consume less supply chain fossil fuel in manufacture than most competing materials, and also avoids the likes of cement process emissions and fossil fuel emissions due to biomass displacement (Sathre and O'Connor, 2010b). Buchanan and Levine (1999) provide one of the early examples of applying the above to specific built environment contexts and show that converting wood to a construction material requires much lower process energy and results in lower carbon emissions than other materials. Their results indicate that a 17% increase in wood usage in the New Zealand building industry could result in a 20% reduction in carbon emissions from the manufacture of all building materials. Similar studies have since been undertaken in other countries. For instance, Upton et al. (2008) found that net greenhouse gas emissions associated with wood-based houses in the United States are 20%–50% lower than emissions associated with thermally similar steel or concrete-based houses and assert that the difference between wood and nonwood building systems could be in the order of 9.6 MT of CO_2 equivalents per year.

Sathre and O'Connor's metaanalysis (Sathre and O'Connor, 2010a,b) draws together 66 existing studies and focuses upon the environmental benefits of using wood to displace other products. The quantitative aspect of their study focused on 21 of the above studies and their findings indicate that wood mostly has a positive displacement factor—as a reasonable estimate of the GHG mitigation efficiency—over a range of product substitutions and analytical methodologies (Sathre and O'Connor, 2010b, p. 106). Even so, a key area of realizing these benefits is that at end-of-life (of a building) wood must be reused, recycled or at least landfilled in a way that prevents GHG emissions from escaping—see for instance modeling undertaken by Börjesson and Gustavsson (2000) and by Petersen and Solberg (2003, 2004, 2005).

In terms of direct comparison with competing forms of nonresidential construction, Ding and Forsythe (2017) present the results of a global warming potential (GWP) study, for a hypothetical office building and apartment building used to model the difference between conventional in situ reinforced concrete framed construction versus prefabricated engineered wood construction. A key finding for the office building was that the RC design resulted in 4.4 times greater GWP than the timber version. For the apartment building, the RC option resulted in 2.1 times more GWP than the timber version. These calculations did not even include the additional amount of carbon stored in the timber which would have added approximately 1962 and 1588 tons of stored carbon, respectively (Ding and Forsythe, 2017).

Benefits also extend back to the very beginning of the supply chain. For instance, authors such as Kauppi et al. (2001) chart the potential role of forestation and wood products from a geoengineering perspective with a view to managing carbon pools within global ecosystems around the world. Logging and its impact on the surrounding flora and fauna clearly represent a related area of concern but nowadays the use of timber is commonly underpinned by the specification of timber drawn from forests that comply with stewardship schemes such as the Forest Stewardship Council and The Programme for the Endorsement of Forest Certification (PEFC). One example of this in the Australian context is the way that the predominant Green Star rating system used for sustainability assessment in nonresidential buildings only allows timber credit points to be amassed based around one of the above stewardship schemes (Green Building Council of Australia, 2016).

At the downstream end of the supply chain, the pathway forward for massive timber construction is inherently linked to prefabrication and supportive digital technologies such as BIM and file-to-factory-to-site methodologies. Life cycle assessment research indicates that such approaches offer improved environmental outcomes relative to traditional onsite methods (Kuittinen et al., 2013). More specifically, Mao et al. (2013) found that a semiprefabricated approach to construction produced a 9% saving in greenhouse gases relative to conventional construction—85% of which was saved embodied energy. Similarly, Monahan and Powell (2011) found that embodied carbon for a semidetached house was 34% less than the comparative traditional methods using a panelized timber framing system.

In interpreting the above discussion, carbon-related studies are complex and variable due the intricacies and assumptions surrounding functional units, system boundaries, and timber energy flows that typically traverse between different economic sectors from forest to factory, to construction, to energy, and eventually, waste management (Kuittinen et al., 2013). Even so, there is a strong case concerning the sustainability credentials of massive timber construction technology. Notably, it is particularly well placed in meeting carbon reduction strategies; it can contribute significantly to the overall materials used in a building and; its applicability to prefabrication processes means it uses less carbon that traditional construction processes.

11.5 CASE STUDY B: ADOPTION IN THE AUSTRALIAN CONTEXT OF MASSIVE TIMBER CONSTRUCTION

Despite the importance of measuring the worth of massive timber construction in sustainability terms, actual adoption is in many ways dependent on nuances concerning the overarching market behavior toward it. This can be seen through the eyes of the previously mentioned technology diffusion—it is the process by which an innovation's acceptance spreads and brings about change. Rogers' (1962) *s*-curve is useful in exploring this idea further in the context of massive timber construction. For instance, timber construction spread across Australia is relatively new and therefore representative of the "first mover" stage of adoption. Even so, impetus for its uptake has been underpinned by European technology, which is more advanced and more in keeping with Rogers' "early majority" category. Consequently, whilst the main focus of this case study focuses on market behavior in Australia, discussion begins with insights about establishment of the technology in Europe.

With regard to this, a large amount of forest resource has created a strong timber culture in the likes of Switzerland, Austria, Northern Italy, and Germany. Here, modern manifestations of massive timber construction began in the 1970s with the advent of Brettstapel construction.[1] Further development came via improved adhesive technology resulting in CLT, GLT, and similar products. Such products are now produced *en masse* by companies such as BinderHoltz, Stora Enso, and KLH who of note have developed significant export markets including the supply of massive timber construction elements into Australia. Even so, the timber culture in this part of the world goes well beyond basic manufactured products and instead provides a far more complete and integrated supply chain in terms of technology adoption. For instance, an innovative university sector actively participates in materials testing and development of new multiengineered systems (examples include

1. http://www.brettstapel.org/Brettstapel/What_is_it.html.

Bern University, Rosenheim University, Technical University of Graz, ETH Zurich); development of advanced digital technologies make the transfer of BIM data to factory fabricatable machine files possible (see for example the widely used CADworks and Dataholz software); advanced manufacturing equipment is capable of enabling digitally intelligent and highly automated processing of timber elements (see for instance Hundegger and Homag[2]); manufacturers of metal connectors facilitate efficient prefabrication of large, high-load structures (see for example Rothoblaas and Sihga[3]); advanced design and fabrication contractors pull the above strands together into a systematic means of delivering large and architecturally complex buildings on and off site (see for example Wiehag, Haring, Erne, and Blumer Lehmann[4]).

Combined, this network is broad spanning enough and exportable enough to provide a holistic technology proposition for countries such as Australia to adopt. It serves to bridge the supply side of the technology gap in terms of providing infrastructure that can accelerate uptake in places like Australia, at a rate that would otherwise not be possible. For instance, the Australian context discussed below utilizes the likes of European-based manufacturing equipment, timber products, connectors, file to factory technology, and design tools.

Notwithstanding this, there is still the need to bridge the Australian side of the gap, which primarily concerns the demand side problem of creating uptake for the technology. Here, focus is placed on a limited number of entrepreneurial "first movers," who have provided the impetus for progress in Australia. Specific instances include the large multinational property and construction company Lendlease, and the quickly growing Sydney-based company Strongbuild. Both have been prepared to take the market risk associated with new technology adoption. Both have utilized European supply chains and associated technology as a (low capital) means of initiating adoption. Both use massive timber construction as part of more expansive business models that involve a somewhat compressed and integrated supply chain that aims to deliver fast, prefabricated and sustainable construction solutions.

In detailing the above, Lendlease have a strategy "committed to creating and delivering innovative and sustainable property and infrastructure solutions for future generations" (Lendlease, 2016). They were the first to adopt massive timber construction in Australia via the 10 story Forte apartment building (completed in Melbourne, 2012). It was groundbreaking in so far as at commencement, there were virtually no timber buildings in Australia beyond two stories in height, and yet upon completion it was the tallest timber building in the world. Much has since been written about the project globally. It is a "landmark" project in terms of technology adoption since it

2. https://www.hundegger.de/de/maschinenbau/unternehmen.html, https://www.homag.com/en/.
3. https://www.rothoblaas.com, https://www.sihga.com.
4. http://en.wiehag.com, https://www.haring.ch/en/, https://www.erne.net/de/, http://www.blumer-lehmann.ch/en/.

is not especially notable for its architectural merit but more so for the synthesis of sustainability outcomes, timber technology and fast, prefabricated construction.

The emphasis is mainly around the building's carbon capture credentials and this has proven to be a selling point among end buyers of apartments. What is less apparent at first glance, is the level of integration in the delivery of the building. Unlike most Australian building projects where delivery often involves a fragmented process coupled with risk divestment down the supply chain, Lendlease took a more vertically integrated and compressed approach. For instance, they had a high degree of direct involvement in the project as property developer, design manager and head construction contractor onsite. The prefabricated CLT structure compressed key parts of the supply chain. Here, KLH in Austria worked with Lendlease using BIM-based technology to provide a means for communicating design information between the contrasting locations and in a way that could readily be converted to machine language for offsite cutting and fabrication of panels. Once shipped to site, the panels were lifted into position using a crane and only a small team of carpenters. There is much to reflect on concerning the fast and quiet onsite construction process. For instance, Lendlease made a comparison between how fast the structure of the Forte building was constructed compared to a similar and nearby in situ concrete building also built by Lendlease (see for instance the comparative online video of the two Lendlease buildings which shows how easily the Forte building out paces the in situ concrete building https://www.youtube.com/watch?v = cqXygHyU5ws).

Following projects by Lendlease concern the Docklands library which completed 2 years later (Melbourne, 2014) and more recently, the six story International House office building, which followed a further 3 years later (Sydney, mid-2017). Unlike the Forte building, these latter buildings are of a more institutional nature and more open in design. They show off more timber aesthetically, which would seem to be a more obvious statement about the sustainability features of timber and the sustainability ideals of those occupying the building.

These projects again represent landmark projects in terms of being highly publicized buildings that promote sustainability and what is possible using timber technology. For instance, the recently completed International House figures strongly in media commentary of the new harbor-front Barangaroo precinct, on the edge of the Sydney CBD. In addition, these buildings still exhibit the same delivery features as the Forte project in terms of prefabricated massive timber elements shipped from Europe, and made using file-to-factory technology (this time via Stora Enso). Lendlease has now shown its ongoing intent via its "Designmake" business which provides internally focused expertise that links their design and advanced offsite manufacturing of prefabricated building components. Gradually, this will likely allow greater capacity for more Australian-based processing of offsite timber construction.

Relative to Lendlease, Strongbuild have come from a much smaller base but have grown very quickly. Their first project to attract similar attention to those by Lendlease, is the MacArthur Gardens affordable housing complex in Sydney. It involves three massive timber construction towers with heights of 6, 7, and 8 stories, respectively, and due for completion in mid-2017. A more recent project which is at an earlier stage of progress involves the 10 story AVEO retirement housing project, also in Sydney. Once completed, it will arguably be the largest volume massive timber project in Australia. Other Strongbuild projects include large but lower rise townhouse projects and institutional buildings.

Like Lendlease, Strongbuild have a high degree of integration and a compressed supply chain. However, unlike Lendlease their involvement stops short of being directly involved in property development and instead focuses within the role of a head construction contractor. For instance, using ostensibly their own internal resources they can deliver the entire timber structure under a single package and can also manufacture and install cabinetry, fitout, stairs, and service pods. The breadth of delivery can be scaled according to project need and if appropriate, also allows them to act as a major subcontractor rather than head contractor. Importantly, this allows them greater freedom to move around in the marketplace in terms of seeking different clients and achieving economies of scale, which seemingly differs from the more internally focused Lendlease scenario.

Even so, similarities still exist in terms of the timber-related aspects of their supply chain which in Strongbuild's case, comes from Binderholz in Austria. For instance, in delivering the previously mentioned projects, panels were shipped from Austria according to the design needs of the projects involved. As prefabrication is a central part of their business model, there is high-level use of BIM and file-to-factory information flows. As with Lendlease, such methods make it possible to mitigate the tyranny of distance between Austria and Australia. Again, such an approach leverages European technology including fabrication softwares and manufacturing equipment. Even so, Strongbuild continue to develop their own offsite capacity, which will presumably allow greater long-term value adding within Australia. The scope of their operation is one of the first to exist in the Australian market and marks a change in the existing supply chain capability.

Yet another issue of significance, which falls outside the previously mentioned factors but is still apparent in the Australia context, is the previous difficulties for massive timber construction in passing efficiently through the regulatory compliance process—under Australia's National Construction Code (Australian Building Codes Board, 2016). Until recently, massive timber construction required a so-called "alternative solution" for proving compliance, which is known to be more expensive and harder to achieve than the simplified and more common "deemed to satisfy" pathway. Of note, this situation changed in 2016 where buildings up to 25 m high (approximately eight stories high) can now be dealt with under the more standardized "deemed to satisfy" provisions. This is likely to make massive timber construction a more realistic and cost-effective option for future projects and will engender greater confidence by clients in choosing it as a construction solution.

What can be observed from the above instances is that massive timber construction technology adoption begins at an organizational level and under a business model that pairs sustainability with prefabrication in delivering buildings. It also involves a close supply relationship with timber supply companies from Europe. Consequently, technology adoption is linked to those who are already at a more advanced stage in the previously discussed Rogers' *s*-curve. In time, it is expected that this path dependency will likely adjust as the technology in Australia moves from "first movers" to more advanced stages of technology adoption. For instance, Australia's first CLT manufacturer, CrossLam, will begin production in the near future and so locally made product will become available. It would therefore seem that there is now sufficient market confidence in adoption of the technology, to merit the required capital investment within Australia.

11.6 CONCLUSION

This chapter has shown that expecting sustainable construction technology adoption to happen serendipitously in a complex and fragmented construction industry is fraught with uptake problems unless situational context is provided. Adoption can be piecemeal and may have limited overall impact. This chapter has served to test, refine, and apply situational context concerning the adoption of sustainable construction technology surrounding massive timber construction. Whilst it clearly has strong sustainability credentials that resonate across the full breadth of the supply chain, its uptake varies as seen through the lens of Rogers' *s*-curve. For instance, central Europe is probably best described as being in the "early majority" category with a well-developed supply chain and export capability. Australia is still in the "first movers" category and this implicates the need for technology supply from Europe and reliance on early risk takers who create demand for adoption of the technology in Australia. For these companies, interest in massive timber construction revolves around more holistic business models that not only incorporate sustainability but an emphasis on design driven prefabrication as well. Related features include use of digital technologies as well as compressed and integrated supply chains.

To date, the gestation period for massive timber construction in Australian has occurred over some 4–5 years. Clearly, it takes time to develop local expertise, networks, regulatory standardization, and client interest given the risks in balancing sustainability benefits against traditional construction methods. Landmark buildings create market momentum but it is still difficult to make use of this momentum where there is insufficient local supply chain development for the client base to fully commit to the new technology. In some ways, this is the danger period for technology adoption as an unwanted lull can stall uptake momentum. There is a need to create traction quickly; otherwise, the cost benefits expected of economies of scale cannot be realized.

The next step in moving massive timber technology adoption forward is likely to be stronger development of typical or standardized construction solutions that makes client decision-making simpler and more automatic in the face of competing with traditional construction methods. Economies of scale will then follow. It would seem that a simplified compliance pathway is a good sign in supporting this cause. Even so, the role of landmark buildings will be ongoing in drawing attention to the new technology, which will itself serve to increase adoption.

REFERENCES

ABS, 2014. Innovation in Australian Business, 2012–13. Australian Bureau of Statistics, Belconnen, ACT, Australia.

Abbott, C., Sexton, M., Barlow, C., 2015. Regulation and innovation in new build housing: insights into the adoption and diffusion of micro-generation technologies. In: Orstavik, F., Dainty, A.R.J., Abbott, C. (Eds.), Construction Innovation.Wiley-Blackwell, pp. 79–89.

Akadiri, P.O., Fadiya, O.O., 2013. Empirical analysis of the determinants of environmentally sustainable practices in the UK construction industry. Constr. Innov. Inform. Process Manage. 13 (4), 352–373.

Alkalbani, S., Rezgui, Y., Vorakulpipat, C., Wilson, I.E., 2013. ICT adoption and diffusion in the construction industry of a developing economy: the case of the Sultanate of Oman. Archit. Eng. Design Manage. 9 (1), 62–75.

Allan, C., Jaffe, A.B., Sin I., 2014. Diffusion of green technology: a survey. Motu working paper 14-04. Motu Economic and Public Policy Research, April 2014, Wellington, New Zealand.

Armstrong R.A., 2010. Sustainability and the tall building: recent developments and future trends. Council on Tall Buildings and Urban Habitat, Research Paper, Illinois, USA.

Arvanitis, S., Peneder M., Rammer, C., Stucki T., Woerter M., 2016. The adoption of green energy technologies: the role of policies in an international comparison. KOF working papers no 411, September 2016, Zurich, Germany.

Australian Building Codes Board, 2016. National Construction Code, vol. 1, Canberra, Australian Government.

Australian Sustainable Built Environment Council, 2016. Low Carbon, High Performance. Australian Sustainable Built Environment Council, Sydney, Australia.

Barlow, J., 2000. Innovation and learning in complex offshore construction projects. Res. Policy 29 (7–8), 973–989.

Berry, S., Sharp, A., Hamilton, J., Killip, G., 2014. Inspiring low-energy retrofits: the influence of 'open home' events. Build. Res. Inform. 42 (4), 422–433.

Bollinger, B.K., 2011. Green technology adoption in response to environmental policies. PhD Dissertation. Stanford University, Stanford, CA.

Börjesson, P., Gustavsson, L., 2000. Greenhouse gas balances in building construction: wood versus concrete from life-cycle and forest land-use perspectives. Energy Policy 28 (9), 575–588.

Brandon, P., Shu-Ling, L., 2008. Clients Driving Innovation. Wiley-Blackwell, Oxford.

Buchanan, A.H., Levine, S.B., 1999. Wood-based building materials and atmospheric carbon emissions. Environ. Sci. Policy 2 (6), 427–437.

CII, 2008. Leveraging Technology to Improve Construction Productivity. Construction Industry Institute, New York.

Christie, L., Donn, M., Walton, D., 2011. The 'apparent disconnect' towards the adoption of energy-efficient technologies. Build. Res. Inform. 39 (5), 450–458.

Cierpicki, S., Wright, M., Sharp, B., 2000. Managers' knowledge of marketing principles: the case of new product development. J. Empirical General. Market. Sci. 5 (3), 771–790.

Ding, G., Forsythe, P., 2017. A Study Comparing the Global Warming Potential of Timber and Reinforced Concrete Construction in Office and Apartment Buildings. Report, Forest and Wood Products Australia, Melbourne, Australia.

Ding, G., Forsythe, P., Forest and Wood Products Australia, 2018. A study comparing the global warming potential of timber and reinforced concrete construction in office and apartment buildings, no. PNA308a – 1 213, pp. 1–16.

Forsythe, P.J., 2014. The case for bim uptake among small construction contracting businesses. In: Proceedings of the 31st International Symposium on Automation and Robotics in Construction and Mining, University of Technology Sydney, Sydney, pp. 480–487.

Gambatese, J.A., Hallowell, M., 2011. Factors that influence the development and diffusion of technical innovations in the construction industry. Constr. Manage. Econom. 29 (4), 507–517.

Goulding, J., Sexton, M., Zhang, X., Kagioglou, M., Aouad, G.F., Barrett, P., 2007. Technology adoption: breaking down barriers using a virtual reality design support tool for hybrid concrete. Constr. Manage. Econom. 25 (12), 1239–1250.

Gourville, J.T., 2005. The Curse of Innovation: A Theory of Why Innovative New Products Fail in the Marketplace. Harvard Business School, Boston, MA.

Green Building Council of Australia (2016) Revised Timber Credit. Retrieved from: <http://www.gbca.org.au/green-star/revised-timber-credit/2693.htm> (accessed 05.01.17).

Hinkka, V., Tätilä, J., 2013. RFID tracking implementation model for the technical trade and construction supply chains. Autom. Constr. 35, 405–414.

Hottenrott, H., Rexhauser, S., Veugelers, R., 2016. Organisational change and the productivity effects of green technology. Resour. Energy Econom. 43 (1), 172–194.

IEA, 2013. Transition to Sustainable Buildings: Strategies and Opportunities to 2050. International Energy Agency Publications, Paris, France.

IPCC, 2014. Working Group III – Mitigation of Climate Change: Technical Summary Final Draft, Intergovernmental Panel on Climate Change, Geneva, Switzerland.

Jones, K., Stegemann, J., Sykes, J., Winslow, P., 2016. Adoption of unconventional approaches in construction: the case of cross-laminated timber. Constr. Build. Mater. 125, 690–702.

Jumadi, H., Zailani S., 2011. Determinants of green technology innovation adoption among transportation companies in Malaysia, Conference paper.

Kauppi, P., Sedjo, RJ., Apps, M., Cerri, C., Fujimori, T., et al. Technical and economic potential of options to enhance, maintain and manage biological carbon reservoirs and geo-engineering. In: Mitigation 2001. Metz B, et al. (Ed.) The IPCC Third Assessment Report, Cambridge, Cambridge University Press; 2001.

Kibert, C.J., 2016. Sustainable Construction: Green Building Design and Delivery, fourth ed. Wiley, New York, United States.

Kremer, P.D., Symmons, M.A., 2015. Mass timber construction as an alternative to concrete and steel in the Australia building industry: a PESTEL evaluation of the potential. Int. Wood Prod. J. 6 (3), 138–147.

Kremer, P.D., Symmons, M.A., 2018. Perceived barriers to the widespread adoption of Mass Timber Construction: An Australian construction industry case study. Journal of Mass timber Construction 1, 1–8.

Kuittinen, M., Ludvig, A., Weiss, G., 2013. Wood in Carbon Efficient Construction: Tools, Methods and Applications. CEI-Bois, Brussels.

Lees, T., Sexton, M., 2012. The domestication and use of low and zero carbon technologies in new homes. In: Smith, S.D. (Ed.), Proceedings 28th Annual ARCOM Conference, September 3–5, 2012, Edinburgh, UK. Association of Researchers in Construction Management, pp. 1389–1398.

Lees, T., Sexton, M., 2014. An evolutionary innovation perspective on the selection of low and zero-carbon technologies in new housing. Build. Res. Inform. 42 (3), 276–287.

Lehmann, S., 2012. Low carbon construction systems using prefabricated engineered solid wood panels for urban infill to significantly reduce greenhouse gas emissions. Sustain. Cities Society 6, 57–67.

Lendlease (2016) Lendlease. Retrieved from: <http://www.lendlease.com/au/company/about-us/> (accessed 05.01.17).

Loosemore, M., 2014. Innovation, Strategy and Risk in Construction: Turning Serendipity into Capability. Routledge, London.

Loosemore, M., Lim, B., 2016. Linking corporate social responsibility and organizational performance in the construction industry. Construction Management and Economics. Retrieved from: <www.tandfonline.com/doi/full/10.1080/01446193.2016.1242762>.

Lu, W., Yuan, H., 2013. Investigating waste reduction potential in the upstream processes of offshore prefabrication construction. Renew. Sustain. Energy Rev. 28 (4), 804–811.

Mao, C., Shen, Q., Shen, L., Tang, L., 2013. Comparative study of greenhouse gas emissions between off-site prefabrication and conventional construction methods: two case studies of residential projects. Energy Build. 66, 165–176.

Monahan, J., Powell, J., 2011. An embodied carbon and energy analysis of modern methods of construction in housing: a case study using a lifecycle assessment framework. Energy Build. 43 (1), 179–188.

Morris, S., 2013. Improving Energy Efficient, Sustainable Building Design and Construction in Australia – Learning from EuropeSpecialised Skills Institute, Melbourne, Australia.

Murphy, M.E., Perera, S., Heaney, G., 2015. Innovation management model: a tool for sustained implementation of product innovation into construction projects. Constr. Manage. Econom. 33 (3), 209–232.

Nahmens, I., Reichel, C., 2013. Adoption of high performance building systems in hot-humid climates – lessons learned. Constr. Innov. Inform. Process Manage. 13 (2), 186–201.

Nelms, C.E., Russell, A.D., Lence, B.J., 2007. Assessing the performance of sustainable technologies: a framework and its application. Build. Res. Inform. 35 (3), 237–251.

New London Architecture, 2017. Tall Buildings Survey. New London Architecture, London.

Petersen, A.K., Solberg, B., 2003. Substitution between floor constructions in wood and natural stone: comparison of energy consumption, greenhouse gas emissions, and costs over the life cycle. Canad. J. Forest Res. 33 (6), 1061–1075.

Petersen, A.K., Solberg, B., 2004. Greenhouse gas emissions and costs over the life cycle of wood and alternative flooring materials. Clim. Change 64 (1–2), 143–167.

Petersen, A.K., Solberg, B., 2005. Environmental and economic impacts of substitution between wood products and alternative materials: a review of micro-level analyses from Norway and Sweden. Forest Policy Econom. 7 (3), 249–259.

Rajagopalan, N., Kelley, S.S., 2017. Evaluating sustainability of buildings using multi-attribute decision tools. Forest Prod. J. 67 (3), 179–189.

Rogers, E.M., 1962. Diffusion of Innovations. Free Press, Glencoe, New York.

Rogers, E.M., 2003. Diffusion of Innovations. Free Press, New York.

Rushton, T., 2010. Opening pandora's box. Build. Survey. J. July–August. RICS, London, pp. 26–30.

Samuelson, O., Björk, B., 2013. Adoption processes for EDM, EDI and BIM technologies in the construction industry. J. Civil Eng. Manage. 19 (Suppl. 1), S172-S187.

Sargent, K., Hyland, P., Sawang, S., 2012. Factors influencing the adoption of information technology in a construction business. Austral. J. Constr. Econom. Build. 12 (2), 72–86.

Sathre, R., O'Connor, J., 2010a. Meta-analysis of greenhouse gas displacement factors of wood product substitution. Environ. Sci. Policy 13 (2), 104–114.

Sathre, R., O'Connor, J., 2010b. A synthesis of research on wood products and greenhouse gas impacts, Technical Report No. TR 19 R, FPInnovations, Vancouver BC.

Skibniewski, M.J., 2014. Information technology applications in construction safety assurance. J. Civil Eng. Manage. 20 (6), 778–794.

Thorpe, D., Wright, C., 2015. Use of advanced and green construction materials by small and medium-sized enterprises. In: Raiden, A. , Aboagye-Nimo, E. (Eds.), Proceedings 31st Annual ARCOM Conference, September 7–9, 2015, Lincoln, UK. Association of Researchers in Construction Management, pp. 227–236.

UNEP-SBCI, 2014. Greening the Building Supply Chain. United Nations Environment Programme, Geneva, Switzerland.

Upton, B., Miner, R., Spinney, M., Heath, L.S., 2008. The greenhouse gas and energy impacts of using wood instead of alternatives in residential construction in the United States. Biomass Bioener. 32 (1), 1–10.

WEF, 2017. Global Risk Report 2017, World Economic Forum, Davos, Switzerland.

Walker, D.H.T., Lloyd-Walker, B., 2011. Profiling Professional Excellence in Alliance Management. Alliancing Association of Australia, Melbourne, Australia.

Widen, K., Atkin, B., Hommen, L., 2008. Setting the game plan – the role of clients in construction innovation and diffusion. In: Brandon, P. Shu-Ling, L. (Eds.), Clients Driving Innovation. Wiley-Blackwell, Oxford, pp. 56–79.

FURTHER READING

ABS, 2009. Innovation in Australian Business. Australian Bureau of Statistics, Belconnen, ACT, Australia.

Actualitix (2016) ActualitixData. Retrieved from: <http://en.actualitix.com/country/che/switzerland-forest-area.php> (accessed 05.01.17).

The Economist, 2013. January 12th–18th, Innovation pessimism: has the ideas machine broken down? The Economist 406 (8818), 19–23.

Walker, E., Redmond, J., Giles, M., 2010. A proposed methodology to promote adoption of green production by small firms. Int. J. Business Stud. 18 (1), 39–48.

Chapter 12

Lean Principles in Construction

Oladapo Adebayo Akanbi[1], Ogunbiyi Oyedolapo[2] and Goulding Jack Steven[3]
[1]School of Engineering, University of Central Lancashire, Preston, United Kingdom, [2]Department of Quantity Surveying, Yaba College of Technology, Lagos, Nigeria, [3]Department of Architecture and Built Environment, Northumbria University, Newcastle, United Kingdom

12.1 INTRODUCTION

There is an ongoing trend in the development of lean culture and the implementation of lean principles within the construction industry. The term "lean" was borrowed and developed from a range of industries and converted to a suitable form for use in the construction industry. Lean construction relies on the production management principles inspired by the Toyota Production System (Howell, 1999). The principle of lean is mainly aimed at eliminating waste in process activities in order to reduce process cycles, improve quality, and increase efficiency. In the lean context, waste includes all forms of overproduction, overprocessing, delay, excess inventory and motions, failure, and defects (Al-Aomar, 2011). Lean construction is a philosophy based on lean manufacturing concepts. The application of lean production principles to construction is based on the transformation (T), flow (F), and value generation (V) theory of production management (Koskela, 1992). The TFV theory identifies transformation (achieved by resources, equipment, workers), materials flow and customer focus as three independent angles to production. Lean construction emerged due to the failure of the traditional project management approach and results in significant improvements in terms of management and project deliverables (Koskela and Howell, 2002). The emerging concept of lean construction is concerned with the application of lean thinking to the construction industry. The ideas of lean thinking within the UK construction industry seem to be predominantly targeted to improving quality and efficiency (Green, 1999).

One of the priorities of lean construction is the elimination of waste as lean construction tools have evolved to contribute to sustainable construction. Similarly, sustainable construction focuses on the removal of waste from the construction process. Therefore, it could be said that both concepts

Sustainable Construction Technologies. DOI: https://doi.org/10.1016/B978-0-12-811749-1.00010-9

share the same goal of waste reduction. However, organizations struggle to integrate the concepts (Koranda et al., 2012). The need for a more sustainable approach or initiative such as lean has been stressed by the UK Government (DTI, 2006). The construction industry is seen as a major threat to sustainable development due to its negative environmental impact. Therefore, there has been the need for widespread implementation of practices and approaches that would reduce the negative impact of construction activities on the environment. This has raised the construction industry's awareness of the lean approach and provided some impetus toward its adoption in the industry. Thus, the lean approach has been implemented within the construction industry as a means of improving construction activities and work place organization.

12.2 AN OVERVIEW OF THE CONSTRUCTION INDUSTRY

Construction industry activities are concerned with the planning, regulation, design, manufacture, construction, and maintenance of buildings and other structures (Burtonshaw-Gunn, 2009). The construction industry is defined by Druker and White (1996) as comprising new construction work, general construction and demolition work, the construction and repair of buildings, civil engineering, the installation of fixtures and fittings, and building completion work. In addition, the construction industry encompasses the building and the engineering sectors and also includes the process-plant industry. However, the demarcation between these areas is often blurred (Ashworth, 2010).

The construction industry is one of the largest industries in the world. It has the largest number of fatal injuries of main industry groups and it is also one of the most dangerous in terms of health and safety (Ashworth, 2010; HSE, 2013). The construction sector in almost every country is under increasing requirement to adopt the principles of lean in its activities and policies (Brandon and Lombardi, 2005). Miyatake (1996) suggested that changes should be made in the way the construction industry undertakes its activities. The industry makes use of energy, materials, and other resources to create buildings and civil engineering products. The end result of all these activities is huge volumes of waste during and at the end of the facility's life. Therefore, changing this process into a cyclic process will bring increased use of recycled, renewed, and reused resources, and a significant decrease in the use of energy and other natural resources.

The construction industry is one of the largest industries in most developed economies. A variety of statistics illustrates the importance of the construction industry to national economies. In terms of output and contribution to employment, the construction industry is immense. In the United Kingdom, for example, the construction industry contributed almost £90 billion to the national economy (or 6.7%) in value added, comprises over

280,000 businesses covering some 2.93 million jobs, which is equivalent to about 10% of total UK employment (Office for National Statistics UK, 2013). The contracting industry is the largest subsector of the construction sector, accounting for about 70% of total value added generated by UK construction and almost 70% of the sector's jobs (Office for National Statistics UK, 2013).

Murdoch and Hughes (2008) stated that most of the people who study the construction industry do so from their respective points of view, which are based on their professions. Because of this, there are many descriptions of the construction sector, drawn from different specialist disciplines. In a broad context, the term construction can include the erection, repair, and demolition of things as diverse as houses, offices, shops, dams, bridges, motorways, home extensions, chimneys, factories, and airports. Many different firms carry out specialist work relating to particular technologies, but a few firms are confined to only one building type or one technology. Barrie and Paulson (1992) affirmed that the construction industry must include general and specialty construction as there is no clear definition as to what the construction industry is. They further stated that to really understand the construction industry, one must extend its scope to include designers of facilities, material suppliers, and equipment manufacturers.

Meyers (2008) clearly identified a range of actors that can be included in a broad definition of the construction industry as suppliers of basic materials such as cement and bricks, machinery manufacturers who provide equipment used on site, such as cranes and bulldozers used on site. Other actors are site operatives who bring together components and materials, project managers and surveyors who coordinate the overall assembly, developers and architects who initiate and design new projects, facility managers who manage and maintain property, and providers of complementary goods and services such as transportation, distribution, demolition, disposal, and clean-up. There are many interpretations given to the construction industry in the literature some of which are narrow or broad. The construction industry has been referred to as all firms involved directly in the design and construction of buildings and other structures (Morton, 2002). This description exempts the broad categorization of the construction industry.

The construction industry is generally characterized by low productivity, overruns in cost and schedule, errors, poor reputation, shortage of skilled labor and poor safety (Dawood et al., 2002; Health and Safety Executive, 2013). In particular, lack of safety is one of the chronic problems in construction, as is evident from the high accident rates. Although the construction industry only accounts for about 5% of employees in England, for example, it accounts for 27% of fatal injuries to employees and 10% of major reported injuries (HSE, 2013). According to HSE (2013), in the United Kingdom, slips and falls account for more than half (56%) of all major injuries and almost a third of over 7 day (31%) injuries to employees, making up 37% of all reported injuries to employees. The construction industry has several unique features, which distinguish it from other industries. Such features include its fragmented nature, one-off projects, the physical nature of its products, and multiplicity of participants (Thomassen, 2004; Fellows et al., 2002; Dubois and Gadde, 2002; de Valence, 2010; Amiri et al., 2014).

12.3 THE CONCEPT OF LEAN CONSTRUCTION AND INNOVATION

The concepts and principles of lean are to generally make the construction process more efficient by the removal of waste, which is regarded as nonvalue generating activities (Koskela, 2000). Lean construction is a new production philosophy, which has the potential of bringing innovative changes in the construction industry. The need for more innovation in construction has been raised as construction faces the challenge of minimizing the environmental impact of its consumption of materials and energy (Sturges et al., 1999). However, complexities within the construction industry make introducing these innovative technologies difficult. For example, each technology may have to be compatible with numerous parties and the residential construction industry contains a particularly high degree of uncertainty in innovative product adoption (Koebel, 2004; Conference Board of Canada, 2004).

Innovation in construction is "the act of introducing and using new ideas, technologies, products and/or processes aimed at solving problems, viewing things differently, improving efficiency and effectiveness, or enhancing standards of living" (CERF, 2000). The construction industry has been tagged with a poor record of innovation when compared with the manufacturing industry. The construction industry was the worst performing industry in five out of six categories of innovation compared to 11 other industries (DTI, 2004). The emergence of lean construction is to bring substantial change to the construction industry to achieve the objectives of sustainability within the built environment in the critical social, economic, and environmental aspects. Lean construction offers new techniques of constructing sustainable projects. It is about costs reduction by cutting waste, innovating by engaging people, and organizing the workplace to be more efficient. Evidence from the literature reveals that innovation through lean improvement in construction processes has provided proof of sustainability outcomes in terms of reduced waste, effort, and time, and increased productivity (Ogunbiyi et al., 2011; Wu and Wang, 2016).

Lean construction is similar to the current practices in the construction industry; both practices pursue meeting customer needs while reducing the wastage of resources. However, the difference between conventional and lean construction is that lean construction is based on production management principles, and it has better results in complex, uncertain, and quick projects. The lean approach identifies seven types of waste: overproduction, overstocking, excessive motion, waiting time, transportation, extra-processing and defects (Womack and Jones, 1996). It seeks to minimize these wastes to provide significant benefits in terms of increased organizational and supply chain communication and integration (Ogunbiyi et al., 2011). The various methodologies for attaining lean production include just-in-time (JIT), total quality management (TQM), concurrent engineering, process redesign, value based management, total productive maintenance, and employee involvement (Womack and Jones, 1996).

Lean construction conceives a construction project as a temporary production system dedicated to three goals of delivering the project, maximizing value, and minimizing waste (Koskela, 2000). The aim for lean construction is to work on continuous improvement, waste elimination, strong user focus, value for money, high quality management of projects and supply chains, and improved communications (OGC, 2000). A study by Eriksson (2010) has shown how the implementation of lean construction impacts the performance of supply chain actors in construction. An investigation of the basic principles of lean construction shows how the various aspects of the approach can be grouped into six core elements: waste reduction, process focus in production planning and control, end customer focus, continuous improvements, cooperative relationships, and systems perspective. Fig. 12.1 presents lean construction in practice. The common elements of lean according to Jørgensen (2006) and Womack and Jones (2003) are:

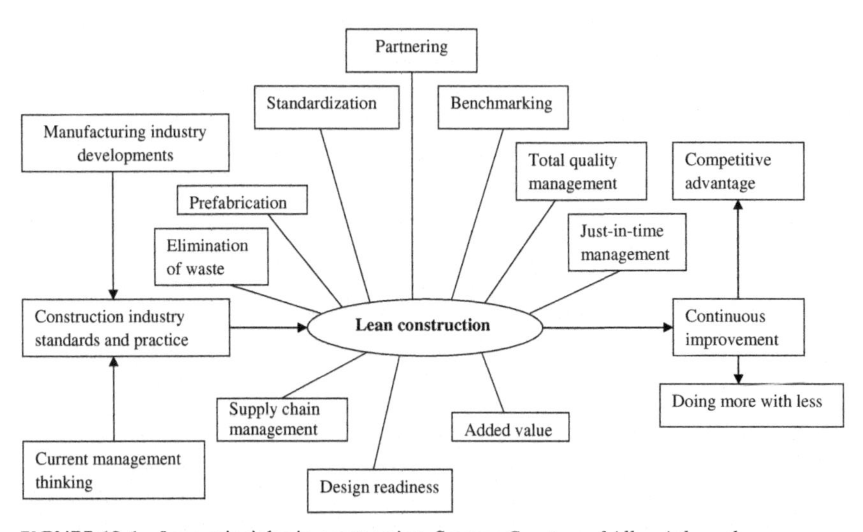

FIGURE 12.1 Lean principles in construction. Source: *Courtesy of Allan Ashworth.*

- A focus on eliminating/reducing waste and sources of waste in relation to the delivery of artifacts or services that represent value to the end customer.
- End customer preference is adopted as the reference for determining what is to be considered value and what is waste.
- Management of production and supply chain from a (customer) demand pull approach.
- Approaching production management through focus on processes and flows of processes.
- An (at least to some degree) application of a system's perspective for approaching issues of waste elimination/reduction.

Lean construction implementation efforts can be divided into three stages. Stage 1 focuses on waste elimination from a technical and operational perspective. The responsibilities and focus are tied to managers rather than individual workers. The essential parts of this stage are elimination of needless movements, cutting out unnecessary costs, optimizing workflow, and sharing the benefits from improved performance (Green and May, 2005). Stage 2 focuses on eliminating adversarial relationships and enhancing cooperative relationships and teamwork among supply chain actors. The essential parts are cooperation, long-term framework agreements, workshops, and facilitators. The workshops and facilitator roles are needed in order to enhance good communication among the project participants, which in turn improves integration and coordination (Pheng and Fang, 2005). Knowledge sharing and joint learning are important in enhancing continuous improvement. Therefore, the understanding of lean concept by projects participants must be improved (Green and May, 2005). This can be facilitated by relevant training in workshops where project participants meet periodically to exchange knowledge and experience and also jointly suggest ideas for the most visible problems in the workplace (Salem et al., 2006). Aspects related to Stage 2, according to Eriksson (2010), are limited bid invitation, soft parameters, long-term contracts, collaborative tools, and broad partnering team. Lean Stage 2 does not go much beyond the concept of partnering since it is about eliminating waste derived from suboptimizations and adversarial relationships through increased integration and collaboration.

Stage 3 involves structural changes to project governance and it is regarded as the most sophisticated. Its essential parts are information technology, prefabrication, Last Planner, bottom-up activities and emphasis on individuals, a rethink of design and construction, decreased competitive forces, long-term contracts, training at all staff levels, and a systems perspective of both processes and the product (Green and May, 2005). Aspects related to lean Stage 3 are joint IT tools, prefabrication, Last Planner, self-control, concurrent engineering, limited bid invitation, soft parameters, long-term contracts, special interest groups, training, suggestions from workers, coherent procurement decisions, large scale contracts, and properly balanced objectives. Only when striving to achieve Stage 3 is a radical change from other types of project governance required (Eriksson, 2010).

12.4 LEAN PRINCIPLES AND LEAN THINKING IN CONSTRUCTION

Womack and Jones (2003) defined five lean principles to eliminate waste in organizations. Lean thinking has been considered to be one potential approach for improving organizational performance in terms of value generation (Womack and Jones, 2003). Lean construction is the application of lean thinking to the design and construction process creating improved project delivery to meet client needs and improve profitability for constructors. It places "optimizing the total value" instead of "minimizing the cost" as the main goal. Within lean, cost cutting has to be seen in the perspective of eliminating nonvalue-adding activities (Womack and Jones, 2003). There are five lean principles, namely, "identify value from the customer perspective," "map the value stream," "achieve flow within the work process," "achieve customer pull at the right time," and "strive for perfection and continuous improvement" (Picchi and Granja, 2004; Fewings, 2013). These principles are referred to as the strategic approach term "lean thinking." Fig. 12.2 represents the five lean principles within which lean construction techniques can be successfully applied and the description of some of these lean techniques are given in the next subsection.

12.4.1 Identifying Value

The principle of value in construction is considered from the point of view of the customer's perception, i.e., specifying value from the customer's perspective. The definition of value in construction is subjective and complex. Koskela (2000) explored the use of the term "value" and deduced that value can be related to either market value or utility value. This perception of value is supported by many other researchers as presented in lean construction papers. Value management and value engineering are the two methodologies used in gaining value knowledge about a construction design. Value Management is described as, "Conceptualization of production (from value viewpoint): As a process where value for the customer is created through fulfillment of his requirements" (Bertelsen and Koskela, 2002, p. 3). Value engineering refers to the analysis of technical building design to reduce cost but maintain fitness for purpose. Value management is concerned with understanding how the brief for a design can be developed so that a client's requirements can be captured in the design (Kelly and Male, 1993) thereby improving the value perception of the client.

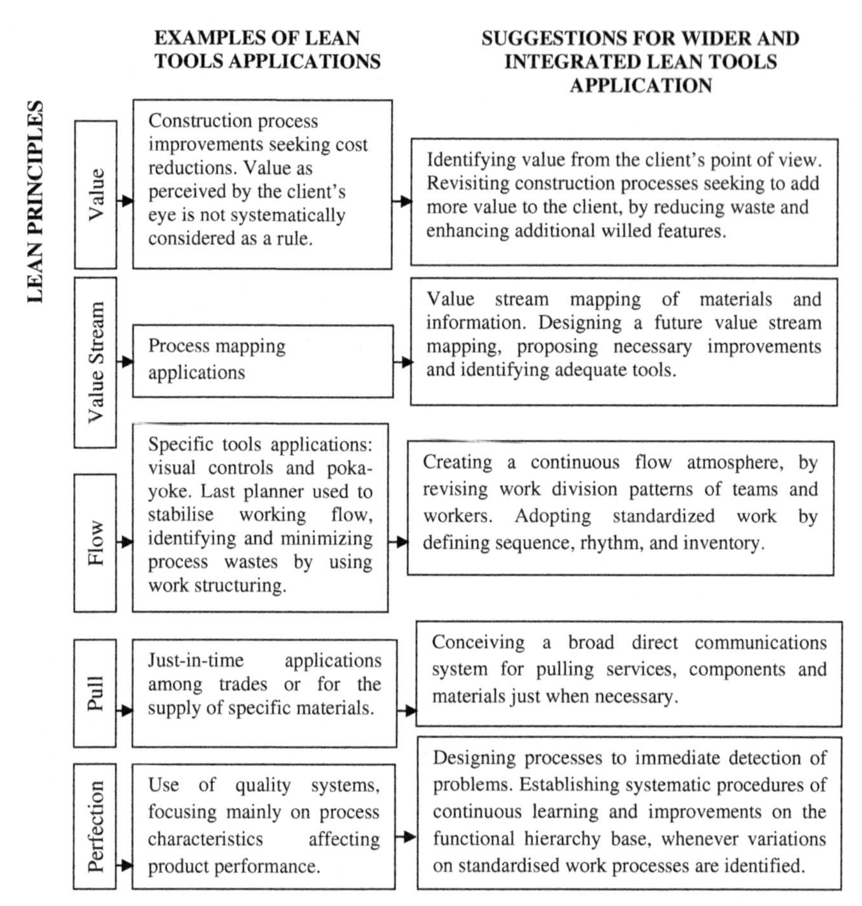

FIGURE 12.2 Examples of lean tools already reported in construction implementation and suggestion for wider and integrated application for sector. Source: *Picchi, F., Granja, A., 2004. Construction sites: using lean principles to seek broader implementations. In: Proceedings of International Group of Lean Construction, 12th Annual Conference, Copenhagen, Denmark, August 3–5, p. 6.*

Ballard and Howell (1998) stated that value is generated through a process of negotiation between customer's ends and means. According to Lindfors (2000), value is the products or services that increase profit, decrease time and cost, and improve quality for the company and generate profit or value for the customer. Leinonen and Huovila (2000) mentioned three different kinds of value: exchange value, use value, and esteem value. The first two can be translated directly into market value and utility value. The third value has a broader scope than only the product–customer perception. Marosszeky et al. (2002) described the importance of working with project culture and values for achieving the desired level of quality. A model for reinforcing the manager's belief is applied, and it is concluded that each organization tends to view quality from its parochial perspective due to the culture.

Value, as defined in lean thinking (Womack and Jones, 2003), refers to materials, parts, or products—something materialistic, which is possible to understand and to specify (Koskela, 2004). Value may be divided into external and internal value (Emmitt et al., 2005); external value being the clients' value and the value which the project should end up with, while internal value is the value that is generated by and between the participants of the project delivery team (contractor, architects, designers, etc.). Emmitt et al. (2005) stated that value is the end goal of all construction projects and therefore the discussion and agreement of value parameters is fundamental to the achievement of improved productivity and client/user satisfaction. Emmitt et al. (2005) view value as an output of the collective efforts of the parties contributing to the design and construction process; central to all productivity; and providing a comprehensive framework in which to work. Value identification is vital in lean construction and must be established as client requires a product that fulfills its purpose, requirement, and value for money (Ballard and Howell, 2004).

12.4.2 Value Stream Mapping

Mapping the value stream is the second principle of lean thinking. A value stream identifies every step necessary to create and deliver a product to the customer (Womack and Jones, 1996). The first step to understanding this is mapping the current state. Thus, identifying and mapping the value stream is a key requirement in the implementation of lean thinking. The value stream map is therefore an outline of operations that lead to valuable achievement of the product and identifies alternative routes to maximize performance in the construction process (Dulaimi and Tanamas, 2001; Forbes and Ahmed, 2011). As noted by Fewings (2013), value stream entails all the value-adding steps required to design, produce, and provide the product. In achieving an effective delivery process in a construction project, all the nonvalue-adding activities must be minimized, i.e., those activities that do not add value to the customer. The nonvalue-adding activities consume resources such as time, space, and money without adding value to the product (Forbes and Ahmed, 2011).

According to Fewings (2013), flow is a key process of perfecting and balancing the interconnected activities through which a product can be developed. The flow aspect has been suggested to be given more attention in construction instead of emphasizing on the transformation aspect (Koskela and Howell, 2002). In managing flow, Koskela (2000) presented seven flows toward the perfect execution of a work package. These include space, crew, previous work, equipment, information, materials, and external conditions such as weather. It should be noted that each of these flows has its own nature and should be managed accordingly. Among these flows, the physical flow of materials is probably the easiest to deal with while the external condition is mostly the flow of unlikely things that may happen. According to Garnett et al. (1998), flow is strategically concerned with achieving a holistic route by which a product is developed. It attacks the fragmentation that is inherent in the construction industry by revealing it as highly wasteful. The basic units of analysis in lean construction are information and resources flow.

12.4.3 Allowing Customer Pull

Pull is the need to be able to deliver the product to the customer as soon as the customer needs it at the strategic level. Pull is the ability to deliver the product to the client at the earliest possible time (Bicheno, 2000). The principle of pull makes use of just-in-time applications to meet client needs and subsequently customizing and delivering them more predictably when the client requires them (Garnett et al., 1998). There are several risks and uncertainties associated with the delivery of construction projects which may hinder the delivery of a product to the client within the specified period and with minimum resources (Dulaimis and Tanamas, 2001).

12.4.4 Pursuing Perfection

This is a key concept at the strategic level because it defines the need for the way of working and organizing to deliver construction products to become a way of life with an inherent culture. To achieve perfection means constantly considering what is being done, how it is being done and harnessing the expertise and knowledge of all those involved in the processes to improve and change it (Womack and Jones, 1996; Dulaimi and Tanamas, 2001). The principle of perfection involves producing exactly what the customer wants in terms of quality and quantity at the right time at a fair price and with minimum waste, with an ultimate target of zero waste (Bicheno, 2000). Perfection can be achieved through a continuous improvement in eliminating all forms of obstacles and nonvalue-adding tasks along the flow process (Dulaimi and Tanamas, 2001). Lean thinking principles have been applied mainly on site activities (i.e., at the production level). However, it should be incorporated at the organizational level to guide senior managers in organizing change (Womack and Jones, 1996).

There is a further level of organization in the construction industry where lean principles can be applied, namely, the construction project. Lean principles can be implemented by adopting the Plan, Do, Check, Act (PDCA) cycle. Koskela (1992) identified a process for implementing lean construction:

1. Process—company's work should be viewed as process with a flow of key elements such as information and material depending whether it is a management process, e.g., design management or an operational process, e.g., constructing a floor slab.
2. Reduce nonvalue-adding activities—each process should be examined to reduce nonvalue-adding activities such as movement of materials to enhance the effectiveness and efficiency of value adding processes.
3. Develop a more effective operating strategy—having developed a more effective operating strategy, the organization of the work force must be taken into account.
4. Change the organization culture—the culture of the organization needs to be changed to support lean construction. Tools need to be developed to facilitate key parts of the new process.

The implementation issues of lean such as barriers and success factors have been identified and discussed by many studies. These barriers need to be overcome in order for the construction industry to reap the benefits of implementing lean construction. Implementing lean construction requires action and good understanding of the drivers and techniques. The implementation of lean is believed to start on projects and spread throughout the organization and suppliers (Howell and Ballard, 1998). The implementation of lean requires a change management strategy. There are four levels of change: event, system, behavior, and mental model.

A number of authors have argued that the construction industry has failed to adopt process improvement techniques such as TQM (Shammas-Thoma et al., 1998), supply chain management (Vrijhoef and Koskela, 2000), and just-in-time (Low and Mok, 1999) that have improved performance in other industries. Organizations in lean transformation report an unexpected occurrence, which calls for larger improvements (Senge et al., 1994). The construction industry should view implementing lean as a system. Howell and Ballard (1998) advised that, "implementing lean thinking will lead to change in almost every aspect of project and company management. No one step-by-step guide can be offered because change at the mental model level is a developmental process. Each principle driven action will reveal new opportunities hidden because people simply could not think in ways that made the change possible. Thinking causes action, action causes deep learning, and learning causes new thinking."

12.5 LEAN CONSTRUCTION TOOLS AND TECHNIQUES

Within a company, there are many lean tools and techniques that can be used. These tools and techniques include value stream, 5S, just-in-time, visual management, preventative maintenance, continuous improvement activities, Kanban, etc. The adoption of the lean approach within a company has potential significance in terms of productivity, service delivery, and quality, which ultimately results in substantial cost savings.

Salem et al. (2005) in their study "Site Implementation and Assessment of Lean Construction Techniques" carried out an evaluation on lean construction tools such as Last Planner, increased visualization, daily huddle meetings, first run studies, 5S process, and fail safe for quality and safety. The effectiveness of the lean construction tools was evaluated through the lean implementation measurement standard and performance criteria. It was found that Last Planner, increased visualization, daily huddle meetings, and first run studies achieved more effective outcomes than expected on the project. However, the results of implementation of 5S process and Fail Safe for quality did not meet the expectations of the tool champions and the research team. It was found that there is a need for behavioral changes and training for effective use of lean tools. Most of the lean construction tools selected for the project are either ready to use or recommended with some modifications.

12.5.1 Last Planner System

Last Planner System (LPS) is a technique that shapes workflow and addresses project variability in construction. It is a system of production control that emphasizes the relationship between scheduling and production control to improve the flow of resources (Ballard, 2000; Fewings, 2013). The Last Planner is the person or group accountable for operational planning, that is, the structuring of product design to facilitate improved work flow, and production unit control, that is, the completion of individual assignments at the operational level (Ballard, 2000). People, information, equipment, materials, prior work, safe space, and safe working environment are the seven flows required to come together at the workplace to enable construction transformation to flow. LPS manages all seven flows by building relationships, creating conversations, and securing commitments to action at the right level at the right time throughout the process (Mossman, 2008). According to Ballard and Howell (1994), the use of lean-based tools like Last Planner reduces accident rates. The aim of LPS, according to Ballard (1997), is to improve productivity by eliminating barriers to workflow. One of the main advantages is that it replaces optimistic planning with realistic planning by assessing the Last Planners' performance based on their ability to achieve their commitments (Salem et al., 2005).

12.5.2 Increased Visualization

According to Moser and Dos Santos (2003), the increased visualization lean tool is about communicating key information effectively to the workforce through posting various signs and labels around the construction site. This is because workers can remember elements such as workflow, performance targets, and specific required actions if they visualize them. Salem et al. (2005) noted that the increased visualization tool makes operations and quality requirements clearer using charts, displayed schedules, painted designated inventory, and tool locations. This tool is similar to the lean manufacturing tool, visual controls, which is a continuous improvement activity that relates to the process control (Abdelhamid and Salem, 2005).

12.5.3 Daily Huddle Meetings

Two-way communication is the key of the daily huddle meeting process in order to achieve employee involvement (Schwaber, 1995, cited by Salem et al., 2005). With awareness of the project and problem solving involvement along with some training that is provided by other tools, employee satisfaction (job meaningfulness, self-esteem, sense of growth) will increase. This is a lean construction tool where a brief daily start-up meeting is conducted. This allows the team members to quickly give the status of what they have been working on since the previous day's meeting, especially if an issue might prevent the completion of an assignment (Salem et al., 2005).

12.5.4 First Run Studies

According to Ballard (1997), First Run Studies are used to redesign critical assignments, part of continuous improvement effort; and include productivity studies and review work methods by redesigning and streamlining the different functions involved. The use of video files, photos, or graphics to show the process or illustrate the work instruction is common with First Run Studies (Abdelhamid and Salem, 2005). The first run of a selected assignment should be examined in detail, bringing ideas and suggestions to explore alternative ways of doing the task. A PDCA cycle is suggested to develop the study (Forbes and Ahmed, 2011). "Plan" refers to select work process to study, assemble people, analyze process steps, brainstorm how to eliminate steps, and check for safety, quality, and productivity (Salem et al., 2006). "Do" means to try out ideas on the first run. "Check" is to describe and measure what actually happens. "Act" refers to reconvene the team, and communicate the improved method and performance as the standard to meet. This tool is similar to the combination of the lean production tool, graphic work instructions, and the traditional manufacturing technique, time and motion study (Abdelhamid and Salem, 2005).

12.5.5 5S Process

The 5S process (sometimes referred to as the Visual Work Place) is about "a place for everything and everything in its place." It has five levels of housekeeping that can help in eliminating wasteful resources (Kobayashi, 1995; Hirano, 1996): "Seiri" (Sort) refers to separate needed tools/parts and remove unneeded materials (trash). "Seiton" (Straighten or set in order) is to neatly arrange tools and materials for ease of use (stacks/bundles). "Seiso" (Shine) means to clean up. "Seiketsu" (Standardize) is to maintain the first 3Ss. Develop a standard 5 S's work process with expectation for the system improvement. "Shitsuke" (Sustain) refers to create the habit of conforming to the rules. This tool is similar to the 5S housekeeping system from lean manufacturing (Abdelhamid and Salem, 2005). The material layout is commonly used for acceleration of 5S implementation on the construction site. Spoore (2003) stated that 5S is an area-based system of control and improvement. The benefits from implementation of 5S include improved safety, productivity, quality, and set-up-times improvement, creation of space, reduced lead times, cycle times, increased machine uptime, improved morale, teamwork, and continuous improvement (kaizen activities).

12.5.6 Fail Safe for Quality and Safety

The "Poka-Yoke" devices as new elements that prevent defective parts from flowing through the process were introduced by Shingo (1986). Fail safe for quality relies on the generation of ideas that alert for potential defects. This approach is opposed to the traditional concept of quality control, in which only a sample size is inspected and decisions are taken after defective parts have already been processed. This is similar to Visual inspection (Poka-Yoke devices) from lean manufacturing. Fail Safe can be extended to safety but there are potential hazards instead of potential defects, and it is related to the safety risk assessment tool from traditional manufacturing practice. Both elements require action plans that prevent bad outcomes. The logic of lean construction implementation requires a certain sequence of initiatives, which progressively reveal additional opportunities for improvement (Ballard, 1997).

12.5.7 Concurrent Engineering

Concurrent engineering has been defined as the parallel execution of different development tasks in multidisciplinary teams with the aim of obtaining an optimal product with respect to functionality, quality, and productivity (Rolstadås, 1995). Concurrent engineering goes beyond diagrams, charts, and algorithms. It demands a multidisciplinary team effort where information sharing and communication are key to identify ideas (Kamara, 2003). According to Gil et al. (2000), the success in lean product process development relies on the involvement of all participants in the early design. Therefore, the relationship with client should not be overlooked as the client may facilitate concurrent engineering efforts that reduce the project's cost. Partnering with subcontractors and suppliers can also influence the outcome of concurrent engineering efforts.

12.5.8 Value Stream Mapping

Value stream, according to Womack and Jones (1996, p. 19), is "the set of all specific actions required to bring a specific product through the three critical management tasks of any business. This is the problem-solving task running from concept through detailed design and engineering to production launch, and the information management task proceeding from raw materials to a finished product in the hand of the customers."

Paez et al. (2005) classified the operative techniques utilized in lean construction into three levels. The different levels are described below, and the classification is summarized in Table 12.1.

1. Level One: Direct application of the techniques from lean manufacturing.
2. Level Two: Modification of the techniques taken from lean manufacturing.
3. Level Three: The all-in-all lean construction specific techniques.

TABLE 12.1 Classification of Lean Methodologies/Tools

Levels	Lean construction technique	Related lean manufacturing technique
Level One	– Material Kanban cards	– Kanban system
Level Two	– Visual inspection – Quality management tools – Concurrent engineering	– Visual inspection (Poka-Yoke devices) – Multifunctional layout – TQM – Standard operations – Single minute exchange of dies (SMED)
Level Three	– Last Planner – Plan conditions of work environment (PCWE) – Daily huddle meetings	– Kanban system – Production leveling – Toyota verification of assembly line (TVAL)

Source: Paez, O., Salem, S., Solomon, J., Genaidy, A., 2005. Moving from lean manufacturing to lean construction: toward a common sociotechnological framework. Human Factors Ergonom. Manuf. Serv. Indust. 15(2), 233–245 (p. 241).

12.6 BENEFITS OF LEAN CONSTRUCTION

The theory of how lean production can work in a construction environment for the purpose of achieving the same benefits as derived in the automotive industry was initiated by Koskela in 1992. Koskela reviewed the theory of lean production in terms of its constituent elements and its conceptual basis as a production philosophy. Construction was defined as a production philosophy and the problems that practitioners would have in adopting the approach was identified (Koskela, 1992). The three layers of lean production identified by Koskela are as follows:

- A production method which was effective and waste free;
- A general management philosophy;
- A set of tools to continuously improve quality.

Koskela (2000) further argued that construction production should not be seen as conversion activities but rather a process flow. Some of the benefits of adopting a process flow viewpoint include the removal of nonvalue-adding activities such as waiting, transporting, and inspection of materials. Two important aspects are identified in the general management philosophy of lean production are the reorganization of the work force to facilitate new operating processes, and the cultural changes that are required within the firm and individuals for the success of the lean production philosophy. It therefore becomes imperative for a company intending to adopt a lean production philosophy to consider what the most suitable organization structure would be for the new way of working because lean implementation in various industries can lead to different results (Wu and Wang, 2016). Similarly, it is essential for the company to adapt existing techniques to suit its own unique environment or create other tools and techniques to support its new operating and management structures. It should be noted that the tools and techniques are developed to support the other two aforementioned elements. The benefits of implementing lean construction can be grouped under environmental, economic, and social aspects. Lean construction is one of the strategies for improving the sustainability of construction, in other words one method of achieving sustainable construction (Ogunbiyi et al., 2014).

Several examples of the application of lean construction techniques were presented by Forbes et al. (2002). This included a Brazilian company which collaborated on a research program with the University of Sao Paulo to improve the integration of design and production processes, and used Last Planner on a 90-day construction project; and the application of the Last Planner Control System on a housing project in Quito, Ecuador. Some of the benefits achieved are presented, respectively: Communication and motivation among the design team influenced the integration of design features with process considerations directly, the implementation of lean construction and control procedures significantly improved production efficiency, in terms of buildability and production cost control and elimination of not only material waste, but nonvalue-adding tasks as well. Additional benefits include a reduction in project duration from 90 to 83 days and reduced rework. The Last Planner facilitated improved quality control and the application of lean methods, and the Percent Plan Complete (PPC) and Performance Factor (PF) improved. It was proven at the construction site that look ahead planning enables one to keep current activities linked with the master pull schedule.

Marzouk et al. (2011) assessed the impact of applying lean principles to design processes in construction consultancy firms to aid in decision-making at early stages of construction projects using a computer simulation tool. It was concluded that applying lean construction principles to the design process significantly helped to improve process efficiency, in terms of reduced process durations and increased resource utilization.

In the review of three case studies on lean principles for rapid construction, Yahya and Mohamad (2011) identified the shortening of order fulfillment leading times, less project downtime, more innovation, and true reduction in the chronicle predecessor as some of the benefits. In the case study of a design and build firm in Colorado, United States, which applied lean construction principles, the reduced project time of up to 30% was attributed to developments of facilitation of innovation in design and assembly, improvement in site work flow by proper definition of production units, and visualization of processes and the use of dedicated design team on any design from beginning to end (Constructing Excellence by Watson, 2004).

The adoption and implementation of lean concepts and tools translated into fast and huge productivity gains for a construction company in Fortaleza, Brazil. According to Jose and Alves 2007, this led to the organization of international seminars and events on innovative practices in Lean Construction in 2004 and 2006 which raised the interest of local and national and national construction companies in Brazil. Gradually, it became established and glaring that the adoption of lean principles facilitated the progress of companies, sustained the innovative practices that had been introduced and implemented, and stimulated the introduction of new ones. The inability of some companies to sustain the benefits arising from the use of lean construction principles this way was attributed to lack of integration of lean construction implementation within their business strategy (Jose and Alves, 2007).

Scherrer-Rathje et al. (2009) stated that despite the significant benefits lean offers in terms of waste reduction and increased organizational and supply chain communication and integration, implementing lean and achieving the levels of organizational commitment, employee autonomy, and information transparency needed to ensure its success is a daunting task. Not every company will be successful in its first attempt to get lean.

12.7 ORGANIZATIONAL CHALLENGES IN THE IMPLEMENTATION OF LEAN CONSTRUCTION

There are many challenges to the application of lean construction practice across different countries. Many studies have identified challenges encountered with the application of lean construction techniques. The adoption of innovative management practices, such as supply chain management and lean thinking, from a manufacturing context to the construction industry, is not without challenges (Höök and Stehn, 2008). Many companies worldwide have tried to implement lean but a majority of them only achieved modest levels of success as the adoption of lean has presented more failure than success among many industries (Mohd-Zainal et al., 2011). The common causes of lean failures include, among other things, poor leadership, poor communication, lack of concrete processes or mechanisms, lack of clear targets or direction, lack of conducive environments, staff resistance to change, and lack of learning that leads to poor understanding of lean (Hines et al., 2004). Barriers to the implementation of lean construction has been classified by many authors including Olatunji (2008), Alinaitwe (2009), Bashir et al. (2010), and Sarhan and Fox (2012). Ogunbiyi (2014) further categorized these barriers into management, process, people, resource, technology, and other related barriers.

12.7.1 Management-Related Barriers

Top management commitment and support is vital in achieving a successful adoption and implementation of innovative strategies such as lean (Salem et al., 2005; Cano et al., 2015). The successful implementation of lean requires adequate management commitment and support. Lack of management commitment and support, poor communication, lack of customer focus and process-based performance management system have been categorized under the management-related barriers (Ogunbiyi, 2014; Forbes and Ahmed, 2004; Sospeter and Kikwasi, 2017; Tezel et al., 2018). Other studies have identified management-related issues such as poor project definition, inadequate resources, and delay in materials delivery (Olatunji, 2008). Koskela (1999) identified inaccurate preplanning while Forbes and Ahmed (2004) presented additional challenges that include delay in decision-making, unsuitable organizational structure, weak administration, and poor procurement selection strategies. In addition, Alinaitwe (2009) and Forbes and Ahmed (2004) identified poor communication as one of the barriers in implementing lean. Poor communication can be divided into poor communication among employees and poor communication between the senior management and the general workforce. Employees are to be involved in the implementation process, but the importance of involving the general workforce is often neglected by senior management. Poor communication can result in no change within the organization as knowledge would have remained within the circle of the senior managers (Achanga, 2007). It is important to establish effective communication among the parties by means of partnering and integrated team working route (Thomas and Thomas, 2005) to increase adaptability to corporate culture change. This helps to improve knowledge sharing and cooperation within the work group for performance improvement (Coyle-Shapiro, 1995; Burnes et al., 2003).

12.7.2 Technology-Related Barriers

Lack of adequate skills and knowledge and lack of fundamental techniques are classified as technology barriers. The successful implementation of the lean concept by an organization depends on the level of commitment, knowledge, and skills. However, organizations do face significant barriers in taking the first steps toward adopting lean. Examples include understanding the underlying concepts of lean (Green, 1999). The construction industry adopted the use of lean thinking principles from the manufacturing sector. Therefore, many lean construction principles and techniques are adapted from the manufacturing sector. This has led to a debate on the extent to which these tools and techniques can be applicable to construction (Green, 1999; Howell and Ballard, 1998). In fact, Eriksson (2009) has argued for a need for some of these techniques and principles to be amended to suit construction. The use of inappropriate tools and techniques has been identified by many researchers (including Bashir et al., 2010; Johansen et al., 2002) as a barrier to the successful implementation of lean. It is imperative to have a full understanding of the lean manufacturing concepts in order to clearly understand the concept of LC. Two of the barriers identified by Tabatabaee et al. (2017) are lack of knowledge and lack of expertise, which reflect the inadequacy of training and education in relevant techniques. This is because a central tenet of lean is that improvements are based on the ideas and knowledge of employees (Found and Harvey, 2006; Van Dun et al., 2008).

12.7.3 Resource-Related Barriers

Inadequate training and financial issues in terms of training cost are categorized as resource-related barriers. Resources in financial terms are required for employee training programs and for the use of external consultants. Another form of financial issue in terms of training cost is the financial incapacity of organizations which has been considered as one of the major barriers to the adoption and implementation of lean (Achanga, 2007). Training or team training is not successful unless reinforced by regular follow-ups of an ongoing systematic change in how work is conducted (Wiklund and Wiklund, 2002). A lack of quality training leads to insufficient implementation of quality methods and quality learning including lean (Sandvik and Karrlson, 1997; Tabatabaee et al., 2017; Dinesh et al., 2017). Education, training, and participation are factors critical in the implementation of a quality improvement process (James, 1996). Training and communication play a crucial role in increasing employee awareness, knowledge and understanding of the adoption of environmental management systems (Zutshi and Sohal, 2004). Required training is necessary for proper implementation of lean across an organization. Effective implementation of an improvement program is about organizational learning and without organizational learning there can be no continuous improvement (Wiklund and Wiklund, 2002). Organizational learning is also critical in the implementation of lean; otherwise, organizations focus on personal mastery rather than "team learning" and a systems view (Senge, 1990).

12.7.4 Process-Related Barriers

Lack of adequate lean awareness and understanding, lack of implementation understanding and concepts, gaps in standards and approaches, and long implementation periods are categorized as process-related barriers. Furthermore, Mossman (2009) identified lack of time for innovation while Alarcón et al. (2002) and Abdullah et al. (2009) found long periods of implementation as the challenge in some organizations. Many organizations have been put off considering the implementation period of the concepts of lean. Lean implementation should not be considered as a quick process but should be viewed as a journey for continuous improvement. It requires training and the adoption of a culture of continuous improvement and developing the system to support lean implementation as well as long-term thinking (Mossman, 2009; Rother, 2010). One of the major threats to the implementation of lean is the fact that there are no standard approaches to how a company should implement lean; this has presented a lot of challenges for organizations which intend to implement lean (Bernson, 2004). Bernson (2004) presented the challenges of a standard approach to lean as selecting the appropriate level of detail, lack of customization at the local level, and a top-down implementation model.

12.7.5 People-Related Barriers

Human attitudinal issues and resistance to change pose challenges to the implementation of lean construction in many organizations. Lack of committed leadership, misconceptions about lean practice, poor leadership, lack of cooperation, lack of team work, and poor understanding of clients' briefs are some of the attitudinal factors (Howell, 1999; Olatunji, 2008; Common et al., 2000; Forbes and Ahmed, 2004; Mossman, 2009). To move toward sustainability, companies need employee involvement in changing corporate culture (Hanna et al., 2000). The success of the adoption of environmentally responsible practices is dependent on employee involvement in cultural change because organizations are viewed as complex systems of individuals and coalitions, each having its own beliefs and culture. It is imperative to change beliefs and values assigned to the environment by all employees in an organization. To achieve this, employees need to understand the need for change and to be in a position to create appropriate responses. A clear understanding of the future direction of business goals make employees commit to their organizations (Walker et al., 2007). Organizational culture is a main element for promoting an innovative environment.

There is a human element in the culture of an organization that cannot be left out and is the determinant in effective business performance and management of change. Moffett et al. (2002) observed that to change an organization's culture, peoples' values, norms, and attitudes must be amended so that they make the right contribution to the collective culture of the organization. Another aspect that must be understood is that each organization requires a different set of cultural values. If an organization is dealing with ambiguous situations that require a variety of insights, then there is a higher need for flexibility.

12.7.6 Other Barriers

Government policy and the fragmented nature of the construction industry are classified as other barriers to the implementation of lean. In some countries, government policy issues such as inconsistency in policies and unsteady price of goods impede the implementation of lean construction (Olatunji, 2008; Alinaitwe, 2009). The fragmented nature of the construction industry is recognized as restricting change within the industry (Myers, 2005). The UK construction industry has been characterized by a complex and fragmented structure and this is conceptualized as a barrier to effective implementation of any process improvement within the construction sector. The traditional construction process is characterized by its fragmented nature with loosely coupled actors who only take part in some of the phases of the process (Johansen et al., 2002). The effect of the fragmented nature of the construction industry has been identified by many studies (Bashir et al., 2010; Mossman, 2009).

12.8 RESEARCH REPORT ON THE IMPACT OF LEAN CONSTRUCTION TECHNIQUES ON SUSTAINABLE CONSTRUCTION IN THE UNITED KINGDOM

The authors conducted a survey of UK construction participants in 2013 to explore the contribution of lean techniques to sustainable construction. All the 100 companies on the directory of United Kingdom top 100 construction companies were contacted and asked to participate in the survey. Seventy out of the firms contacted indicated their interest and were willing to participate and complete the questionnaire. Fifty-five of the firms eventually completed and returned the questionnaires.

As shown in Table 12.2, 70 questionnaires were distributed to construction professionals in various construction firms. Fifty-five were returned and analyzed.

TABLE 12.2 Survey Return

	Number	Percentage (%)
Total number of questionnaire received	55	79
Total number of questionnaire unreturned	15	21
Total number of questionnaire distributed	70	100

Source: Ogunbiyi, O., Oladapo, A., Goulding, J., 2014. An empirical study of the impact of lean construction techniques on sustainable construction in the UK. J. Constr. Innov. Inform. Process, Manage. 14 (1) 96.

TABLE 12.3 Ranking of the Benefits of Synchronizing Lean and Sustainability

Benefits of synchronizing lean and sustainability	Mean	Std. deviation	Rank
Improved corporate image	3.29	.533	1
Increased productivity	3.27	.525	2
Reduction in waste	3.24	.543	3
Reduction in energy consumption	3.22	.567	4
Improvement in sustainable innovation	3.20	.558	5
Improved process flow	3.20	.558	5
Reduction in material usage	3.20	.590	7
Reduced cost and lead time	3.20	.678	8
Improvement in health and safety	3.18	.580	9
Improvement in environmental quality	3.16	.601	10
Reduction in water usage	3.16	.601	10
Increased sustainable competitive advantage	3.11	.567	12
Increased compliance with customers' expectation	3.16	.660	13
Increased employee morale and commitment	3.05	.731	14

Source: Ogunbiyi, O., Oladapo, A., Goulding, J., 2014. An empirical study of the impact of lean construction techniques on sustainable construction in the UK. J. Constr. Innov. Inform. Process, Manage. 14 (1) 98.

12.8.1 Benefits of Synchronizing Lean and Sustainability

The results (presented in Table 12.3) show that improved corporate image which is ranked (1) is the most important benefit of synchronizing lean and sustainability while increased employee morale and commitment (ranked 12) is the least.

12.8.2 Lean Principles/Techniques for Enabling Sustainability

The results presented in Table 12.4, show the level of use of lean principle/techniques for enabling sustainability in respondents' organizations. 4 = high use, 3 = medium use, 2 = low use, and 1 = don't use. The most used lean techniques are just-in-time, visualization tool, value analysis, daily huddle meetings, and value stream mapping while six sigma is the least used techniques.

TABLE 12.4 Lean Principle/Techniques for Enabling Sustainability

Lean principles/techniques for enabling sustainability	Mean	Std. deviation	Rank
Just-in-time	2.75	.440	1
Visualization tool	2.67	.818	2
Daily huddle meetings	2.60	.564	3
Value analysis	2.60	.830	4
Value stream mapping	2.51	.690	5
Total quality management	2.49	.605	6
Fail safe for quality	2.47	.742	7
5S	2.44	.714	8
Total preventive maintenance	2.38	.828	9
First run studies	2.29	.567	10
Last planner	2.29	.875	11
Concurrent engineering	2.09	.752	12
Pull approach	2.04	.543	13
Kanban	1.91	.823	14
Kaizen	1.91	.845	15
Six sigma	1.53	.742	16

Source: Ogunbiyi, O., Oladapo, A., Goulding, J., 2014. An empirical study of the impact of lean construction techniques on sustainable construction in the UK. J. Constr. Innov. Inform. Process, Manage. 14 (1) 99.

12.8.3 Areas of Linkage Between Lean and Sustainability

The results (presented in Table 12.5) show that waste reduction, environmental management, and value maximization is the most important area of linkage between lean construction and sustainability while cost reduction is the least.

The results of the survey indicated that there were several benefits associated with the implementation of lean construction and sustainable construction. The overall perspective of professionals within the construction industry, according to the questionnaire survey, showed that benefits such as improved corporate image and sustainable competitive advantage, improved process flow and productivity, improvement in environmental quality, and increased compliance with customer's expectations were realized from the integration of principles of lean construction and sustainable construction within construction industry. Just-in-time, visualization tool, value analysis, daily huddle meetings, and value stream mapping are the most common lean tools/techniques employed to enhance sustainability. The study also identified several areas of linkage between lean and sustainability, including waste

TABLE 12.5 Ranking of Areas of Linkage Between Lean Construction and Sustainability

Area of linkage between LC and sustainability	Mean	Std. deviation	Rank
Waste reduction	3.55	.503	1
Environmental management	3.55	.503	1
Value maximization	3.53	.539	3
Health and safety improvement	3.49	.605	4
Performance maximization	3.47	.504	5
Design optimization	3.45	.571	6
Quality improvement	3.44	.572	7
Resource management	3.38	.527	8
Energy minimization	3.31	.577	9
Elimination of unnecessary process	3.31	.635	10
Continuous improvement	3.29	.658	11
Cost reduction	3.20	.711	12

Source: Ogunbiyi, O., Oladapo, A., Goulding, J., 2014. An empirical study of the impact of lean construction techniques on sustainable construction in the UK. J. Constr. Innov. Inform. Process, Manage. 14 (1), 100.

reduction, environmental management, value maximization, and health and safety improvement among others. Full details of the survey results can be found in Ogunbiyi et al. (2014).

12.9 SUMMARY

This chapter has drawn from literature on the construction industry relating to lean principles and their application. Lean tools and techniques have been examined and an overview of the application of lean principles in construction is provided. Organizational barriers to the implementation of lean construction as well as the benefits of lean construction were also discussed. These barriers were grouped into process, people, cost, management, technology, and other related barriers. Lean practices can lead to environmental benefits; inversely environmental practices often lead to improved lean practices. The chapter also provides a summary of the results of an empirical study on the impact of lean construction techniques on sustainable construction in the United Kingdom. It is concluded that the contribution of lean construction to sustainable construction goes beyond the environmental aspect but also extends to the social and economic aspects. The lean approach also has a positive influence on sustainable construction in terms of improved safety. Lean implementation can exist at both levels of strategic and operational management, therefore the implementation issues can be viewed from both perspectives. The lean approach has delivered significant economic benefits to companies. However, the implementation of lean is still facing difficulties due to the variability of construction processes and products. A better understanding of lean concepts by the construction industry can contribute to improvement in all aspects of construction organization.

REFERENCES

Abdelhamid, T., Salem, S., 2005. Lean construction: a new paradigm for managing construction projects. In: The International Workshop on Innovations in Materials and Design of Civil Infrastructure, Cairo Egypt. http://www.researchgate.net/publication/242085758_LEAN_CONSTRUCTION_A_NEW_PARADIGM_FOR_MANAGING_CONSTRUCTION_PROJECTS, accessed 02/05/2013.

Abdullah, M.M.B., Uli, J., Tari, J.J., 2009. The relationship of performance with soft factors and quality improvement. Total Quality Manage. Busi. Excel. 20 (7), 735–748.

Achanga, P. (2007). Development of an impact assessment framework for lean manufacturing within SMEs. Ph.D. thesis, Cranfield University (submitted).

Al-Aomar, R., 2011. Handling multi-lean measures with simulation and simulated annealing. J. Franklin Inst. 348 (7), 1506–1522.

Alarcón, L.F., Diethelm, S., Rojo, O. 2002. Collaborative implementation of lean planning systems in Chilean construction companies. In: Tenth Annual Conference of the International Group for Lean Construction, August, Brazil.

Alinaitwe, H.M., 2009. Prioritising lean construction barriers in Uganda's construction industry. J. Constr. Devel. Countr. 14 (1), 15–29.

Amiri, M., Ardeshir, A., Fazel Zarandi, M.H., 2014. Risk-based analysis of construction accidents in Iran during 2007-2011-meta analyze study. Iranian J. Public Health 43 (4), 507–522.

Ashworth, A., 2010. Cost Studies of Buildings, fifth ed. Pearson Education Limited, England, UK.

Ballard, G., 1997. Lookahead planning: the missing link in production control. In: Proceedings of the 5th Annual Conference of the International Group for Lean Construction. Griffith University, Gold Coast, Australia.

Ballard, G., Howell, G., 1994. Implementing lean construction: Stabilizing work flow. Proceedings of the 2nd Annual Meeting of the International Group for Lean Construction, Santiago, Chile.

Ballard, G., Howell, G., 1997. Implementing lean construction: improving downstream performance. Lean Constr. 111–125.

Ballard, G., Howell, G., 1998. Shielding production: essential step in production control. J. Constr. Eng. Manag. 124 (1), 11–17.

Ballard, G., Howell, G.A., 2004. Competing construction management paradigms. Lean Constr. J. 1 (1), 38–45.

Ballard, H.G. (2000). The last planner system of production control. Doctoral dissertation, the University of Birmingham, Birmingham, UK Available at: <http://www.leanconstruction.org> (accessed 10.12.12).

Barrie, D.S., Paulson, B.C., 1992. Professional Construction Management. McGraw-Hill International Edition, Toronto, Ont.

Bashir, M.A., Suresh, S., Proverbs, D.G., Gameson, R., 2010. Barriers towards the sustainable implementation of lean construction in the United Kingdom, ARCOM Doctoral Workshop, vol. 25. University of Wolverhampton.

Bernson, M. (2004). The value of a common approach to lean. Master thesis, Massachusetts Institute of Technology. Available at: <http://dspace.mit.edu/bitstream/handle/1721.1/34753/56607252.pdf?sequence = 1> (accessed 7.9.12).

Bertelsen, S., and Koskela, L. (2002). Managing the three aspects of production in construction. In: Proceedings of the 10th Annual Conference of the International Group for Lean Construction, Gramado, Brazil.

Bicheno, J., 2000. The lean toolbox, 2nd ed PICSIE books, Buckingham.

Brandon, P.S., Lombardi, P., 2005. Evaluating Sustainable Development in the Built Environment. Blackwell Science Ltd., Oxford.

Burnes, B., Cooper, C., West, P., 2003. Organizational learning: the new management paradigm? Manage. Deci. 41 (5), 452–464.

Burtonshaw-Gunn, S.A., 2009. Risk and Financial Management in Construction. Gower Publishing Limited, England, UK.

Cano, S., Delgado, J., Botero, L., and Rubiano, O., 2015. Barriers and success factors in Lean Construction's implementation - Survey in pilot context. In: Proceedings of 23rd Annual Conference of the International Group for Lean Construction. Perth, Australia, July 29-31, pp. 631-641.

CERF (2000). Guidelines for moving innovations into practice. Working Draft Guidelines for the CERF International Symposium and Innovative Technology Trade Show 2000, CERF, Washington, DC.

Common G., Johansen E., and Greenwood D. (2000) A Survey Of The Take-Up Of Lean Concepts Among UK Construction Companies: The 7th International Group for Lean Construction Annual Conference. Brighton, July 2000.

Conference Board of Canada (2004). Sixth Annual Innovation Report – Progress Report. The Canada Project, Ottawa.

Coyle-Shapiro, J., 1995. The Impact of a TQM Intervention on Teamwork: a Longitudinal Assessment. Employee Relat. 17 (3), 63–74.

de Valence, G., 2010. Defining an industry: what is the size and scope of the Australian building and construction industry. Constr. Econ. Build. 10 (1-2), 119–131.

DTI (2004) Detailed Results From the Third UK Community Innovation Survey. Available at: <http://www.dti.gov.uk/iese>. (accessed 25.07.11).

DTI, 2006. Review of Sustainable Construction. DTI, London.

Dawood, N., Akinsola, A., Hobbs, B., 2002. Development of automated communication of system for managing site information using internet technology. Automat. Constr. 11 (5), 557–572.

Dinesh, S., Sethuraman, R., Sivaprakasam, S., 2017. The review on lean construction an effective approach in construction industry. Int. J. Eng. Res. Mod. Edu., (Special Issue) 119.

Druker, J., White, G., 1996 Managing People in Construction, Institute of Personnel and Development, London.

Dubois, A., Gadde, L.E., 2002. The construction industry as a loosely coupled system: implications for productivity and innovation. Constr. Manage. Econom. 20 (7), 621–631.

Dulaimi, M.F., Tanamas, C., 2001. The principles and applications of lean construction in Singapore. Proceedings of the 9th Annual Conference of the IGLC. Kent Ridge Crescent, Singapore.

Emmitt, S., Sander, D. and Christoffersen, A.K. (2005). The value universe: defining a value based approach to lean construction, In: Proceedings of the 13th Annual Conference of the International Group for Lean Construction, Sydney.

Eriksson P.E. (2009). A case study of paternering in lean construction. In: The 5th Nordic Conference on Construction Economics, Reykjavík University, Iceland.

Eriksson, P.E., 2010. Improving construction supply chain collaboration and performance: a lean construction pilot project. Suppl. Chain Manage. Int. J. 15 (5), 394–403.

Fellows, R., Langford, D., Newcombe, R., and Urry S. (2002). Construction Management in Practice. Oxford, Blackwell Science

Fewings, P., 2013. Construction Project Management: An Integrated Approach, second ed. Spon Press, London.

Forbes, L.H., Ahmed, S.M., 2004. Construction Integration and Innovation through lean methods and e-business applications. Construction Research Congress: Wind of Change: Integration and Innovation. American Society of Civil Engineers Press, pp. 1–10.

Forbes, L.H., Ahmed, S.M., 2011. Modern Construction: Lean Project Delivery and Integrated Practices. CRC Press, Boca Raton.

Forbes, L.H., Ahmed, S.M., and Barcala, M. (2002). Adapting lean construction theory for practical application in developing countries. In: Proceedings of the first CIB W107 International Conference: Creating a Sustainable Construction Industry in Developing Countries (Eds. Division of Building Technology, CSIR), November, Stellenbosch, South Africa, pp. 11–13.

Found, P.A., Harvey, R., 2006. The role of leaders in the initiation and implementation of manufacturing process change. Int. J. Knowledge Cult. Change Manage. 6, 35–46.

Garnett, N., Jones, D.T. and Murray, S. (1998). Strategic application of lean thinking. In: Proceedings of the fifth Annual Conference of the International Group for Lean Construction, Guaruja, Brazil.

Gil, N., Tommelein, I.D., Kirkendall, R.L., Ballard, G. (2000). Contribution of specialty contractor knowledge to early design. In: Proceedings of the Eighth Annual Conference of the International Group for Lean Construction (IGLC-8), 17-19 July, Brighton, UK.

Green, S., May, S., 2005. Lean construction: arenas of enactment, models of diffusion, and the meaning 'leanness.'. Build. Res. Inform. 33 (6), 498–511.

Green, S.D., 1999. The missing arguments of lean construction. Constr. Manage. Econom. 16, 133–137.

Hanna, M.D., Newman, W.R., Johnson, P., 2000. Linking Operational and Environmental Improvement through employee involvement. Int. J. Operat. Prod. Manage. 20 (2), 148–165.

Health and Safety Executive (HSE) (2013). Construction Industry. Available at: <http://www.hse.gov.uk/statistics/industry/construction/> (accessed 14.12.13).

Hines, P., Holweg, M., Rich, N., 2004. Learning to evolve: a review of contemporary lean thinking. Int. J. Oper. Prod. Man. 24 (10), 994–1011.

Hirano, H., 1996. 5S for Operators: 5 Pillars of the Visual Workplace. Productivity Press, Portland, OR.

Höök, M., Stehn, L., 2008. Applicability of lean principles and practices in industrialized housing production. Constr. Manage. Econom. 26 (10), 1091–1100.

Howell, G. (1999). What is Lean Construction. In: Proceedings of the Seventh Annual Conference of the International Group for Lean Construction, Berkeley, California, USA.

Howell, G., and Ballard, G. (1998). Implementing lean construction: understanding and action. In: Proceedings of the Sixth Annual Conference of the International Group for Lean Construction.

James, P.T.J., 1996. Total Quality Management: An Introductory Text. Prentice Hall, Englewood Cliffs, NJ.

Johansen, E., Glimmerveen, H.,. and Vrijhoef, R. (2002). Understanding lean construction and how it penetrates the industry: a comparison of the dissemination of lean within the UK and the Netherlands. In: Proceedings of the 10th annual conference of the International Group for Lean Construction, Brazil.

Jose, P.B.N. and Alves, C.L. (2007). Strategic issues in lean construction implementation. In: Proceeding of IGLC-15, Michigan, USA.

Jørgensen, B. (2006). Integrating lean design and lean construction: processes and methods. Ph. D. thesis, Department of Civil Engineering (BYG.DTU), Technical Unviversity of Denmark (DTU) (submitted).

Kamara, J. (2003). Enablers for concurrent engineering in construction. In: Proceedings of the 11th Annual Conference of the International Group for Lean Construction, Salford, UK: University of Salford.

Kelly, J., Male, S., 1993. Value Management in Design and Construction. E& FN Spon, London.

Khaba, S., Bhar, C., 2017. Modeling the key barriers to lean construction using Interpretive structural modeling. J. Model. Man. 12 (4), 652–670.

Kobayashi, I., 1995. 20 Keys to Workplace Improvement. Productivity Press, Portland, OR.

Koebel, C.T. (2004). Residential construction diffusion research for the NSF-PATH program. In: Syal, M., Hastak, M., Mullens, M. (Eds), Proceedings of the NSF Housing Research Agenda Workshop, February 12–14, Orlando, USA.

Koranda, C., Chong, W., Kim, C., Chou, J.S., Kim, C., 2012. An investigation of the applicability and sustainability of lean concepts to small construction projects. J. Civil Eng. 16 (5), 699–707.

Koskela, L. (1992). Application of the New Production Philosophy toConstruction, Technical Report # 72, Center for Integrated Facility Engineering, Department of Civil Engineering, Stanford University, CA. Available at: <http://www.ce.berkeley.edu/~tommelein/Koskela-TR72.pdf> (accessed 10.12.12).

Koskela, L. (1999) Management of production in construction: a theoretical view. In: Proceedings 7th Annual Conference of the International Group for Lean Construction, University of California, Berkeley, USA. Available at: <http://usir.salford.ac.uk/9429/1/1999_Management_of_production_in_construction_a_theoretical_view.pdf> (accessed 10.12.12).

Koskela, L. (2000). An exploration towards a production theory and its application to construction. VTT Technical Research Centre of Finland, Espoo.

Koskela, L., 2004. Moving on-beyond lean thinking. Lean Constr. J. 1, 24–37.

Koskela, L., Howell, G., 2002. The underlying theory of project management is obsolete. Proceedings of PMI Research Conference, 293–302.

Leinonen, J. & Huovila, P. (2000). The house of the rising value. In: Proceedings of the Eighth Annual Conference of the International Group for Lean Construction, Brighton, England.

Lindfors, C. (2000). Value chain management in construction: controlling the housebuilding process. In: Proceedings of the Eighth Annual Conference of the International Group for Lean Construction, Brighton, England.

Low, S.P., Mok, S.H., 1999. The application of JIT philosophy to construction: a case study in site layout. Constr. Manag. Econ. 17, 657–668.

Marosszeky, R.T.M., Karim, K., Davis, S. and McGeorge, D. (2002). The importance of project culture in achieving quality outcomes in construction. In: Proceeding of the 10th Annual Conference of the International Group for Lean Construction, Gramado, pp. 1-13.

Marzouk, M., Bakry, I., El-Said, M., 2011. Application of lean principles to design processes in construction consultancy firms. Int. J. Constr. Supply Chain Manage. 1 (1), 43–55.

Meyers, D., 2008. Construction Economics A New Approach, second ed. Taylor & Francis, London.

Miyatake, Y., 1996. Technology development and sustainable construction. J. Manage. Eng. 12 (4), 23–27.

Moffett, S., McAdam, R., Parkinson, S., 2002. Developing a model for technology and cultural factors in knowledge management: a factor analysis. Knowl. Proces Manag. 9 (4), 237–255.

Mohd-Zainal, A., Goodyer, J., and Grigg, N. (2011). Organisational learning to sustain lean implementation in New Zealand manufacturing companies. In: Proceedings of the 3rd International Conference on Information and Financial Engineering, IPEDR, 12, pp. 151-156.

Morton, R., 2002. Construction UK: Introduction to the Industry. Blackwell Publishing, Oxford.

Moser, L., and Dos Santos, A. (2003). Exploring the role of visual controls on mobile cell manufacturing: a case study on drywall technology. In: Proceedings of the Eleventh Annual Conference of the International Group for Lean Construction, Blacksburg, VA, pp. 418–426.

Mossman, A. (2008). Last planner five crucial conversations for reliable flow and project delivery. Available at: <http://www.thechangebusiness.co.uk/TCB/LPSBenefits_files/Last_Planner_5_crucial_conversations.pdf> (accessed 12.11.13).

Mossman, A., 2009. Why isn't the UK Construction Industry going lean with gusto. Lean Constr. J. 5 (1), 24–36.

Murdoch, J., Hughes, W., 2008. Construction Contracts Law and Management. Taylor & Francis, London.

Myers, D., 2005. A review of construction companies' attitudes to sustainability. Constr. Manage. Econom. 23, 781–785.

OGC (2000) Achieving Sustainability in Construction Procurement, Produced by the Sustainability Action Group of the Government Construction Clients' Panel (GCCP). Available at: <http://www.ogc.gov.uk/documents/Sustainability_in_Construction_Procurement.pdf> (accessed 7.11.10).

Office for National Statistics UK (2013) ONS Labour Force Survey. Available at: <http://www.ons.gov.uk/ons/guide-method/method-quality/specific/labour-market/labour-market-statistics/index.html> (accessed 14.12.13).

Ogunbiyi, O. (2014). Implementation of the lean approach in sustainable construction: a conceptual framework. Ph.D. thesis, University of Central Lancashire, Preston, UK (submitted).

Ogunbiyi, O., Oladapo, A., and Goulding, J. (2011). Innovative value management: assessment of lean construction implementation. In: RICS Construction and Property Conference, Salford, U.K., p. 696.

Ogunbiyi, O., Oladapo, A., Goulding, J., 2014. An empirical study of the impact of lean construction techniques on sustainable construction in the UK. J. Constr. Innov. Inform. Process, Manage. 14 (1).

Olatunji J. (2008). Lean-in-Nigerian construction: state, barriers, strategies and "goto-gemba" approach. In: Proceedings of 16th Annual Conference of the International Group for Lean Construction, Manchester, United Kingdom.

Paez, O., Salem, S., Solomon, J., Genaidy, A., 2005. Moving from lean manufacturing to lean construction: toward a common sociotechnological framework. Human Factors Ergonom. Manuf. Serv. Indust. 15 (2), 233–245.

Pheng, L.S., Fang, T.H., 2005. Modern-day lean construction principles: some questions on their origin and similarities with Sun Tzu's Art of War. Manage. Decision 43 (4), 523–541.

Picchi, F., Granja, A., 2004. Construction sites: using lean principles to seek broader implementations. Proceedings of International Group of Lean Construction, 12th Annual Conference, Copenhagen, Denmark, August 3–5. pp. 1–12.

Rolstadås, A., 1995. Planning and control of concurrent engineering projects. Int. J. Product. Econom. 38, 3–13.

Rother, M., 2010. Toyota Kata: management people for improvement, adaptativeness, and superior results. McGraw Hill, New York.

Salem, O., Solomon, J., Genaidy, A., Luegring, M., 2005. Implementation and assessment of lean construction techniques. Lean Constr. J. 2 (2).

Salem, O., Solomon, J., Genaidy, A., Minkarah, I., 2006. Lean construction: from theory to implementation. J. Manage. Eng. 22 (4), 168–175.

Sandvik, W.P., Karrlson, S., 1997. Critical aspects on quality method implementation. Total Qual. Manage. 8, 55–66.

Sarhan, J., Xia, B., Fawzia, S., Karim, A., Olanipekun, A., 2018. Barriers to implementing lean construction practices in the Kingdom of Saudi Arabia (KSA) construction industry. Constr. Innov. 18 (2), 246–272.

Sarhan, S. and A. Fox, (2012). Trends and challenges to the development of a lean culture among UK construction organizations. Available at: <http://www.iglc20.sdsu.edu/papers/wp-content/uploads/2012/07/116%20P%20015.pdf> (accessed 02.3.13).

Scherrer-Rathje, M., Boyle, T.A., Deflorin, P., 2009. Lean, take two! Reflections from the second attempt at lean implementation. J. Business Horizon 52 (1), 79–88.

Senge, P., 1990. The Fifth Discipline – The Art and Practice of the Learning Organization. Doubleday, New York.

Senge, P., Kleiner, A., Roberts, C., Ross, R., Smith, B., 1994. The Fifth Discipline Fieldbook. Doubleday, New York.

Shammas-Toma, M., Seymour, D., Clark, L., 1998. Obstacles to implementing total quality management in the UK construction industry. Constr. Manag. Econ. 16, 177–192.

Shingo, S., 1986. Zero Quality. Control: Source Inspection and the Poka-Yoke System. Productivity Press, Cambridge, MA, pp. 57–69.

Sospeter, N.G., Kikwasi, G.J., 2017. The state of adoption of lean construction in the Tanzanian construction industry. J. Constr. Proj. Manage. Innov 7 (1), 1689–1711. 1.

Spoore, T. (2003). Five S (5S): The Key to Simplified Lean Manufacturing. The Manufacturing Resources Group of Companies (MRGC). The article was originally written for the Durham Region Manufactures Association (DRMA), February 2003, Newsletter. Available at: <http://mrgc.org> (accessed 03.09.2013).

Sturges, J.L., Egbu, C.O. and Bates M.B. (1999) Innovations in construction. construction industry development in the new millennium. In: Proceedings of the 2nd International Conference on Construction Industry Development, and 1st Conference of CIB TG 29 on Construction in Developing Countries, The Pan Pacific, Singapore.

Tabatabaee, S., Mahdiyar, A., Yahya, K., Marsono, A.K., Sadeghifam, A.N., 2017. Level of awareness on lean thinking concept in construction among higher learning students in Malaysia. Malays. J. Civil Eng. 29 (1).

Tezel, A., Koskela, L., Aziz, Z., 2018. Lean thinking in the highways construction sector: motivation, implementation and barriers. Prod. Plan.Control 29 (3), 247–269.

Thomas, G., Thomas, M., 2005. Construction Partnering and Integrated Teamworking. John Wiley & Sons.

Thomassen, M. A. (2004). The economic organisation of building processes On specialization and coordination in interfirm relations, PhD Dissertation, Technical University of Denmark. Available at <http://orbit.dtu.dk/en/publications/the-econonic-organisation-of-building-processes-on-specialisation-and-coordination-in-interfirm-relations(f53db032-05ba-44b5-8909-a41a32d482b5).html> (accessed 14.12.13).

de Valence, G., 2010. Innovation, procurement and construction industry development. Constr. Econom. Build. 10 (4), 50–59.

Van Dun, D.H., Hicks, J.N., Wilderom, C.P.M., and Van Lieshout, A.J.P. (2008). Work values and behaviors of middle managers in lean organizations: a new research approach towards lean leadership. In: Proceedings of the 11th Bi-annual, Conference of the International Society of the Study of Work and Organizational Values, Singapore.

Vrijhoef, R., Koskela, L., 2000. The four roles of supply chain management in construction. Eur. J. Purch. Supply Manage. 6, 169–178.

Walker, H.J., Armenakis, A.A., Bernerth, J.B., 2007. Factors influencing organizational change efforts: an integrative investigation of change content, context, process and individual differences. J. Organ. Change Manage. 20 (6), 761–773.

Watson, M. (2004). Constructing Excellence. Construction Lean Improvement Programme. Lean Institute, UK.

Wiklund, H., Wiklund, S., 2002. Widening the Six Sigma concept: An approach to improve organizational learning. Total Quality Manage. 13 (2), 233–239.

Womack, J., Jones, D., 1996. Lean Thinking: Banish Waste and Create Wealth in Your Corporation. Simon & Schuster, New York.

Womack, J., Jones, D., 2003. Lean Thinking: Banishing Waste and Create Wealth in Your Corporation. Free Press, New York.

Wu, P. and Wang, X. (2016). A critical review of the factors affecting the success of using lean to achieve green benefits. In: 24th Annual Conference of the International Group for Lean Construction, Boston, USA, July 20–22, 2016.

Yahya, M.A., Mohamad, M.I., 2011. Review on lean principles for rapid construction. J. Teknol. 54, 1–11.

Zutshi, A., Sohal, A.S., 2004. Adoption and maintenance of environmental managementsystems: critical success factors. Manage. Environ. Quality Int. J. 15 (4), 399–419.

FURTHER READING

White, R.E., Prybutok, V., 2001. The relationship between JIT practices and type of production system. Omega 29 (2), 113–124.

Chapter 13

BIM-Enabled Sustainable Housing Refurbishment—LCA Case Study

Ki Pyung Kim
School of Natural and Built Environments, University of South Australia, Australia

13.1 INTRODUCTION

The construction industry has long been regarded as one of the major contributors toward a negative environmental impact due to the high amount of waste and energy consumption generated from activities related to demolition and construction (Fadiya et al., 2014). In particular, 10%–15% of construction materials were delivered to the construction site and ended up as wastes (HM Government, 2010). As public awareness about the importance of sustainability has grown, sustainability has emerged as a corporate strategy and an integral part of business plans and sustainable construction has become a major consideration in the construction industry (Aigbavboa et al., 2017; Akadiri et al., 2012). Currently, sustainable construction is broadly defined in the literature by various researchers, but it shares the common characteristics such as the minimization of resource consumption, the maximization of resource reuse, the use of renewable or recyclable resources, and the protection of the natural environment (Yin et al., 2018; Tabassi et al., 2016). As refurbishment renders reduction of demolition waste and landfill usage and reuses structures, components and materials of existing building where possible, it can be considered as sustainable construction in a broad sense. According to a recent research, there is also the benefit of social sustainability as the outcome of refurbishment brings positive social effects such as improved conditions for local communities and transport, schools, and facilities (Sodagar, 2013).

The UK government legislated the Climate Change Act 2008 aiming at 80% CO_2 reduction by 2050 against 1990 CO_2 emission levels. Indeed, the United Kingdom possesses the oldest housing stock among developed countries with 8.5 million properties over 60 years old (EST, 2010). The housing

Sustainable Construction Technologies. DOI: https://doi.org/10.1016/B978-0-12-811749-1.00019-5

sector alone accounts for 27% of the total United Kingdom CO_2 emissions (BEIS, 2017a), and 87% of those housing responsible for the 27% CO_2 emission will still be standing in 2050 (BEIS, 2017a). As energy prices continue to increase and operational energy costs become more expensive, it has become more important to keep operational energy costs lower in the use phase of the entire building life cycle. Since major CO_2 emission, approximately 80%, is generated at the use phase of a house (BIS, 2011), it is essential to carry out sustainable housing refurbishment to achieve the targeted CO_2 reduction effectively and efficiently. The Committee on Climate Change (CCC) forecasted that energy efficiency improvements through refurbishment could reduce annual CO_2 emissions from the UK housing sector by about 17 million tonnes by 2020 (CCC, 2013). New builds have been constructed with higher energy performance standards with low operational energy costs while existing buildings—in particular, existing housing with inefficient energy performance such as solid wall housing—are vulnerable to increasing operational energy costs. Eighty-three percent of total energy consumption in a house—space heating alone consumes 66% of total domestic energy and the energy for generating hot water consumes 17%—has a direct impact on energy efficiency and CO_2 emissions (BEIS, 2017b). Therefore, it is an overarching matter to improve energy efficiency in the housing sector to reduce CO_2 emission.

Refurbishment is considered a better option than demolish-and-rebuild because of the financial and environmental benefits from refurbishment of existing buildings. Furthermore, environmental benefits in terms of conservation of embodied energy in existing housing can be achieved when refurbishing a house rather than demolish-and-rebuild (Riley and Cotgrave, 2011). According to Building and Social Housing Foundation (BSHF, 2008), refurbished homes can make an initial saving of 35 tonnes of embodied CO_2 per property compared to new build homes. New builds are responsible for 50 tonnes of embodied CO_2 while refurbishment is responsible for 15 tonnes, although new homes eventually make up for the high embodied CO_2 through lower operational CO_2. Thus, housing refurbishment can enhance the conservation of embodied energy. As the operational energy efficiency improves by refurbishing a house up to the modern energy standard, the impacts and importance of embodied energy from existing housing refurbishment are increasing since new build housing from 2016 must have zero operation energy consumed and zero CO_2 emission (EST, 2017). The UK government (2013) recommends that embodied energy analysis should be included in the decision-making processes at the design stage, and recently, Brighton and Hove City Council (2011) made the cradle-to-gate embodied CO_2 assessments a mandatory process when planning applications are prepared, and without complete calculation of embodied CO_2, applications cannot be registered. Indeed, Sartori and Hestnes (2007) found that the amount of embodied CO_2 in a low carbon building becomes three times higher than in a conventional building. Thus, the LCA should be the primary methodology to compare various refurbishment solutions and to make proper decisions based on environmental impacts. Various researchers assert that life cycle assessment (LCA) enables design and construction professionals to conduct a cost-benefit analysis or a life cycle cost analysis to satisfy the customers' requirements financially and environmentally because LCA quantifies the environmental impacts in terms of CO_2 emission, and supports professionals to decide the most energy efficient and environmental friendly solutions throughout the life cycle (Assiego de Larrivaa et al., 2014; Burton, 2012).

13.2 LIFE CYCLE ASSESSMENT FOR HOUSING REFURBISHMENT

As all types of buildings including housing leave an environmental footprint over the lifecycle, the environmental impacts of a building should be appreciated based on the systems perspective (Burton, 2012). A systems approach to buildings is essential, in particular for housing, because buildings continue to adaptively change its purpose and entities based on the various needs and requirements of occupants and owners. Thus, in order to considerately plan the environmental impacts of a building from the outset of a project, a building should be considered as a whole-correlated organism living and growing to fit to purpose rather than a final product, which remains as built statically. Consequently, the building industry has growing interest in utilize LCA as a method to assess and evaluate the environmental impacts of an entire building as well as building materials through its lifecycle from the construction phase to the end of life, although LCA is initially developed and applied to the manufacturing industry (Guinee et al., 2011). Furthermore, LCA methodology is gaining increased international attention, and particularly the European Union is focusing on LCA to quantify the environmental impacts depending on refurbishment measures (Monteiro and Freire, 2012). LCA has been used to assess the environmental performance of different building materials including concrete, insulation materials, wood structure, and other building elements such as window and door (Brás and Gomes, 2015; Dylewski and Adamczyk, 2014; Pargana et al., 2014; Salazar, 2014). LCA methodology can guide and support key project stakeholders to make an informed decision about building designs and materials based on the environmental impacts assessment outcomes. However, LCA could be time consuming if a construction professional is not skilled and knowledgeable enough to interpret the LCA outcomes and make informed decisions. Indeed, currently the construction industry is suffered from a skill shortage problem as there is a lack of personnel who can consider the sustainability of refurbishment options regarding life cycle CO_2 emission and other environmental impacts holistically (CIOB, 2011).

Furthermore, researchers and practitioners emphasize that life cycle performance analysis including LCA should be utilized as decision-making supporting criteria within a larger decision-making framework in the situation where design and construction decisions on a building need to be made. As changes in the later phases of a construction project tend to increase costs (PMI, 2017), consideration should be given when selecting construction materials and building designs. As the building must fit for purpose, and in particular for housing, the refurbished house should accommodate the occupant's needs and requirements. If refurbishment cannot provide substantial and tangible benefits and satisfaction to customers, a refurbished house can be considered as functionally failed. In order for design and construction professionals to manage the costs and environmental impacts associated with housing refurbishment through the entire life cycle of a house, the LCA should be utilized as a decision-making support tool to maximize the value of investment in terms of low operational cost and life cycle CO_2 emission.

This trade-off relationship is essential to determine a sustainable refurbishment solution because this holistic life cycle approach should provide the most sustainable refurbishment solution by identifying the intersecting point between the construction phase and operation phase. Beccali et al. (2013) and Blengini and Di Carlo (2010) proved that operational energy and embodied energy are in inverse proportion to each other as shown in Fig. 13.1. Life cycle cost (LCC) and LCA methods are regarded as two distinct methods as LCC mainly focuses on life cycle financial implications and LCA focuses on life cycle environmental impacts. LCC mainly does not consider factors such as operational CO_2 emission and embodied CO_2 for

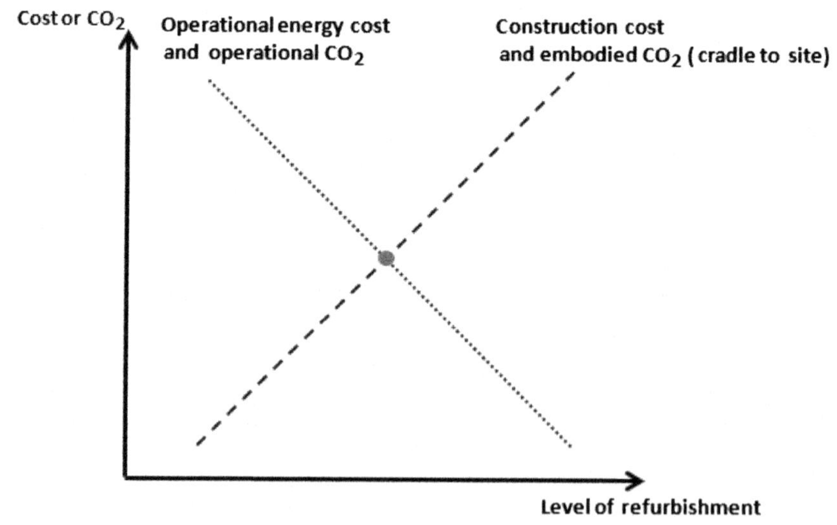

FIGURE 13.1 Inverse proportion between operation phase and construction phase (Beccali et al., 2013; Blengini and Di Carlo, 2010).

constructions that do not have financial impacts on a building or a project throughout the life cycle. In contrast, LCA does not take financial implications such as construction costs and operational energy costs into consideration when it assesses and evaluates environmental impacts of building materials and a whole building. Nevertheless, LCC and LCA can be applied to building materials and buildings since there is an essential trade-off relationship between them. Indeed, operation and maintenance costs are five times larger than initial construction cost (Hughes et al., 2004), and more than 80% of the energy is consumed during a home's operation and maintenance phase (Hacker et al, 2008). In order to achieve lower operational energy cost and CO_2 emission in the use phase of a house, a higher amount of embodied CO_2 with larger capital investment is required in the construction phase. Minimizing the energy demands in the use phase results in an increase in the embodied CO_2 of a building due to the increase in materials and other installations, and, according to Sartori and Hestnes (2007), the amount of embodied CO_2 in a low carbon building is three times higher than in a conventional building. As a result, various researchers assert the simultaneous use of LCC and LCA for sustainable housing refurbishment is highly relevant since decision-making for construction materials and refurbishment measures at each phase of a refurbishment project life cycle influence one another (Construction Products Association, 2014; Bin and Parker, 2012; Ardente et al., 2011).

Despite the importance of the simultaneous use of LCA and LCC, the life cycle studies in the housing sector have been limitedly conducted as shown in Table 13.1, and even the life cycle approach is mainly considered for new builds rather than existing housing refurbishment projects.

Moreover, LCC and LCA methodologies are not easy to use for construction projects because proper LCC and LCA datasets for construction materials and building are not available at the early design phase, and it is challenging to retrieve the necessary data from various project stakeholders due to the fragmented nature of the construction industry (Kovacic et al., 2018; Pal et al., 2017; Assiego de Larrivaa et al., 2014; Flanagan and Jewell, 2005). Furthermore, without a proper dataset for LCC and LCA, informed decisions on refurbishment solutions cannot be made (Dukanovic et al., 2016; Ferreira et al., 2015; Bribian et al., 2009). Thus, these two methodologies require information sharing and collaboration among project stakeholders. If a company or organization starts using LCC and LCA, the data can be reused and updated as they continue with similar projects. Eventually, all the data they have been using will become reliable historical data. Therefore, a proper dataset for LCC and LCA should be prepared in conjunction with the improvement of current fragmented refurbishment processes. Through the studies, researchers commonly argued that the LCA-based decision-making in selecting the most environmentally responsible construction materials or building elements can be further enhanced by integrating the

TABLE 13.1 Life Cycle Impact Studies in the Housing Sector Around the World

Researchers	LCA study scope	Building type	Location	Findings
Peuportier (2001)	Cradle to Grave (LCA)—New Build	Typical French House	France	Timber buildings and wood-based products can improve environmental performances
Adalberth et al. (2001)	Cradle to Grave (LCA)—Existing Houses	Multifamily Buildings	Sweden	Prefabrication and timber-framed house can improve environmental performances
Wang et al. (2005)	Refurbishment and New Build (LCC)	Residential Building	Canada	Life cycle analysis methodology can improve design alternative selection processes economically and environmentally
Asif et al. (2007)	Construction Materials (LCA)—New Build	Three Bedroom Semidetached House	United Kingdom	Concrete, timber and ceramic tiles among wood, aluminum, glass, concrete, and ceramic tiles are revealed as the three major energy expensive materials
Martinaitis et al. (2007)	Refurbishment (LCC)	Residential Building	Lithuania	Simultaneous consideration of cost and energy efficiency improvement is required for better appraisal of building renovation options
Hacker et al. (2008)	Construction and Use Stages (LCA)—New Build	Two Bedroom Semidetached House	United Kingdom	Thermal mass can achieve better passive control of heating and cooling energy demands
Hammond and Jones (2008)	Construction Materials (LCA)—Existing Houses	Apartments	United Kingdom	Little difference between embodied energy and carbon for houses is observed due to external construction works
Blengini (2009)	Cradle to Grave (LCA)—Existing Houses	Residential Building	Italy	Trade-off relationship between energy efficiency and environmental impact (LCA)
Juan et al. (2009)	Refurbishment (LCC)	Residential Building	Taiwan	Optimal refurbishment actions can be decided by considering the trade-off between cost and quality
Bribian et al. (2009)	Construction and Operation Stages (LCA)—New Build	Single Family House	Spain	The top cause of energy consumption in residential building is heating, and the second is the building materials

(*Continued*)

TABLE 13.1 (Continued)

Researchers	LCA study scope	Building type	Location	Findings
Ortiz et al. (2009)	Construction and Operation Stages (LCA)—New Build	Single Family House	Spain	Appropriate combinations of building materials and improvement in behaviors can reduce environmental loads over a life cycle of a building
Monahan and Powell (2011)	Construction Stage (LCA)—New Build	Three Bedroom Semidetached House	United Kingdom	34% reduction in embodied carbon can be achieved when the modern methods of construction (MMC) in housing construction
Tsai et al. (2011)	Refurbishment and New Build (LCC)	Residential Building	Taiwan	CO_2 emission costs are embedded into building project costs based on LCA
Rosa et al. (2012)	Cradle to Grave (LCA)—Existing Houses	Residential Building	United Kingdom	Recycling the building materials can reduce the environmental impacts—3% reduction of CO_2 emission for the detached and semidetached houses and 2% for the terraced house
Assiego de Larriva et al. (2014)	Refurbishment (LCA)	Apartments	Spain	Passive measures or the implementation of active systems such as insulation, changing the windows, and a controlled mechanical ventilation can reduce gross energy requirement (4.4%–9%), and global warming potential (2.6%–4.3%)
Pacheco-Torres et al. (2014)	LCA (Construction Stage)—New Build	Single Family Detached House	Spain	Foundation and structures of a three-floor single-family detached house are responsible up to 39% of the total CO2 emissions due to building construction
Kmeťková and Krajčík (2015)	Cradle to Grave (LCC)—Existing Houses	Apartments	Slovakia	Cost-optimal levels of energy performance need to consider life-cycle costs of the building
Rauf and Crawford (2015)	Cradle to Grave (LCA)—Existing Houses (Services)	Three Bedroom Detached House	Australia	Service life of buildings can have a significant effect on the life cycle embodied energy, and a 29% reduction of the embodied energy can be achieved with extending services life from 50 to 150 years

(*Continued*)

TABLE 13.1 (Continued)

Researchers	LCA study scope	Building type	Location	Findings
Schwartz et al. (2016)	LCC and Carbon Footprint—Refurbishment	Housing Complex	United Kingdom	Environmental footprints can be minimized by insulating thermal bridges and using different heating systems and fuels
Soust-Verdaguer et al. (2016)	Cradle to Grave (LCA)—Existing Houses	Single Family House	Spain	Simplification of LCA methods is challenging due to various assessment standards, and simplified strategy for LCA method is required
Tam et al. (2017)	LCC—Construction Material (Timber)	Residential Building	Australia	LCC was conducted on timber as a construction material under three types of weather conditions, and the most suitable timbers are recommended based on the weather conditions
Syngros et al. (2017)	Construction Materials (LCA)—Existing Houses	Three-Story Multifamily Dwelling	Greece	A practical baseline for construction materials in terms of embodied CO2 has been identified such as concrete and steel
Pal et al. (2017)	LCC and Carbon Footprint—New Build	Residential Building	Finland	Various design variables such as insulation thicknesses, window types, heating systems, and PV area are explored to minimize carbon footprints
Kovacic et al. (2018)	Cradle to Grave (LCA)—New Build	Residential Building	Austria	Environmental impact of a passive house is slightly better than a traditional house due to the behavioral impacts of occupants
Rodrigues et al. (2018)	Refurbishment (LCC)	Residential Building	Spain	Preventive maintenance is important to reduce life-cycle cost and increasing materials service life

LCA method with an effective ICT tool such as building information modeling (BIM) (Ronning and Brekke, 2014). Currently, there are various software packages that can calculate the LCA in the market as listed in the Table 13.2.

TABLE 13.2 LCA Software List

Names	Developers	Remarks	Countries
LCAidTM	University of Western Australia	LCA evaluation tool among design options	Australia
Eco-Quantum	Interfaculty Environmental Science Department (IVAM), University of Amsterdam	LCA evaluation tool for residential building	Netherlands
BEES (Building for Environmental and Economic Sustainability) 4.0	National Institute of Standards and Technology (NIST)	LCA evaluation and economic performance assessment	United States
Envest	BRE	Environmental and financial tradeoffs among design options	United Kingdom
Athena	Athena Sustainable Materials Institute	Comparative analysis tool among various design alternatives and Calculating environmental impacts	Canada
eToolLCD	eTool	Web-based whole building LCA design software	Australia
SimaPro	Preconsultants	LCA calculation process in this software can be edited and expanded	Netherlands
GaBi	University of Stuttgart with PE and IKP	LCA calculation process is user-defined and not fixed. Mainly used for the automobile industry	German
IES IMPACT (Integrated Material Profile And Costing Tool)	BRE with Integrated Environmental Solution (IES) and Technology Strategy Board (TSB)	LCC and LCA tradeoffs calculation. Used in conjunction with IES VE (virtual environment) within a BIM system	United Kingdom

Note: Various LCA tools are available in the market in addition to the tools listed in this table.

Most software is designed to study the LCA of a building depending on the LCA dataset of the software developers' country. The LCA dataset should be carefully selected because the LCA study can indicate different outcomes with the same construction materials based on a specific location and the appropriate climate (Sartori and Hestnes, 2007). For example, the United Kingdom and Sweden are located north of Spain and France where the climate is colder and where, as a result, a house uses fewer and lighter construction materials, such as insulation materials, depending on the climate. The Envest and VE/IMPACT have been developed by the BRE, a United Kingdom-based organization, and they are suitable for this research since they consider LCC and LCA simultaneously. In particular, the BRE developed the VE/IMPACT software that has a capability to calculate LCC and LCA simultaneously in partnership with the IES and Technology Strategy Board based on the Envest, which use the specific database developed for LCC and LCA calculation in the UK construction environment. Furthermore, the use of VE/IMPACT is encouraged by the BREEAM assessment manual since the use of IES VE/IMPACT can get additional BREEAM credits. Currently, the GaBi and SimaPro are well-known LCA calculation tools in the industry (AIA, 2010); however, there are potential issues to calculate the LCA outcomes because both software packages enable a user to customize and redefine the LCA calculation processes that cause different LCA study outcomes and the interpretation could be varied and different among construction professionals (Jankovic, 2017). As a result, the LCA outcomes can be different depending on the user, and users of LCA data must fully understand the implication of utilization of different LCA data sources and cautiously conduct a comparative analysis of results. Therefore, the IES VE/IMPACT will be adopted in this research to formulate the LCA and LCC of refurbishment solutions.

13.3 IMPLICATION OF BUILDING INFORMATION MODELING FOR SUSTAINABLE HOUSING REFURBISHMENT

The UK government released the whole-house refurbishment strategy named as the "2050-ready Home" aiming at refurbishing 80% of the housing stock by adopting whole-house refurbishment approach by 2020 (EST, 2017; HM Government, 2011). According to researchers (McMullan, 2017; Thorpe, 2014), a partial refurbishment such as double-glazed window or cavity wall insulation is capable of achieving limited CO_2 reduction, which is between 25% and 35%, while a whole-house refurbishment can achieve 80% CO_2 reduction. Whole-house refurbishment requires a large amount of capital investment at the beginning, and the investment will be compensated from reduced energy bill over a building life cycle. To understand the financial and environmental implications of various refurbishment alternatives, and determine a well-integrated refurbishment solution among them based on the understandings, the UK government and researchers recommend the integrated use of LCA and LCC methodologies to determine the most affordable refurbishment solution among various alternatives (ISO, 2008; Hacker et al., 2008; HM government, 2010). Yet, the actual implementation of LCA for the whole-house refurbishment projects is not widely adopted and practiced. It is challenging since the whole-house refurbishment requires skills and knowledge to understand the trade-off relationship between the capital investment and energy efficiency improvement for providing affordable refurbishment, and the offered refurbishment solutions should be feasible to customers both economically and environmentally (Menassa, 2011; Konstantinoua and Knaack, 2013; Thuvander et al., 2012). Moreover, refurbishment solutions are proposed at the end of design phase in the current refurbishment process when flexibility of refurbishment solutions and opportunities to explore various refurbishment alternatives are significantly limited (Ma et al., 2012; Thuvander et al., 2012). As a result, many researchers point out the absence of an integrated decision-making framework to estimate the financial and environmental impact of a refurbishment solution from the early design stage. Furthermore, they emphasize the importance of proper decision-making and the necessity of using proper information and communications technology (ICT) tools that support construction professionals considering various refurbishment solutions at the early design phase (Eastman et al, 2011; Hannele et al., 2012). Thus, it is essential to cope with diverse information such as the effectiveness of refurbishment measures, financial feasibility, and environmental impact simultaneously from the early design phase (Killip, 2008), when there are more alternatives to select the most affordable refurbishment solution. In response to current issues and limitations, many researchers state the potential and importance of BIM for informed decision-making on refurbishment solutions at the early design stages. It is because BIM is capable of enhancing collaboration among stakeholders, and the improvement in the integration of project information by exploring and comparing various refurbishment alternatives at the early design stages can lead to better refurbishment solution (Basbagill et al., 2013; HM Government, 2012). Juan et al. (2009), Jenkins et al. (2012), and Rysanek and Choudhary (2013) and Konstantinoua and Knaack (2013) suggested whole-building simulation model with the list of all the potential refurbishment options that can examine various refurbishment alternatives at the early design phase prior to determining a final refurbishment solution. The proposed models by the researchers commonly emphasize the granularity of information regarding refurbishment materials and options that can be examined and tested at the early design stage. According to Scheneider and Rode (2010), 50% of possible refurbishment alternatives that can render better outcomes of refurbishment are neglected due to a lack of collaboration among key project stakeholders at the early design stage. Poor decision-making at the early design phase results in significant changes in the time and cost of a project which leads to reworks, and the cost for reworks and changes become five times larger as changes occur at later phases (Doran et al., 2009).

Currently, BIM is regarded as a major paradigm shift in the construction industry as it is a catalyst for changes of process and culture that requires more integrated approach than before (Hannele et al., 2012; Succar, 2009), and among other advantages to utilize BIM, there is one specific benefit—sustainability enhancement—that is essential to determine the most environmentally friendly refurbishment solution. BIM supports informed decisions regarding sustainability issues such as energy performance and embodied CO_2 of a building at the early design phase by assessing the energy performance and the embodied CO_2 of a building where most of the level of sustainability is determined (Redmond et al., 2012; Krygiel and Nies, 2008). However, in a traditional way, it is challenging to secure sustainable attributes of a building from the early design phase because essential information regarding the construction materials that determine embodied CO_2 is not readily available. In addition, it is challenging to exchange and integrate essential information amongst project stakeholders, in particular between design and construction professionals due to the fragmented nature of the construction industry. In addition, BIM can achieve sustainability in the construction supply chain by making construction and procurement processes more effective and efficient, and construction waste materials throughout a project life cycle can be reduced (HM Government, 2012; Crosbie et al., 2011). In response to the current challenges in utilizing the LCA methods within a BIM environment, there have been various efforts to utilize BIM for the LCA calculation to develop and select more sustainable design solution at the early design stage. Anton and Diaz (2014) recognized that LCA method can be fitted into BIM for assessing the environmental performance of a building, and found that BIM is capable of facilitating an integrated design practice and rendering enhanced construction information flow and coordination among diverse stakeholders. Ladenhauf et al. (2016) suggested a BIM-based LCA algorithm to detect an optimized LCA input data for energy analysis for a building. Shoubi et al. (2015) and Ham and Golparvar-Fard (2015) also utilized BIM to minimize the operational energy use of a building by assessing thermal performance and environmental impacts of various construction materials and generating a various combination of construction materials for more energy efficient building. Both studies focused on capturing more accurate LCA information using BIM as a decision-making supporting tool.

Although there has been various studies to integrate the LCA method with BIM, the studies are limited in its scope of the study to a partial component of a building. Only a few studies have integrated BIM with the LCA methodology to examine environmental impacts of a building holistically (Anton and Diaz, 2014). Hong et al. (2012) and Kim et al. (2013) proposed an integrated model to assess the LCC and LCA based on a well-structured life cycle information datasets in the South Korean construction context. Researchers assert the importance of the availability of life cycle information of various refurbishment alternatives at the early design phase, and this research is supported by Kim and Park (2013) as researchers conducted a BIM feasibility study for housing refurbishment in the United Kingdom context whether BIM is feasible for an information management system for housing refurbishment, and consequently, BIM is recognized as feasible when sufficient BIM datasets including the LCA is available. Ajayi et al. (2015) conducted a BIM-based LCA for a two-story building based on a 30-year life cycle period and found out that a timber structure house is the most environmentally responsible type of a building. Ferreira et al. (2015) and Dukanovic et al. (2016) apply the LCA to a residential building refurbishment in Italy and Serbia, respectively, and researchers recognize that BIM is capable of managing the life cycle information of structural components of buildings. Overall research findings explicitly indicate that BIM is capable of enabling design and construction professionals to integrate the LCA, and facilitate stakeholders' early involvement as it has been emphasized by researchers in sustainable refurbishment. Researchers share the same perspectives that the integration of life cycle information of each refurbishment alternative is essential to generate the most affordable refurbishment solution in a BIM environment using the LCA calculation. Researchers commonly pointed out that BIM can only be an enabler when proper and reliable datasets are available. Indeed, the LCA datasets are still maintained, calculated, and compared separately within a BIM system, and eventually it fails to achieve the seamless updates on LCA calculations depending on different selections of refurbishment alternatives (Shadram et al., 2016). Currently, the International Council for Local Environmental Initiatives developed the SMART SPP Guide to integrate LCC and LCA for the sustainable procurement (ICLEI, 2016). However, the tool is still calculating the LCC and LCA separately. To overcome the unintegrated practice, researchers argue that the improvement on data exchange format and interoperability of LCC and LCA datasets are essential (Shadram et al., 2016). Bueno et al. (2017) argue that a universal data exchange protocol such as IFC is required to be developed further in order to calculate and integrate the LCA study results simultaneously. Soust-Verdaguer et al. (2017) also share the similar idea that an information-enriched BIM model can be utilized as a single source data repository for the LCA. Furthermore, researchers asserted that proper classification and well-defined LCA datasets are vital to enable designers and engineers to examine various design alternatives without data distortions among different BIM tools using different data formats such as gbXML or IFC. Furthermore, the researchers emphasized the importance of seamless data connection between LCA datasets and BIM elements for more accurate thermal performance and environment impact assessment. Indeed, the importance of seamless data connection is advocated by Diaz et al. (2016) as researchers asserted that manual establishment of data relationship between the BIM elements and an LCA database is prone to generate errors in the situation where a researcher quantifies the environmental impacts corresponding to different categories.

Consequently, the BIM standards including PAS 1192 series and BS 8536 series have been developed and they provide level of 3D building model definitions and the level of details of a federated BIM model that contains individual BIM models as a whole such as architectural, MEP, structural, and energy simulation models from the design stage to the asset management stage. Both standards advise that a BIM model should be constructed based on the common data environment and well-defined BIM library to minimize human errors and maximize efficiency in construction project management in a BIM environment. According to CIC (2013), all of these requirements for whole-house refurbishment can be fulfilled by adopting BIM, and BIM can provide the opportunity for homeowners or potential home buyers to customize their homes as they wish using the 3D visualization that does not require an in-depth knowledge of 2D drawings. Therefore, this chapter investigated the financial and environmental life cycle impacts of refurbishment solutions using BIM tools including IES VE/IMPACT, and examine the capability of BIM to formulate LCA and LCC for housing refurbishment by a case study with a BIM simulation.

13.3.1 Case Study: Methodology and Scope of Building Information Modeling Simulation

This research adopts a case study with BIM simulation approach to formulate LCC and LCA of housing refurbishment options within a BIM environment. This research focuses on a contemporary event questioning how a BIM can support proper decision-makings for housing refurbishment by formulating LCA and LCC, and why BIM is recognized as an enabler for the construction industry to achieve higher effectiveness and efficiency. As this research investigates the why and how regarding current issues in the construction industry, a case study is the most relevant strategy than others (Yin, 2003). For a case study, Autodesk Revit 2016 is adopted for basic housing model development and IES VE for operational energy cost calculation and IES IMPACT, which is operated within the IES VE simulation platform, for LCC and LCA calculation. Autodesk Revit was selected because it is one of the most commonly used BIM tools, and comparable with AutoCAD platform that is a dominant 2D tool still widely utilized in the UK construction industry (NBS, 2017). The IES VE/IMPACT can manage a 3D BIM model and has been evidenced by a number of researches for energy simulations in refurbishment and has a capability to simulate all possible building energy assumptions compared to other tools (Jankovic, 2017; Murray et al., 2012; Crawley et al., 2008). Since there is no "one-size-fits-all" solution for housing refurbishment in the United Kingdom (Jenkins et al., 2012), the tool must be capable of coping all possible alternatives and this requirement makes the IES VE/IMPACT relevant for this research. In order to develop a basic BIM model for simulation, The average housing condition data published by the UK government, practitioners (Brinkley, 2017; Palmer and Cooper, 2013; Riley and Cotgrave, 2013; Neufert, 2012) were used to build up a case building model in a BIM system hypothetically because the condition of housing indicates a wide range of variation in its characteristic such as year built, construction types physical dimensions, extra retrofitted measures, and construction materials, which cannot be generalized (Jenkins et al., 2012). For energy simulation, LCC and LCA calculation within a BIM environment, various supplementary data sources have been used as follows:

- LCC and LCA—IES IMPACT dataset provided by BRE (2013);
- Cost for materials and labor—SMM7 Estimating Price Book 2013 (BCIS, 2012);
- Operational energy cost: Inherited Default from IES VE;
- Embodied CO_2 for materials—University of Bath (Hammond and Jones, 2011);
- Embodied CO_2 for construction Works—Black Book (Franklin and Andrews, 2010).

This research examined the whole-house fabric refurbishment, as the approach should be the first stepping stone to improve a whole-house and various researchers and construction professional organizations have argued that the whole-house fabric should be improved first rather than upgrade services or renewable energy systems (Liang et al., 2017; NIHE, 2015; National Refurbishment Centre, 2012). In particular, the Zero Carbon Hub, which is a United Kingdom-based organization supported by the Department of Energy and Climate Change, release a series of energy improvement guideline aiming at the 80% CO_2 reduction and it is strongly recommended to improve energy efficiency of the whole-house fabric of a building (Liang et al., 2017; BRE, 2014; Institute for sustainability, 2011). Thus, this research adopted a whole-house fabric refurbishment for a case study.

13.4 WHOLE-HOUSE FABRIC REFURBISHMENT OPTIONS

The current housing refurbishment options, i.e., refurbishment practices for whole-house fabric refurbishment were identified as shown in Table 13.3.

The identified refurbishment options are currently prevalent among construction professionals and recognized as the most cost and carbon effective practice by various construction organizations and actual refurbishment

TABLE 13.3 Current Best Refurbishment Practices for Whole-House Fabric Refurbishment

Elements	Construction types	Best refurbishment measures	Capital costs	Disruptions	Carbon cost effectiveness
Roof	Pitched roof	Rafter insulation	££1,000–£5,000	Low	£100–£500/ 1 tonne CO_2
		Loft insulation	£100–£1,000	Moderate	Repay more than initial capital cost
Wall	Solid wall	External wall insulation	Over£10,000	Moderate	Repay more than initial capital cost
		Internal wall insulation	£5,000–£10,000	Significant	Repay more than initial capital cost
Floor	Suspended timber floor	Underfloor insulation (insulation between joists)	£100–£1,000	Significant	Repay more than initial capital cost
		Surface insulation (insulation over the floor board)	£100–£1,000	Significant	Repay more than initial capital cost
Window	Single glazing	Double glazing	£1,000–£5,000	Moderate	£100–£500/ 1 tonne CO_2
		Triple glazing	Over£10,000	Moderate	£100–£500/ 1 tonne CO_2

projects such as the Retrofit for the Future (TSB, 2014; Construction Products Association, 2014).

13.5 ENERGY PERFORMANCE STANDARDS

Building Regulation Part L 2010, Building Regulation Part L 2013, and the Fabric Energy Efficiency Standard (FEES) will be adopted for energy simulations. The Building Regulation Part L 2010 and 2013 mandate the minimum energy efficiency standard for housing fabric as shown in Table 13.4 in the first column. In addition to the minimum standard, Building Regulation Part L 2010 and 2013 advise construction professionals to consider further energy efficiency by providing a notional energy efficiency standard aimed at a 25% CO_2 reduction as shown at the second and third columns. The U-values has been updated between 2010 and 2013.

The Fabric Energy Efficiency Standard has been recently introduced to the Building Regulation Part L 2013 aimed at achieving zero carbon homes by 2016. These energy efficiency standards have been adopted because there is no energy efficiency standard for housing refurbishment and these are the most reliable standards at present. Furthermore, FEES has been introduced specifically for zero carbon in existing housing to achieve the required 80% CO_2 reduction. This research will identify the most financially and environmentally feasible energy standard for the whole-house fabric refurbishment by comparing the different outcomes LCC and LCA based on these different energy standards. For comparative analysis on financial and environmental impacts, the fiber glass and expanded polystyrene (EPS) were used as refurbishment materials for insulation because the initial cost is the most influential factor for customers to select refurbishment materials (Park and Kim, 2014). Furthermore, only information regarding these two insulation

TABLE 13.4 Current Energy Efficiency Standards (BR Stands for Building Regulations Part L)

Housing elements	Energy standards (W/m^{2}K)			
	BR 2010/2013 (Minimum)	BR 2010 (Notional)	BR 2013 (Notional)	FEES (Maximum)
Wall	0.3	0.22	0.18	0.15
Floor	0.25	0.18	0.13	0.13
Roof	0.2	0.15	0.13	0.13
Window	2.0	1.4	1.4	1.2
Door	2.0	1.2	1.2	1.0

Note: The standards stands for the U-value (W/m2K) of each housing element.

TABLE 13.5 Thickness of Insulation Materials for Each Energy Standard (mm)

Housing elements	Insulation materials	Energy performance standards			
		BR 2010/2013 (Minimum)	**BR 2010 (Notional)**	**BR 2013 (Notional)**	**FEES (Maximum)**
Wall	Fiber glass	120	170	210	260
	EPS	100	140	175	215
Floor	Fiber glass	145	170	260	260
	EPS	120	140	215	215
Roof	Fiber glass	190	260	300	300
	EPS	155	215	250	250
Door	Wooden door	45	90	90	105
Window	Timber framed	Double glazing : 24 mm	Triple glazing: 42 mm	Triple glazing: 42 mm	Triple glazing: 42 mm
		(U-value frame: 2.71, glazing: 1.75)	(U-value frame: 3.1, glazing: 1.27)	(U-value frame: 3.1, glazing: 1.27)	(U-value frame: 0.85, glazing: 1.27)

materials is commonly available in both data sources—Autodesk Revit and IES VE/IMPACT. The thickness of refurbishment materials varies in the energy standards as shown in Table 13.5.

13.6 BASIC INFORMATION FOR HOUSE MODELS

The general information about the detached solid wall housing in the United Kingdom is provided in Fig. 13.2 and Table 13.6. The gross internal floor area (GIFA) was used for the calculation of LCC and LCA and energy performance simulation. For the information including basic dimensions, floor plans and descriptions of rooms about other housing types—semidetached/end terraced and terraced house—are provided in Appendix 13.1.

13.7 DETAILED INFORMATION OF HOUSES FOR ENERGY SIMULATION

The basic house models for IES VE/IMPACT are visualized as shown in Fig. 13.3. These models were transferred from the Autodesk Revit models.

The information regarding air permeability and thermal bridging has been inherited from IES VE/IMPACT simulation software because this

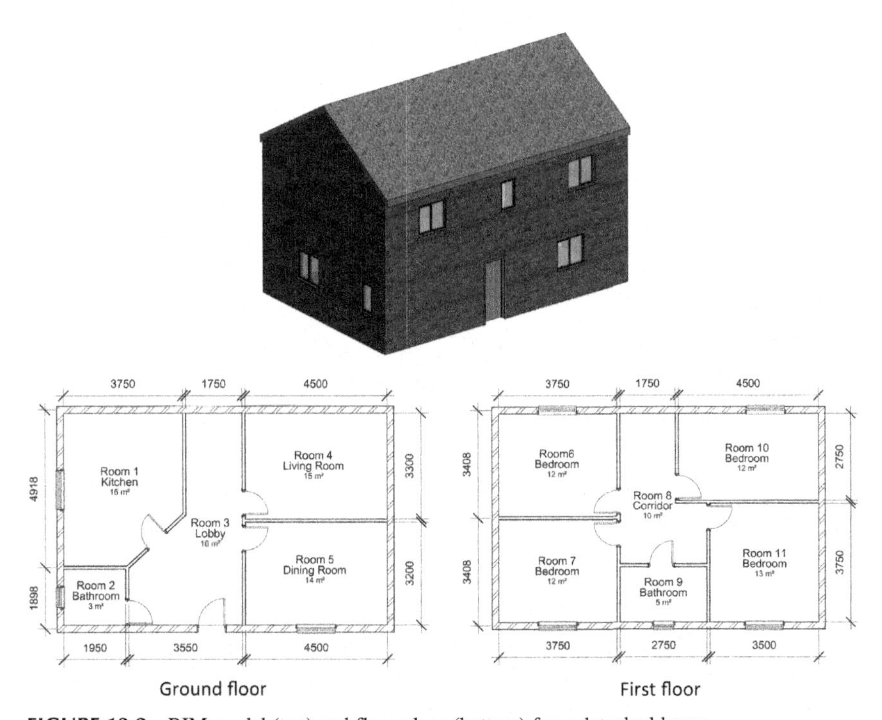

FIGURE 13.2 BIM model (top) and floor plans (bottom) for a detached house.

TABLE 13.6 Detached House—Room and Space Information

Floors	Rooms	Descriptions	Areas (m^2)
Ground floor	Room 1	Kitchen	16
	Room 2	Bathroom	3
	Room 3	Lobby	16
	Room 4	Living room	15
	Room 5	Dining room	14
First floor	Room 6	Bedroom	12
	Room 7	Bedroom	12
	Room 8	Corridor	10
	Room 9	Bathroom	5
	Room 10	Bedroom	12
	Room 11	Bedroom	13
Total usable floor area			130

Note: Roof (Loft) was not included into the usable floor area.

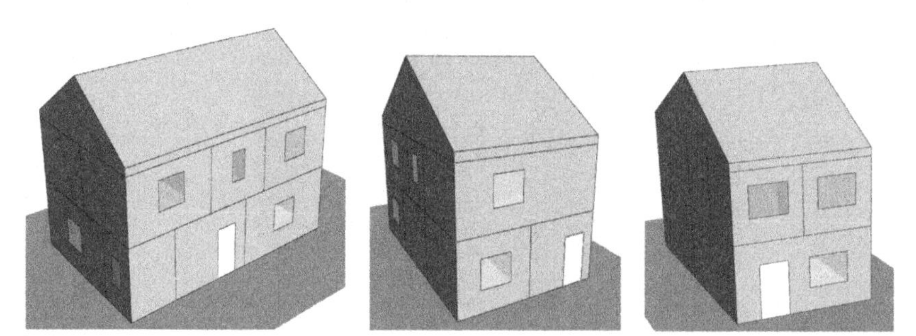

FIGURE 13.3 Energy simulation model in IES VE/IMPACT transferred from Autodesk Revit. From left, detached, semidetached/end terraced, and terraced house.

information cannot be generalized as a typical information since various housing conditions exist. The energy simulation was conducted based on the weather dataset of London, United Kingdom, because the differences in energy demand reduction based on the location—Edinburgh, Manchester, London—were not significant as the differences were located between 1% and 3%, when an 80% CO_2 reduction refurbishment scenario was adopted (Mohammadpourkarbasi and Sharples, 2013). Thus, the weather data based

TABLE 13.7 Detailed Construction Information

Elements	Construction types	Components	Thickness (mm)	U-values (W/m^2K)
Roof	Pitched roof with timber joist and rafter	Roofing tile	25	0.8
		Wood (batten)	25	
		Roofing felt	5	
		Timber structure	140	
External wall	Solid brickwork	Dense gypsum plaster finish (external)	13	2.1
	Masonry wall (single leaf)	Solid brickwork	220	
Party wall	Timber stud partition wall	Gypsum wall board	12.5	0
		Air infiltration barrier	75	
		Gypsum wall board	12.5	
Floors	Suspended timber floor	Timber joist structure	225	0.7
		Chipboard	25	
		Carpet	10	
Ceiling	Generic ceiling	Gypsum wall board	12.5	N/A
Windows	Double glazing	Double glazing, timber frame	6 mm Glazing	2.0
Exterior door	Wooden door	Wooden door	44	3.0

Note: U-value and thickness adopted from RdSAP 2012 (BRE, 2012).

on London was used, and there could be a minimal discrepancy in the outcomes of simulation. The detailed information regarding construction types of a house and energy simulation including ventilation and heating (CIBSE, 2016) are provided as shown in Tables 13.7 and 13.8.

A 60-year life cycle studies of LCC and LCA have been conducted by following the international standards—ISO 14040 and ISO 15686 (ISO, 2006; ISO, 2008). The purpose of this research leads to a greater focus on the quantified energy use and CO_2 emission in the use phase of a building after refurbishment, and the LCA study adopts a cradle-to-grave approach

TABLE 13.8 Detailed Information for Energy Simulation

Information	Detached	Semidetached/end terraced	Terraced
Number of floors	2	2	2
Construction type	Solid brick wall	Solid brick wall	Solid brick wall
Ventilation	Natural ventilation	Natural ventilation	Natural ventilation
Heating (using water)	Radiator	Radiator	Radiator
Main energy source	Natural gas	Natural gas	Natural gas
Household size (number of people)	Single family (2.3)	Single family (2.3)	Single family (2.3)
Indoor temperature	19–23 °C	19–23 °C	19–23 °C
Usable floor area	130 m^2	90 m^2	60 m^2
Ceiling heights	2.7 m	2.7 m	2.7 m

Reference: Palmer and Cooper (2013).

with the exclusion of the recycle, reuse, and/or disposal stage as this contributes minimal percentages of CO_2 impact throughout the entire life cycle of a house (Rosa et al., 2012).

13.8 LIFE CYCLE ENVIRONMENTAL IMPACTS AND FINANCIAL IMPLICATIONS

13.8.1 Environmental and Economic Feasibility for Housing Refurbishment

When LCA study outcome is used as a decision-making criterion for refurbishment, it is important to compare the LCA of a refurbished building to an existing building. When the LCA of a refurbished building is lower than an existing building, refurbishment can be considered to be environmentally feasible, and it also can be considered economically beneficial for customers to refurbish their homes when the LCC of a refurbished building is lower than an existing building (Gustafsson, 1992). Thus, this research investigates whether refurbishment is environmentally and financially feasible for each housing type as shown in Figs. 13.4–13.6. More detailed LCA and LCC study results for each housing type are provided in Appendix 13.4 and 13.5.

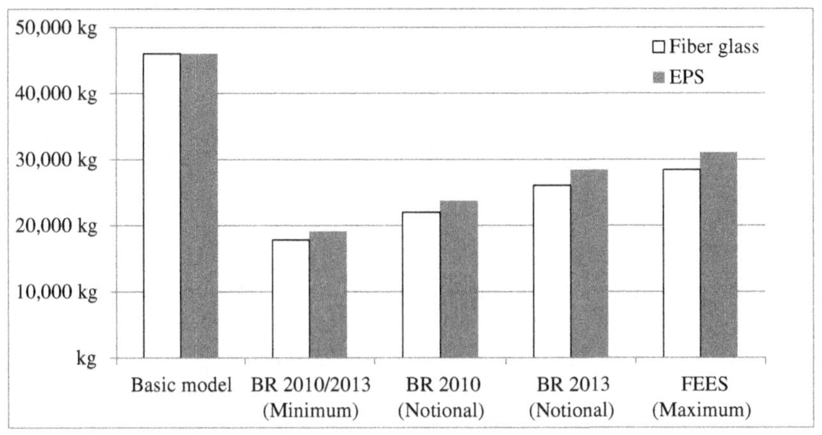

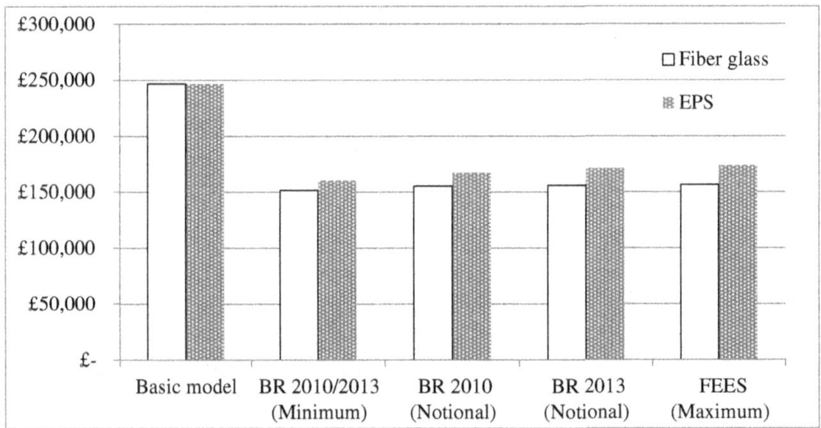

FIGURE 13.4 Life cycle assessment (LCA, top) and life cycle cost (LCC, bottom) of existing and refurbished house—detached house.

The LCA results for each housing type continue to increase as more insulation materials are installed for energy efficiency improvement (see Figs. 13.4–13.6) which require more embodied CO_2 associated with the materials and construction works although the operation energy continues to decrease. All the LCC results of refurbishment with different energy standards are also lower than the LCC of the existing house. Thus, it is confirmed that refurbishment is environmentally and economically beneficial for each housing type to improve energy efficiency compared to the new build.

The difference of LCA study outcomes based on the fiber glass and the EPS are caused by the involvement of fossil fuels in the raw material, i.e., the embodied CO_2 for the fiber glass (4.25 kg/m^3) is lower the EPS (12 kg/m^3). There are also differences in the LCC results between using the

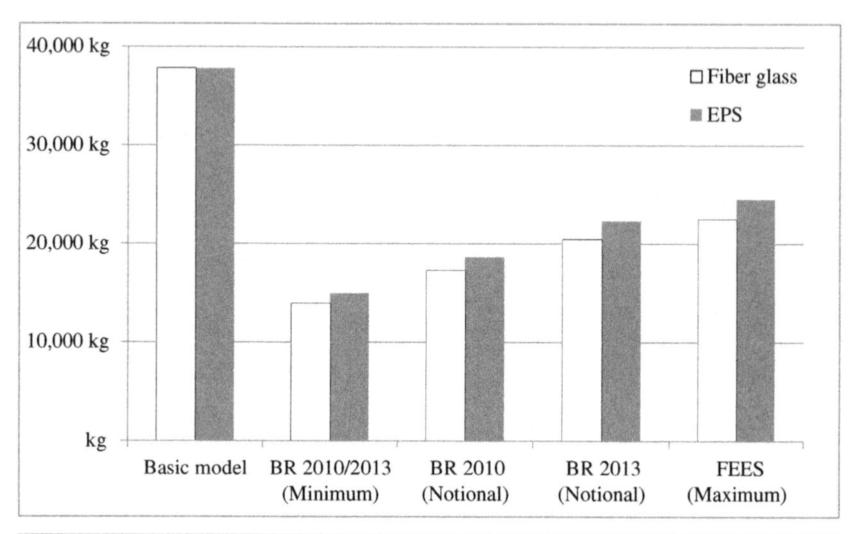

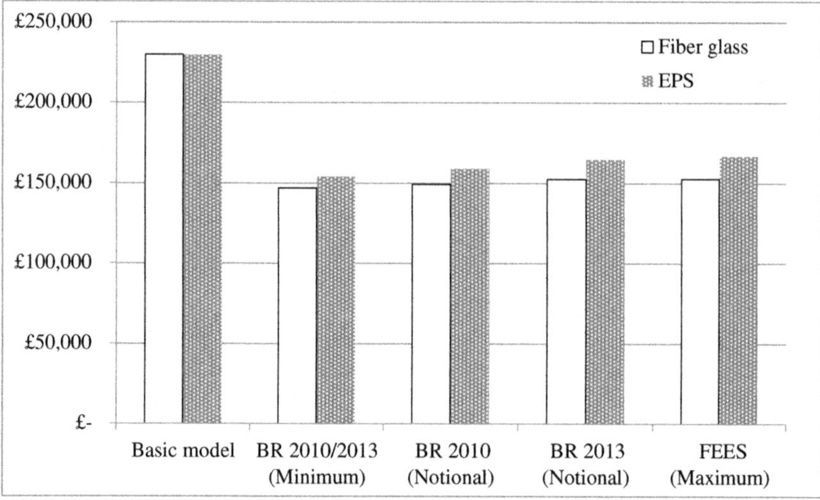

FIGURE 13.5 Life cycle assessment (LCA, top) and life cycle cost (LCC, bottom) of existing and refurbished house—semidetached/end terraced house.

fiber glass and the EPS because the initial material cost of fiber glass (£5.25/m^2) is less than the EPS (£9.88/m^2). This difference in material costs impacts on construction costs and operating costs such as major and minor repairs. Overall, the differences of the LCA and LCC depending different energy standards are identified, and the root cause for the difference is revealed as the initial attributes of materials in terms of embodied CO_2 and initial costs of insulation materials.

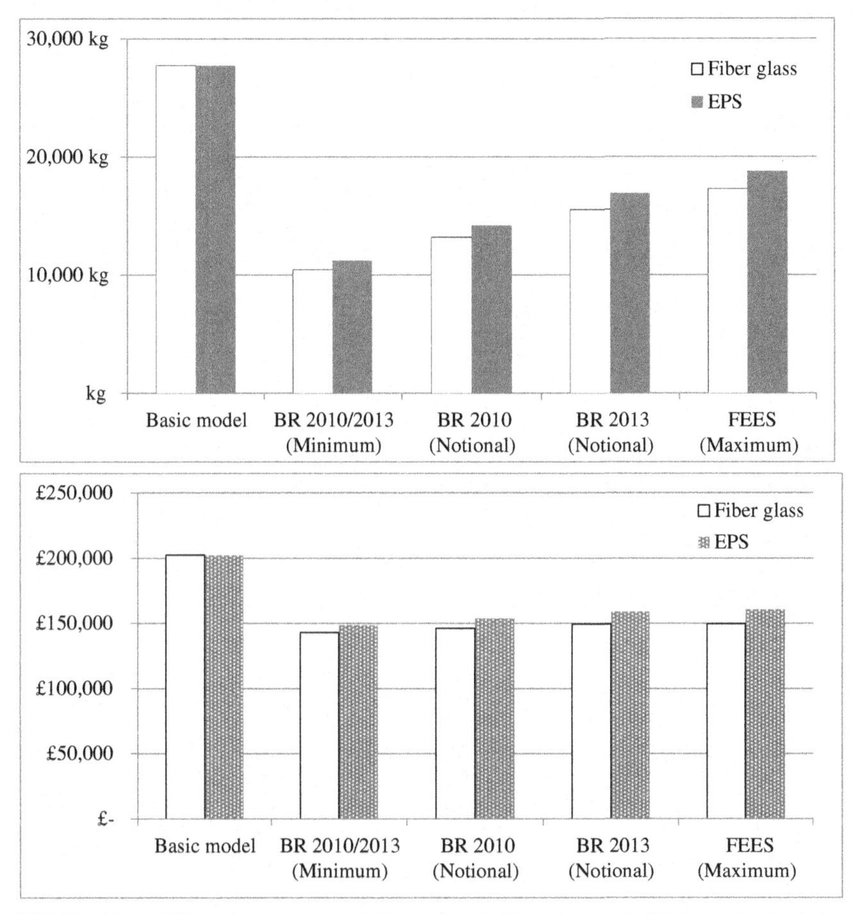

FIGURE 13.6 Life cycle assessment (LCA, top) and life cycle cost (LCC, bottom) of existing and refurbished house—terraced house.

13.9 COMPARATIVE ANALYSIS FOR LIFE CYCLE ASSESSMENT AND LIFE CYCLE COST AMONG DIFFERENT HOUSE TYPES

When the embodied CO_2 for reducing 1 kg CO_2 is considered as a decision-making criterion for refurbishment, the minimum energy standard is the most economical option for all three types of houses as shown in Figs. 13.7–13.10. The semidetached/end terraced house is identified as the most environmentally responsible and economically affordable house type for refurbishment, and the detached house is identified as the largest amount of embodied energy consuming housing type to be refurbished.

The construction costs for reducing 1 kg CO_2 between different types of houses are identified as shown in Figs. 13.9 and 13.10. The terraced house is

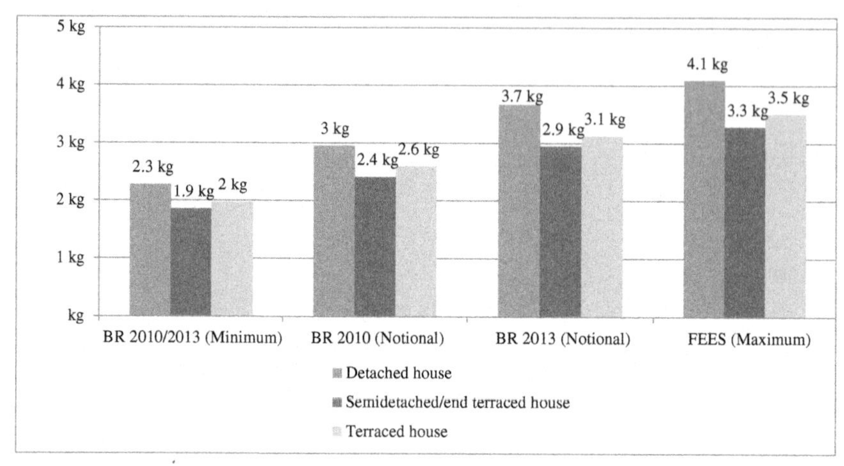

FIGURE 13.7 Embodied CO_2 for reducing 1 kg of operational CO_2—fiber glass.

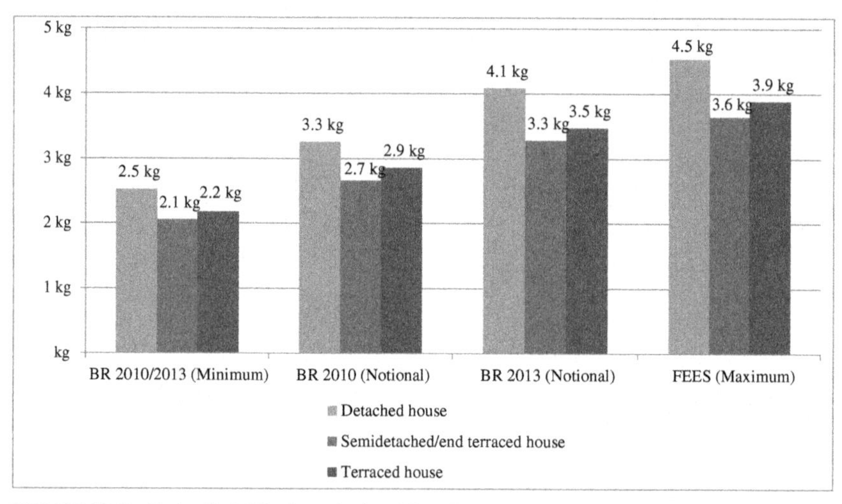

FIGURE 13.8 Embodied CO_2 for reducing 1 kg of operational CO_2—EPS.

the most expensive house type to reduce 1 kg of operational CO_2 when the fiber glass is used. The detached house is the most expensive when EPS is used. For both materials, the semidetached/end terraced house is the most economical house type to achieve 1 kg CO_2 reduction.

If only the LCA results including the amount of embodied CO_2 related to construction works and transportation are considered or only the LCC results including construction costs are considered, it cannot be environmentally feasible to adopt a higher energy standard from the beginning as identified. Only the BR 2010/2013 energy standard, which requires the minimum energy efficiency, should be adopted and could be the only option. However,

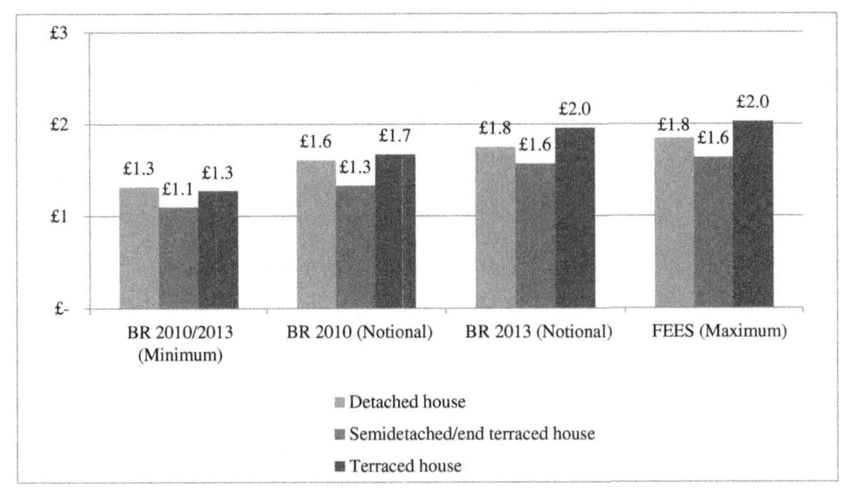

FIGURE 13.9 Construction cost for reducing 1 kg of operational CO_2—fiber glass.

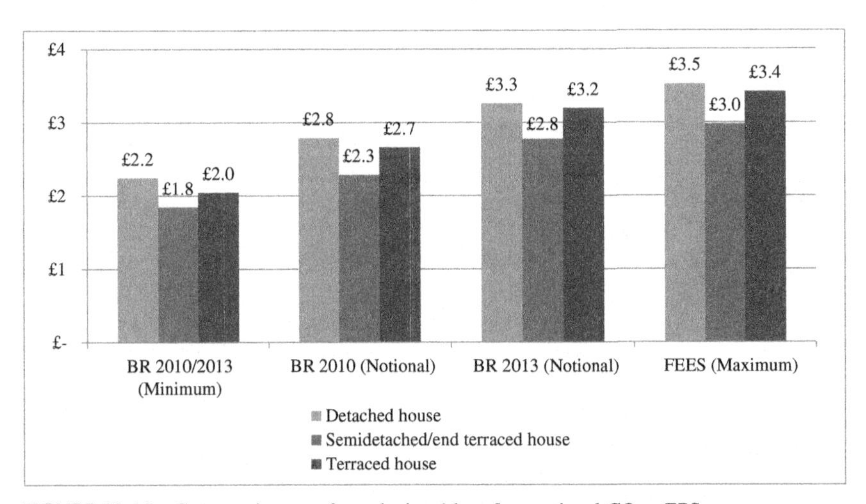

FIGURE 13.10 Construction cost for reducing 1 kg of operational CO_2—EPS.

when the LCC and LCA study results are considered in an integrated way, more financially affordable and environmentally responsible refurbishment option with higher energy standard can be implemented. As shown in Fig. 13.11, it is revealed that the operating costs of the EPS has the same pattern as the construction cost, while the operating costs of the fiber glass do not continuously increase for all housing types.

More importantly, the operating costs with the BR 2013 (Notional) and the FEES (Maximum) are less than the operating costs with the BR 2010 (Notional) for a detached house. Semidetached and terraced house also

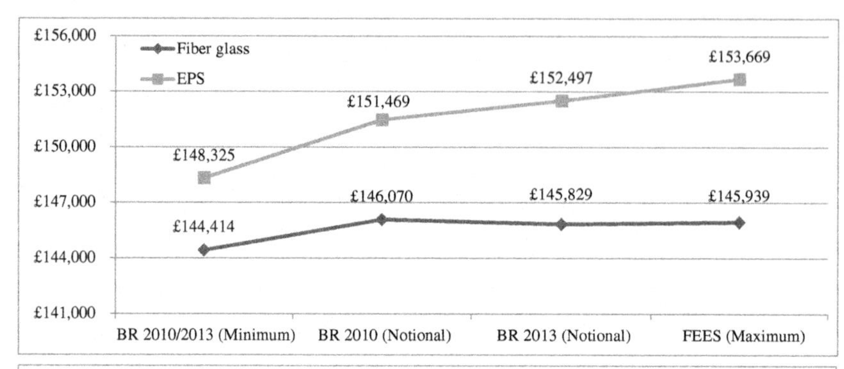

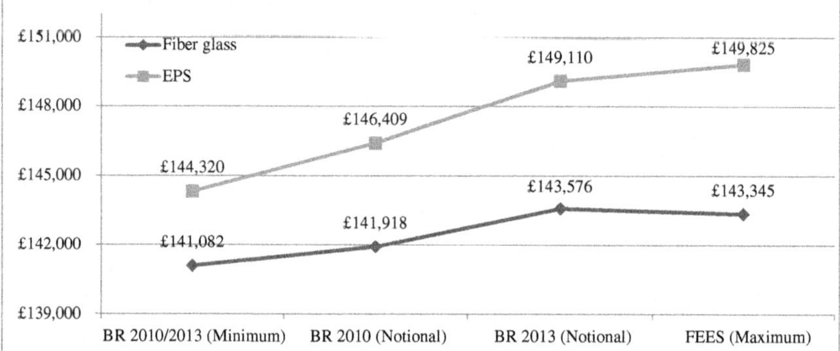

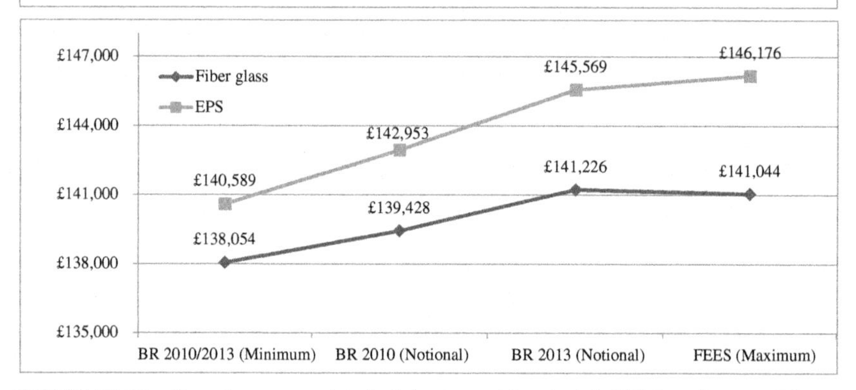

FIGURE 13.11 Operating costs—detached (top), semidetached (middle), and terraced house (bottom).

renders lower operating costs than the BR 2013 (Notional) energy standard. Based on this result, construction professionals can advise customers that a higher energy standard such as BR 2013 (Notional) or FEES (Maximum) is more beneficial than BR 2010 (Notional), when they wish to achieve high energy efficiency. The fluctuation of operating costs is caused by the inverse proportion relationship between construction cost and energy cost, which echo with the findings of preceded researches identified through literature review. The construction cost continues to increase for applying higher energy standard, while the operating energy costs continue to decrease as energy performance continues to be improved. Thus, this relationship needs to be considered at the early design stage with design professionals being responsible for identifying the optimum point where the total cost of construction cost and energy cost result are at the minimum level.

13.10 CO_2 EMISSION REDUCTION AND CO_2 PAYBACK PERIOD

Since housing refurbishment is highly relevant and aligned with the UK government CO_2 reduction target, it is important to consider the amount of CO_2 reduction depending on different energy standards. The amount of CO_2 emissions per year also continues to diminish as higher energy standards are adopted as shown in Fig. 13.12—although there is no significant CO_2 reduction after BR 2010 (Notional) adoption. The average percentage of CO_2 reduction with the adoption of the FEES standard is 58% (10,985−5,328.3 kg), which is similar to the results of other research and practical experience with the maximum of 60% CO_2 reduction (Construction Products Association, 2014; Sodagar, 2013).

Furthermore, as a considerable amount of energy is required for housing refurbishment as adopting higher energy standard, it should be considered if the invested embodied energy can be recovered through a life cycle of a refurbished house. Thus, the CO_2 payback period is important to be considered and it is calculated as shown in the equation below (Genchi et al., 2002). The CO_2 payback period stands for a duration when the amount of

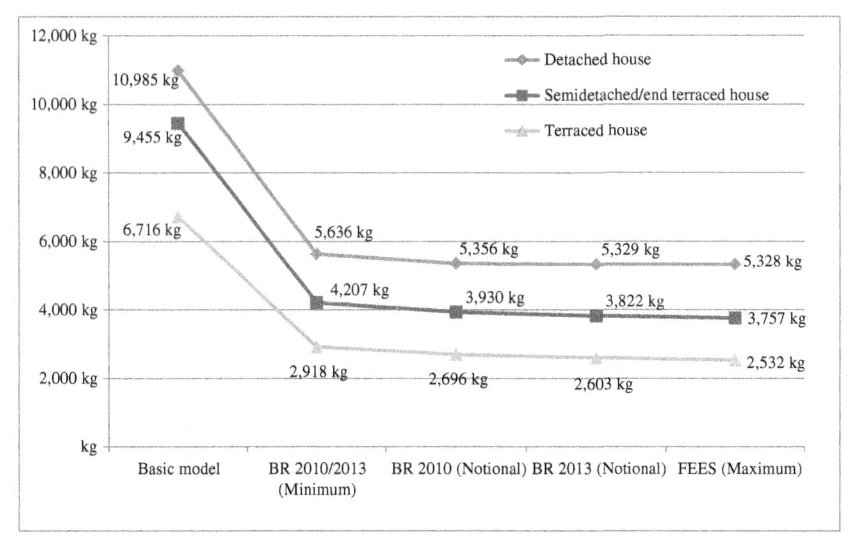

FIGURE 13.12 CO_2 emission per year.

operational CO_2 reduction after refurbishment compensates the embodied CO_2 invested in the refurbishment.

$$\frac{\text{Total amount of embodied } CO_2}{\text{Annual } CO_2 \text{ emission reduction after refurbishment}} = CO_2 \text{ Payback Period}$$

The CO_2 payback periods for each house types are shown in Fig. 13.13. The BR 2010/2013 (Minimum) energy standard indicates the shortest CO_2 payback period among the other energy standards. However, it should be noted that the gap between the BR 2010/2013 (Minimum) and the FEES (Maximum) is nearly 2 years for both the fiber glass and the EPS. If the CO_2 payback period is considered in conjunction with the embodied CO_2, the payback period for retrieving the embodied CO_2 by reducing CO_2 emissions is not significantly different among different energy standards.

As it is proven previously, the FEES should be recommended and adopted for the maximum benefits of life cycle financial implication. Thus, design and construction professionals can implement the highest energy standard, which is FEES, for the maximum benefits of embodied CO_2 payback by reducing the largest amount of operational CO_2 reduction. However, it must be noted that there is always the possibility that the outcome can be changed as the input information is changed since this research used a hypothetical case study with the best possible information that is currently available. Therefore, it is worth emphasizing that the outcome of this research should be used as a supporting tool for decision-making, not as definitive decision-making criteria.

13.11 DISCUSSIONS ABOUT THE LIMITATION OF BUILDING INFORMATION MODELING TOOLS

Through the case study, BIM is proven its relevance and capability to be utilized as an information management platform for housing refurbishment by formulating the LCA and LCC studies, and energy simulations. It is also revealed that BIM can be feasible for housing refurbishment as a decision-supporting tool when relevant housing condition and construction data is provided to construct an information-enriched model. Although BIM can provide various computational and visual aids, there are still challenges existed to fully utilize BIM for housing refurbishment and other construction projects. Two major issues associated with the utilization of BIM are identified: Data Exchange and Interoperability and Unstandardized Specification System Between Different Data Source below described.

13.11.1 DATA EXCHANGE AND INTEROPERABILITY

Existing CAD files, which is currently prevalent 2D tool for the construction industry, cannot utilized when the file is imported into a BIM tool since the 2D CAD data has no parametric information and cannot be converted into an

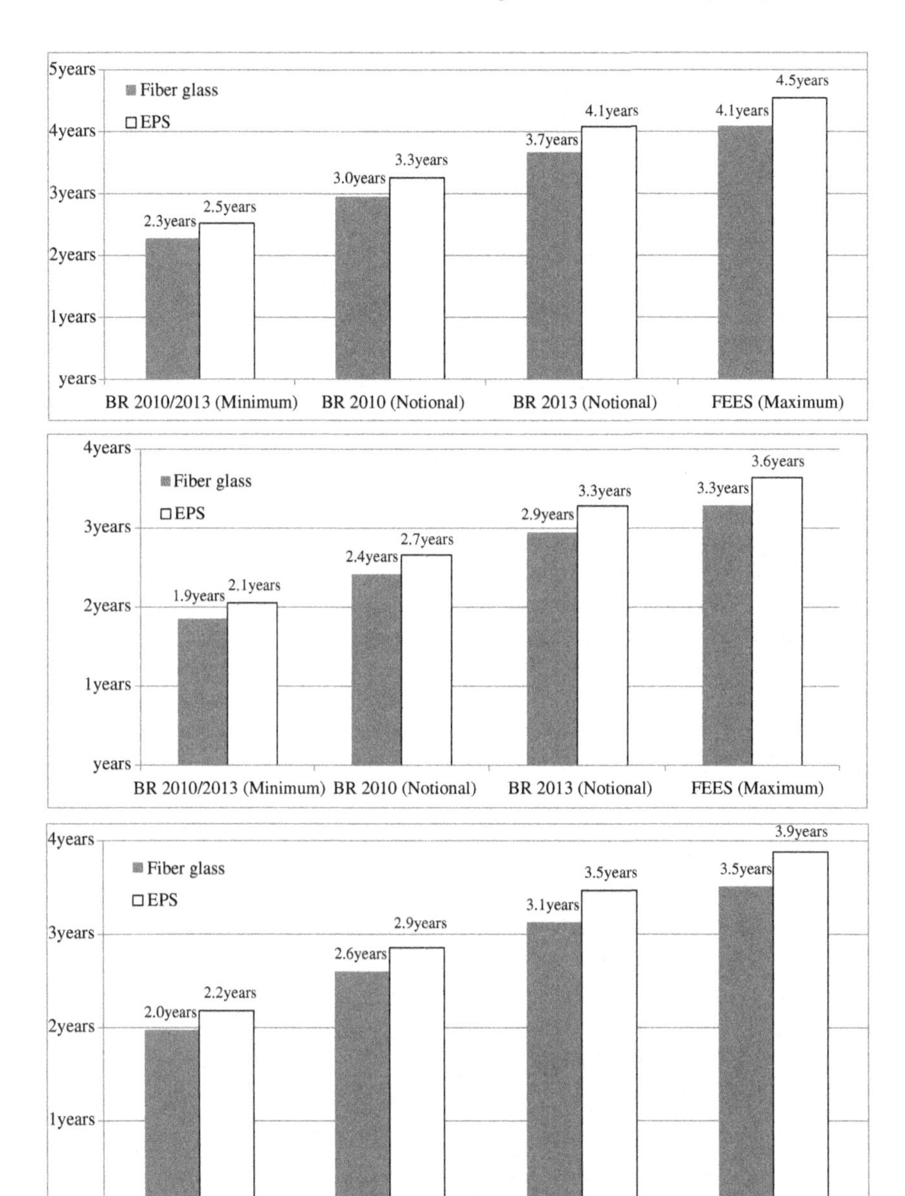

FIGURE 13.13 CO_2 payback period—detached (top), semidetached (middle), and terraced house (bottom).

individual object. Although the 2D CAD file can be used as a reference line, it has less capability than a BIM model. All the elements, structures, and service systems are presented as one mass without parametric information such as dimensions, thickness, and thermal information because they are not

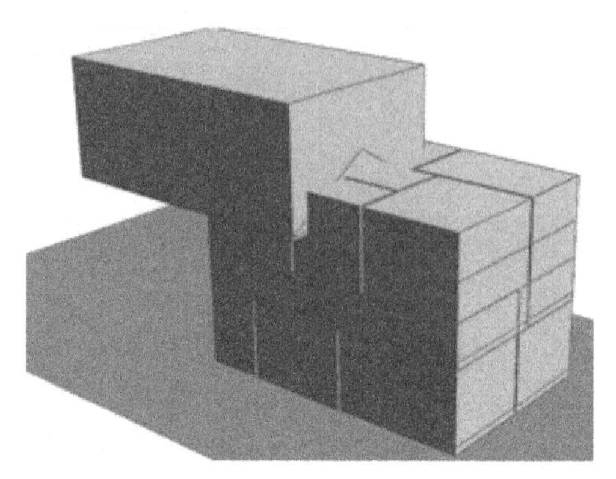

FIGURE 13.14 Basic model import result in IFC data format.

object oriented and cannot be separately converted into a BIM object with parametric information. Consequently, additional steps to reconstruct a BIM model from 2D CAD data is inevitable in the situation where only 2D floor plans with CAD data are available for housing refurbishment.

Furthermore, the study found interoperability issues with data exchange format. The IFC format cannot be exchanged between Autodesk Revit and IES VE/IMPACT as shown in Fig. 13.14. The geometric information is not inherited from the initial BIM model when IFC data is used for data import to the IES VE/IMPACT, while gbXML format transfers a dimensionally intact BIM model.

Interoperability between different BIM software is a critical technical barrier, yet the interoperability issues are still not resolved although the concept of IFC and gbXML data formats within BIM system should exchange necessary data without any conflicts. Thus, the gbXML file format instead of IFC format was used in this research.

13.11.2 UNSTANDARDIZED SPECIFICATION SYSTEM BETWEEN DIFFERENT DATA SOURCE

Not all building elements are classified under the same code and the general code of buildings components are currently changing from SMM7 to NRM. A standardized data format/template in the shared classification should make the datasets of a housing information modeling more reliable and efficient in calculating the bill of quantity and generating a cost management plan. Thus, construction professional organizations such as the NBS, which provide or plan to provide open BIM objects library, should take into consideration a standardized coding system for further BIM object development. The information on costs and CO_2 of construction materials has different classification systems and they are categorized by unclassified coding systems as shown in Table 13.9.

TABLE 13.9 Unstandardized Specifications Among Different Data Sources

Specifications	Building element and construction materials				References
	Roof–pitched roof, timber	Roof–insulation, fiber glass	Lowest floor–suspended timber floor	Lowest floor–wood chipboard	
NRM 1	2.3.1	2.3.1	1.1.3	1.1.3	IES IMPACT
Assembly code	B1020400	B1020400	B1010375	B1010375	Autodesk Revit
SMM7 (materials costs)	G20055	3015103 A	G20052	K20002	SMM7
SMM6 (embodied CO_2 for construction work)	G202911S	PA003	G202102F	K111308D	Black Book
None (embodied CO_2 for raw construction material)	None	None	None	None	University of Bath

A triple-glazed window is essential to achieve the FEES for the maximum CO_2 reduction and energy efficiency according to government mandates. However, insufficient data about costs (material and labor costs) and CO_2 impacts (embodied CO_2 and CO_2 associated with construction works) for the triple-glazed window are not publically available. At present, the only information that can be obtained is from the manufacturers or special traders. As a result, the information on cost and CO_2 impact can vary between suppliers. In this research, the costs and embodied CO_2 are calculated by using the database provided by reliable professional construction bodies such as SMM7 and the Black Book. In regard of the manufacturing industry standpoint, more collaboration can be initiated to develop industry standard life cycle inventory (LCI) data and make it publically accessible to appreciate more accurate LCA impactions of building materials and a building as a whole. Indeed, in order to generate more reliable information for LCC and LCA, data sources provided by well-known highly-rated construction organizations have been used in this research as inputs at the beginning to avoid a situation known colloquially as "garbage in, garbage out." Although BIM tools have a built-in dataset such as basic dimensions, density, and U-value of materials, these datasets are not sufficient to formulate LCC and LCA. Furthermore, BIM software such as Autodesk Revit uses a BIM object library, which is known as "Family Library," and the library can be provided any third parties who have a capability to make BIM objects. Thus, proper datasets with LCA and LCC information should be equipped with BIM objects for more reliable usage of BIM tools, while specifications for construction materials are manually matched in this research.

13.12 CONCLUSION

This chapter not only addresses major benefits to consider LCC with LCA for better and informed decisions for housing refurbishment materials and solutions, but also identifies the feasibility of BIM utilization for the simultaneous LCA and LCC studies of existing housing refurbishment. Through this chapter, the implication of LCC and LCA trade-off relationships of a whole-house fabric refurbishment based on the different energy standards are fully explored and appreciated. Based on the LCA outcomes, design and construction professionals can understand that the embodied CO_2 continues to increase as higher energy standards adopted. Consequently, design and construction professionals can considerately plan for reuse of existing building materials and structure, and consider materials with low embodied energy from the outset of a project. The use of recycled materials can be considered and promoted as well when refurbishment materials are chosen. Furthermore, if the LCC results are appreciated in conjunction with LCA, it can be concluded that the most environmentally responsible and financially affordable energy standard and refurbishment materials can be determined by considering the trade-off relationship. This chapter can provide insights for other researchers and construction professionals to understand the implication of LCA and LCC of housing refurbishment by providing the following findings.

1. Regardless of housing types, the construction costs continue to increase in order to achieve higher energy standards and larger operational CO_2 reduction.
2. As the construction costs increase, the LCC and the embodied CO_2 increase.
3. In order to achieve larger operational energy cost savings, more construction costs and embodied CO_2 are required. Thus, it is inevitable to put more insulation materials and construction costs to achieve high energy efficiency in a house.
4. Refurbishment material is the most critical factor to influence on the construction costs and embodied CO_2. In order to reduce the embodied CO_2, recycled materials should be considered for the insulation material.
5. When a customer wants high energy standards, it is beneficial to adopt the FEES (Maximum) energy standard. In the case of the detached house, the BR 2013 (Notional) is also beneficial in addition to the FEES (Maximum).
6. Currently, the Building Regulation is the best energy efficiency standard, which renders the most financially and environmentally feasible refurbishment solution. However, the FEES needs to be applied for achieving 60% CO_2 reduction through whole-house fabric refurbishment prior to any service or renewable energy installation.
7. The semidetached/end terraced house has the shortest discounted payback period and CO_2 payback period compared to detached and terraced houses, regardless of the construction materials used, in this case fiber glass and EPS.

The integration of LCC and LCA studies outcomes can significantly improve decision-makings among key stakeholders including architects, engineers, and contractors, and support them to understand and practice more sustainable construction practices. Whilst the case study has not considered replacement of existing building services and installation of renewable energy systems, the research outcomes prove that 60% CO_2 reduction is technically feasible, and also substantiate that the FEES can be environmentally and financially beneficial. Since this research is confined to whole-house fabric refurbishment, the difference and allowance between simulated and actual results also should be monitored and managed. Moreover, the substantial contribution to the current knowledge regarding BIM adoption for the housing sector is made because the case study proves that BIM is feasible to be utilized to generate and determine proper refurbishment solution when reliable datasets regarding LCA and LCC studies are available. The findings of this research contribute to shed light on examining potential BIM use and provide a better understanding the implication of BIM utilization for housing refurbishment. More importantly, the current challenges—Data Exchange and Interoperability and Unstandardized Specifications for Construction Materials—are identified and the current remedial actions are also provided, although it is not a permanent solution for the challenges. However, these findings will provide opportunities for other researchers and practitioners to tackle current issues and move forward to the more digitally prepared construction industry.

REFERENCES

AIA, 2010. AIA Guide to Building Life Cycle Assessment in Practice. AIA, United States.

Adalberth, K., Almgren, A., Petersen, E.H., 2001. Life cycle assessment of four multi-family buildings. Int. J.Low Energy Sustain. Build. 2 (2001), 1–21.

Aigbavboa, C., Ohiomah, I., Zwane, T., 2017. Sustainable construction practices: a lazy view of construction professionals in the South Africa construction industry. Energy Proced. 105 (2017), 3003–3010.

Ajayi, S.O., Oyedele, L.O., Ceranic, B., Gallanagh, M., Kadiri, K.O., 2015. Life cycle environmental performance of material specification: a BIM-enhanced comparative assessment. Int. J. Sustain. Build. Technol. Urban Develop. 6 (1), 14–24.

Akadiri, P.O., Chinyio, E.A., Olomolaiye, P.O., 2012. Design of a sustainable building: a conceptual framework for implementing sustainability in the building sector. Buildings 2 (2), 126–152.

Anton, L.A., Diaz, J., 2014. Integration of life cycle assessment in a BIM environment. Proced. Eng. 85 (2014), 26–32.

Ardente, F., Beccali, M., Cellura, M., Mistretta, M., 2011. Energy and environmental benefits in public buildings as a result of retrofit actions. Renew. Sustain. Energy Rev. 15 (1), 460–470.

Asif, M., Muneer, T., Kelley, R., 2007. Life cycle assessment: a case study of a dwelling home in Scotland. Build. Environ. 42 (2007), 1391–1394.

Assiego de Larrivaa, R., Rodríguez, G.C., Lopez, J.M.C., Raugei, M., Palmer, P.F., 2014. A decision-making LCA for energy refurbishment of buildings: conditions of comfort. Energy Build. 70 (2014), 333–342.

BEIS, 2017a. Local Authority Carbon Dioxide Emissions Estimates 2015. Department for Business, Energy & Industrial Strategy, London, UK.

BEIS. 2017b. Energy Consumption in the UK. Department for Business, Energy & Industrial Strategy, London, UK.

BRE. 2012, The Government's Standard Assessment Procedure for Energy Rating of Dwellings 2012. BRE, Watford, UK.

BRE, 2014. Low Carbon Domestic Refurbishment. Watford, UK.

BRE, 2013. Green Guide Specification, Building Research Establishment. Watford, UK.

BSHF, 2008. New Tricks with Old Bricks, Building and Social Housing Foundation. London, UK.

Basbagill, J., Flager, F., Lepech, M., Fischer, M., 2013. Application of life-cycle assessment to early stage building design for reduced embodied environmental impacts. Build. Environ. 60, 81–92. BCIS. 2012. SMM7 Estimating Price Book 2013, 18th ed. Building Cost Information Service, London, UK.

Beccali, M., Galatioto, A., Leone, G., Longo, S., 2013. Is the NZEB benchmarking approach suitable for assessing energy retrofit design? Appl. Mech. Mater. 361–363, 402–407.

Bin, G., Parker, P., 2012. Measuring buildings for sustainability: comparing the initial and retrofit ecological footprint of a century home—the REEP house. Appl. Energy 2012 (93), 24–32.

Blengini, G.A., 2009. Life cycle of buildings, demolition and recycling potential: a case study in Turin, Italy. Build. Environ. 44 (2), 319–330.

Blengini, G.A., Di Carlo, T., 2010. The changing role of life cycle phases, subsystems and materials in the LCA of low energy buildings. Energy Build. 42 (2010), 869–880.

Brás, A., Gomes, V., 2015. LCA implementation in the selection of thermal enhanced mortars for energetic rehabilitation of school buildings. Energy Build. 92 (2015), 1–9.

Bribian, I.Z., Uson, A.A., Scarpellini, S., 2009. Life cycle assessment in buildings: state-of the-art and simplified LCA methodology as a complement for building certification. Build. Environ. 44 (12), 2510–2520.

Brighton and Hove City Council, 2011. Brighton and Hove City Council Climate Change Strategy, UK.

Brinkley, M., 2017. The Housebuilder's Bible, 12th ed. Ovolo Publishing, Cambridge, UK.

Bueno, C., Fabricio, M.M., 2017. Methodological discussion of insertion and exploration of LCA data embedded in BIM elements. WIT Trans. Built Env. 169 (2017), 101–110.

Burton, S., 2012. Handbook of Sustainable Refurbishment – Housing. Earthscan, Abingdon, Oxon.

CCC, 2013. Reducing the UK's Carbon Footprint and Managing Competitiveness Risks. Committee on Climate Change, London, UK.

CIBSE, 2016. CIBSE Guide B: Heating, Ventilating, Air Conditioning and Refrigeration. Norfolk, UK.

CIC, 2013. Growth Through BIM. Construction Industry Council. London, UK.

CIOB, 2011. Buildings Under Refurbishment and Retrofit. London, UK.

Construction Products Association, 2014. An Introduction to Low Carbon Domestic Refurbishment. London, UK.

Crawley, D.B., Handb, J.W., Kummertc, M., Griffith, B.T., 2008. Contrasting the capabilities of building energy performance simulation programs. Build. Environ. 2008 (43), 661–673.

Crosbie, T., Dawood, N., Dawood, S., 2011. Improving the energy performance of the built environment: the potential of virtual collaborative life cycle tools. Automat. Constr. 20 (2011), 205–216.

Diaz, J., Anton, L.A., Reitschmidt, G., 2016. Integrating BIM and LCA. Comparison of different approaches for the integration of lifecycle information in a digital planning process. In: Proceedings of the 16th International Conference on Computing in Civil and Building Engineering, Osaka, Japan, pp. 397–402.

Doran, D., Douglas, J., Pratley, R., 2009. Refurbishment and Repair in Construction. Whittles Publishing, FL.

Dukanovic, L., Radivojevic, A., Rajci, A., 2016. Potentials and limitations for energy refurbishment of multi-family residential buildings built in Belgrade before the World War One. Energy Build. 115 (2016), 112–120.

Dylewski, R., Adamczyk, J., 2014. Life cycle assessment (LCA) of building thermal insulation materials. Eco-Efficient Construction and Building Materials. Woodhead Publishing, pp. 267–286.

EST. 2010. Fabric first, Focus on Fabric and Services Improvements to Increase Energy Performance in New Homes, CE320. Energy Saving Trust, London, UK.

EST, 2017. The Clean Growth Plan: A "2050-ready" New-Build Homes Policy. Energy Saving Trust, London, UK.

Eastman, C., Teicholz, P., Sacks, R., Liston, K., 2011. BIM Handbook: A Guide to Building Information Modeling for Owners, Managers, Architects, Engineers., Contractors, and Fabricators, 2nd ed John Wiley and Sons, US.

Fadiya, O., Georgakis, P., Chinyio, E., 2014. Quantitative analysis of the sources of construction waste. J. Constr. Eng. 2014, 1–9.

Ferreira, J., Pinheiro, M.D., de Brito, J., 2015. Economic and environmental savings of structural buildings refurbishment with demolition and reconstruction—a Portuguese benchmarking. J. Building Eng. 3, 114–126.

Flanagan, R., Jewell, C., 2005. Whole Life Appraisal for Construction. Blackwell Publishing Ltd, Garsington Road, Oxford, UK.

Franklin and Andrews, 2010. UK Building Black Book: The Cost and Carbon Guide. Franklin and Andrews, London, UK.

Genchi, Y., Kikegawab, Y., Inabaa, A., 2002. CO2 payback–time assessment of a regional-scale heating and cooling system using a ground source heat–pump in a high energy–consumption area in Tokyo. Appl. Energy 71 (2002), 147–160.

Gustafsson, S.-I., 1992. Optimization of building retrofits in a combined heat and power network. Energy 17 (2), 161–171.

HM Government, 2010. Low Carbon Construction, Innovation & Growth Team Final Report. London, UK.

HM Government, 2011. The Carbon Plan: Delivering our Low Carbon Future. London, UK.

HM Government, 2012. Building Information Modelling—Industrial Strategy: Government and Industry in Partnership, London, UK.

Hacker, J., De Saulles, T.P., Minson, A.J., Holmes, M.J., 2008. Embodied and operational carbon dioxide emissions from housing: a case study on the effects of thermal mass and climate change. Energy Build. 40 (2008), 375–384.

Ham, Y., Golparvar-Fard, M., 2015. Mapping actual thermal properties to building elements in gbXML-based BIM for reliable building energy performance modeling. Automat. Constr. 49 (2015), 214–224.

Hammond, G. and Jones, C., 2011, Embodied Carbon, The Inventory of Carbon and Energy (ICE). BSRIA BG 10/2011. BSRIA, London, UK.

Hammond, G.P. and Jones, C.I., 2008. Embodied energy and carbon in construction materials. Proc. Inst. Civil Eng. Energy 161(2008), 87–98.

Hannele, K., Reijo, M., Tarja, M., Sami, P., Jenni, K., Teija, R., 2012. Expanding uses of building information modeling in life-cycle construction projects. Work 41 (2012), 114–119.

Hong, T., Kim, J., Koo, C., 2012. LCC and LCCO2 analysis of green roofs in elementary schools with energy saving measures. Energy Build. 45 (2012), 229–239.

Hughes, W.P., Ancell, D., Gruneberg, S., Hirst, L., 2004. Exposing the myth of the 1:5:200 ratio relating initial cost, maintenance and staffing costs of office buildings. In: Proceedings of the 20th Annual ARCOM Conference, HeriotWatt University, Edinburgh, pp. 373–381.

ICLEI, 2016. 2014 Procurement Directives, European Sustainable Procurement Network.

ISO, 2006. Environmental management – Life cycle assessment – Requirements and Guidelines. BSI, London, UK.

ISO, 2008. Standardized Method of Life Cycle Costing for Construction Procurement. BSI, London, UK.

Institute for Sustainability, 2011. Sustainable Retrofit Guides, Technology Strategy Board. London, UK.

Jankovic, L., 2017. Designing Zero Carbon Buildings – Using Dynamic Simulation Methods. Routledge, Abingdon, Oxon, UK.

Jenkins, D.P., Peacok, A.D., Banfill, P.F.G., Kane, D., Ingram, V., Kilpatrick, R., 2012. Modelling carbon emissions of UK dwellings—the Tarbase domestic model. Appl. Energy 2012 (93), 596–605.

Juan, Y.-K., Kim, J.H., Roper, K., Castro-Lacouture, D., 2009. GA-based decision support system for housing condition assessment and refurbishment strategies. Automation Constr. 18 (2009), 394–401.

Killip, G., 2008. Building a Greener Britain; Transforming the UK's Existing Housing Stock. Federation of Master Builders, London, UK.

Kim, K.P., Park, K.S., 2013. BIM feasibility study for housing refurbishment projects in the UK. Organ. Technol. Manage. Constr. Int. J. 5 (3), 756–774.

Kim, M.H., Song, H.B., Song, Y., Jeong, I.T., Kim, J., 2013. Evaluation of food waste disposal options in terms of global warming and energy recovery: Korea. Int. J. Energy Environ. Eng. 4 (1), 1–12.

Kmeťková, J., Krajčík, M., 2015. Energy efficient retrofit and life cycle assessment of an apartment building. Energy Proced. 78 (2015), 3186–3191.

Konstantinoua, T., Knaack, U., 2013. An approach to integrate energy efficiency upgrade into refurbishment design process, applied in two case-study buildings in Northern European climate. Energy Build. 59 (2013), 301–309.

Kovacic, I., Reisinger, J., Honic, M., 2018. Life cycle assessment of embodied and operational energy for a passive housing block in Austria. Renew. Sustain. Energy Rev. 82 (2018), 1774–1786.

Krygiel, E., Nies, B., 2008. Green BIM: Successful Sustainable Design with Building Information Modeling. Wiley Publishing, Indianapolis.

Ladenhauf, D., Battisti, K., Berndt, R., Eggeling, E., Fellner, D.W., Gratzl-Michlmair, M., et al., 2016. Computational geometry in the context of building information modeling. Energ. Build. 115 (2016), 78–84.

Liang, X., Wanga, Y., Royapoora, M., Wub, Q., Roskilly, T., 2017. Comparison of building performance between conventional house and passive house in the UK. Energy Proced. 142 (2017), 1823–1828.

Ma, Z., Cooper, P., Daly, D., Ledo, L., 2012. Existing building retrofits: methodology and state-of-the-art. Energy Build. 55 (2012), 889–902.

Martinaitis, V., Kazakevicius, E., Vitkauskasb, A., 2007. A two-factor method for appraising building renovation and energy efficiency improvement projects. Energy Policy 35 (2007), 192–201.

McMullan, R., 2017. Environmental Science in Building, 8th ed Palgrave Macmillan, Basingstoke, UK.

Menassa, C., 2011. Evaluating sustainable retrofits in existing buildings under uncertainty. Energy Build. 43 (2011), 3576–3583.

Mohammadpourkarbasi, H., Sharples, S., 2013. The eco-refurbishment of a 19th century terraced house: energy and cost performance for current and future UK climates. Buildings 2013 (3), 220–244.

Monahan, J., Powell, J.C., 2011. A comparison of the energy and carbon implications of new systems of energy provision in new build housing in the UK. Energy Policy 39, 290–298.

Monteiro, H., Freire, F., 2012. Life cycle assessment of a house with alternative exterior walls: comparison of three impact assessment methods. Energy Build. 47 (2012), 572–583.

Murray, S.N., Rocher, B., O'Sullivan, D.T.J., 2012. Static simulation: a sufficient modelling technique for retrofit analysis. Energy Build. 47 (2012), 113–121.

NBS, 2017. National BIM Report 2017. National Building Specification, London, UK.

NIHE, 2015. Energy Efficiency Good Practice Guide for Refurbishment of the Residential Sector. Northern Ireland Housing Executive.

National Refurbishment Centre, 2012. Refurbishing the Nation Gathering the evidence. National Refurbishment Centre, Watford, London, UK.

Neufert, E., 2012. Neufert Architects' Data. Blackwell Publishing, Oxford, UK.

Ortiz, O., Bonnet, C., Bruno, J.C., Castells, F., 2009. Sustainability based on LCM of residential dwellings: a case study in Catalonia, Spain. Build. Environ. 44 (2009), 584–594.

PMI, 2017. Guide to the Project Management Body of Knowledge. Project Management Institute, US.

Pacheco-Torres, R., Jadraque, E., Roldán-Fontana, J., Ordónez, J., 2014. Analysis of CO2 emissions in the construction phase of single-family detached houses. Sustain. Cities Society 12 (2014), 63–68.

Pal, S.K., Takano, A., Alanne, K., Siren, K., 2017. A life cycle approach to optimizing carbon footprint and costs of a residential building. Build. Environ. 123 (2017), 146–162.

Palmer, J. and Cooper, I., 2013. United Kingdom Housing Energy Fact File 2013. London, UK.

Pargana, N., Pinheiro, M.D., Silvestre, J.D., de Brito, J., 2014. Comparative environmental life cycle assessment of thermal insulation materials of buildings. Energy Build. 82 (2014), 466–481.

Park, K.S., Kim, K.P., 2014. Essential BIM input data study for housing refurbishment: homeowners' preferences in the UK. Buildings 2014 (4), 467–487.

Peuportier, B.L.P., 2001. Life cycle assessment applied to the comparative evaluation of single family houses in the French context. Energy Build. 33 (2001), 443–450.

Rauf, A., Crawford, R.H., 2015. Building service life and its effect on the life cycle embodied energy of buildings. Energy 79 (2015), 140–148.

Redmond, A., Hore, A., Alshawi, M., West, R., 2012. Exploring how information exchanges can be enhanced through Cloud BIM. Automat. Constr. 24 (2012), 175–183.

Riley, M., Cotgrave, A., 2011. Construction Technology 3: The Technology of Refurbishment and Maintenance. Palgrave, MacMillan.

Riley, M., Cotgrave, A., 2013. Construction Technology 1: House Construction, 3rd ed Palgrave Macmillan, London.

Rodrigues, F., Matosa, R., Alvesc, A., Ribeirinhoa, P., Rodrigues, H., 2018. Building life cycle applied to refurbishment of a traditional building from Oporto, Portugal. J. Build. Eng. 17 (2018), 84–95.

Ronning, A., Brekke, A., 2014. Life cycle assessment (LCA) of the building sector: strengths and weaknesses. Eco-Efficient Construction and Building Materials. Woodhead Publishing, pp. 63–83.

Rosa, M., Franca, C., Azapagic, A., 2012. Environmental impacts of the UK residential sector: Life cycle assessment of houses. Build. Environ. 54 (2012), 86–99.

Rysanek, A.M., Choudhary, R., 2013. Optimum building energy retrofits under technical and economic uncertainty. Energy Build. 57 (2013), 324–337.

Salazar, J., 2014. Life cycle assessment (LCA) of windows and window materials. Eco-Efficient Construction and Building Materials. Woodhead Publishing, pp. 502–527.

Sartori, I., Hestnes, A.G., 2007. Energy use in the life cycle of conventional and low energy buildings: a review article. Energy Build. 39 (3), 249–257.

Scheneider, D., and Rode, P. (2010).Energy Renaissance. High Performance Building, US.

Schwartz, Y., Raslan, R., Mumovic, D., 2016. Implementing multi objective genetic algorithm for life cycle carbon footprint and life cycle cost minimisation: a building refurbishment. Energy 97 (2016), 58–68.

Shadram, F., Johansson, T.D., Lu, W., Schade, J., Olofsson, T., 2016. An integrated BIM- based framework for minimizing embodied energy during building design. Energy Build. 128 (2016), 592–604.

Shoubi, M.V., Shoubi, M.V., Bagchi, A., Barough, A.S., 2015. Reducing the operational energy demand in buildings using building information modeling tools and sustainability approaches. Ain Shams Eng. J. 6 (1), 41–55.

Sodagar, B., 2013. Sustainability potentials of housing refurbishment. Buildings 3 (2013), 278–299.

Soust-Verdaguer, B., Llatas, C., García-Martínez, A., 2016. Simplification in life cycle assessment of single-family houses: a review of recent developments. Build. Environ. 103 (2016), 215–227.

Soust-Verdaguer, B., Llatas, C., García-Martínez, A., 2017. Critical review of BIM-based LCA method to buildings. Energy Build. 136 (2017), 110–120.

Succar, B., 2009. Building information modelling framework: a research and delivery foundation for industry stakeholders. Automat. Constr. 18 (3), 357–375.

Syngros, G., Balaras, C.A., Koubogiannis, D.G., 2017. Embodied CO2 emissions in building construction materials of Hellenic dwellings. Proced. Environ. Sci. 38 (2017), 500–508.

TSB, 2014. Retrofit for the Future Reducing Energy Use in Existing Homes, A Guide to Making Retrofit Work. Technology Strategy Board, London, UK.

Tabassi, A.A., Roufechaei, K.M., Ramli, M., Bakar, A.H.A., Ismail, R., Pakir, A.H.P., 2016. Leadership competences of sustainable construction project managers. J. Cleaner Product. 124 (2016), 339–349.

Tam, V., Senaratne, S., Le, K., Shen, L., Perica, J., Illankoon, C., 2017. Life-cycle cost analysis of green-building implementation using timber applications. J. Cleaner Product. 147 (2017), 458–469.

Thorpe, D., 2014. Energy Management in Buildings—The Earthscan Expert Guide. Routledge, London, UK.

Thuvander, L., Femenoas, P., Mjornell, K., Meiling, P., 2012. Unveiling the Process of Sustainable Renovation. Sustainability 4 (6), 1188–1213.

Tsai, W.-H., Lin, S.-J., Liu, J.-Y., Lin, W.-R., Lee, K.-C., 2011. Incorporating life cycle assessments into building project decision-making: an energy consumption and CO2 emission perspective. Energy 36 (2011), 3022–3029.

Wang, W., Zmeureanu, R., Rivad, H., 2005. Applying multi-objective genetic algorithms in green building design optimization. Build. Environ. 40 (15), 12–25.

Yin, B.C.L., Laing, R., Leon, M., Mabon, L., 2018. An evaluation of sustainable construction perceptions and practices in Singapore. Sustain. Cities Society 39 (2018), 613–620.

Yin, R.K. 2003. Case Study Research, Design and Methods, 3rd ed. Thousand Oaks, Sage.

FURTHER READING

BSI, 2008. Standardised Method of Life Cycle Costing for Construction Procurement. A Supplement to BS ISO 15686-5. Buildings and Constructed Assets, London, UK.

Berg, B., 2014. Using BIM to Calculate Accurate Building Material Quantities for Early Design Phase Life Cycle Assessment. Master Thesis, Victoria University of Wellington, New Zeland.

Boardman, B., 2007. Home Truths, A Carbon Strategy to Reduce UK Housing Emission by 80% by 2050. ECI Research Report 34. University of Oxford's Environmental Change Institute, Oxford.

Brinkley, M., 2008. The Housebuilder's Bible, 7th ed. Ovolo Publishing, Cambridge, UK.

Cavieres, A., Gentry, R., Al-Haddad, T., 2011. Knowledge-based parametric tools for concrete masonry walls: conceptual design and preliminary structural analysis. Automat. Constr. 20 (6), 716–728.

Cheung, F.K., Rihan, J., Tah, J., Duce, D., Kurul, E., 2012. Early stage multi-level cost estimation for schematic BIM models. Automat. Constr. 27 (2012), 67–77.

Diaz, J. and Anton, L.A., 2014. Sustainable constuction approach through integration of LCA and BIM tools. In: Proceedings of the 2014 International Conference on Computing in Civil and Building Engineering, Orlando, Florida, June 23–25, pp. 283–290.

Reeves, A., 2009. Achieving Deep Carbon Emission Reductions in Existing Social Housing: The Case of Peabody. Ph.D thesis, De Montfort University.

Sathre, R., González-García, S., 2014. Life cycle assessment (LCA) of wood-based building materials. Eco-Efficient Construction and Building Materials. Woodhead Publishing, pp. 311–337.

Tobias, L. and Vavaroutsos, G., 2009. Retrofitting Office Buildings to be Green and Energy-Efficient: Optimizing Building Performance, Tenant Satisfaction, and Financial Return, Urban Land Institute, Washington, DC, US.

APPENDIX 13.1 BASIC INFORMATION FOR SEMIDETACHED/ END TERRACED HOUSE MODEL

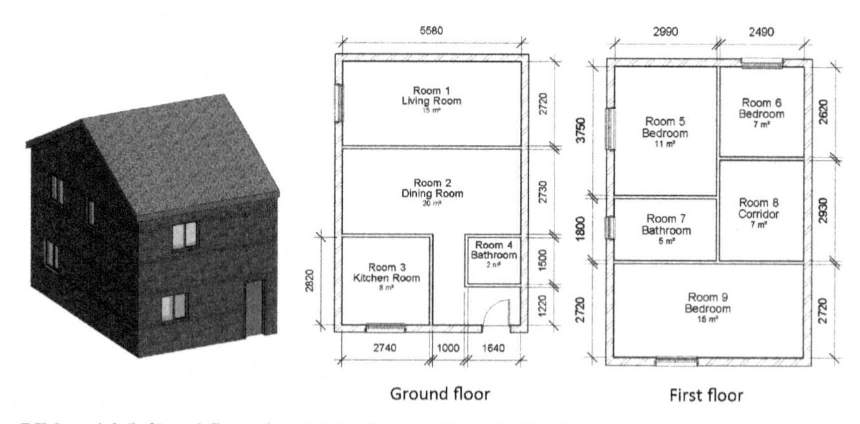

BIM model (left) and floor plans (right) for a semidetached/end-terrace house

Room and space information— semidetached/end terraced house

Floors	Rooms	Descriptions	Areas (m^2)
Ground floor	Room 1	Kitchen	15
	Room 2	Bathroom	20
	Room 3	Lobby	8
	Room 4	Living room	2
First floor	Room 5	Dining room	11
	Room 6	Bedroom	7
	Room 7	Bedroom	5
	Room 8	Corridor	7
	Room 9	Bathroom	15
Total usable floor area			90

APPENDIX 13.2 BASIC INFORMATION FOR TERRACED HOUSE MODEL

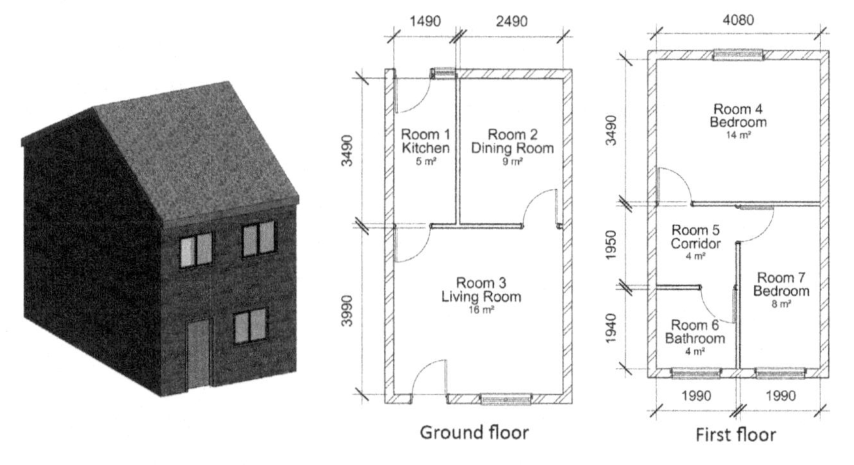

BIM model (left) and floor plans (right) for a terraced house

Room and space information—terraced house

Floors	Rooms	Descriptions	Areas (m^2)
Ground floor	Room 1	Kitchen	5
	Room 2	Dining room	9
	Room 3	Living room	16
First floor	Room 4	Bedroom	14
	Room 5	Corridor	4
	Room 6	Bathroom	4
	Room 7	Bedroom	8
Total usable floor area			60

APPENDIX 13.3 LIFE CYCLE ASSESSMENT AND LIFE CYCLE COST STUDY RESULTS FOR DETACHED HOUSE

Detached house			Current basic model	Energy standards			
				BR 2010/2013 (Minimum)	BR 2010 (Notional)	BR 2013 (Notional)	FEES (Maximum)
Energy demand (KWh/year/m^2)			209.8	52.5	44.8	41.5	39.3
CO_2 emission (kg/year/m^2)			84.5	43.4	41	41	41
Energy demand (MWh/year)			38.4	9.6	8.1	7.6	7.2
CO_2 emission (kg/year)			10,985	5,635.5	5,355.6	5,328.5	5,328.3
Energy cost (£/year)			1,150	295	252.54	234.75	224.75
Life cycle cost (£)	Construction cost	Fiberglass	41,371.35	7,065.57	9,055.37	9,899.04	10,425.47
		EPS		12,004.63	15,690.73	18,419.59	19,917.36
	Operating cost	Fiber glass	205,359.48	144,414.43	146,069.73	145,829.16	145,938.91
		EPS		148,325.25	151,469.01	152,497.49	153,668.72
	Total cost	Fiber glass	246,730.83	151,480.0	155,125.10	155,728.2	156,364.38
		EPS		160,329.88	167,159.74	170,917.08	173,586.08
Life cycle assessment (kg)	Embodied CO_2 (cradle-to-site)	Fiber glass	34,994.9	12,197.25	16,624.06	20,750.04	23,140.86
		EPS		13,505.52	18,336.66	23,114.4	25,689.4
	Total (cradle-to-grave)	Fiber glass	45,979.9	17,832.75	21,979.66	26,078.54	28,469.16
		EPS		19,141.02	23,692.26	28,442.9	31,017.7

APPENDIX 13.4 LIFE CYCLE ASSESSMENT AND LIFE CYCLE COST STUDY RESULTS FOR SEMIDETACHED/END TERRACED HOUSE

Semidetached/end terraced			Current basic model	Energy standards			
				BR 2010/2013 (Minimum)	BR 2010 (Notional)	BR 2013 (Notional)	FEES (Maximum)
Energy demand (kWh/year/m^2)			274.3	66.3	54.8	48.7	45
CO_2 emission (kg/year/m^2)			105.1	46.7	43.7	42.5	41.7
Energy demand (MWh/year)			36	8.7	7.2	6.4	5.9
CO_2 emission (kg/year)			9,454.8	4,207.2	3,929.5	3,822.2	3,756.5
Energy cost (£/year)			1,078.7	266.3	223.4	200	184.43
Life cycle Cost (£)	Construction cost	Fiber glass	32,794.94	5,742.68	7,343.17	8,845.20	9,313.88
		EPS		9,666.66	12,633.56	15,626.69	16,950.57
	Operation cost	Fiber glass	197,053.94	141,081.55	141,917.82	143,575.71	143,344.72
		EPS		144,319.75	146,409.23	149,109.73	149,824.78
	Total cost	Fiber glass	229,848.88	146,824.23	149,261	152,420.9	152,658.6
		EPS		153,986.41	159,042.79	164,736.42	166,775.35
Life cycle assessment (kg)	Embodied CO_2 (cradle-to-site)	Fiber glass	28,302.39	9,725.01	13,346.31	16,601.90	18,710.29
		EPS		10,771.70	14710.09	18,486.26	20,756.82
	Total (cradle-to-grave)	Fiber glass	37,757.19	13,932.21	17275.81	20,424.1	22,466.79
		EPS		14,978.9	18639.59	22,308.46	24,513.32

APPENDIX 13.5 LIFE CYCLE ASSESSMENT AND LIFE CYCLE COST STUDY RESULTS TERRACED HOUSE

Terraced house			Current basic model	Energy standards			
				BR 2010/2013 (Minimum)	BR 2010 (Notional)	BR 2013 (Notional)	FEES (Maximum)
Energy demand (kWh/year/m^2)			307.6	78.2	64.5	57.2	51.2
CO_2 emission (kg/year/m^2)			112	48.6	45	43.4	42.2
Energy demand (MWh/year)			26.3	6.7	5.5	4.9	4.4
CO_2 emission (kg/year)			6,716.2	2,917.9	2,695.5	2,602.6	2,532.4
Energy cost (£)			789.6	207.6	173.8	155.6	141.5
Life cycle cost (£)	Construction cost	Fiber glass	23,949.07	4,830.93	6,711.13	8,040.43	8,479.04
		EPS		7,773.90	10,696.91	13,133.80	14,279.80
	Operation cost	Fiber glass	178,400.73	138,054.12	139,428.08	141,225.9	141,044.1
		EPS		140,588.7	142,952.93	145,569.1	146,175.7
	Total cost	Fiber glass	202,349.8	142,885.05	146,139.22	149,266.33	149,523.16
		EPS		148,362.6	153,649.84	158,702.92	160,455.46
Life cycle assessment (kg)	Embodied CO_2 (cradle-to-site)	Fiber glass	20,967.9	7,495.79	10,456.70	12,867.58	14,694.12
		EPS		8,286.12	11,483.02	14,285.32	16,245.98
	Total (cradle-to-grave)	Fiber glass	27,684.1	10,413.69	13,152.2	15,470.18	17,226.52
		EPS		11,204.02	14,178.52	16,887.92	18,778.38

Chapter 14

Bridging Sustainable Construction Technologies and Heritage: Novel Approach to the Conservation of the Built Environment

Cazacova Liudmila and Yapicioglu Balkiz
Arkin University of Creative Arts and Design, Kyrenia, Cyprus

14.1 INTRODUCTION

The certainty of global warming and, consequently, climate change urges for planners' actions to create low carbon cities and, therefore, reduce the carbon emissions (Yung and Chan, 2012). The building industry, according to Horvath (2004) and Yung and Chan (2012), due to its large scale, economic strength, and social importance, is considered the leading consumer of materials and energy, one of the most significant polluters and producers of carbon emissions. According to Organization for Economic Co-operation and Development report (OECD, 2010), buildings consume more than 40% of the total amount of the globally produced energy and generate one-third of global greenhouse emissions. Due to the building or construction sector's processes, approximately 136 million tonnes of waste are annually created (Yung and Chan, 2012; Tam and Tam, 2006). Given the above listed accounts, the building industry should reposition itself and take the responsibility of creating low carbon cities and, consequently, contributing to the reduction of carbon emissions on global scale. Horvath (2004) accents the vitality of learning and better understanding the building industry and its processes through research, which will assist in transforming the construction processes via innovations, make them sustainable oriented, and ultimately reduce their environmental impacts.

Tam and Tam (2006) define construction industry as a vast complex mechanism composed of numerous parts (e.g., building materials manufacturing,

Sustainable Construction Technologies. DOI: https://doi.org/10.1016/B978-0-12-811749-1.00011-0

building climate control systems, service and maintenance) and with an immense number of members. For delivering complete construction project, mutual efforts of architects, designers, engineers (civil, mechanical, electrical, electronic, software, etc.), miners, manufactures (construction materials, processes, equipment, building systems, etc.), maintenance and built environment facilities' management companies, and others are required. As put by Tam and Tam (2006) "construction by its nature is not an environmental-friendly activity." Nevertheless, it is quite clear that construction industry has to support a world of continuing population growth and economic development and cannot afford to apply a cradle-to-grave approach (where the raw materials are extracted, manufactured, used for a lifespan, and finally disposed). Construction has mostly been a profit-driven industry, and it has been shown that it is one of the industries least likely to invest in innovation and research and development (Dale, 2007). The industry, henceforward, must pay heed to the widespread social interest in environmental preservation. It cannot further increase its environmental impact because it is not socially and environmentally acceptable (Horvath, 2000) and must find ways of changing and adapting to increasing concern for the environment. Therefore, to adjust this giant mechanism to the sustainability notion, each part engaged in this process should follow sustainability's doctrines (Hussin et al., 2013).

Kuhlman and Farrington (2010), referring to Brundtland Report (1987), define sustainability as meeting society's present needs without compromising the ability of future generations to meet their own needs. Sustainable construction, on the other hand, has no solitary definition that is accepted worldwide. Several schoolers, e.g. Kibert (1994, 2013), Hussin et al. (2013), Saleh (2015), SBC (2003), have defined sustainable construction in their works, but the simplest definition was given by the European Union and described by Du Plessis (2012) as: "the use and/or promotion of (a) environmentally friendly materials, (b) energy efficiency in buildings, and (c) management of construction and demolition waste." Given that all the definitions have an almost exclusive focus on environmental impact, the mission of sustainable development is to adopt appropriate methods in construction industry to protect the environment (Du Plessis, 2012).

One of the methods recently adopted by the construction industry is green construction (sustainable construction, sustainable, or green buildings), which is defined by US Green Building Council (Kriss, 2014), as "planning, design, construction, and operations of buildings with several central, foremost considerations: energy use, water use, indoor environmental quality, material selection and the building's effects on its site." These efforts aim to amplify the positive and mitigate the negative of these effects throughout the entire life cycle of a building—sitting, design, construction, operation, maintenance, renovation, and demolition.

To make a construction sustainable it should be minimized the waste of materials by design, proper selection of quantity and quality of materials, and application of sustainable (green) materials. A sustainable or green is considered a material that does not deplete nonrenewable (natural resources) and whose use has no adverse impact on the environment (Srinivas, 2015). Furthermore, to reduce the impact on environment and energy consumption, reusable, recyclable at the end-of-life materials, with low embodied energy,[1] and available in a proximate distance from the construction site (to minimize the consumption of fuel, emissions, and road congestion) should be used (Wahlström et al., 2014).

Another effective method is recycling, which besides minimizing the construction waste also reduces the requests for new resources, lessens the amount of energy for transportation and production, and use waste which would otherwise be lost to landfill sites (Pullen, 2013; Tam V. & Tam C., 2006). Recalling the statement made by Yung and Chan (2012), nearly half of the total amount of waste created by the construction industry comes from buildings' demolition. Henceforth, the most environmentally friendly building is the one that do not have to be built because it already exists, according to Murray (2012). He considers that "Keeping buildings out of landfill dumps by not demolishing them is, *ipso facto*, an environmental benefit."

Building adaptive reuse, as suggested by Conejos et al. (2012), is a noble alternative to traditional demolition. He defines it as "a significant change to an existing building function when the former function has become obsolete." Bullen and Love (2010) describe adaptive reuse as "a process that changes a disused or ineffective item into a new item that can be used for a different purpose" or a process that "involves converting a building to undertake a change of use required by new or existing owners" (Bullen and Love, 2011). This method, especially when it applies to the adaptive reuse of heritage (vernacular) building, is a synergy of methods listed (which are (1) sustainable construction, (2) employment of green materials for waste minimization, and (3) recycling). Adaptive reuse of buildings, as described by Yung and Chan (2012), bypasses the wasteful process of demolition, therefore is considered as an element of sustainable development process that has several environmental benefits—it saves energy, saves resources, minimizes construction waste, and reduces carbon emissions, and consequently, contributes to the environmental (materials and resources efficiency), economic (construction cost reductions), and social (retention) sustainability improvement (Bullen and Love, 2011). Adapting an old building and improving its performance is considered by Conejos et al. (2012) an

1. Energy consumption produces carbon dioxide, which contributes to greenhouse emissions, therefore how green the building material is determined by the amount of its embodied energy, which is the total amount of energy employed in a construction of a building (raw materials extraction and transportation, building materials manufacturing and transportation to the building site) (Bergham, 2012). Embodied energy is also an indicator of the overall impact of the materials and building on the environment ("Sustainable building and construction,"2003).

eco-friendlier approach than constructing a new energy efficient building. By reusing the function of the building, its components, materials, the amount of embodied energy, which is usually required for creation of new functional facilities demanded by the community, is reduced (Judson and Iyer-Raniga, 2010; Bullen and Love, 2010; Pullen, 2013). It also benefits the communities via conserving the existing green spaces, and therefore, improving the air quality, maintaining the microclimate and habitat, ecosystem and water quality. According to Bullen and Love (2010; 2011), adaptive reuse improves the performance of the building and increases its financial, environmental, and social value. Moreover, it assists communities and governments in slowing down the continuous process of urban expansion and minimizes its environmental, social, and economic costs.

Referring to Conejos et al. (2012) adaptive reuse is already widely applied in the United States, Australia, and across the Asia Pacific region for refunctioning facilities and buildings including important heritage monuments. Bullen and Love (2011) also describes ongoing experiments in adaptive reuse applied for heritage buildings conservation and urban context regeneration in Australia. The worldwide recognition of buildings adaptive reuse can significantly contribute to communities' sustainable development and to the improvement of the built environment quality.

The built environment consists of a variety of buildings in different shapes and sizes, old and new. Sustainable construction innovation, therefore, is not only about new construction but also inclusive of existing buildings. Our aim in this chapter, therefore, is to examine how the construction industry can reposition itself to increase focus on the adaptive reuse of existing buildings as an alternative to demolition and replacement by introducing a new integrated ICOMOS (International Council on Monuments and Sites) + C2C (Cradle-to-Cradle) life cycle approach to sustainable Built Heritage conservation. A vernacular dwelling in old area of Mirbat town of Dhofar region in Sultanate of Oman has been selected as a case study to depict our integrated ICOMOS + C2C approach.

14.2 CURRENT APPROACHES IN SUSTAINABLE BUILT HERITAGE CONSERVATION

Built Heritage—in the case of this chapter, old historic buildings—strengthen the history of the place, while creating friendly neighborhoods that are unique and yet familiar for a certain community. These buildings are bridges connecting tangible places and intangible cultural traditions (VHF, 1995), therefore, should be conserved as society's heritage. Just recently, in September 2015, the international community has agreed on new global framework for sustainable development, the 2030 Agenda for Sustainable Development, which contains an explicit heritage target, Target 11.4 that calls "for making cities and human settlements inclusive, safe, resilient and sustainable by strengthening efforts to protect and safeguard the world's cultural and natural heritage" (UN General Assembly, 2015).

Sustainability is distinct by meeting society's present needs without compromising the ability of future generations to meet their own needs. These needs, which are formed by social, economic, and environmental factors, could be met through the conservation of the built environment (VHF, 1995), which anticipates transferring the heritage to the future generations and prolong buildings' life (Cinieri and Zamperini, 2013). Even though sustainable development and heritage conservation are different movements, they share similar intentions and both aim to enhance the relationship between people and the built environments. As Godwin (2011) puts it, "There is no inherent conflict between the retention of historic buildings and the principles of sustainability." However, as Bullen and Love (2010) states, it is important to acknowledge how and the ways a building can make a positive contribution to sustainability, especially the community the building serves and represents. Over the past decade there has been a great desire with both sustainable and heritage conservation movements to align their agendas, for example, the International Green Construction Code (IgCC) and US Green Building Council's LEED O + M; however, both of these codes are short of addressing historical buildings (WBDG, 2014; Roberts, 2007).

Many old heritage buildings prove to be able to conserve energy while creating a comfortable and healthy indoor environment with a low environmental footprint (Madhumathi et al., 2014; Hines, 2011), or we can even associate them being less brutal toward environment. This can be attributed to the bioclimatic means used in architecture to achieve thermal comfort, where embedded energy in materials used is a significant sustainable aspect of Built Heritage (WBDG, 2014; Sayigh and Marafia 1998). Because historical buildings might not perfectly meet energy conservation criteria, have restrictions in construction techniques and typological and functional features (Cinier and Zamperini, 2013), the contemporary society do hesitate to fully recognize the value of the embodied energy of existing buildings and benefit from the passive design techniques of the Built Heritage. Consequently, and unfortunately, quite a few of historical building have been demolished globally on the base that "the costs of renovating and adapting these buildings for new uses are too high" (Shipley et al., 2006). Regardless, the practice shows that buildings conservation can be sustainable.

Recently, in the field of sustainable Built Heritage conservation research has been encouraged, where the researchers show curiosity in built cultural heritage conservation methods, wherein as a classification criterion building historical–cultural value is included using the concepts of life cycle approach (WBDG, 2014; Cinier and Zamperini, 2013; Godwin, 2011; Roberts, 2007). For example, Parks Canada's Standards and Guidelines for the Conservation of Historic Places in Canada agrees retaining and reusing heritage buildings by introducing energy saving alterations during conservation, even though some character-defining elements of the Built Heritage cannot be altered to save energy (HRB, 2007).

Another example in the field of green cultural Built Heritage conservation is the study conducted by Italian researchers along with the Italian National Agency for New Technologies, Energy and Sustainable Economic development, which aimed to develop a methodology for historical buildings energy improvement (Cinier and Zamperini, 2013).

Such conservation methods preserve value of the building while prolonging its lifespan (VHF, 1995), sustainability, in this case, is about being ecologically conscious where the embodied energy is preserved and about understanding the existing sustainable features of the Built Heritage. Cultural Built Heritage does not need new land, needs less extraction and production and processing of new materials, and most importantly, as Cinier and Zamperini (2013) emphasizes, "heritage buildings often have morphological and technological features appropriate to environment and climate."

According to Mora et al. (2011), a life cycle to the buildings, which is being actively encouraged now, is a foundation of sustainable development. Traditional architecture and building engineering sciences have accumulated vast knowledge in construction materials and techniques and environmental design, which unfortunately are not copiously applied in contemporary projects. The application of this knowledge along with the building materials' durability analysis will positively contribute to the sustainability of the construction (Mora et al., 2011). On the other hand, Silvestre et al. (2014) considers that waste flow during all stages as production, construction, use, and end-of-life of building materials should be controlled and understood the effect of their variation on C2C environmental performance of building materials improvement.

Given the above cited statements, and as mentioned by Cazacova (2015), the application of the life cycle approach, which considers all the aspects of sustainability, toward built cultural heritage conservation makes the conservation process green while preserving the value of the building. However, it is also pivotal to state that preserving historical merit of a building is as important as conserving a building to meet society's needs. To put it another way, sustainable building design is not only just about performance, but also about how users respond to the building. The quality of the direct emotional impact they have on us (soft to the touch, warm, and able to grow old gracefully) are living, human qualities (Hines, 2011) and should also be valued in sustainable conservation of Built Heritage. With the burgeoning sustainability rhetoric and field of conservation, Built Heritage in our modern industrial society should be valued not only as society's heritage, but for being inherently sustainable.

Therefore, traditional conservation of cultural Built Heritage (which is further explained in Section 14.3.1) should be integrated in a larger strategy of sustainable management of the embodied energy, which in return should be considered as integrated part of cultural heritage.

14.3 INNOVATIVE APPROACH TO BUILT HERITAGE CONSERVATION: ICOMOS + C2C

This section will explore the concept of ICOMOS (International Council on Monuments and Sites principles) and the C2C life cycle approach and will present our innovative integrated approach. Our innovative approach is based on the philosophy that Built Heritage conservation can promote cultural heritage sustainability along with social and environmental sustainability.

14.3.1 ICOMOS for Built Cultural Heritage Conservation

ICOMOS is one of the three formal advisory bodies to the World Heritage Committee[2] and is the professional and scientific advisor to the Committee on all aspects of the cultural heritage. The meaning of cultural heritage, as explained by Burra Charter (ICOMOS, 2013), is no longer just about "single monuments identified as objects of art to cultural landscapes, historic cities, and serial properties." The concept of heritage extends beyond "tangible heritage," to the intangible dimensions of heritage as well where economic, social, and environmental developments are linked to it.

The Charter for the Conservation of Places of Cultural Heritage Value (ICOMOS, 2010) states that built cultural heritage's (place) value means possessing historical, archeological, architectural, technological, aesthetic, scientific, spiritual, social, traditional, or other special cultural significance, associated with human activity. Furthermore, the same Charter (ICOMOS, 2010) acknowledges the places of cultural heritage as places retaining lasting values, teaching about the past and culture of the predecessors, providing the context for community's identity, providing variety and contrast in the modern world and appreciated in their own right. The Burra Charter (ICOMOS, 2013), on the other hand, defines cultural heritage as historical records that reflect the diversity of our societies forming the countries' identities, therefore are valuable and inimitable. Article 15 of Burra Charter (ICOMOS, 2013) states that cultural heritage (place) must be identified and protected (conserved) and that the demolition of significant fabric of a place is generally not acceptable.

The Charter for the Conservation of Places of Cultural Heritage Value (ICOMOS, 2010) defines conservation as a process of caring for a place with the aim of safeguarding its cultural heritage value. Furthermore, the purpose of conservation is not only to care for places, but also for their structures, materials, and cultural meaning. On the other hand, Article 4 of Burra Charter (ICOMOS, 2013) declares conservation as responsible for making

2. The other two advisory bodies to the World Heritage Committee are IUCN (International Union for Conservation of Nature) and ICCROM (International Centre for the Study of the Preservation and Restoration of Cultural Property).

TABLE 14.1 ICOMOS— The Conservation Process—Steps and Tasks

Conservation process	
Understanding significance	Identify place and association—secure the place and make it safe
	Gather and record information about the place sufficient to understand its significance—documentary, oral, physical
	Assess significance
	Prepare statement of significance
Develop policy	Identify obligations arising from significance
	Gather information about other factors affecting the future of the place—owner/manager's needs, external factors, physical conditions
	Develop policy—identify and consider opinions and test their impact on place significance
	Prepare a statement of conservation policy
Manage	Manage the place in accordance to the policy—develop and implement strategies through a management plan, record sthe place prior ant changes
	Monitor and review any of the processes

use of all the knowledge, skills, and disciplines, which can contribute to the study and care for the place wherein traditional techniques and materials are preferred for the fabric safeguarding. Henceforth, Article 5 of Burra Charter (ICOMOS, 2013) necessitates that conservation of a place should identify and take into consideration all aspects of cultural and natural significance without unwarranted emphasis on any single value at the expense of others. Conservation process consists of three parts: (1) understanding significance, (2) developing conservation policy, and (3) manage conservation process (see Table 14.1) (ICOMOS, 2013).

Conservation process could be executed via various methods listed below in increasing degrees of intervention (ICOMOS, 2013):

1. Preservation,[3] through stabilization, maintenance, or repair;
2. Restoration,[4] through reassembly, reinstatement, or removal;

3. Preservation of a place involves as little intervention as possible, to ensure its long-term survival and the continuation of its cultural heritage value (ICOMOS, 2013).
4. Restoration is returning the existing fabric of a place to a state known earlier without the introduction of new material (Article 1, Burra Charter, ICOMOS, 2013).

3. Reconstruction[5]; and
4. Adaptation.

However, Article 7 of Burra Charter (ICOMOS, 2013) recommends adaptation (which means modifying a place to suit the current use or a proposed use, involving the least possible loss of cultural heritage value) as one of the potential methods of cultural heritage conservation. The Nara Document on Authenticity (ICOMOS, 1994) states that the new use (function) should involve minimal change to the original fabric and use, and should respect associations and meanings. As described in Burra Charter (ICOMOS, 2013) adaptation might be executed by two means, restoration or reconstruction, depending on the physical condition of the place.

14.3.2 Cradle-to-Cradle for Built Cultural Heritage Conservation

The concept "Cradle-to-Cradle" or "C2C," based on life cycle approach, states that everything we have and will ever have is, in one form or another, on the planet now and C2C industrial system copies natural cyclical systems allowing for a fully sustainable environment. The Earth's resources do not get replenished from outside, therefore, we must never use them up faster than the Earth's own ecosystems can replenish them (Bergman, 2012). C2C questions the finite lifetime outlook on materials and the products and advocates materials and the products to be part of an endless cycle that replicate the nature's (Braungart and McDonough, 2002). C2C utilizes a "waste is food" principle, where any waste produced goes to feed another system, and a biometric approach to the design of products and systems that models human industry on nature's processes viewing materials as nutrients circulating in healthy, safe metabolisms.

According to Mulhall and Braungart (2010), a C2C building should contain measurable elements that add value and celebrates innovation and enjoyment by significantly enhancing the quality of materials, biodiversity, air, and water; using current solar income; being deconstructible and recyclable, and performing diverse practical and life-enhancing functions for its stakeholders. For a building to achieve a C2C, it has to fulfill the following three basic principles (Mulhall and Braungart, 2010):

1. Define materials and their intended use pathways, integrate biological nutrients, enhance air and climate quality, and enhance water quality.
2. Integrate renewable energy.

5. Reconstruction is returning a place to a state known earlier through introduction of new recycled and salvaged material from other places, into the fabric. In some cases, when is beneficial for conservation, modern techniques, and materials could be introduced (Article 1, Burra Charter, ICOMOS, 2013).

3. Actively support biodiversity, celebrate conceptual diversity with innovation, add value and enhance quality for stakeholders, and enhance stakeholder well-being and enjoyment.

However, with the modern technologies applied in construction the probability to achieve a 100% C2C building, which has 0% negative and 100% positive impact on its surroundings is very low (Lilletorget, n.d.). Therefore, instead of trying to achieve a 100% C2C building it is better to attain as many as possible building's C2C inspired elements. Suggested by source process of obtaining C2C building comprises three phases: (1) analysis, (2) evaluation, and (3) optimization. Analysis phase (1) establishes a statement of the project's context, its challenges, opportunities, and actors, wherein the C2C intensions are formulated and the C2C building inspired elements as focus area are selected. During the evaluation phase (2) the selected C2C building inspired elements are evaluated and the materials' health, recycling potential, and beneficial content assessment are implemented. Finally, the optimization phase (3) organizes the process of continuous work toward the goal of 100% positive impact (Lilletorget, n.d.).

14.3.3 ICOMOS + C2C

Built cultural heritage and vernacular architecture, which employ local materials and construction techniques and passive design for indoor thermal comfort facilitation, obeys C2C concepts requiring the product/materials to be either biodegradable, or reusable/recyclable. The practice shows that building conservations can be sustainable when it is executed via adaptation—changing building's function and reusing it (VHF, 1995). The comparison of ICOMOS and C2C conservation processes show similar aims, steps, and tasks as shown in Table 14.2. Therefore, build cultural heritage reuse with its renovation/restoration, therefore, provides an opportunity of C2C concepts application for conservation process enhancement (Sayigh and Marafia, 1998).

The integrated approach—ICOMOS + C2C—for vernacular built cultural heritage conservation suggested by Cazacova (2016) is where the connection of tangible and intangible heritage are linked and applied in the selected case study in this chapter. This method considers build cultural heritage as a product, which is not demolished or recycled, but reused. By such integration, ICOMOS legislations stimulate vernacular and cultural heritage conservation; while C2C principles incorporated into conservation process, assist in vernacular heritage integration into planet biological cycle, where anything that would otherwise have resulted in waste is conceived as a nutrient for other technical or biological processes.

The guidelines for integrated ICOMOS + C2C approach are divided into three steps, which are described below (Cazacova, 2016):

TABLE 14.2 ICOMOS and C2C Building's Conservation Processes Comparison

Conservation process			
ICOMOS		C2C	
Understanding significance	Identify place and association—secure the place and make it safe	Statement of project's context establishment	**Phase 1 - Analysis**
	Gather and record information about the place sufficient to understand its significance—documentary, oral, physical	Challenges, opportunities, and actors identification	
	Assess significance	Formulation of C2C intensions	
	Prepare statement of significance	C2C building inspired elements selection	
Develop policy	Identify obligations arising from significance	Evaluation of the selected C2C building inspired elements	**Phase 2 - Evaluation**
	Gather information about other factors affecting the future of the place—owner/manager's needs, external factors, physical conditions	Assessment of materials' health and recycling potential	
	Develop policy—identify and consider opinions and test their impact on place significance	Beneficial content assessment	
	Prepare a statement of conservation policy	Implementation	
Manage	Manage the place in accordance to the policy—develop and implement strategies through a management plan, record the place prior to changes	Organizes the process of continuous work toward the goal of 100% positive impact	**Phase 3 - Optimization**
	Monitor and review any of the processes		

Step 1: Understanding significance and analysis. This step identifies the project aim and its context through understanding of the significance of the selected building and C2C inspired elements. The building significance and its value are assessed using ICOMOS guidelines and grading scale (ICOMOS, 2011), the statement of significance is prepared and the impact of changes to the building is defined. C2C intentions of the project are stated, building inventory is executed, building's C2C inspired elements identified, and each construction material pathways are traced.

Step 2: Developing policy and evaluation. Here the cultural significance policy is developed, the intensions of the conservation project are stated and conservation plan is designed, which states the goal of the conservation, its objectives and strategies. The Built Heritage physical condition assessment is executed, C2C inspired elements are evaluated, detailed inventory is prepared indicating each element ID number, its location within the building, its significance, overall impact of changes and materials' health, recycling potential and content assessment is implemented. During this step are also considered owner/management needs and the decision to conserve the building by adaptation/reuse is made. And finally, the conservation plan, where the goals, objectives, and strategies are stated, is prepared and its statement of implementation is explained.

Step 3: Manage and optimization. At this last step the conservation process is managed (from the beginning to the very end—from restoration/reconstruction to maintenance) in accordance to the previously developed conservation policy. Here C2C principles introduce the integration of innovative experienced contractors, and stakeholders' partnership to the conservation process toward the goal of 100% positive impact through conservation policy management. This stage also suggests technical resources and administrative staff selection, individual responsible for decision-making appointment, preparation of events sequences and financial resources allocation.

The tools to be used during the conservation process, which are required by the ICOMOS and C2C, are listed in Table 14.3.

14.3.4 International Case Study—Mirbat

As a case study a vernacular dwelling in old area of Mirbat town of Dhofar region in Sultanate of Oman has been selected where the innovative ICOMOS + C2C approach to Built Heritage conservation is applied, as was explained in the previous section. Previously conducted research by Cazacova et al. (2014) in the old area of Mirbat town, Dhofar region, Sultanate of Oman, where vernacular houses are located, identify the place as possessing cultural heritage value.

TABLE 14.3 ICOMOS and C2C Tools for Sustainable Heritage Conservation

	ICOMOS	C2C
Tools	Adaptation, reuse, compatible use	Actively beneficial qualities of the materials
	Detailed examination and recording	Defined product recycling/reusing content
	Use of culture knowledge, skills and prefer traditional techniques and materials	Defined use pathways
		Design for assembly, disassembly and reverse logistics
	Demolishing is not acceptable	Materials pooling
	Reserve place cultural value	Preferred ingredients list
		Integrate built environment's features to support cultural and conceptual diversity

Mirbat, a town of Dhofar province, which was founded in forth millennium BC, is located on the coast of the Indian Ocean. As all ports of the province, Mirbat was a chain in frankincense trade along with other ports of Dhofar and very well known also due to the export of Arabian horses and slaves (ICOMOS, 2000; Cazacova et al., 2014). Trading activities influenced the lifestyle of local population and the city's architecture, in which can still be traced the inspiration of Hadramaut style (see Fig. 14.1). Mirbat old town with its port, old bazar, castle, forts, mosques, and traditional housing is an outstanding example of Dhofari Islamic style in architecture.

As mentioned in the ICOMOS World Report 2000 on Monuments and Sites in Danger (ICOMOS, 2000) Sultanate of Oman, in 1970, started its great movement toward a most modern nation and as a result major changes occurred in Omanis' lifestyle. The desire of living a modern and more comfortable life suggested leaving traditional houses and customary lifestyle and move to newly built modern houses, well equipped and air-conditioned. The consequences of this great movement are the desertion of almost all traditional settlements with houses built of mud brick (or stone) on stone foundations with beautifully carved windows/doors and evolvement of new ones built of concrete with modern features and finishes.

The problem with many young and developing countries such as Oman is that traditional housing is seen as a regressive past and is not highly valued by the public. Hence, traditional settlements are not yet protected heritage places (Al-Belushi, 2013). The owners of the houses do not maintain them and, therefore these settlements appear as ghost cities with empty streets,

FIGURE 14.1 Collapsing Bayt al Siduf, residence in Mirbat, 2013. Bayt al Siduf is a traditional house built by a merchant in traditional Yemeni-style, mud-brick walls and hand-carved doors and windows. Source: *Photograph taken by author.*

collapsing roofs and walls, fading with wind and rain windows and doors. Little of the traditional houses are still withstanding, but in maximum 10 years will be eroded to a mass of clay and stone and disappear. The decay of the abandoned historic buildings of Mirbat were first mentioned and illustrated in Heritage at Risk 2000 issue (ICOMOS, 2000) and later on in Heritage at Risk (ICOMOS, 2006) and illustrated by Fig. 14.2. ICOMOS World Report 2000 on Monuments and Sites in Danger (ICOMOS, 2000; Cazacova, 2014) named Mirbat settlement and its traditional houses as vanishing heritage that are a record of the country's history, should be documented, and the most unique buildings and fabrics should be preserved for the coming generations.

According to Cazacova (2014) description, though the area of Mirbat town currently has been drastically expanded, its old town can still be recognized by the traditional residences. Initially, Mirbat consisted of a port, customhouse, a castle, two forts, a mosque, and 901 traditional buildings. At present 459 of the traditional residences were demolished and 311 modern residences build as a replacement. Furthermore, 39 of the old houses are ruined, 241 are abandoned, and only 162 are inhabited. Very little of the old

FIGURE 14.2 Abandoned and decaying vernacular residences, Mirbat, 2013. Source: *Photograph taken by author.*

houses are inhabited by their owners and maintained, otherwise the other residences are inhabited by foreign labors (see Fig. 14.3).

A typical Mirbat's traditional residence is of unique style and could be depicted as one/two storied building of sandy color, with an inner courtyard, and a modest facade, variously shaped and richly carved wooden windows/doors/shutters/screens, roof parapet crowned with stepped crenels/merlons, roof parapet's spots of various shape and projecting roof drains. The residences were built in the beginning of 20th century with the employment of local materials and local traditional construction skills, which were

FIGURE 14.3 View from the harbor to the old town of Mirbat, where the traditional houses are located, 2013. Source: *Photograph by author.*

developed by the local builders since ancient times and passed down from one to another generation.

The residence selected for this study is shown in Fig. 14.4 and identified as shown in the following map (see Fig. 14.5).

According to the designed methodology, this research consist of three steps: (1) understanding significance and analysis; (2) developing policy and Evaluation; (3) manage and optimization, wherein each step follows ICOMOS legislation requirements and incorporates C2C concepts and each step is detailed as follows (see Table 14.2).

14.3.4.1 Step 1: Understanding Significance (ICOMOS) and Analysis (C2C)

14.3.4.1.1 ICOMOS: Understanding building significance

Previously conducted research (Cazacova et al., 2014) by the students of Dhofar University describes the recognition of Mirbat town's old area as a historical environment, lists its attributes, states its cultural significance and suggests its further conservation.

The statement of significance prepared by the Cazacova et al. (2014) declares that the old area of Mirbat town which stretches along the sea and its designated by its organic street pattern, public places such as castle, two forts, port, customhouse, bazar, mosques, and vernacular dwellings of unique local Islamic style built in the beginning of 20th century, demonstrates the development of the local community and its society achievements in urban planning design, construction, and crafting as well as cultural traditions and inhabitants' life style at the certain period of time. Therefore, the place is considered to have historic and cultural value. On the other hand, the beauty of the structures indicates its aesthetic value. Mirbat vernacular houses' cultural heritage value is also appreciated by the Reports Heritage at Risk (ICOMOS, 2000, 2006).

One of the vernacular buildings from above-described area of Mirbat, which was selected for this study, has been assessed according to the criteria

FIGURE 14.4 Residence No.108, which was selected as case study, 2013. Source: *Photograph taken by author.*

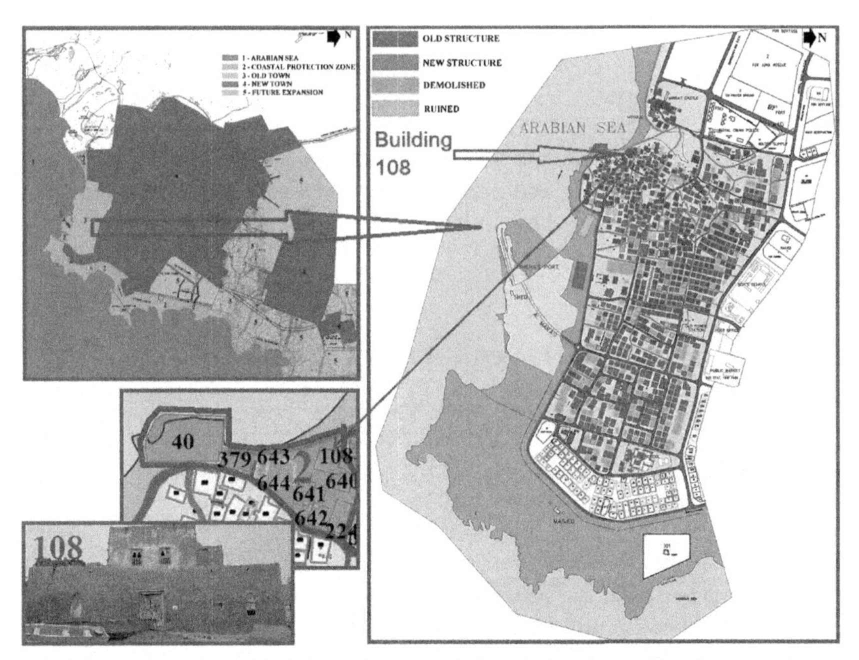

FIGURE 14.5 Residence No.108 location on Mirbat's map location and its view from the sea. Source: *Photographs and map by author.*

suggested by Kerr (2013). Its inventory form (see Table 14.4) was prepared, where the Built Heritage was given an identification number—BHR 108, and its statuary designation, brief description, location, photographs, physical condition, authenticity, integrity, interrelationship, and sensitivity were described. The value of the Built Heritage was defined using the following grading scale:

- Very high
- High
- Medium
- Low
- Negligible
- Unknown potential

BHR 108 is undesignated structure, it is of clear national importance and located next to the nationally designated structures (castle, forts, and residences) with standing remains, therefore is considered of high importance.

The development magnitude of impact assessment was executed according to the scale or severity of impacts or changes that are judged taking into account their direct and indirect effects and whether they are temporary or permanent, reversible or irreversible:

- No change
- Negligible change
- Minor change
- Moderate change
- Major change

Since BHR 108's elements were changed and appear slightly different and the building itself was noticeably changed, the magnitude of impact on BHR 108 is considered to have minor, temporary, reversible changes with indirect impact.

14.3.5 C2C: Inspired Elements Identification

With the aim of holistic quality achievement building's C2C inspired elements incorporate various goals and features, therefore is not compulsory for entire building or development to be C2C. As an alternative a focus on few elements that integrate desired goals to maximize their effectiveness is made. Selection of building C2C inspired elements considers the goals to be achieved (Mulhall and Braungart, 2010; Hussin et al., 2013).

According to the C2C concept for built environment each building is composed of several layers such as site, structure, skin, services, space plan, and stuff as shown in Fig. 14.6.

TABLE 14.4 Mirbat Selected Dwelling Inventory Form

Feature	Inventory description	Location/appearance
Unique ID number	BHR 108	Location—Mirbat old town
Asset name	Built Heritage	Map
Type of asset	Residence	
Date	March 22, 2013	
Statutory designation	108—Mirbat Municipality	
Brief description	The single story, circa 1960s local Islamic-style dwelling is characterized by sandy color, modest façade, variously shaped and richly carved wooden windows, carved wooden doors with and opening, roof parapet crowned with stepped crenels/merlons, roof parapet's spots and projecting coconut palm drains Heritage attributes, i.e., character-defining elements of the property include, but are not limited to: • Two stories scale • Islamic style • No foundations • Limestone with timber reinforcement walls • Flat roof • Various shapes and size wooden carved windows with shutters • Carved wooden doors with copper pins • Roof parapet with stepped crenels/merlons • Roof parapet spots • Projecting roof drains	South and East Façades
Condition	Partially ruined	
Authenticity	Historical, artistic, social	
Integrity	Residential quarters, castle, forts, market, mosques	

(Continued)

TABLE 14.4 (Continued)

Feature	Inventory description	Location/appearance
Inter-relationship	The dwelling is part of fabric (old residential quarters of the town bordered with a castle, two forts, and a market)	
Sensitivity	Area of built cultural heritage	
Importance	High	
Development magnitude of impact	Moderate changes	
Development significance of effect	Indirect impact, reversible changes	

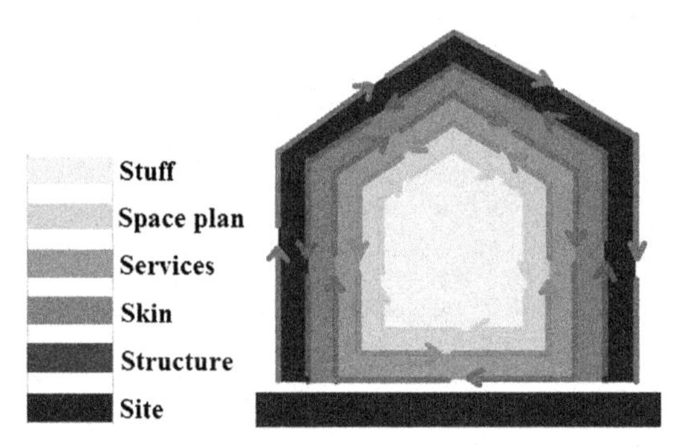

FIGURE 14.6 Cradle-to-cradle concept—building layers. Source: *Diagram by author.*

Hence, the goals of the research project, which are building conservation, building's structure, and skin, are studied, wherein the C2C inspired elements are identified. Building's C2C inspired elements' materials are examined, their pathways are traced, ingredients, and the amount of embodied energy are defined. Mirbat traditional building's structure and skin was fragmented into several components foundations, walls, roof/ceiling, floors, openings, windows, doors, interior and exterior drains and finishes (Fig. 14.7). Each component was studied independently and its materials and ingredients were identified, where the techniques of construction was also assessed as follows:

Foundations: No foundations for the house were built. Usually, the soil was extracted until the hard rock was reached, then the walls were built on it.

Walls: Exterior and interior walls were built of limestone, which was cut on the nearby mountains and carried by camels or donkeys to the construction site. The rocks were shaped into blocks of certain size using hand

FIGURE 14.7 Mirbat dwelling's structure and skin components. Source: *Photographs taken by author.*

tools—hammer and chisel. Mortar mud, simply made of the clay soil mixed with water, was used for wall construction. The clay soil was extracted nearby the construction site using grab hoe. The walls were also reinforcement by timber every third course. The branches of Tamarind trees were used for the reinforcement of stonewalls which grow in Hashir Mountains near Mirbat. The cut trees were brought to the side by camels or donkey, then the timber was cut using hand tools such as wood axe—locally refereed as Al Jarz—and crosscut saw.

Roof, ceiling, floor: For roof and ceilings spanning beams and purlins were used, which were made for local trees Samir and Tamarind. The cut trees or branches from the mountains were brought to the side by camels or donkeys. Crosscut saw, axe, hand planes, and daggers were used to make beams and purlins from the cut trees and branches. Coconut palm tree fiber was laid over the purlins and mixture of sand with local limestone plaster (norah) was applied. The floor finish on the ground floor was a compacted earth and on the upper level was plastered with norah.

Openings: The openings in the walls were supported by lintels, mostly made by local wood, which spanned at the bottom and the top of the openings. The arches were constructed using a frame where centering was made of palm fronds upon which the stones were laid to interlock and form a desired shape. The stones were fixed with mud mortar and the centering was removed after the mortar was cured. For small triangular arches two lime stone plates were arranged at 45 degree angle.

Finish: Finishing of the walls were very modest and both the interior and the exterior surfaces, the roof and the ceiling were finished by norah plaster.

Windows, doors, and shutters: The doors, windows, shutters, and screens were made from local wood by the local craftsman. Each window was made of fixed frame from outside and shutters that opened inside the house. All these features were hand made using similar tools used for wall construction (crosscut saw, hand planes, and carving knives, mallets). The hinges and pins were made of copper by the local craftsmen.

Sewerage drain canals and roof drains: The toilet was usually located on the upper floor and connected to a drain canal build of limestone and plastered with norah. The drain penetrated the exterior wall and run down along it. The roof drains were similarly constructed, but were made of coconut palm trunk, which was cut alongside and hollowed in order to obtain a pipe. The slope of the roof toward the drain canals was created using layers of norah. The coconut palm trees, which grow in Mirbat, were cut and brought to the side by camels or donkeys and then were hand crafted by locally made hand tools (wood axe, crosscut saw, and chisel).

Roof parapet merlons/crenels: The roof of the house was decorated with stepped parapet (limestone made merlons/crenels) and with spots protruding the parapet. These spots could be of triangular or rectangular shape (similar method as windows) and were used for observing the outside view or for draining the rainwater.

Table 14.5 is a summary of C2C inspired elements' materials examination, their pathways, ingredients and the amount of embodied energy identification.

14.3.5.1 Step 2: Developing Policy (ICOMOS) and Evaluation (C2C)

At this stage (a) a conservation policy is developed and the conservation plan is prepared, where the goal is stated, the objectives and strategies are listed; (b) the evaluation of the selected C2C building inspired elements is executed, the assessment of materials' health and recycling potential is done, the beneficial content is assessed; and finally, the statement of conservation policy is implemented.

14.3.5.1.1 Conservation Policy and Conservation Plan

The main goal of this project is Mirbat dwelling conservation by adaptation (the building is not demolished but reused). The objectives and strategies for conservation plan implementation are given in Table 14.6.

Reuse or Compatible Use?

a. Collaboration with the residence owner, local community, and authorities suggested giving a compatible use to the dwelling.

In the case of old Mirbat, the residents have already deserted the old houses and moved into newly built residences nearby. The owners of the

TABLE 14.5 Built Heritage BHR 108 Construction Materials, Their Pathways and Embodied Energy Identification

	Structural element	Materials	Origins	Manufacturing process	Transportation process	Embodied energy, MJ
1	Foundations	No	–	–	–	–
2	Columns	Limestone	Mirbat	Handmade	Camel/donkey	0
3	Walls	Limestone	Mirbat	Handmade	Camel/donkey	0
4	Mortar	Mud	Mirbat	Handmade	Camel/donkey	0
5		Reinforcement: Timber	Mirbat	Handmade	Camel/donkey	0
6	Plaster	Norah	Mirbat	Handmade	Camel/donkey	0
7	Ceiling/roof	Beams: Timber	Mirbat	Handmade	Camel/donkey	0
		Purlins: Timber	Mirbat	Handmade	Camel/donkey	0
		Coconut palm net	Mirbat	Handmade	Camel/donkey	0
		Mixture:				
		Sand +	Mirbat	Handmade	Camel/donkey	0
		Norah	Mirbat	Handmade	Camel/donkey	0
		Norah plaster	Mirbat	Handmade	Camel/donkey	0
8	Stairs	Limestone	Mirbat	Handmade	Camel/donkey	0
9	Openings	Lintels: Timber	Mirbat	Handmade	Camel/donkey	0
		Arch: Limestone	Mirbat	Handmade	Camel/donkey	0
		Centering: Timber	Mirbat	Handmade	Camel/donkey	0

(*Continued*)

TABLE 14.5 (Continued)

	Structural element	Materials	Origins	Manufacturing process	Transportation process	Embodied energy, MJ
10	Finish	Plaster: Norah	Mirbat	Handmade	Camel/donkey	0
11	Windows	Wood	Mirbat	Handmade	Camel/donkey	0
		Copper hinges	Mirbat	Handmade	Camel/donkey	0
12	Shutters	Wood	Mirbat	Handmade	Camel/donkey	0
		Copper hinges	Mirbat	Handmade	Camel/donkey	0
13	Doors	Wood	Mirbat	Handmade	Camel/donkey	0
		Copper pins	Mirbat	Handmade	Camel/donkey	0
		Copper hinges	Mirbat	Handmade	Camel/donkey	0
		Lock: wood, copper	Mirbat	Handmade	Camel/donkey	0
14	Screens	Wood	Mirbat	Handmade	Camel/donkey	0
		Copper hinges	Mirbat	Handmade	Camel/donkey	22
15	Drains canals	Limestone +	Mirbat	Handmade	Camel/donkey	0
		Norah plaster	Mirbat	Handmade	Camel/donkey	0
16	Roof drains	Coconut palm trunk	Mirbat	Handmade	Camel/donkey	0
17	Roof crenels and merlons	Limestone	Mirbat	Handmade	Camel/donkey	0
	Total Embodied Energy					22

TABLE 14.6 Built Heritage BHR 108 Conservation Plan, Its Goal, Objectives, and Strategies

Goal: Conservation by adaptation	
Objectives	**Strategies**
1. To reuse or compatible use	a. Collaborate with building's owner, community, and local authority b. Consider local, regional, and country legislations c. Reflect the association of the place
2. To assess residence physical condition	a. Involve experts in built cultural heritage assessment b. Invite local traditional builders c. Consult local craftsmen
3. To minimize the impact on building significance	a. Decide restoration or reconstruction b. If reconstruction—with or without the introduction of new materials c. If reconstruction with or without the introduction of modern materials or techniques
4. To execute detailed inventory of attributes and materials	a. List all attributes b. List each attribute material c. Identify materials pathways d. Identify and list each material's ingredients
5. To develop a conservation policy	a. Identify obligations arising from significance b. Involve local community—citizens, authorities, industry, funders
6. To identify options and test their impact on significance	a. In case of reconstruction with the introduction of new materials make sure there are similar and the same value salvaged components, attributes or materials b. In case of reconstruction with the introduction of modern materials and techniques select materials with the ingredients similar to the original one and that comply with C2C requirements c. Select a technique than will not diminish built cultural heritage significance d. Ensure the modern selected materials are locally available e. Ensure the new techniques can executed by the local builders
7. To prepare a statement of conservation policy	a. Communicate the conservation policy to the stakeholders

old deserted houses were suggested to have their old residence reconstructed for renting it during the khareef period as holiday home for tourists.[6]

b. At this stage, the local community and authorities' opinion was considered as well as the constraints of local legislations.[7] The visitors can spend their holiday in an old traditional house and experience the local cultural traditions, thus its cultural significance retains, the associations and meanings are respected and continuation of practice is provided (ICOMOS, 2013).

c. The new given function reflects the association of the place and makes the building accessible for public. This will promote local cultural Built Heritage, conserve it and keep it well maintained.

Residence Physical Condition Assessment

a. For the "residence physical condition assessment," experts in built cultural heritage assessment are involved. Building structure and skin's elements are examined. Alterations to the original structure and skin elements as well as those which are missing are identified.

Each element/attribute (e.g., exterior wall, interior wall, roof, window, door) is given and identification number, assessed independently, and its significance of the effect of change (i.e., the overall impact) are defined. The changes and the overall impact on attribute might be adverse or beneficial and is measured according to the following scale (Kerr, 2013):

- Major beneficial
- Moderate beneficial
- Minor beneficial
- Negligible beneficial
- Neutral
- Negligible adverse
- Minor adverse
- Moderate adverse
- Major adverse

---

6. It is important to mention here that Dhofar region is the single place on the Gulf peninsula, which experiences monsoon periods (khareef) that lasts from the end of June to the end of August. During this period thousands of tourists from Gulf countries, as well as from other nearby regions, visit Dhofar. However, the hospitality sector at the moment cannot handle these numbers and quite a few tourists end up sleeping outdoors. Consequently, the local population created another economy by renting their houses during the khareef season, which complements both this shortage of accommodation and the local residents' income.

7. In this particular case, the legislations of Sultanate of Oman and Dhofar Governorate are considered.

Building's plans were drawn, where each attribute was positioned and its identification number indicated (Figs. 14.8 and 14.9). The changes and the overall impact on each attribute assessment determine its conservation method: (a) no need of any interaction, (b) renovation needed, (c) restoration required, (d) reconstruction, or (e) replacement needed.

The assessment of each individual Built Heritage's attribute is recorded in a table where the changes and impacts are identified and methods of evaluation (qualitative/ quantitative) are indicated. As an example of such a report is Table 14.7, where the attribute W15-II (window, second floor) is evaluated.

b. The contribution of local traditional builders in building "physical conditions assessment" is essential. They can easily identify the alterations to the original elements and materials and can suggest the methods of renovation, restoration, and reconstruction with the application of the same as original materials and methods, same as the original.

c. Consultation with local craftsmen assists in recording the alternated and missing building elements such as windows, doors, shutters. The craftsmen also suggest the materials and methods of renovation, restoration, or reconstruction using original materials and methods. They also assist in substitution of missing elements.

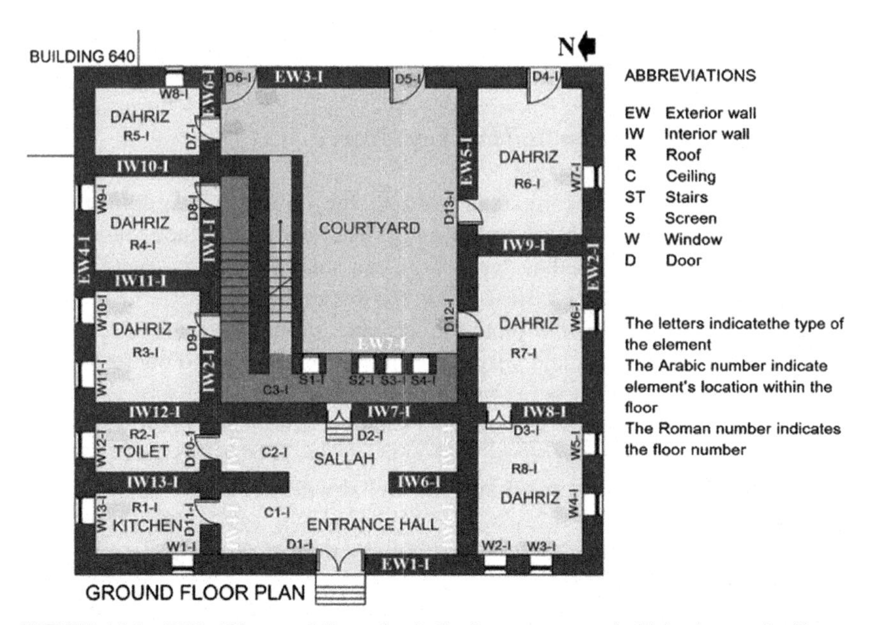

FIGURE 14.8 BHR 108 ground floor plan indicating structure and skin's elements/attributes, abbreviations, ID numbers and their location. Source: *Drawing by author.*

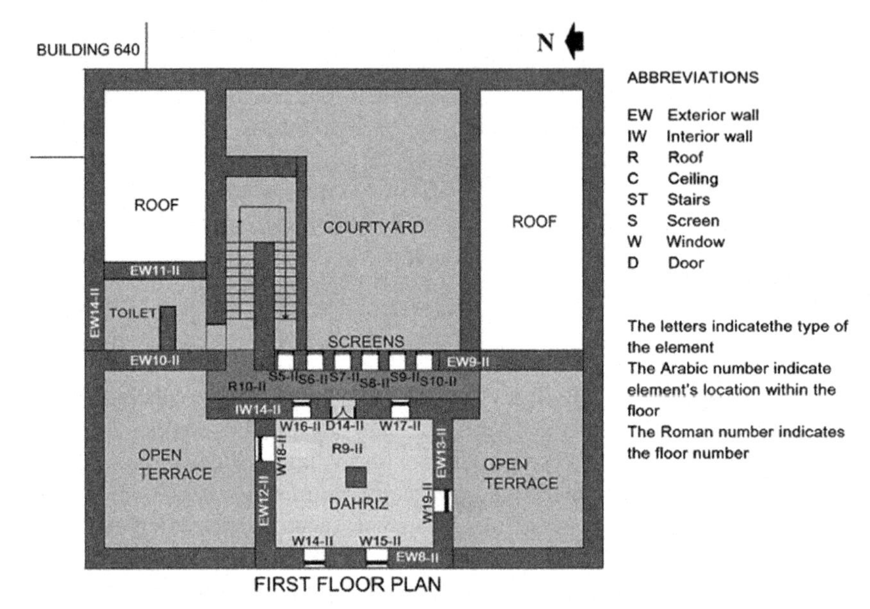

FIGURE 14.9 BHR 108 first floor plan indicating structure and skin's elements/attributes, abbreviations, ID numbers, and their location. Source: Drawing by author.

"Building and its attributes physical condition summary" describes that the impact of changes on some exterior walls are very large (demolished); interior walls—slight; ceiling and roofs—moderate/slight; stairs—neutral; doors—slight; on windows—moderate/large; screens—moderate/large.

Minimizing the Impact on Building Significance

a. Restoration or reconstruction.

It is decided to restore or reconstruct the building according to the results of the assessment of building physical condition. Since the impact of changes were assessed as "very large" on some major construction elements of BHR 108 a reconstruction is recommended.

b. Reconstruction—with or without the introduction of new materials.

Since there is large and very large impact of changes to on the building's attributes, the introduction of new materials is necessary (exterior wall cladding, windows screens and shutters, wall's screens). The new introduced materials are to be the same as the original, with the same ingredients and manufactured and assembled with traditional construction techniques.

c. Reconstruction with or without the introduction of modern materials or techniques.

The introduction of new modern material—window's glass (originally the windows were unglazed) between the outer screen and inner shutter—is

TABLE 14.7 Built Heritage BHR 108—Attribute W15-II Evaluation Report

Element	Window						
W15-II	Significance of impact					Evaluation methods	Photograph
	Neutral	Slight	Moderate/ slight	Moderate/ large	Large/very large		
Lintel	✓					Executed with the involvement of local builders	
Frame		✓				Executed with the involvement of local carpenters and craftsmen	
Exterior pane—screen			✓			Executed with the involvement of local carpenters and craftsmen	
Interior pane—shutters			✓			Executed with the involvement of local carpenters and craftsmen	
Finish				✓		Executed with the involvement of local carpenters and craftsmen	
	Suggested conservation method					Remarks	
	No any	Renovation	Renovation	Re construction	Re placement	Suggested reconstruction with the application of new material	

suggested for improving indoor comfort and air quality. Herein the employment of modern construction techniques is required. The added glass will not be seen from outside and, therefore, this will not affect the appearance of the building.

Detailed Inventory of Build Heritage BHR 108 Attributes and Materials' Execution

a. Listing all attributes.

The attributes of BHR 108 building can be listed as: exterior walls (EW), interior walls (IW), floors (F), ceilings (C), roof (R), stairs (ST), openings (O), windows (W), doors (D), screens (S), floor drains (FD), roof drains (RD) and roof's stepped parapet (RP).

b. Listing each attribute material.

All BHR 108 building materials are listed and recorded in Table 14.8. The list of materials is required to ensure that the new introduced materials and techniques for reconstruction are fabricated with the employment of identical to the original materials and with identical techniques and complies with C2C concepts.

c. Identifying materials' pathways.

As shown in Table 14.8 the materials employed in building construction are limited to limestone, wood/timber (Tamarind, Samir, and coconut palm), clay, water, dates molasses, coconut palm tree fiber, and sand.

BHR 108's main construction material is limestone, which is used for interior and exterior walls construction, stairs, drains, and roof decorative parapet and, as shown in Fig. 14.10, comes from the mountains in the region of Mirbat. The valleys (wadi) around the town are the source of clay, sand and water, which are mixed for mortar making, ceiling/roof construction and plaster mixture. The copper, that is also locally, available is used to make pins and hinges for the doors, windows, and shutters.

Timber is widely used in BHR 108's and other Built Heritage building's constructions: windows, doors, shutters, screens, lintels, walls' reinforcement, ceiling/roof structure, beams, and as centerings for the openings. Tamarind, Samir, and coconut palm trees are available in Mirbat. Coconut palm tree fiber is also applied for ceiling construction and for insulation for windows and doors. The molasses, which is used for plaster mixture, is provided by Date palm trees growing locally in Mirbat (Fig. 14.11).

d. Identifying and listing each material ingredients.

Here, as requested by C2C concepts, each material's ingredients are defined and their life cycle is traced. To comply with C2C concepts the dwelling has to be deconstructable and its materials have to be either decomposable or recyclable. In order to assess whether the selected C2C inspired elements—structure and skin—fulfill these requirements, a

TABLE 14.8 BHR 108 Attributes and Their Materials, Ingredients

ID	Attribute	Materials						
EW	Exterior walls	Limestone	Timber—Tamarind	Mortar—clay, water	Plaster—limestone, wood, water, dates molasses			
IW	Interior walls	Identical to EW						
F-I	FloorsGround floor	Compacted earth	–	–	–			
F-II	First floor	–	–	–	Plaster—limestone, wood, water, dates molasses			
C	Ceiling	–	Timber—Samir, tamarind	–	Plaster—limestone, wood, water, dates molasses	Coconut palm fiber	Sand	
R	Roof	Identical to C						
ST	Stairs	Limestone						
O	Openings	Limestone	Timber—Tamarind					
D	Doors		Wood—Samir					Copper
W	Windows	Identical to D						
S	Screens		Wood—Samir					
FD	Floor drain	Limestone			Plaster—limestone, wood, water, dates molasses			
RD	Roof drains		Timber—Coconut palm					
RP	Roof parapet	Identical to EW						

FIGURE 14.10 (A) Mirbat mountains—the source of limestone; (B) Mirbat valley—the source of clay, sand, and water; and (C) mountains—the source of copper. Source: *Diagrams by author.*

FIGURE 14.11 Tamarind, Samir, Coconut palm, and Date palm trees—sources of timber, fiber, and dates molasses. Source: *Diagram by author.*

life cycle diagram is drawn for each material and its ingredients (Figs. 14.12–14.16).

As seen from Fig. 14.12, the limestone is extracted from the nearby mountains, transported by camels to the construction site, where the cladding blocks for walls are cut. With limestone walls, the building can be easily deconstructed and the blocks could be recycled or returned to the nature.

Fig. 14.13 shows the life cycle of mud mortar and its ingredients—clay and water. During the mortar curing process the water evaporates, and in the case of building disassembly the clay returns to its biological cycle in the form of dust.

Fig. 14.14 explains the life cycle of Tamarind, Samir, and coconut palm tree timber. Wood products such as beams, lintels, windows, doors, shutters, and screens can be reused in the case of building disassembly or can serve as fertilizer for new trees. Copper, which is extracted locally, is used for pins, hinges, and fabrication locks, and can be recycled when the building is disassembled as shown in Fig. 14.15.

For exterior and interior finishes a special type of plaster—norah—invented by the local builders is used. Its production is a complex process where crashed limestone and wood are moisturized, heated, and left for long periods to make mortar powder. For plaster production, water and dates molasses are added. During plaster curing the water evaporates and when the building is disassembled the plaster is decomposed (can be used for tress fertilizer) (Fig. 14.16).

Developing the Conservation Policy

a. Identification of obligations arising from BHR 108 significance.

The building is a part of the place of cultural heritage and any interventions to it will have result as impact on its significance. The intervention to the surrounding area of the built cultural heritage should be managed in such a way that its impact is sympathetic with the value of the resources. When such intervention is inevitable for evading direct or indirect impacts on built cultural heritage mitigation actions are recommended. Mitigation can be defined as a process of minimizing expected adverse impacts to cultural

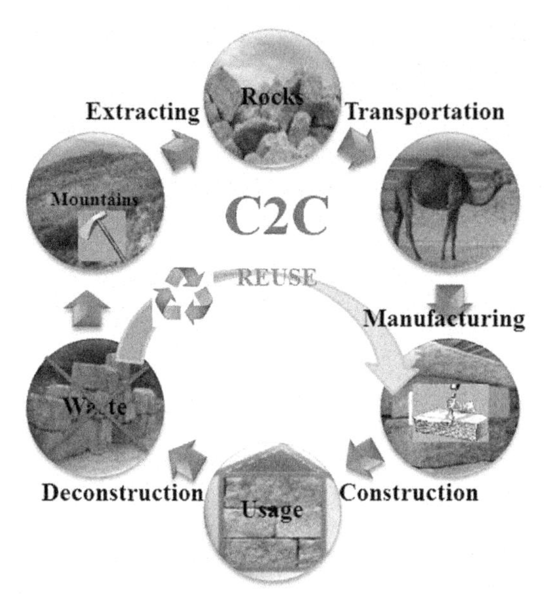

FIGURE 14.12 Limestone cycle. Source: *Diagram by author.*

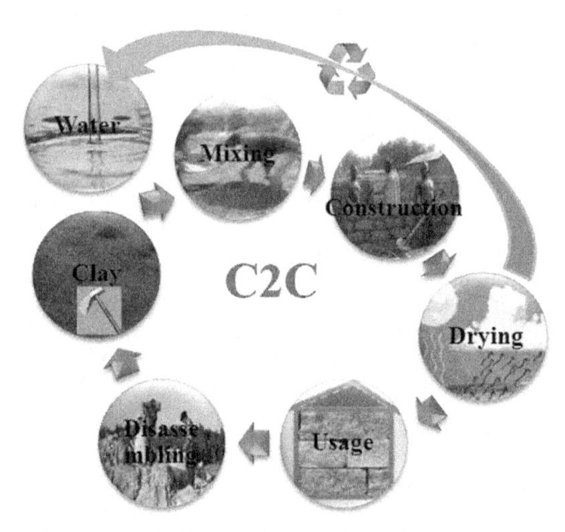

FIGURE 14.13 Mud mortar cycle. Source: *Diagram by author.*

heritage resources. This might include actions such as avoidance, monitoring, protection, relocation, corrective landscaping, documentation of the Built Heritage resource if to be demolished or relocated, and salvage of building materials.

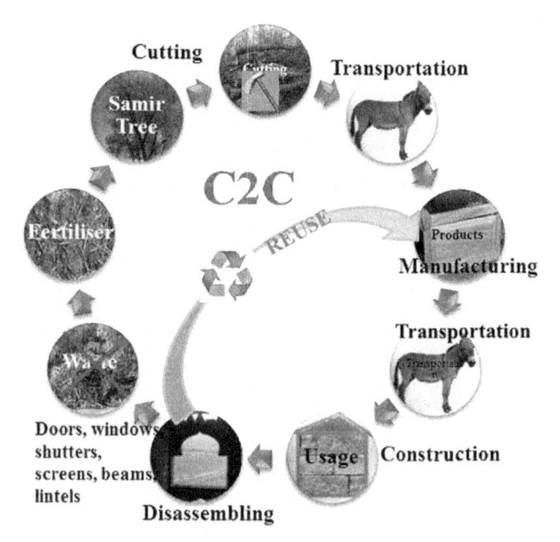

FIGURE 14.14 Wood cycle. Source: *Diagram by author.*

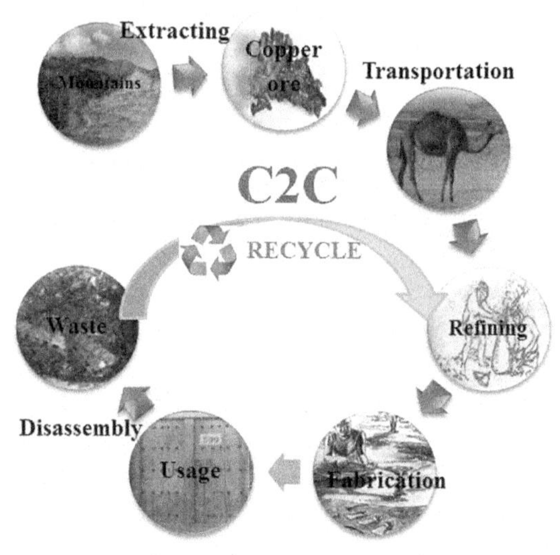

FIGURE 14.15 Copper cycle. Source: *Diagram by author.*

No any predicted direct impacts on BHR 108 are identified. Improvements of the roads and parking on the area are identified as indirect impacts. The old area of Mirbat town was planned for walking and for camels, horses, and donkeys, but not for cars. Therefore, the dirt roads, which are very narrow, are currently being widened and asphalted. There is also lack of parking space.

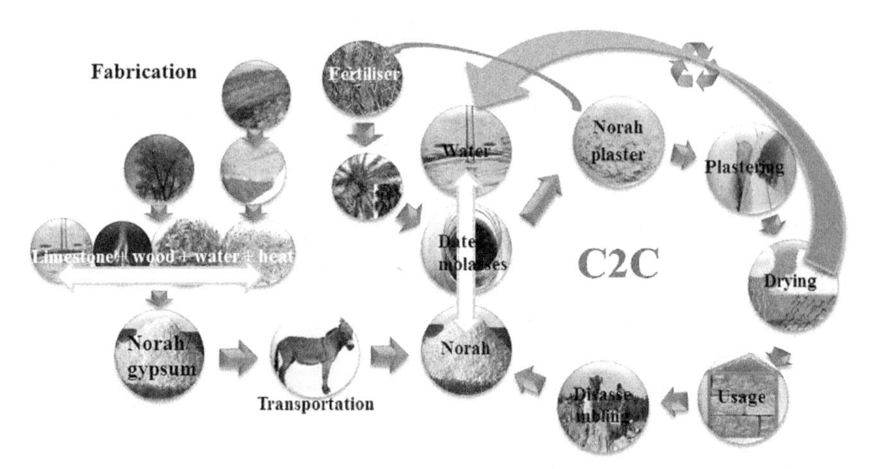

FIGURE 14.16 Norah plaster cycle. Source: *Diagram by author.*

An assessment of possible impacts to the BHR 108 which will negatively impact its significance was prepared and is listed as the following:

- Demolishing of other buildings of the place;
- Substitution of old buildings by new ones;
- Shadows created by possible new structures that might be higher will alter the appearance or change the visibility;
- Changes in landscaping of the surrounding environment might cause direct or indirect obstruction of dwelling view;
- Isolation of a heritage attribute from it surrounding environment, context, or a significant relationship;
- Change in land use as such to provide open space for residential use, allowing new development or site alteration to fill in the formerly open spaces;
- Improvements for the existing roads and parking;
- Excavations of the streets for draining system improvement.

For minimizing the impacts on BHR 108 significance, the following mitigation actions are recommended:

- Avoid demolishing of other buildings of the same fabric to substitute old buildings by new ones. The application of integrated ICOMOS + C2C approach guarantees the Built Heritage conservation by adaptation or new compatible use instead of demolishing. Thus, instead of generation construction waste that results from buildings demolition, the residences will continue serving local community.
- Avoid construction in the surrounding building area.
- Avoid changes in landscaping of the surrounding environment that might cause direct or indirect obstruction of dwelling view. Instead, a

suitable landscaping of the surrounding building area should be executed as a buffer. The new landscaping plan and all planted material and other should comply with C2C principles—only local vegetation should be planted.

- Do not isolate the building from it surrounding environment and context (fabric).
- Avoid change of land use and new buildings construction on the open spaces of the fabric.
- Improve the roads, parking, and draining system in a way that the impact on building significance is minimal. Only materials that comply with C2C principles should be employed. The techniques selected for improvement should have minimal impact on place significance.
- Keep clear the area of coastline in front of the building.

b. Involving local community

Conservation process and its management, which is explained by Articles 12, 26, and 29 (ICOMOS, 2013), engage participants for whom the place has special associations and meanings, or social/spiritual/cultural values. Residences' owners, citizens, authorities, industries, and funders are provided opportunity to contribute and are involved in conservation and management and take responsibilities for it. The organizations and individuals responsible for decisions are appointed and named in the statement of conservation policy, where specific responsibility for each decision is defined.

Identifying the Options and Test Their Impact on Significance

a. Reconstruction with the introduction of new materials and its impact on BHR 108 significance.

In this particular case, a reconstruction is an appropriate method of conservation. There are damaged attributes and alterations, but there is also enough evidence to reproduce the fabric to its earlier state, thus its cultural significance retains the same value (Article 20, ICOMOS, 2013). This can be achieved by taking advantages of all the local traditional builders' knowledge/skills and other disciplines, which can contribute to conservation of the fabric.

The impact on fabric significance can result from reconstruction with the (a) application of new material and (b) application of new techniques.

There are two options for reducing the impact resulting from reconstruction with the application with new materials:

1. Use similar and the same value recycled or salvaged from other places components, attributes or materials. Since all buildings of the place are considered as cultural heritage this option is not applicable. Disassembling other buildings of the place will lead to the detriment of its cultural significance, thus is not acceptable (Article 1, ICOMOS, 2013).

2. To reproduce the missing components or attributes using the same materials. This option is preferred and achievable. The involvement of local experienced builders (for walls, roof/ceilings, and drains), craftsmen (for windows, doors, shutters, screens, roof drains) and application of traditional techniques along with C2C scientists will benefit the conservation (Article 4, ICOMOS, 2013). The list of decomposing and recycling materials applied in BHR 108 structure and skin given in Fig. 14.17 assists in selection of identical materials, thus fabric's form, scale, character, color, and texture remain indistinguishable and its value is not diminished. However, the visual setting of the fabric and relationship remains the same, and all changes to the fabric are reversible and considered temporary (Article 15, ICOMOS, 2013).

Table 14.9 specifies the guidelines to be followed in selecting materials and their ingredients for achieving reconstruction's C2C compliance.

b. Reconstruction is acceptable only with a condition that (new) materials of similar ingredients to the original one that comply with C2C requirements are selected. New materials and techniques for reconstruction selection must comply with the guidelines given in Table 14.9.

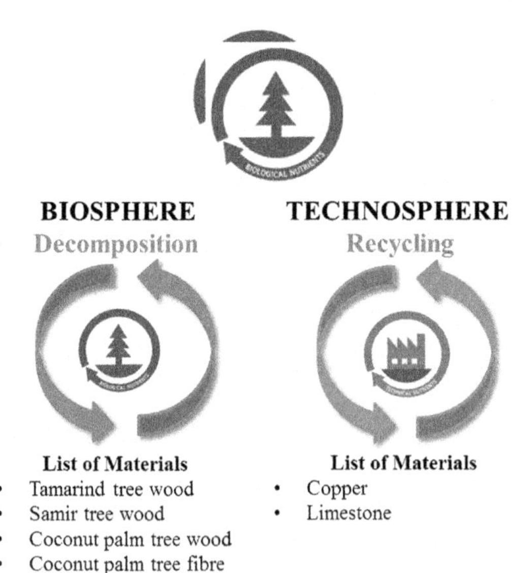

FIGURE 14.17 Mirbat dwelling's list of decomposing and recycling materials. Source: *Diagram by author.*

TABLE 14.9 BHR 108 Conservation Process—C2C Element and Its Relation to the Goal and Milestones (According to point 1, which was described above)

Element	
Define building (structure and skin) materials and their pathways	
Description	New materials employment: – With quality and contents that are measurable defined in biological or technical pathways through use and recovery – With impacts that are measurable beneficial for human health and the environment
Strategies	List the beneficial materials: – Preferred ingredients list – Defined use pathways and periods – Think of assembly, disassembly, (reverse logistics) reuse, recycle – Dedicated recycling lines – Actively beneficial qualities
Goal	The building conservation project to be carried out using the most advanced available level of defined materials and beneficial pathways
Milestones	The most advanced level of defined materials that can be reliable implemented to be identified by the planners prior the building conservation tenders delivery. Ensure that the leading companies in conservation field are aware of tendering process for this project

In this case study, to add a new material is suggested to the building's skin—glazing to the windows. The glass will be added in-between the exterior screen and the interior shutter. This will positively affect the indoor comfort (minimize the heat and dust entering the interior) and since the glass will be behind the exterior screen, this will not negatively affect the exterior of the building. As suggested in Table 14.6, glass cycle (Fig. 14.18) is followed and this assures that the material is recyclable and complies with C2C concepts.

c. Selecting a technique that will not diminish built cultural heritage significance.

For reproduction of ruined ceilings/roofs and walls, local builders are involved. The reproduction will follow the same to the original construction techniques. The employment of identical materials is ensured by the supervision of C2C scientists. Such integration of knowledge and skill will substantially benefit the conservation process.

d. Confirmation of availability of contemporary materials selected locally.

Only one new contemporary material is introduced to the construction process—glass. Glass manufacturers are available in Salalah, Dhofar region, which is only 75 km from Mirbat, therefore low embodied

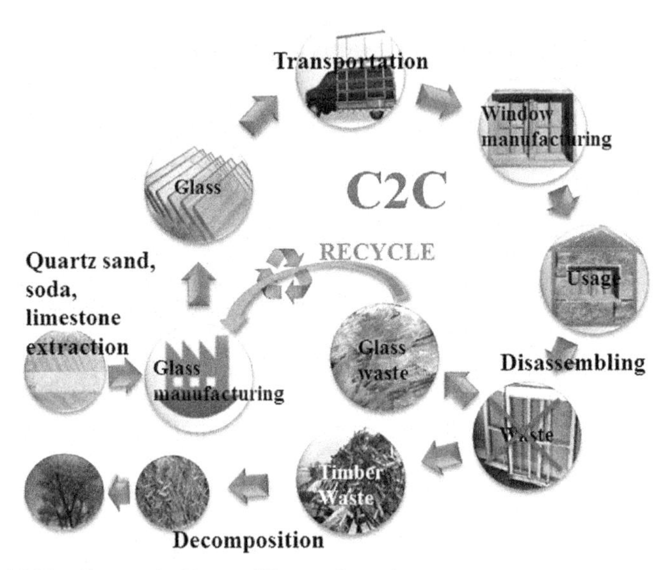

FIGURE 14.18 Glass cycle. Source: *Diagram by author.*

energy. The glass will be cut to the size at the plant and delivered by a truck to the site. The thickness of the glass and number of layers are suggested by C2C scientists, who will also supervise the installation. The installation will be executed by the local builders (Article 4, ICOMOS, 2013).

e. Confirmation of local builders' skills for new techniques execution.

A local construction company is selected that will provide a carpenter for glass installation. The installation is executed under the supervision of traditional craftsmen (carpenter) and C2C professional.

Preparing the Statement of Conservation Policy Considering the factors affecting the future of the place (owner, community and administrative bodies' needs, resources available and physical conditions) (Article 6, ICOMOS, 2013) a statement of Built Heritage—Mirbat dwelling BHR 108—conservation policy is prepared (Article 3, ICOMOS, 2013) which states how the short- and long-term conservation process will be carried out and includes the description of fabric and setting, use, interpretation, management, control into fabric intervention, future developments, and adaptation and review. Statement of conservation policy is prepared with the involvement of stakeholders for whom the place has special associations and meaning (Article 12, ICOMOS, 2013) and its brief is given below (ICOMOS, 2010):

- Fabric and setting description arises from the statement of significance and states that as fabric is considered BHR 108 and as setting (place) is the old area of Mirbat town. The map of the town where the location of the fabric is indicated is necessary. The conservation method suggested is adaptation (compatible use), which proposes minimal changes to the fabric and setting.

- A new compatible use of the fabric as holiday home is suggested. The place (setting) is suggested to be a touristic place. These uses will not diminish the value of the cultural significance and will make the fabric and setting accessible for the public.
- The cultural significance of the fabric and setting is not altered, indeed involves minimal changes and respects the associations and meaning of the place. The place remains as residential quarters with old traditional residences reflecting the local culture and life style during the period of the first half of 20th century.
- The management structure for conservation policy implementation is formed, wherein each body or person understands the significance of the heritage. For succeeding conservation and its management decisions, a cultural heritage department is formed under the supervision of the local municipality and ministry of cultural heritage. The department of cultural heritage takes responsibilities for place conservation management, while the owner of the building, which reports to the department of cultural heritage, is responsible for day-to-day management of the fabric (BHR 108). The management plan identifies the mechanism for decision-making and recording, as well as place and fabric security and maintenance provision.
- Control of intervention in the place and fabric is executed by a person from the department of cultural heritage, which is appointed by the ministry of cultural heritage. The conservation policy clearly specifies inevitable for conservation process interventions, the degree and nature of interventions, the impact of the intervention on cultural heritage significance, conservation proposals assessment, process of materials tests, and methods of information delivery to the stakeholders.
- Constrains that could limit the accessibility or investigation of the place and fabric are listed as land use and buildings' ownership, local culture (life style) and local legislations.
- The guidelines for place future development resulting from use change are designed.
- Provisions for review, which are essential part or conservation process, are made.
- Consequences of conservation policy are stated, where it is explained how the conservation policy will or will not impact the place and setting, its significance, locality and its amenity, owner and user, other involved stakeholders (Article 3, ICOMOS, 2013).

a. Communicate the conservation policy to the stakeholders.

All stakeholders (house owner, representative of local community, department of cultural heritage, municipality along with professionals in built cultural heritage conservation, sponsors, and local professionals) are members of conservation policy statement committee development. The members of the committee are responsible to deliver the conservation policy to all the stakeholders for whom the place has special associations and meaning.

14.4 STEP 3: MANAGE (ICOMOS) AND OPTIMIZATION (C2C)

Managing the conservation plan is the third and last step in the conservation process. The management is executed in accordance with the conservation policy prepared during the second step and includes building restoration or reconstruction, followed by its future maintenance. C2C principles introduce the third step, the integration of innovative experienced contractors, and stakeholders' partnership to the conservation process.

A strategy for conservation policy implementation is arranged, wherein the building owner opinion and its financial resources are considered, and the technical and other staff needed are selected. The sequences of events are listed and the conservation management structure is designed (ICOMOS, 2010). People with appropriate knowledge and skills should maintain the direction, supervision, and implementation of conservation plan at all stages. Therefore, the organizations and individuals responsible for the decisions are named during this stage and their responsibilities are specified. The resources needed for conservation are specified and the agreements with the suppliers are made. A log of conservation events and decisions should be kept and placed in a permanent archive, which is publicly accessible when is culturally appropriate.

14.5 DISCUSSIONS AND CONCLUSIONS

The high humans' demands for planet's resources are beyond planets' capacity to replenish them, and as a result of humans' actions the present-day environmental unbalance was created. To prevent the growth of environmental unbalance in the future, the present generations must well manage the planet's resources (Atia and De Herde, 2010). Therefore, Ortiz et al. (2009) urges governments and agencies to apply construction codes and policies for improving sustainability in construction and create environmental, social, and economic indicators for constructions' sustainability assessment. Furthermore, Khalaf (2015) calls the researchers to adopt a strategy for solving the problems related to the built environment for transforming "architecture into a celebration of a human ecological footprint with wholly positive effects."

For the last few decades, research focused on environmental design for building performance improvement over its lifespan are actively conducted (Yung and Chan, 2012). Whilst some researchers are working on methods to close the energy cycle of the building, others suggest methods to close the land, water, and building's materials cycle (Atia and De Herde, 2010). As a result of researchers collaborative efforts, indicators such as eco-footprint, life cycle and building adaptive reuse assessment have been introduced. Tools such LEED and the Sustainable Project Rating Tool (SPiRiT) are already applied for environmental performance of design proposals' assessment (Yung and Chan, 2012). Life cycle assessment (LCA) method also adds its contribution to sustainable constructions via assessing all stages of the building life cycle (Ortiz et al, 2009).

On the other hand, Bullen and Love (2011) considers that, due to the current existing buildings rules and conservation legislations, which might confine building's functioning, owners, developers are inactively reusing the heritage buildings. To find a solution to this problem, Property Council of Australia, for example, encourages heritage legislation to preserve only the best features of the building but not preserve the entire heritage building (Bullen and Love, 2011).

Given all the above-mentioned, green adaptive reuse of buildings (which embeds environmental design and technologies into existing buildings) turn out to be an imperative method (Yung and Chan, 2012) for sustainable construction, built environment improvement, and heritage conservation. The following examples show that latterly in the field of building adaptive reuse's assessment the researches are actively conducted. Conejos et al. (2012), for example, in his research suggests a model for buildings adaptive reuse's assessment called adaptSTAR. AdaptSTAR, similarly to Green Building Council's Green Star or LEED, measures the performance of adapted buildings based on following design criteria: technological, environmental, physical, functional, and, additionally, sociocultural (to be added for heritage buildings). Furthermore, Langston's (2012) research proposes a model for validation of the adaptive reuse potential (ARP) using iconCUR that identifies and ranks adaptive reuse potential in existing buildings based on estimation of building expected physical life, current age of the building, assessment of physical, economic, functional, technological, social, legal, and political conditions. According to Mora et al. (2011), tools as LCA can be successfully applied for heritage buildings' performance assessment and, as stated by Judson et al. (2014), the researches have to move toward approaches with integrated (combined) methods for performance assessment.

We have, in this chapter, shown that by recycling buildings instead of erecting new ones the architecture takes a new direction toward humans' ecological negative footprints minimization and demonstrated how sustainability can come to the fore as an important concept in Built Heritage conservation by illustrating a new integrated approach ICOMOS + C2C using a case study in Oman. Our comparison of ICOMOS and C2C conservation processes in this chapter highlighted that ICOMOS and C2C share similar aims, steps, and tasks, which inspired the suggested integrated ICOMOS + C2C approach to Built Heritage conservation. Our new approach follows ICOMOS legislation requirements and incorporates C2C concepts. ICOMOS + C2C considers build cultural heritage as a product, which is not demolished or recycled, but reused. By such integration ICOMOS legislations stimulate vernacular and cultural heritage conservation; while C2C principles incorporated into conservation process, assist in vernacular heritage integration into planet biological cycle.

Our case study shows that Built Heritage conservation can be culturally and environmentally sustainable, respectively, via adaptive reuse by applying and integrated ICOMOS + C2C approach and is a link between tangible and intangible heritage. The integrated ICOMOS + C2C approach is an important tool to manage conservation of Built Heritage. It explores and supports a life cycle approach in conservation where traditional Built Heritage is integrated in a larger strategy of sustainable management of the embodied energy, which in return is considered as integrated part of cultural heritage.

This research focuses on only two of the several layers (building's structure, building's skin, services, space plan and stuff) defined by C2C concept for the built environment, which are building's structure and skin (see Fig. 14.6) and takes them through the sustainable conservation process with integrated approach ICOMOS + C2C. However, for future work and research this method might be also applied for other layers (e.g., services, space plan) for building performance improvement (better indoor air quality, energy, and water self-sufficiency).

REFERENCES

Al-Belushi, M.A.K., 2013. The heritage prospective and urban expansion in capital cities: old defence sites in Muscat, Oman. WIT Trans. Built Environ. 131, 551–562.

Attia, S. & De Herde, A., 2010. Towards a definition for zero impact buildings. In: Proceedings of Sustainable Buildings CIB 2010, Maastricht, Netherlands, 2010. Retrieved from: <http://hdl.handle.net/2268/167561>.

Bergman, D., 2012. Sustainable Design: A Critical Guide. Princeton Architectural Press, New York, pp. 18–21.

Braungart, M., McDonough, W., 2002. Cradle to Cradle: Remaking the Way We Make things. North Point Press, New York.

Brundtland, G.H., 1987. Report of the World Commission on Environment and Development: Our Common Future. Oslo, 20 March 1987. Retrieved from: <http://www.un-documents.net/our-common-future.pdf>.

Bullen, P.A., Love, P.E.D., 2010. The rhetoric of adaptive reuse or reality of demolition: views from the field. Cities 27, 215–224. Retrieved from: https://doi.org/10.1016/j.cities.2009.12.005.

Bullen, P.A., Love, P.E.D., 2011. Adaptive reuse of heritage buildings. Struct. Surv. 29 (5), 411–421. Retrieved from: https://doi.org/10.1108/02630801111182439.

Cazacova, L., 2014. Traditional Housing Features Shaped by the Defensive City: Case study Mirbat. Unpublished Doctoral thesis, European University of Lefke.

Cazacova, L., 2015. Cradle to cradle concept for built cultural heritage protection. Recent advances in earth sciences, Environment and Development. Energy, Environmental and Structural Engineering Series 37, 127–134.

Cazacova, L., 2016. Sustainable built cultural heritage conservation with cradle to cradle. Int. J. Ener. Environ. 10, 61–69.

Cazacova, L., Hidalgo, M., Farhan, M., Al-Masikhi, F., Al-Hafidh, F., Qatan, H., et al. (2014). Recognition of Mirbat town's old area as historical environment. In: Proceedings of REHAB 2014 Conference. Tomar, Portugal, March, 19–21, 2014, Vol. II, pp. 581–591. doi: 10.14575/gl/rehab2014.

Cinieri, V. & Zamperini, E. (2013). Lifecycle oriented approach for sustainable preservation of historical Built Heritage. In: Built Heritage 2013 Monitoring Conservation Management, pp. 465–474. Retrieved from: <http://www.bh2013.polimi.it/papers/bh2013_paper_325.pdf>.

Conejos, S., Langston, C., Smith., J., 2012. AdaptSTAR model: a climate-friendly strategy to promote built environment sustainability. Habitat Int. 1 (9), 1–13. Retrieved from: <http://epublications.bond.edu.au/sustainable_development/93>.

Dale, J. (2007). Innovation in Construction: Ideas Are the Currency of the Future. CIOB Survey. Retrieved from: <http://www.ciob.org/sites/default/files/Innovation%20in%20Construction.pdf>.

Du Plessis, C., 2012. Towards a regenerative paradigm for the built environment. Build. Res. Inform. 40 (1), 7–22.

Godwin, P.J., 2011. Building conservation and sustainability in the United Kingdom. Proced. Eng. 20, 12–21. Retrieved from: https://doi.org/10.1016/j.proeng.2011.11.135.

Hines, M. (2011). The building conservation directory. Conservation in the age of sustainability. Retrieved from: <http://www.buildingconservation.com/articles/sustainable-conservation/sustainable-conservation.htm>.

Historic Resources Branch (HRB). (2007). Make History: Preserve Manitoba Past. Green Guide to Heritage Conservation. Report by Canada's Manitoba Culture, Heritage and Tourism. Retrieved from: <https://members.museumsontario.ca/sites/default/files/members/green-guide2010.pdf>.

Horvath, A. (2000). Construction for Sustainable development—A Research and Educational Agenda. Berkeley-Stanford Construction Engineering & Management Workshop: Defining a Research Agenda for AEC Process/Product Development in 2000 and Beyond, August 26–28, 1999. Retrieved from: <http://faculty.ce.berkeley.edu/tommelein/CEMworkshop/Horvath.pdf>.

Horvath, A., 2004. Construction materials and the environment. Annu, Rev. Environ. Res. 29, 181–204. Available from: https://doi.org/10.1146/annurev.energy.29.062403.102215.

Hussin, J.M., Rahman, I.A., Memon, A.H., 2013. The way forward in sustainable construction: issues and challenges. Int. J. Adv. Appl. Sci. 2 (1), 15–24. Retrieved from: <http://iaesjournal.com/online/index.php/IJAAS>.

ICOMOS. (1994). The Nara Document on Authenticity. Retrieved from: <http://www.icomos.org/charters/nara-e.pdf>.

ICOMOS. (2000). H@R 2000: Heritage at Risk. ICOMOS World Report 2000 on Monuments and Sites in Danger. Germany. Retrieved from: <http://www.icomos.org/risk/world_report/2000/oman_2000.htm>.

ICOMOS. (2006). H @ R 2006/07: Heritage at Risk. ICOMOS World Report 2006/2007 on Monuments and Sites in Danger. Paris, France. Retrieved from: <http://www.icomos.org/risk/world_report/2006-2007/pdf/H@R_2006-2007_web.pdf>.

ICOMOS. (2010). New Zealand Charter for the Conservation of Places of Cultural Heritage Value. Retrieved from: <http://www.icomos.org/charters/ICOMOS_NZ_Charter_2010_FINAL_11_Oct_2010.pdf>.

ICOMOS. (2011). Guidance on Heritage Impact Assessments for Cultural World Heritage Properties. A Publication of the International Council on Monuments and Site. Retrieved from: <http://www.icomos.org/world_heritage/HIA_20110201.pdf>.

ICOMOS. (2013). The Australia ICOMOS Charter for Places of Cultural Significance. The Burra Charter. Retrieved from: <http://australia.icomos.org/wp-content/uploads/The-Burra-Charter-2013-Adopted-31.10.2013.pdf>.

Judson, E.P., Iyer-Raniga, U., Horne, R., 2014. Greening heritage housing: understanding homeowners' renovation practices in Australia. J. Housing Built Environ. 29 (1), 61–78. Retrieved from: https://doi.org/10.1007/s10901-013-9340-y.

Judson, P. & Iyer-Raniga, U. (2010). Reinterpreting the value of built heritage for sustainable development. Retrieved from: <http://www.academia.edu/1475400/Judson_P._and_Iyer-Raniga_U._2010_Reinterpreting_the_value_of_built_heritage_for_sustainable_development>.

Kerr, J.S. (2013). Conservation Plan: A Guide to the Preparation of Conservation Plans for Places of European Cultural Significance, seventh ed. ICOMOS: Australia. Retrieved from: <http://australia.icomos.org/publications/the-conservation-plan/>.

Khalaf, R.W., 2015. The reconciliation of heritage conservation and development: the success of criteria in guiding the design and assessment of contemporary interventions in historic places. Int. J. Architect. 9 (1), 77–92. Retrieved from <http://www.archnet-ijar.net/index.php/IJAR/article/viewFile/504/pdf_28>.

Kibert, C.J. (1994). Sustainable construction. In: Proceedings of the First International Conference of CIB TG 16, November 6–9, 1994, Tampa, Florida. Centre for Construction and Environment, M.E. Rinker Sr. School of Building Construction, College of Architecture, University of Florida.

Kibert, C.J., 2013. Sustainable Construction: Green Building Design and Delivery, third ed. John Wiley & Sons, New Jersey.

Kriss. J. (2014). United States Green Building Council. LEED. What is Green Building? August 6, 2014. Retrieved from: <http://www.usgbc.org/articles/what-green-building>.

Kuhlman, T., Farrington, J., 2010. What is sustainability? Sustainability 2, 3436–3448. Retrieved from: https://doi.org/10.3390/su2113436.

Langston, C., 2012. Validation of the adaptive reuse potential (ARP) model using iconCUR. Facilities 30 (3–4), 105–123. Retrieved from: <http://epublications.bond.edu.au/sustainable_development/160>.

Lilletorget. (n.d.) Cradle to cradle objectives. Lilletorget—multi-disciplinary, design—and innovation competition. Retrieved from: <http://www.duurzaamgebouwd.nl/bookstore>.

Madhumathi, A., Vishnupriya, J., Vignesh, S., 2014. Sustainability of traditional rural mud houses in Tamilnadu, India: an analysis related to thermal comfort. J. Multidiscipl. Eng. Sci. Technol. 1 (5), 302–311. Retrieved from: <http://www.jmest.org/wp-content/uploads/JMESTN42350265.pdf>.

Mora, R., Bitsuamlak, G., Horvat, M., 2011. Integrated life-cycle design of building enclosures. Build. Environ. 46, 1469–1479.

Mulhall, D., Braungart, M., 2010. Cradle to Cradle® criteria for the built environment. Ekonomiaz 75 (3), 122–132.

Murray, C. (2012). The Heritage Council. February. Retrieved from: <https://is.muni.cz/el/1421/podzim2015/DU2727/Built_Heritage___Sustainable_Development.txt>.

Organisation for Economic Co-operation and Development—OECD. (2010). Better Policies for Better Lives. Report by OECD—The OECD at 50 and Beyond. Retrieved from: <www.oecd.org>.

Ortiz, O., Castells, F., Sonnemann, G., 2009. Sustainability in the construction industry: a review of recent developments based on LCA. Constr. Build. Mater. 23, 28–39. Retrieved from: https://doi.org/10.1016/j.conbuildmat.2007.11.012.

Pullen, T., 2013. The Sustainable Building Bible: Building Homes for a Greener World. Ovolo, UK, pp. 64–85.

Roberts, T., 2007. Historic preservation and green building: a lasting relationship. Building Green 16 (1), Retrieved from: <https://www2.buildinggreen.com/article/historic-preservation-and-green-building-lasting-relationship>.

Saleh, M.S., Alalouch, C., 2015. Towards sustainable construction in Oman: challenges and opportunities. Proced. Eng. 118, 177–184. Retrieved from: https://doi.org/10.1016/j.proeng.2015.08.416.

Sayigh, A., Marafia, A.H., 1998. Thermal comfort and the development of bioclimatic concept in building design. Renew. Sustain. Ener. Rev. 2 (1), 3–24.

Shipley, R., Utz, S., Parsons, M., 2006. Does adaptive reuse pay? A study of the business of building renovation in Ontario, Canada. Int. J. Heritage Study 12 (6), 505–520. doi.org/10.1080/13527250600940181. Retrieved from: <http://www.tandfonline.com/doi/abs/10.1080/13527250600940181?src = recsys&journalCode = rjhs20>.

Silvestre, J.D., De Brito, J., Pinheiro, M.D., 2014. Environmental impacts and benefits of the end-of-Life of building materials—calculation rules, results and contribution to a "cradle to cradle" life cycle. J. Cleaner Product. 66 (1), 37–45. Retrieved from: https://doi.org/10.1016/j.jclepro.2013.10.028.

Srinivas, H. (2015). What is a Green or Sustainable Building? Green Construction. Concept Note Series E-029, June. Retrieved from: <http://www.gdrc.org/uem/green-const/1-whatis.html>.

Sustainable Building and Construction (SBC): Facts and Figures. (2003). UNEP Industry and Environment. Sustainable Building and Construction, April–September 5. Retrieved from: <http://www.uneptie.org/media/review/vol26no2-3/005-098.pdf>.

Tam, V.W.Y. & Tam, C.M. (2006). A Review on the Viable Technology for Construction Waste Recycling. Resources, Conservation, Recycling 47 (6), 209–221. Retrieved from: https://doi.org/10.1016/j.resconrec.2005.12.002.

The Vancouver Heritage Foundation (VHF). (1995). New Life of Old Buildings: Your Green Guide to Heritage Conservation. Still Creek Press, Vancouver. Retrieved from: <http://www.vancouverheritagefoundation.org/wp-content/uploads/2014/11/VHF-GreenGuide-webbook.pdf>.

UN General Assembly. (2015). Transforming our world: the 2030 Agenda for Sustainable Development. Available from: <http://www.un.org/ga/search/view_doc.asp?symbol = A/RES/70/1&Lang = E>.

Wahlström, M., Laine-Ylijoki, J., Järnström, H., Kaartinen, T., Erlandsson, M., Cousins, A.P., et al.(2014). Environmentally Sustainable Construction Products and Materials – Assessment of Release and Emissions. Nordic Innovation Report 2014:03//March 2014. Nordic Innovation, Oslo. Retrieved from: <http://www.nordicinnovation.org/Global/_Publications/Reports/2014/Environmentally%20Sustainable%20Construction%20Products%20and%20Materials_Final_report.pdf>.

Whole Building Design Guide (WBDG). (2014). Sustainable Historic Preservation. Retrieved from: <https://www.wbdg.org/resources/sustainable_hp.php>.

Yung, E.H.K., Chan, E.H.W., 2012. Implementation challenges to the adaptive reuse of heritage buildings: towards the goals of sustainable, low carbon cities. Habitat Int. 36, 352–361. Retrieved from: https://doi.org/10.1016/j.habitatint.2011.11.001.

FURTHER READING

UN General Assembly. (1997). Report of the World Commission on Environment and Development: Our Common Future. Retrieved from: <http://www.un-documents.net/our-common-future.pdf>.

Workshop: Defining a Research Agenda for AEC Process/Product Development in 2000 and Beyond, August 26–28, 1999. Retrieved from: <http://www.ce.berkeley.edu/~tommelein/CEMworkshop/Horvath.pdf>.

Chapter 15

Conclusions

Vivian W.Y. Tam[1,2], Cuong N.N. Tran[1], Laura Almeida[1] and Khoa N. Le[1]
[1]*School of Computing, Engineering and Mathematics, Western Sydney University, Penrith, NSW, Australia,* [2]*College of Civil Engineering, Shenzhen University, Shenzhen, China*

It is not overexaggerating if someone says the entire world is surfing on a rising tide of construction sustainability. Thanks to the energetic progress of science and technology, the "green" approach in this industry has been established in the past few years by several methodologies. As stated in this book, the key of this revolution is not merely ideology changing but rather technology modernization in every stage of the building life.

That is a long and fantastically journey since the very first time man used the natural caverns to secure themselves from cruel climate and environment. This book basically presents the next chapter of this voyage that will let people synchronize in optimistic harmony with the environment. The future of the world's construction will be formed with the modernistic management in any phase of the project such as procurement, quality control, etc., or neoteric cost-effective technologies that manipulate sustainably in any green projects.

As many typical construction projects, green buildings also are developed in the similar implemented stages of a project and have a continuously increasing life span due to the reason that many progressive and modern technologies, respectively, have been applying in almost all of a project's phases. Besides the cost-effective achieving purpose, which is the particular aspect of any project, developers can derive "green" aspect for their project by utilizing sustainable material, emissions reduction, efficient land use as well as energy use, which is mentioned in Chapter 2.

The preliminary chapters demonstrate all the aspects that involve with the project implementation from the initial step such as planning, designing to the applicableness of revolutionized sustainable technologies in construction.

Current sustainable development approach which is discussed in Chapter 2, Current Management Approach, covers many aspects of human being such as social, environmental, economic pillars (Elkington, 1994) or

Sustainable Construction Technologies. DOI: https://doi.org/10.1016/B978-0-12-811749-1.00012-2

even includes institutional, cultural, human-oriented, and ecological components (Valentin and Spangenberg, 2000; Nurse, 2006; Herva et al., 2011).

In the direction of this approach, several sustainability indicators that are developed within the construction field include LEED (Leadership in Energy and Environmental Design, United States), BREEAM (The Building Research Establishment's Environmental Assessment Method, United Kingdom), and Green Star (Green Building Council, Australia and New Zealand) (Kov, 2013). These life cycle assessment (LCA) tools cover a range of construction projects from new-born buildings to renovation ones, and they are used for evaluating every phases from cradle to grave of a project (Graedel and Guth, 1990). Although LEED (USGBC, 2016) and Green Star (GBCA, 2015) use evaluation score assessment and BREEAM utilizes weighting measure method, these three rating systems have eight identical major evaluating credit criteria: location and transportation, energy, water, indoor atmosphere, material, waste and pollution, management, and others. Energy category occupies uppermost percentage in total score of these criteria for all three tools.

Along with building scheme, there are also some existing rating systems for infrastructure projects such as BEST-in-Highways, Envision, GreenLITES, Greenroads, I-LAST, and Invest. These systems comprise more sustainable aspects than construction field's measurements. However, again, energy consumption plays a significant proportion in both building and infrastructure evaluation systems.

The innovative management approach—as can be seen in Chapter 2, Current Management Approach, as well—will emerge from three key scientific advancements: modern methods of construction (MMC), which also called as prefabrication construction, Building Information Modeling (BIM), and the integrated project delivery process (IPD).

Prefabrication construction could be a ground-breaking movement by creating environmental attributes structure for building (Tam et al., 2015). The aggravated obstacle in any MMC-used projects in the early stages of the prefabrication innovation is the lack of appropriately training and the tremendous market overdemand, which lead total construction cost more expensive than conventional measures (Mao et al., 2016; Tam et al., 2015). Synchronization among all parties that involve to the project from planning to operating phase would be another critical challenge to apply prefabrication into the current developing construction stream owing to their fragmented responsibility and experience that contribute to each project's stage is different and hard collaborative (Molavi and Barral, 2016).

In order to acquire the comprehensive analysis of the environmental impacts of sustainability measures as well as cooperating harmoniously all parties involve with a project's process, the merely current option is use the combination of cloud computing platform, integrated project delivery, and

BIM technology with the aim to provoke an effectively comprehensive and sustainable project management panorama.

Therefore, in the green building context it is crucial to analyze the social–economic and environmental impacts a building will have in all the different steps of its life cycle, from "cradle to grave" (ISO, 2006), by means of an LCA that has as major stages: In Chapter 2, the goal and scope definition, where all the inputs and outputs of the construction process, are defined; life cycle inventory (LCI) analysis, accounting with all the resources and disposals (by means of waste or subproducts) involved during the construction process workflow; life cycle impact assessment (LCIA), establish potential environmental impact indicators related to features such as land use, energy consumption, climate change; and interpretation, which will allow the support of decision-making (Bayer et al., 2010).

Because sustainable buildings are using the approach to reduce energy consumption and provide a better environment for occupants, the trend of sustainable development in building design is irreversible (Liu et al., 2011). Consequently, a change of perspective in traditional construction management methodologies is eminent, where cross-disciplinary teams start working collaboratively during the early stages of a building project and implement a charrette in which all participants share a vision, the goals, and define the next steps for the project, by means of merged brainstorming, discussion and strategic development (Robichaud and Anantatmula, 2011; USGBC, 2011).

Hence, LCA is an useful tool that can be used in the design stages of a building project which will support decisions related with design options, structural systems, products and/or assemblies, environmentally life cycle impact stages, etc. (Bayer et al., 2010). LCADesign is an available tool to evaluate the environmental impacts of materials at the design assessment (Seo et al., 2007) and BIM is expected to become a motor force for the implementation of a LCA perspective (Bayer et al., 2010) a building construction project.

However, sustainable development has a long way to go and the need of a different management approach, not only focused in costs but also in social and environmental characteristics of a project is fundamental (Shen et al., 2010). Some challenges have yet to be overcome in sustainable development, such as high costs, technical difficulties, lack of communication, bigger extension of time in green features approval, and construction practices (Tagaza and Wilson, 2004). But, as referred previously, the existence of an integrated team (with everyone involved in the building project, such as architects, engineers, owners) working together in a like mind context, since the early stages of a project, is the key element to achieve the desired sustainability goals and the success of a project (USGBC, 2011).

In conjunction with financial and policy support from governments, green development plan will be utilizing these pioneering methods in the bid to reduce energy consumption as well as greenhouse gas (GHG) emissions in the life cycle of building (Hoffert et al., 2002).

Management approaches are also analyzed more deeply in sub Chapter 2, Management, in order to achieve the goals of sustainable revolution. In comparison with conventional management methods, which are quite shattered and overlapping in a construction project organization, sustainable management strategies create the effective and efficient links amongst relevant parties in a project that improve continuously in this project's lifetime.

In this Chapter 2, Current Management Approach, five major strategies for sustainable management are presented to cover all stages of a building life cycle from commencement to maintenance phase. In order to ignite attention of both authorities and entrepreneurs towards sustainability, "Sustainable Procurement" improves the opportunity to obtain the appropriate equilibrium among three sustainable development pillars via the cooperation of governance policies, business mechanism, and project management.

Communication and connection improvement between parties in a project will make a form of new collaboration and lead to reduce time and energy in the early phase of this project. This integrative process builds up a combination of activities that implement sustainability methods. Environmental management system is also one of dimensions that prevent massive destructive effects of the construction industry on environment. ISO 14000 family of standards is a system that used by construction organizations in their sustainability efforts. An update version of ISO 14001 has been published in 2015 and provided a sophisticated structure management system for green building project.

LCA becomes a trendy propensity as well as an inevitable demand of construction projects recently. This demand derives from the fact that the environmental effects of construction have become hazardous to human habitat. In terms of CO_2 emission reduction and cost management, LCA supports designers and engineers in a "cradle-to-grave" approach for deciding which solution will be the optimum energy efficiency and ecological choice for their project (Graedel and Guth, 1990). Discussions in Chapter 2 (Current Management Approach) elucidate that not merely new construction projects are applied LCA but also housing refurbishment sector could be considered to utilize this method to evaluate the environmental impacts of every elements related to an entire house throughout its lifespan.

Fig. 15.1 shows that life cycle costing (LCC) and LCA simultaneously contribute to BIM by using and updating data of financial and environmental impacts through the life span of projects gradually. Both financial implications and ecological impacts that are ignited from production phase to demolition stage of the building have a substantial and coherent relationship in order to build reliable datasets for robust BIM development in both brand new as well as refurbished projects.

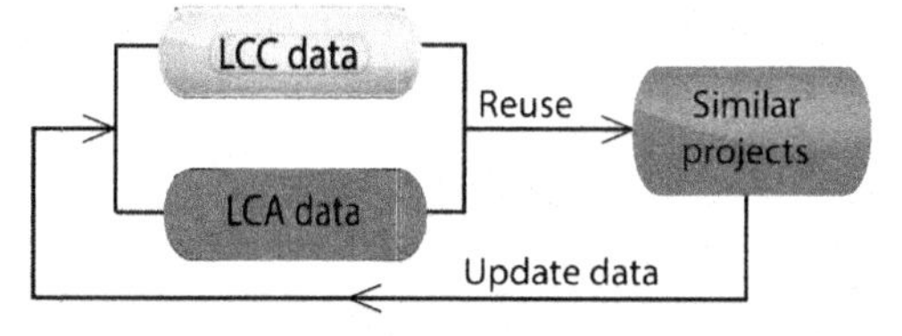

FIGURE 15.1 Life cycle assessment (LCA) and life cycle costing (LCC) in Building Information Modeling (BIM).

As mentioned in Chapter 13, BIM is applied as a whole-building tool in United Kingdom for renovated building projects. It is emphasized that in the early design phase, BIM plays a vital and significant role for decision makers to choose the appropriate refurbishment option. This is due to the reason that the more project information gathers from the beginning of renovation project, the better alternative solutions are recommended perfectly.

Sustainability enhancement benefit is the other benefit that BIM generates for refurbishment project. Energy performance and embodied carbon emission are determined easier in the early stage of project by BIM than traditional approach caused by recently BIM can supply more detail and fundamental analysis material for designers than any conventional project development methods. Furthermore, the consonance amongst every professionals involved with the project will make the procurement and construction processes smoothly and effectively as well as reduce the unnecessary contribution throughout the project's lifespan.

Chapter 13 introduces a housing refurbishment case study using BIM simulation environment in order to estimate LCC and LCA. The 3D BIM model using IES VE/IMPACT platform can satisfy all possible building energy variations that required in renovation projects by utilizing a range of sustainable data sources (Hammond and Jones, 2011; BRE, 2013). This research has recommended the best refurbished effective option is the most cost and carbon effectiveness approach that consensually granted by professional practitioners. The comparison amongst LCC and LCA results caused by diverse energy standards leads to the optimum method for building envelope renovation.

Analysis on different house types also shows that the semidetached/end terrace building is considered as the most "green-building" option for refurbishment while the detached house type consumes the most considerable embodied energy.

Certainly, there is no perfect choice for everything even though BIM is the best selection at this moment by stimulating LCA and LCC, as well as BIM is one of the options to calculate life-cycle assessment, life-cycle cost and life-cycle energy consumption. The difficulties to utilize BIM in reality are data integration and harmonized standard system between different data source.

Architecture and engineering historical data has been built in 2D design schemes for a long time. This is the challenge for anyone who desires to translate this enormous database into BIM. Likewise, each organization has its own standard structure to manage construction projects. That will make difficult to merge database from one's system to the other.

There are sustainable products and technologies, addressed in Chapter 12, Lean Principles in Construction, that are currently being explored in the construction sector in order to reduce the GHG emissions either related with embodied energy as with the use of energy, such as:

- Wood, biofuel;
- Substitute or combine conventional materials with lower environmental impact and natural materials;
- Implement heat recovery, renewable energies, and energy efficiency strategies in the production process of cement, reduce the need of carbon-intensive cement products, as well as use low-carbon cement by adding supplementary cementitious materials (fly ash, silica fume, sewage, etc.), replace carbonate-containing materials with noncarbonate materials and change the clinker ratio in cement production;
- Through energy efficient processes and equipment in the production of steel, as well as by adopting a 3 R (reduce, reuse, recycle) approach in the iron and steel sector;
- New and better technologies like Ultra-Low Carbon Dioxide Steelmaking (ULCOS), development of phase change materials (PCMs) and carbon capture and storage.

Chapter 9 presents the strategies that are currently being used to reduce GHG emissions are:

- Use low impact materials;
- Extent building life span;
- Maximize design of building structures;
- Improve project delivery onsite;
- Increase reuse and recycling of materials.

In a sustainable construction paradigm materials have tremendous relevance. From their physical and thermal characteristics to their source location and its recyclable properties are issues that have to be accounted in the context of green buildings. Furthermore, materials have to promote resources efficiency and in their life cycle be energy efficient and be associated with low carbon emissions. Moreover, materials have to be thermally efficient, be energy efficiently produced and used, and have environmentally low toxicity.

From the LCA perspective the impacts that materials have environmentally are measured in terms of indexes for energy and resources use as well as carbon emissions. Another important factor is the embodied energy and carbon emissions inherent to materials' production, extraction, and transport processes. In order to assess the sustainability of material and building construction processes there are already some existing tools, such as (Cole, 2003):

- Environmental impact assessment (EIA);
- Building environmental rating systems;
- Life cycle assessment.

The impacts materials have in the environment, in terms of an LCA and as addressed in Chapter 10, Emissions, may be as defined in terms of toxicity levels as per the United States Environmental Protection Agency (USEPA, 2006), Fig. 15.2, and in terms of energy use and carbon emissions associated with each different phase of the life cycle of materials.

As result of energy efficiency measures in the operational phase of a building, the embodied energy has been suffering an increment (Sartori and Hestnes, 2007). In terms of materials, the main intensive energy and carbon components in a building are the substructure, superstructure, envelope, roofing, and finishes (RICS, 2012). Therefore, new approaches to a sustainable construction include:

- Improvement in the production of materials;
- Recycling of waste and therefore contribute to its reduction;

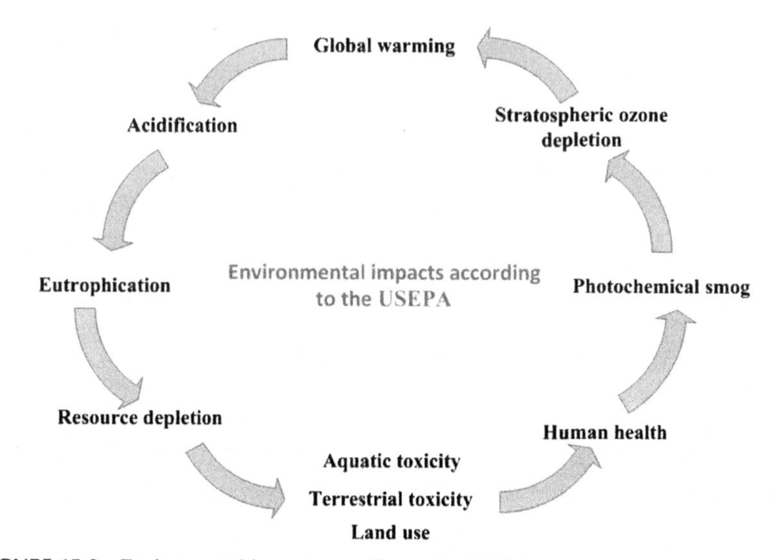

FIGURE 15.2 Environmental impacts according to the USEPA.

- Substitution of materials;
- New construction methods and use of building materials;
- Use of new high-performance materials;
- Use of alternative and renewable building materials.

Material procurement decisions, as mentioned in Chapter 8, Sustainable Water Use in Construction, start early at design and planning stages of a building and, due to the impact materials have in the building's life cycle, these decisions have to be carefully conscious. This fact will instigate the search for alternative materials either new or improved ones, more environmentally friendly, and consequently, the innovation of products in all industries, including the construction one (Renz and Solas, 2016).

In order to implement a sustainable procurement and transportation assessment in a construction project, it will be necessary to fully understand the sustainability concepts and goals in order to minimize possible difficulties and uncertainties relating with the implementation of new technologies and changes. These uncertainties can be classified in Chapter 5, in terms of:

- Time horizon;
- Information;
- Evaluating process;
- Decision makers;
- Material supply.

Furthermore, the demand and the deep impacts that the construction sector has on the environment led to a transformation in this sector's policies and the implementation of sustainable technologies and practices.

The search for alternative sources of energy, besides fossil fuels, and the growing environmental awareness "opened the door" for the renewable energy market. Moreover, specifically in Australia, this fact created an increasing interest in the photovoltaic (PV) and the geothermal heat pump (GHP) systems that were addressed in Chapter 4.

In a building integrated photovoltaics (BIPV) system the PV modules may be mounted in the roof or façade of a building, are considered a functional part of the building structure that enhances the building's appearance and may be used for curtain walls, awnings, windows, and skylights (Peng et al., 2011). BIPV systems are divided in PV foils, tiles, modules, and solar cell glazing, as well as wafer-based technologies and thin-film technologies (Jelle et al., 2012). These systems may be used as a stand-alone or integrated with the grid.

GHP is the most energy efficient renewable technology and as a heat pump is able to provide both heating and cooling by taking advantage of the stable temperatures in the earth. There are two available GHP systems configurations, which allow the system to be assembled vertically or horizontally and depend directly on the type of rock and soil:

- Open-looped (ground-coupled) system that is recommended in climates with moderate temperature variations and the existence of surface water.
- Close-looped (groundwater) system that is allowed transferring the ground temperature to the GHP system by underground continuous piping loops (Cui et al., 2015).

Previous technologies have potential barriers that need to be overcome before their implementation and therefore it is essential to do a previous comprehension research in relation with product and design considerations, installation constrains, high initial costs, social barriers, and lack of support from governments.

In the sustainable development sequence, Chapter 8, Sustainable Water use in Construction, and Chapter 9, Materials describes water as an element that has a vital role in both direct construction activities as well as the production process of construction materials. To obtain green building goals, besides saving energy, water should be used effectively during the life cycle of a building. From construction phase to operation stage of a project, water needs to be carefully conserved and utilized.

In construction phase, pollutants could be flown out with water from the building sites to surroundings via the stormwater system. Therefore, construction runoff should be managed by site management methods, which include but not limited stockpile management, drainage control, erosion control, and sediment control.

This chapter also mentions some measures that preserve as well as reduce water use in construction and mining industries. Different to water use in residential buildings which is primarily related to human activities, amount of water usage in mining industries depends on the mine's habitat and type of mining operations.

Rainwater and wastewater are sources of water that could be used for consumptions. Both of them should be considered to be a part of the planning and design of a building construction. Although the water treatment system, which includes collection, treatment, storage, and distribution, is a financial matter of a house project, investors could be optimistic about the monetary benefit of this system through life cycle cost assessment (LCCA).

The relationship between GHG emission and water used within a building is very strong due to the energy utilization for the water treatment system operation. Saving energy for water treating means less GHG emission.

Buildings have a major impact in the environment not only in terms of materials and energy use but also they represent 19% of the global GHG emissions. The reduction of GHG emissions in the operation stage of buildings put the embodied missions "on spot" which are directly related with materials (Chapter 10, Emissions).

In the context of an LCA, GHG emissions are measured in terms of global warm potential (GWP) and carbon dioxide (CO_2) is used as a reference for GHG effects (Cole, 1999) and all other gases are related to an equivalent mass of CO_2 and therefore referred as CO_2 equivalent ($CO_{2\text{-}e}$) for a 100-year timeframe (IPCC, 2014), as referred in Chapter 12, Lean Principles in Construction.

In order to mitigate the GHG emissions, policies were created but focused essentially on the operating GHG emissions related with energy use in lighting, ventilation, heating, cooling and equipment, leaving the embodied emissions as a gap. Therefore, the International Organization for Standardization (ISO) developed the 14000 standard series to fill up this gap by creating guidelines and framework for the LCA study. The latest 14000 series (ISO 14064 and 14065) defined requirements for GHG validation and verification bodies for accreditation and three different parts to evaluate the GHG emissions:

- Part 1: is related with GHG emissions quantification at organizational level;
- Part 2: quantifies emissions at a project level;
- Part 3: validates and verifies the assertion of GHG emissions.

This goes in line with the GHG protocol, which aims to measure, manage, and report GHG emissions in products and services, as well as establish a framework and requirements for public reporting on GHG emissions. Furthermore, in order to specify the assessment of goods and services in 2008 the Publicly Available Specification (PAS) 2050 was created by the British Standard Institution (BSI, 2011; Sinden, 2009).

In order to evaluate the carbon footprint (CF), which is another approach to measure GHG emissions at several scales (micro to macro), the World Resources Institute (WRI) developed three scopes to define the boundaries of CF assessment (Ranganathan et al., 2004).

- Scope 1 Direct Emissions—Accounts the emissions linked to the combustion of fossil fuels produced in-boundary.
- Scope 2 Indirect Emissions—Includes the indirect emissions associated with energy produced out of the limits of the defined boundary (electricity from the grid).
- Scope 3 Indirect Emissions—Not only includes the previous two scopes, but also adds the whole indirect emission from processes of the supply chain life cycle for materials and energy carriers produced outside the boundary.

In order to calculate the GHG emissions and the consequent environmental impact the existing methods develop an inventory analysis of the activities performed in a building in its life cycle (initial raw material extraction and production, transportation, construction, operation and demolition at the end of the life cycle (Yan et al., 2010)). Emissions may be determined in three different methods:

1. Process-based, also known as a bottom-up approach, where for each process involved in a building life cycle the energy and the materials used are accounted and converted into GHG emissions.
2. Economic input–output (EIO) analysis-based, also known as a top-down approach, considering the direct and indirect impacts of a product or service of the entire economic supply chain (Gerilla et al., 2007; Omar et al., 2014; Yan et al., 2010).
3. Hybrid method that combines the previous two methods in which the total GHG emissions are the sum of both methods.

In line with an LCA approach, there are already available software tools to calculate the GHG emissions, such as GaBi, SimaPro, Athena Eco Calculator, eTool (de Wolf et al., 2016). The LCA process framework of a building accounts emissions from the following stages:

- Initial emissions for all the processes of materials/products (embodied emissions), including transportation;
- Emissions of fuel consumption for plants and equipment used for construction activities onsite;
- Embodied emissions of materials/products used and emission due to energy use in the operating stage;
- Emissions of fuel for plants and equipment used in the demolition of buildings at the end of life stage;
- Emissions of fuel used for the transport of materials/products from distribution centers to building sites, or for disposal of construction and demolition waste to landfill or recycling facilities.

In the end-of-life stage in the construction industry, from an LCA perspective referred as cradle-to-grave, emissions related to reuse and recycling are not included.

The construction industry has some challenges to overcome mainly related with performance and surplus use of resources (Coal Industry, 2002; Oxford Economics, 2015). Being the biggest consumer of materials in the world (Van Odijk and Van Bovene, 2014), its impacts in the environment are also extremely significant representing around 30% of the total GHG emissions worldwide which the major part is related to embodied carbon. Due to the fact that low energy buildings are using more 20%–30% of embodied energy (Sartori and Hestnes, 2007; Thormark, 2007), it is imperative to implement more efficient and environmental friendly material procurement approaches and, consequently, its management. As transportation is the second contributor to GHG emissions, it is important for a good planning of the construction materials transportation. Moreover, a sustainable materials management should analyze economically, environmentally, and socially the impacts materials have in the life cycle of a building in order to be able to reduce waste in the construction sector. A sustainable material procurement approach will require that materials are carefully chosen taking into account embodied energy and carbon as well as the depletion of natural resources.

One of the misunderstandings of sustainability is that it is only related purely with environmental goals (Robinson, 2004; Seghezzo, 2009) while actually reconciles economic, social, and environmental objectives (Lozano, 2008). Moreover, from a new perspective approach, all the previous objectives and its interactions have to be contextualized in a specific a timeframe of a system with predefined boundaries.

Built Heritage, which is named after old historic buildings, should be preserved in sustainable plan (Chapter 14: Bridging Sustainable Construction Technologies and Heritage). The civilizing bridge between future generations and us cannot connect to each other as the absence of antiquated architecture preservation.

In Chapter 11, ICOMOS and C2C (cradle-to-cradle) concept are presented as environmental approaches to preserve cultural heritage. Whilst ICOMOS heritage conservation approach consider the value of legitimate standards to conserve memorial symbols, innovative C2C movement is defined as a biometric process of using one stage's waste as the other stage's material. The integration of these two measures creates a tremendously powerful tool that nourishes construction waste for other phases of life cycle of heritage buildings.

A case study in Chapter 11 describes a practical application of ICOMOS + C2C integration to a vernacular residential building in Mirbat, Dhofar. This sequence is divided into three steps: (1) understanding significance and analysis, (2) developing policy and evaluation, (3) manage and optimization which is demonstrated in Fig. 15.3.

Due to the dimension and the economic significance construction industry has worldwide, it is essential to become aware of questions such as health and the lack of safety, as result of high casualty figures (Ashworth, 2010; Health and Safety Executive, 2013), low productivity, exceeding costs and times, mistakes, little specialized work; as well as energy and other resources that are used during the construction process. The construction industry represents from 5 to 10% of the gross domestic product (GDP) and covers around 10% number of jobs that correspond to around 50% of the gross fixed capital formation (Lopes, 2012). Besides, there are a lot of different actors directly and indirectly related to the construction industry and the uniqueness of its features makes this industry subject to be studied and analyzed individually in order to improve its quality and become more efficient.

The aim of having a more sustainable construction in a sector that has a huge impact in the environment led to the implementation of a lean approach, in order to improve the construction industry. Basically, lean principles intend to turn construction a more efficient process by removing waste in its activities and consequently decrease its process cycles with better

FIGURE 15.3 Three steps of ICOMOS + C2C integration approach.

quality and efficiently (Al-Aomar, 2011). The analogy in the construction industry of the lean production is funded in three major stages, namely, (1) transformation, (2) flow, and (3) value generation (Koskela, 1992); and its aim is focused in better organization and activities. Moreover, by reducing the environmental impact in terms of energy and materials consumption it will allow the construction industry to innovate its processes, technologies, and resources. Also, with the lean approach costs are reduced due to less waste, time, effort, and higher productivity levels (Ogunbiyi et al., 2011; Wu and Wang, 2016).

The lean construction philosophy which is presented in Chapter 12, is founded in the production management concepts of a very demanding and perfectionist industry: the automotive industry. Originally, created by the Toyota industries, the Toyota Production System (TPS) or "just-in-time production" consists in identifying seven wastes in the production process that will improve the overall process, namely: waste of production, of stock, of defects and rework, of movement, of waiting time, of transportation, and of extra processing (Womack and Jones, 1996).

Therefore, as a constant improvement tool that strives for waste elimination, maximizing its value, better quality in management, supply chains, and communications (OGC, 2000), lean construction is based in the application of lean thinking (identifying value, value stream mapping, achieving flow in process, allowing customer pull and pursuing perfection, and continuous improvement) to the design and construction process by meeting client needs and optimizing the total value of a project; and its basic principles are: eliminating/reducing waste and sources of waste, focus in production planning and control, end customer focus, continuous improvements, cooperative relationships, and systems perspective. These principles are implemented in three different stages in which the first stage intends to eliminate any unnecessary tasks or operations, the second aims to enhance cooperation, and the last stage involves crucial alterations to the project such as IT tools, rethink of design and construction, bottom-up activities (Eriksson, 2010).

In order to implement the lean philosophy, there are several tools and techniques available that require training for effective usage and ultimately will improve quality, productivity, and save costs.

- The Last Planner System (LPS) improves workflow productivity, addresses project variability in construction, and reduces accident rates (Ballard, 2000; Fewings, 2013);
- Increased Visualization by foment effectiveness in communicating information;
- Daily Huddle Meetings consist in a brief daily start-up meeting to set up working status;
- First Run Studies by using the PDCA cycle (Plan, Do, Check, Act) in order to redesign crucial construction phases and improve productivity;
- 5 S Process, implementing also the Kaizen activities or the PDCA cycle, improves safety, quality, teamwork, reduces cycle times, etc.;
- Fail Safe for Quality and Safety formulates ideas to prevent possible defects or safety risks in the construction process;
- Current Engineering aims the involvement of an interdisciplinary team, sharing information and ideas related to the project in early stages of the design project;
- Value Stream Mapping involves precise actions in a production chain, from raw materials to the end product that are based in direct application, modification, and all-in-all techniques (Paez et al., 2005).

The aim of a lean philosophy is to establish the analogy of the whole stages involved in construction as a production process, analyzing the effectiveness of all its phases and the elimination of waste, with a general management philosophy and a constant improvement in quality (Koskela, 1992). By seeing the construction as a process flow, it will allow the elimination of superfluous activities that add no specific value to the final project (extra time spending in waiting, transportation, and materials inspection). Furthermore, the earliest lean principles are implemented the more effective the whole construction process will be. Nonetheless, there are some challenges that need to be overcome related to the implementation of an originally manufacturing philosophy to the construction industry. These challenges consist basically in problems associated with poor management and communication, shortage of skills and knowledge in new technologies, inadequate financial resources for training purposes, lack of broader lean specification processes, people unwillingness to change, and other barriers, such as governmental policies and the fragmented nature of the construction industry (Ogunbiyi, 2014). Finally, as lean practices lead to environmental, social, and economic benefits, as well as the improvement of safety in a more sustainable construction, it is beneficial to the construction industry its implementation and a broader comprehension. In conclusion, the book provides designers and practitioners with a valuable tool to apply modern and sustainable development techniques to construction projects. Every aspect of life-cycle assessment is covered in the book such as project management, procurement, transport or construction materials using in projects, and more. Case studies and international regulations and standards in life-cycle assessment are also discussed to help practitioners choosing appropriately sustainable construction technologies for their projects.

REFERENCES

Al-Aomar, R., 2011. Handling multi-lean measures with simulation and simulated annealing. J. Franklin Inst. 348 (7), 1506–1522.

Ashworth, A., 2010. Cost Studies of Buildings, fifth ed Pearson Education Limited, England, UK.

Bayer, C., Gamble, M., Gentry, R., Joshi, S., 2010. A Guide to Life Cycle Assessment of Buildings. The American Institute of Architects, 194.

BRE 2013. Green Guide Specification. Watford, UK.

BSI 2011. Specification for the Assessment of the Life Cycle Greenhouse Gas Emissions of Goods and Services. PAS 2050: 2011. British Standards Institution, London, UK.

Ballard, H.G. 2000. The Last Planner System of Production Control. Doctoral dissertation, The University of Birmingham, Birmingham, UK. Available at: <http://www.leanconstruction.org> (accessed 10.12.12).

Coal Industry, 2002. Industry as a partner for sustainable development. Air Transport Action Group (ATAG), Geneva, Switzerland.

Cole, R.J., 1999. Energy and greenhouse gas emissions associated with the construction of alternative structural systems. Build. Environ. 34 (3), 335–348.

Cole, R., 2003. 'Building environmental assessment methods: a measure of success. Int'. Electronic Journal Construct, Special Issue on the Future of Sustainable Construction, 1–81.

Cui, P., Man, Y., Fang, Z., 2015. Geothermal heat pumps. In: Handbook of Clean Energy Systems. John Wiley & Sons, pp. 1–22.

De wolf, C., Bird, K., Ochsendorf, J. 2016. Material quantities and embodied carbon in exemplary low carbon case studies. In: Habert, G., Schlueter, A. (Eds.), Proceedings: Sustainable Built Environment (SBE) Regional Conference, Expand Boundaries: System Thinking for the Built Environment, Zurich, June 15–17, pp. 726–733.

Elkington, J., 1994. Towards the sustainable corporation: Win-win-win business strategies for sustainable development. Calif. Manage. Rev. 36, 90–100.

Eriksson, P.E., 2010. Improving construction supply chain collaboration and performance: a lean construction pilot project. Supply Chain Manage. Int. J. 15 (5).

Fewings, P., 2013. Construction Project Management: An Integrated Approach, second ed. Spon Press, London, UK.

GBCA 2015. Green Star—Design & As Built. Green Building Council of Australia, Sydney.

Gerilla, G.P., Teknomo, K., Hokao, K., 2007. An environmental assessment of wood and steel reinforced concrete housing construction. Build. Environ. 42 (7), 2778–2784.

Graedel, T.E., Guth, L.A., 1990. The impact of environmental issues on materials and processes. AT&T Tech. J. 69, 129–140.

HEALTH and Safety Executive, H. 2013. Construction Industry. Available at: <http://www.hse.gov.uk/statistics/industry/construction/> (accessed 14.12.13).

Hammond, G.P., Jones, C.I., 2011. Embodied energy and carbon in construction materials. Proc. Inst. Civil Eng. Ener. 161, 87–98.

Herva, M., Franco, A., Carrasco, E.F., Roca, E., 2011. Review of corporate environmental indicators. J. Clean. Prod. 19, 1687–1699.

Hoffert, M.I., Caldeira, K., Benford, G., Criswell, D.R., Green, C., Herzog, H., et al., 2002. Advanced technology paths to global climate stability: energy for a greenhouse planet. Science 298 (5595), 981–987.

IPCC 2014. Climate Change 2014: Impacts, Adaptation, and Vulnerability. Part A: Global and Sectoral Aspects. Contribution of Working Group II to the Fifth Assessment Report of the Intergovernmental Panel on Climate Change. (Field, C.B., Barros, V.R., Dokken, D.J., Mach, K.J., Mastrandrea, M.D., Bilir, T.E., Chatterjee, M., Ebi, K.L., Estrada, Y.O., Genova, R.C., Girma, B., Kissel, E.S., Levy, A.N., MacCracken, S., Mastrandrea, P.R., White, L.L (Eds.). Cambridge University Press, Cambridge, UK/New York, NY.

ISO 2006. ISO 14040:2006 Environmental Management—Life Cycle Assessment—Principles and Framework.

Jelle, B.P., Breivik, C., RØKenes, H.D., 2012. Building integrated photovoltaic products: a state-of-the-art review and future research opportunities. Solar Energy Mater. Solar Cell. 100, 69–96.

N.O. Kov, 2013. Regulations and certificates regarding energy efficiency in buildings. In: Proceedings of the 4th International Youth Conference on Energy (IYCE), June 6–8, 2013, pp. 1–6.

Koskela, L. 1992. Application of the New Production Philosophy to Construction, Technical Report # 72. Center for Integrated Facility Engineering, Department of Civil Engineering, Stanford University, CA. Available at: <http://www.ce.berkeley.edu/~tommelein/Koskela-TR72.pdf> (accessed 10.12.12).

Liu, Z., Osmani, M., Demian, P. & Baldwin, A.N. 2011. The potential use of BIM to aid construction waste minimalisation. In: Proceedings of the CIB W78-W102, International Conference, Sophia Antipolis, France.

Lopes, J. 2012. Construction in the economy and its role in socio-economic development: role of construction in socio-economic development. In: New perspectives on construction in developing countries (pp. 60–91). Routledge.

Lozano, R., 2008. Envisioning sustainability three-dimensionally. J. Cleaner Product. 16, 1838–1846.

Mao, C., Xie, F., Hou, L., Wu, P., Wang, J., Wang, X., 2016. Cost analysis for sustainable off-site construction based on a multiple-case study in China. Habitat Int. 57, 215–222.

Molavi, J., Barral, D.L., 2016. A construction procurement method to achieve sustainability in modular construction. Proced. Eng. 145, 1362–1369.

Nurse, K., 2006. Culture as the fourth pillar of sustainable development. Small States Econom. Rev. Basic Stat. 11, 28–40.

OGC 2000. Achieving sustainability in construction procurement. Produced by the Sustainability Action Group of the Government Construction Clients' Panel (GCCP). Retrieved from: <http://www.ogc.gov.uk/documents/Sustainability_in_Construction_Procurement.pdf> (accessed 07.11.10).

Ogunbiyi, O.E., 2014. Implementation of the lean approach in sustainable construction: a conceptual framework (Doctoral dissertation, University of Central Lancashire).

Ogunbiyi, O., Oladapo, Adebayo & Goulding, Jack 2011, Innovative Value Management: assessment of lean construction implementation, In: RICS Construction and Property Conference, United Kingdom, p. 696.

Omar, W.W., Doh, J., Panuwatwanich, K., 2014. Variations in embodied energy and carbon emission intensities of construction materials. Environ. Impact Assess. Rev. 49, 31–48.

Oxford Economics. 2015. Construction and engineering: Analysis and Forecasts for the Construction Sector. Oxford Economics Aggregates.

Paez, O., Salem, S., Solomon, J., Genaidy, A., 2005. Moving from lean manufacturing to lean construction: toward a common sociotechnological framework. Human Factors Ergonom. Manufact. Serv. Indust. 15 (2), 233–245.

Peng, C., Huang, Y., Wu, Z., 2011. Building-integrated photovoltaics (BIPV) in architectural design in China. Ener. Build. 43 (12), 3592–3598.

RICS 2012. Methodology to Calculate Embodied Carbon of Materials. RICS Information Paper. Royal Institute of Chartered Surveyors, IP32/2012. Available from: <www.ricsbooks.com> (accessed 29.10.14).

Ranganathan, J., Corbier, L., Bhatia, P., Schmitz, S., Gage, P., Oren, K., 2004. The Greenhouse Gas Protocol: A Corporate Accounting and Reporting Standard, revised ed. World Resources Institute and World Business Council for Sustainable Development, Washington, DC.

Renz, A., Solas, M.Z. 2016. Shaping the Future of Construction Industry Institute: A Breakthrough in Mindset and Technology. Industry Agenda: World Economic Forum, Geneva, Switzerland.

Robichaud, Anantatmula, 2011. Greening project management practices for sustainable construction. J. Manage. Eng. 27 (1), 48–57.

Robinson, J., 2004. Squaring the circle? Some thoughts on the idea of sustainable development. Ecol. Econom. 48, 369–384.

Sartori, I., Hestnes, A.G., 2007. Energy use in the life cycle of conventional and low-energy buildings: a review article. Ener. Build. 39, 249–257.

Seghezzo, L., 2009. The five dimensions of sustainability. Environ. Polit. 18, 539–556.

Seo, S., Tucker, S., Newton, P., 2007. Automated material selection and environmental assessment in the context of 3D building modelling. J. Green Build. 2 (2), 51–61.

Shen, L., Tam, V.W.Y., Tam, L., Ji, Y., 2010. Project feasibility study: the key to successful implementation of sustainable and socially responsible construction management practice. J. Clean. Product. 18 (2010), 254–259.

Sinden, G., 2009. The contribution of PAS 2050 to the evolution of international greenhouse gas emissions standards. Int. J. Life Cycle Assess. 14, 195–203.

Tagaza, E., Wilson, J.L. 2004. Green buildings: drivers and barriers e lessons learned from five Melbourne developments. Report Prepared for Building Commission by University of Melbourne and Business Outlook and Evaluation.

Tam, V.W., Fung, I.W., Sing, M.C., Ogunlana, S.O., 2015. Best practice of prefabrication implementation in the Hong Kong public and private sectors. J. Clean. Product. 109, 216–231.

Thormark, C. Energy and resources, material choice and recycling potential in low energy buildings. In CIB Conference SB07 Sustainable Construction, Materials, and Practices, 2007 Lisbon, Portugal.

USEPA 2006. Life Cycle Assessment: Principles and Practice. Available from: <http://www.epa.gov./ORD/NRMRL/lcaccess> (accessed 12.12.12).

USGBC 2011. LEED Green Associate Study Guide, Washington.

USGBC 2016. LEED. U.S. Green Building Council.

Valentin, A., Spangenberg, J.H., 2000. A guide to community sustainability indicators. Environ. Impact Assess. Rev. 20, 381–392.

Van Odijk, S., Van Bovene, F., 2014. Circular Construction. The foundation under a renewed sector. ABN. AMRO. Available at: <https://www.slimbreker.nl/downloads/Circle-Economy_Rapport_Circulair-Construction_05_2015.pdf> (accessed 27.04.17).

Womack, J., Jones, D., 1996. Lean Thinking: Banish Waste and Create Wealth in Your Corporation. Simon & Schuster, New York.

Wu, P., Wang, X. 2016. A critical review of the factors affecting the success of using lean to achieve green benefits. In: Proceedings of the 24th Annual Conference of the International Group for Lean Construction, Boston, MA.

Yan, H., Shen, Q., Fan, L.C., Wang, Y., Zhang, L., 2010. Greenhouse gas emissions in building construction: a case study of one Peking in Hong Kong. Build. Environ. 45 (4), 949–955.

Index

Note: Page numbers followed by "*f*" and "*t*" refer to figures and tables, respectively.

D

E

H

I

J

K

L

Q

R

Printed in the United States
By Bookmasters